CHANGING FAMILIES

Changing Families

JUDY ROOT AULETTE
University of North Carolina at Charlotte

WADSWORTH PUBLISHING COMPANY
Belmont, California
A Division of Wadsworth, Inc.

Editor: Serina Beauparlant
Development Editor: John Bergez
Editorial Assistant: Susan Shook
Production: Ruth Cottrell
Production Services Coordinator: Debby Kramer
Print Buyer: Karen Hunt
Permissions Editor: Peggy Meehan
Designer: Kaelin Chappel
Copy Editor: Sheryl Rose
Photo Researchers: Sarah Bendersky and Stephen Forsling
Technical Illustrator: Weimer Graphics, Inc.
Cover: William Reuter
Signing Representative: Jack Fox
Compositor: Weimer Graphics, Inc.
Printer: R. R. Donnelley

This book is printed on acid-free recycled paper.

I(T)P™

International Thomson Publishing
The trademark ITP is used under license.

Printed in the United States of America

2 3 4 5 6 7 8 9 10—98 97 96 95 94

Library of Congress Cataloging-in-Publication Data

Aulette Root, Judy.
Changing families / Judy Root Aulette.
p. cm.
Includes bibliographical references and index.
ISBN 0-534-21306-5
1. Family—United States. 2. United States—Social conditions—1980- I. Title.
HQ536.A93 1994
306.85′0973—dc20 93-36560
CIP

ISBN 0-534-21306-5

BRIEF CONTENTS

DETAILED CONTENTS

CHAPTER 2 *A HISTORY OF U.S. FAMILIES WITH A FOCUS ON EURO-AMERICANS* 27

CHAPTER 3 *HISTORY OF U.S. FAMILIES WITH A FOCUS ON AFRICAN AMERICANS* 61

CHAPTER 4 *FAMILIES & THE ECONOMIC SYSTEM* 87

CHAPTER 13 CHILDREN 379

CHAPTER 14 FAMILIES, FAMILY POLICY & THE STATE 415

CREDITS

Page 2: © Veronica Buege, Harrison Township, MI
Page 10: © Stock, Boston
Page 13: © Michelle Handler, Davidson, NC
Page 26: Harry Rahn, Lincoln Park, MI
Page 45: © Culver Pictures
Page 49: © FPG International
Page 60: © Archive Photos
Page 73: © Brown Brothers
Page 81: © FPG International
Page 86: © Daymon J. Hartley
Page 94: © Alan Carey/The Image Works
Page 104: © *Daily Calumet*
Page 118: © Ted Soqul/Impact Visuals
Page 134: © Burns Archive
Page 139: © Bob Daemmrich/The Image Works
Page 152: © Bob Daemmrich/The Image Works
Page 163: © Spencer Grant/Photo Researchers
Page 172: © Christopher Takagi/Impact Visuals
Page 184: © Carrie Boretz/The Image Works
Page 194: Table 7.4 from "Women's Work Is Never Done: the Division of Domestic Labor," Shelley Coverman in *Women: A Feminist Perspective*, Jo Freeman, ed., Mayfield Publishing Company, 1989. Reprinted by permission of the publisher.
Page 195: © Paul Fortin/Stock, Boston
Pages 198–199: Box 7-1 from "The Politics of Housework," Pat Mainardi in *America's Families: A Documentary History*, D. Scott and B. Wishy, eds., HarperCollins Publishers, 1970.
Page 200: © FPG International
Page 214: © Spencer Grant/Photo Researchers
Page 221: Figure 8.1 from Love in *America: Gender and Self Development*, Francesca Cancian, Cambridge University Press, 1987. Reprinted with the permission of the publisher.
Page 237: © Nita Winter/The Image Works
Page 239: Figure 8.2 from "Thinking Sex: Notes from a Radical Theory of the Politics of Sexuality," Gail Rubin in *Pleasure and Danger: Exploring Female Sexuality*, C. Vance, ed., HarperCollins Publishers, 1984.
Page 244: Table 8.9 from *Abortion Factbook 1992 Edition: Readings, Trends, and State and Local Data to 1988*, Stanley K. Henshaw and Jennifer Van Vort, April 1992. Reprinted by permission of The Alan Gutmacher Institute.
Page 246: Hazel Hankin/Stock Boston
Page 250: © Alain McLaughlin/Impact Visuals
Page 254: © Bob Daemmrich/The Image Works
Page 261: © Spencer Grant/Photo Researchers
Page 280: © Comstock, Inc.
Page 287: Figure 10.1 from *Population, Introduction to Concepts and Issues*, Fifth Edition, John Weeks. Adapted from K. Davis in *Demographic and Social Aspects of Population Growth*, C. Westoff and R. Parke, Jr., eds., 1972. © Wadsworth, Inc. Reprinted by permission.
Page 291: © Nina Winter/The Image Works
Page 300: © Piet van Lier/Impact Visuals
Page 312: © Tom McKitterick/Impact Visuals
Page 316: Figure 11.1 from *The Washington Post*, September 1, 1992, p. WH5. © 1993 The Washington Post. Reprinted by permission.
Page 320: © Donna Ferrato/Black Star
Page 339: © Robert Fox/Impact Visuals
Page 342: © Robert A. Isaacs/Photo Researchers
Page 356: Figure 12.1 from The New York Times, May 26, 1992. Copyright © 1992 The New York Times Company. Reprinted by permission.
Page 361: © Lionell Delevingne/Stock, Boston
Page 367: © S. Gazin/The Image Works
Page 378: © Harry Wilks/Stock, Boston
Page 390: Poem from *Color* by Countee Cullen. Copyright © 1925 Harper & Brothers; copyright renewed 1953 Ida M. Cullen. Reprinted by permission of GRM Associates, Inc., Agents for the Estate of Ida M. Cullen.
Page 391: © Catherine Smith/Impact Visuals
Page 393: Figure 13.1 from "Marital Satisfaction over the Family Life Cycle," Boyd C. Rollins and Kenneth L. Cannon, *Journal of Marriage and the Family*, 36 (1974): 271–84. Copyright © 1974 the National Council on Family Relations. Reprinted by permission.
Page 404: © *People's Tribune*, Chicago, IL
Page 414: © Kirk Condyles/Impact Visuals
Page 426: © Daymon J. Hartley
Page 439: © Cynthia Bussey

PREFACE

My goal in teaching sociology has been to offer students an alternative to the individualistic psychological view of the world that tends to dominate our culture. I entered my sociology of family classes with this goal and proceeded to address issues related to families from a framework that emphasized social structure, asking questions like How is our government organized? and How does our economic system operate?

My students, however, persisted in approaching the issues at the micro level. Perhaps because of the familiar character of families, they had even greater difficulty keeping their attention on large social institutions and social structure than they did in other sociology courses. They wanted to approach the issues in ways that relate directly to their everyday experience, zooming in on the face-to-face interactions of people in the interiors of families and asking questions like How can I have a happy family? and Why did my parents divorce?

The textbooks available in sociology of families added to my frustration because most were written from a micro-level perspective that inhibited my ability to convince students to shift their attention to social structure. The few textbooks that were written from a macro-level perspective were more consistent with my point of view, but they intensified the barrier between my approach and that of my students because they did not sufficiently acknowledge the personal features of family experience. In response to this dilemma, I decided to write a book that explicitly bridges the gap between these two levels of analysis, a book with a central focus on the interplay between the big picture—a society and its large structural features—and the everyday personal experiences of individuals interacting with one another.

My goal is to show how massive social institutions like the economy, the government, and stratification systems are organized (and have changed historically). But equally important, I want to show the ways in which that social structure plays out in the lives of people interacting with each other in families—how their personal experiences, the way they feel about one another, and the way they treat one another—are influenced and sometimes even determined by what happens in social institutions far beyond the control of any individual person. And I especially want to explain the space between social structure and individual social experience, addressing questions like How exactly did the economic system of slavery shape the families of slaves living within the system? and How do laws about divorce create relationships between divorcing men and women?

This insight, which Mills called *the sociological imagination,* is of course a core principle of the discipline. This book has made that idea the central organizing principle of the text and has consistently and explicitly integrated the concept throughout every chapter. The idea of the sociological imagination is laid out in four components in the first chapter, and each of these components is explicitly featured in every subsequent chapter. First, the micro level of family experience is examined. Second, the macro level, or social context, of families is described. Third, the effect of the macro level on the micro level—the way in which the social structure affects family life—is investigated. Fourth, the effect of the micro-level activity on the macro level—individuals interacting and sometimes organizing into social movements shaping and reshaping the larger society—is examined. Each chapter concludes with a section titled "The Micro Macro Connection" that summarizes and re-emphasizes what the chapter has revealed about the links.

THE IMPORTANCE OF AGENCY

One of the comments that sociology professors often hear is, "You sociologists are so critical. All I learn about is how everything is screwed up. It's depressing." While I want students to take a critical view of their social world, my goal is not to depress them but to help them learn to understand the issues so that they can formulate changes to transform society and solve the problems. Like other textbooks that stress a macro-level approach, this book emphasizes social context and the way in which family issues are determined by social structure or the economically and politically powerful. In addition, this book does something else—it reverses this relationship and calls the reader's attention to the way in which macro-level social institutions are shaped, resisted, and challenged by individuals seeking change.

The title of the book, *Changing Families,* refers to two processes. First, families are dynamic social institutions that change, sometimes dramatically, over time, and certainly they are moving through some important changes now. Second, people can and do seek to change families and the social forces that affect them. Change isn't something that just happens to us; we also create change.

Every chapter in the book is laid out according to the following formula: (1) description: the issues related to the topic are described. For example, how many people are married and divorced, and what happened to families during the industrial revolution? (2) Theory: one or two theoretical debates that exist within the field are presented. For example, is sexuality inborn or is it socially structured, and is violence in families caused by socialization, stress, or patriarchy? (3) Social movements: the actions people have taken to try to cope with, resist, or change the family problems created for them by the organization of our society are discussed. For example, what social movements have emerged, what have they done, what are their goals, and what difficulties have they faced? Chapter 14, the last chapter, further emphasizes this theme of making change by covering the issue of family policy and the government as a critical arena of change.

BUILDING BRIDGES IN THE FIELD OF THE SOCIOLOGY OF FAMILIES

In addition to having distinctions of level of analysis, literature in the field of sociology of families is currently split into many competing perspectives. David Morgan (1985) argues that one cleavage is especially important because it creates two nearly exclusive paradigms, each with its own researchers, writers, journals, and even professional organizations. One perspective might be called *mainstream family sociologists* and the other perspective might be called *feminist* or *women's studies scholars.* The mainstream group has been around longer and has changed over the past few decades. It now includes a diversity of opinion within its parameters. While mainstream family sociologists address issues concerned with gender, this is not a central concern and often is not problematized. Furthermore, Morgan (1985) notes that within the mainstream perspective, little attention is paid to feminist theories or to the importance of the women's movement.

Feminist scholars, on the other hand, have problematized gender—that is, they have moved beyond thinking about similarities, differences, and relationships between women and men as variables and have sought to reveal the ways in which these patterns are unacceptable and the ways in which we might alter them. Feminists have made gender the core of their work, and they have been especially conscious of the link between their ideas, research, and writing, and the ongoing political struggles for equality. Because the social organization of family is central to gender and women's lives, the development of feminist studies in the past two decades has created a large body of literature on issues related to families.

Morgan points out that these two perspectives have been relatively unaware of each other in spite of the fact that they are often looking at the same issues. When I began this project I decided to approach the issue of families from a feminist perspective, and this point of view is one that remains an important feature of the book. During the review process, chapters were read by both mainstream and feminist scholars. The mainstream authors introduced me to their perspective, and I began to more fully integrate their ideas, research, and insights into the work. Sometimes I found similarities and other times I found differences to contrast or to create critiques of one point of view by the other.

In addition to discovering the importance of mainstream sociology, the research that went into writing this book has reinforced my understanding of the diversity within feminist analysis. I have also increased my appreciation of the importance of scholars other than feminists who have sought to revision family sociology, especially those whose work centers around race ethnicity. The integration of material from these many strands was difficult, but I believe the final product is much richer because of it. The book provides readers with a broad view of the field. The contrasts among perspectives often make the issues and ideas stand out in a clearer and more interesting fashion, and I hope the synthesis will help us to come a little closer to a fuller understanding of our social world.

SPECIAL FEATURES

Three important features help readers to deepen their knowledge of family studies and sociology in general: (1) the use of materials from a broad range of disciplines in addition to sociology, (2) the incorporation of explicit discussions of methodology used in the research cited, and (3) the identification of various analysis as parts of different theoretical models.

First, the book pushes the borders of the discipline, bringing information from many fields including philosophy, history, economics, and anthropology.

Second, each chapter includes sections that ask the reader to step back for a moment and look at the way in which data were collected in some pertinent piece of research. For example, the discussion of housework in Chapter 7 refers to the interview methodology used by Ann Oakley in her research. In addition to learning about the way in which housewives perceived their work, she also noted the way in which the methodology of having women researchers interview women housewives created relationships that were like personal conversations and therefore may have produced more valid data than other techniques could have. The discussion of violence in families in Chapter 11 notes the importance of using sampling techniques that allow researchers to argue that their data are representative and not unique to the study but indicative of the problems in the society as a whole.

The third feature is the explicit identification of theoretical models within the text. In every chapter theories are described, contrasted, and critiqued. In addition, when possible they are identified as part of a larger body of social theory. For example, in Chapter 4 the discussions of the variation in coping strategies among families in which a breadwinner becomes unemployed is analyzed as a result of the different way in which the families understand and explain the layoff. This process is then identified as an example of symbolic interactionism, and then symbolic interactionism itself is described and discussed. In Chapter 7 on housework, rational choice models and socialist feminist models are contrasted as ways of understanding the division of labor in families, and then the two theories themselves are described. In Chapter 13 the discussion on children and gender contrasts and describes socialization theories and social structural theory.

ORGANIZATION OF THE BOOK

This book is divided into five sections. The first chapter looks at recent changes in the study of families and introduces the concept of the *sociological imagination,* explaining how it is useful to understanding families.

The second section, Chapters 2 and 3, provides a historical overview of family issues from the Colonial period up to the early 1960s. These two chapters are divided by race ethnicity; Chapter 2 emphasizes the history of Euro-Americans and Chapter 3 focuses on the history of African Americans. Both chapters include historical information about Latinos, Asian Americans, and Native Americans.

The third section of the book focuses on contemporary families and the importance of their social context. Chapter 4 investigates the relationships between families and the economy during the 1970s, 1980s, and 1990s. Chapter 5 looks at the stratification systems of gender, race, and class and the way in which they affect families. Chapter 6 reviews the relationship between work and families.

The fourth section of the book examines relationships and conflicts within families, including: Chapter 7, housework; Chapter 8, love and sex; Chapter 9, marriage; Chapter 10, divorce and remarriage; Chapter 11, violence; Chapter 12, parents; and Chapter 13, children.

The last chapter of the book examines a social institution that is particularly important for those seeking change—the government.

ACKNOWLEDGMENTS

Many people helped me write this book. My family—Albert Aulette, Anna Aulette-Root, and Elizabeth Aulette-Root—have all read chapters, offered advice, and have taken care of me while I wrote. Many others have reviewed manuscripts, helped me find information, and taught me what I need to know to write this. They include Veronica Buege; the Charlotte Public Library librarians; Lynda Ann Ewen, West Virginia Institute of Technology; Walda Katz Fishman, Howard University; Iris Carlton-Laney, Shelly Crisp, Karen Piskurich, and Teresa Scheid, the University of North Carolina at Charlotte; Carol Axtell Ray, San Jose State University; George Root; Ruth Root; Barrie Thorne, University of Southern California; Edith Ward, the University of North Carolina Charlotte librarians, Mary Kay Yarak, and Assata Zurai. I also want to thank Joyce Arditti, Virginia Polytechnic Institute and State University; Denzel Benson, Kent State University; Catherine White Berheide, Skidmore College; Kathleen Blee, University of Kentucky, Gary Hampe, University of Wyoming; Pamela Hewitt, University of Northern Colorado; Masako Ishii-Kuntz, University of California, Riverside; Rachel Kahn-Hut, San Francisco State University; David M. Klein, University of Notre Dame; Annette Lareau, Temple University; Ginger Macheski, Valdosta State College; Lorraine Mayfield-Brown, Montclair State University; Helen Mederer, University of Rhode Island; Ann Moylan, California State University, Sacramento; Cynthia Negrey, University of Louisville; David Pratto, University of North Carolina at Greensboro; Essie Rutledge, Western Illinois University; D. Ann Squier, University of Kansas; Randy Stoecker, University of Toledo; Michael Thornton, University of Wisconsin; and Debra Umberson, University of Texas. Finally, I would also like to thank the people at Wadsworth for allowing me this opportunity and helping me on the way—Serina Beauparlant, Susan Shook, Marla Nowick and Debby Kramer, as well as Ruth Cottrell, Sheryl Rose, Bob Cocke, Sarah Bendersky, and Stephen Forsling.

CHANGING FAMILIES

Weddings are experienced as personal, but they are affected by an array of social forces, including the economy, the government, and religion.

1 HOW TO STUDY FAMILIES IN THE 1990s

Michael and Kimberly are getting married next week. They have spent the past two years getting to know each other and planning their future. They met at the office supply company where Michael is a salesperson and Kimberly is a secretary in the front office. Michael just finished his degree in business and he hopes that, if his career goes like his father's, he will move up the ladder in his company. Kimberly still has a year to go before she finishes her degree in computer science, and she also has high expectations that she will be able to establish a career and eventually have a well-paid and secure job. They expect to have two children, but they plan to wait before they start a family, because they would like to have Kimberly stay home with a new baby, at least for the first year or two. After the wedding, they will be moving into an apartment, but they are sure they will be able to save enough money in two or three years to put a down payment on a house. Michael's parents were divorced when he was in grade school, and he believes that they took the easy way out. Michael is convinced, and Kimberly agrees, that divorce is not in their future.

- *This young couple has made many choices about marriage and family. But how well have Michael and Kimberly predicted their future?*
- *Has our society, for example, the media and the schools, shaped the way they think about these issues?*
- *Do social institutions like the legal system, the government, and places of employment set limits on the options from which Kimberly and Michael have chosen?*
- *What might interfere with their choices, causing their lives to move in a dramatically different direction?*
- *Kimberly and Michael are very much in agreement about the choices they have made. What could happen that would cause them to discover that they had serious differences about the best way to organize their family life?*

These are the kinds of questions that sociologists ask. They are also important questions, however, even for those of us who are not sociologists. Most of you have families with whom you interact, and many of you are thinking about forming relationships or having children to begin a new generation of families. The better we understand what we expect from our families and what helps us in—or prevents us from—establishing and maintaining happy relationships, the more likely we will be to attain our goals.

ANALYZING SOCIAL LIFE

In order to answer the questions we have posed about Michael and Kimberly's plans, we need to ensure that our information about families is accurate and that the concepts we use to make sense of that information are capable of interpreting reality.

This book will provide the most accurate information possible about the reality of family organization in contemporary American society. The book will attempt to make sense of that reality by analyzing it in a conscious and systematic way.

Although you may be unaware of it, you are constantly analyzing your social environment. You predict events, make assumptions, interpret behavior, correlate activities and feelings with other activities and feelings, and assign significance to relationships and interactions among people. All of these activities—predicting, assuming, interpreting, correlating, and assigning significance—are aspects of creating sociological analysis. You probably do these activities, however, without giving them much thought, and if someone were to ask you your analysis of a social issue you might have a hard time answering.

This book will help you to learn a formal method for analyzing your experiences and your relationship to the social world. One tool that is especially important in this analysis is called the Sociological Imagination.

THE SOCIOLOGICAL IMAGINATION: BRIDGING THE GAP BETWEEN THE INDIVIDUAL AND SOCIETY

C. Wright Mills, one of the most important twentieth century American sociologists, called individuals' lives *biographies* and the "lives" of societies, *histories*. He taught that the relationship between biography and history should be the focus of sociology. Mills argued that our goal as sociologists should be to examine and understand individual activities and the way in which they add up to make societies. He maintained, at the same time, that we must also examine and understand the way in which society affects relationships among individuals. Mills named the ability to bridge this gap between society and the individual social experience the Sociological Imagination.

Mills noted that using the Sociological Imagination demands a lot of practice. People do not naturally think like sociologists. Most people in our society, like Kimberly and Michael, believe that individuals *choose* their own destiny. Although it is true that individuals do make choices, they make them within the limits of the society in which they live. Sociologists attempt to understand what those limitations are and how they affect individual choices.

C. Wright Mills explained this task in his book *The Sociological Imagination* (1959) by stating:

> Neither the life of an individual nor the history of a society can be understood without understanding both. Yet men do not usually define the troubles they endure in terms of historical change and institutional contradiction. . . . The sociological imagination enables its possessor to understand the larger historical scene in terms of its meaning for the inner life and the external career of a variety of individuals. . . . The first fruit of this imagination—and the first lesson of the social science that embodies it—is the idea that the individual can understand his own experience and gauge his own fate only by locating himself within his period, that he can know his own chances in life only by becoming aware of those of all individuals in his circumstances. . . . We have come to know that every individual lives, from one generation to the next, in some society; that he lives out a biography, and that he lives it out within some historical sequence. (Mills, 1959, pp. 3–10)

Like Michael and Kimberly, we all will make decisions about whom to live with and establish intimate relationships, whether to marry, whom to marry, whether to have children, and how to divide work within our households. Many people negotiating these choices with one another create a society. But beyond these individuals and their interaction is a social structure that shapes the decisions that individuals make and the experiences that individuals have. Social structure includes, for example, the laws that allow some people to marry but not others; the ideas we have about the acceptable way for husbands to treat wives or parents to treat children; the technology available for housework; and the pay scale for wage work.

Social structure even includes language, as we can see in the quotation above. In the 1950s when Mills was writing, authors used all masculine pronouns. In the 1990s this language seems odd and if Mills were writing today he would undoubtedly use more generic words like persons. These social structural factors—like laws, ideologies, technologies, pay scales, and language—shape, limit, and sometimes even determine the activities and choices of individuals as they interact with one another.

Micro and Macro Levels of Analysis

Using the Sociological Imagination means that we will be analyzing social experience at two levels: the micro and macro levels of analysis. The *micro* level focuses on the family life of individual families or individual family members in face-to-face relationships. The changes in micro-level issues over time are referred to as biography by Mills. Analyzing the micro level of social experience means we are examining society in its smallest expression. Here we would observe social life up close among people who are in contact with each other on a regular basis. This might be called the everyday life of people.

The *macro* level of analysis focuses on social structure, which includes ideologies, technologies, and social institutions like the government, the economy, and social classes. Analyzing social life at the macro level means that we try to understand society, or at least a large segment of society, as a whole. The way in which these macro issues change over time is called history by Mills.

SOCIETY IS A HUMAN INVENTION

C. Wright Mills observed not only that biography is affected by history, but that history is affected by biography. The society in which we live was created by individual people working together. The laws, ideas, technology, and ways of carrying out day-to-day tasks that characterize our society did not drop from the sky or emerge as a fact of nature, but were invented and implemented by humans, and they are constantly being reinvented and recreated. Mills wrote of every individual: "By the fact of his living he contributes, however minutely, to the shaping of this society and to the course of its history, even as he is made by society and by its historical push and shove" (Mills, 1959, p. 11).

This statement has two important implications. First, it means that our families and the social contexts in which they are embedded are the subjects of constant debate and struggle among all the different people in our society who are either seeking to maintain or seeking to change the status quo. Second, it means that *we* have the opportunity to shape the course of history—our own and our society's—by entering into these disputes and the social movements that have emerged around them.

In the past two decades, the contest over who should decide what a family is, what makes a family good or bad, and how best to solve the problems families face has been especially intense. If we discover that our family problems are caused by the limitations placed on us by our society and we choose to alter those limitations through social action, we will discover a flurry of activity and organizations already operating to address family issues. Activists from the political right to the left are putting forward their vision of the problems and their solutions on many questions related to families, like childcare, abortion, housing, welfare, sexuality, and violence.

For example, suppose that Michael and Kimberly have a child before they feel financially ready. Although they would prefer to have Kimberly stay home to take care of the baby, they may find themselves unable to survive without her paycheck. They will then find that daycare is expensive. If they decide to become involved in some political action to convince the federal government to fund daycare for working parents like themselves, they would find numerous organizations already lobbying in Congress. Some of those groups petitioning and rallying would be arguing for federal funding of daycare and others would be arguing that childcare—including how it is paid for—is a matter that is best left to parents and that the government should not interfere.

The Micro/Macro Connection

In sum, C. Wright Mills's model for understanding the social organization of families points toward four sets of issues that we will examine:

1. We will investigate the micro level, the family experience of individuals and individual families.
   - How do different family members use their time?
   - How do they treat each other?
   - What are the patterns of conflict? How do family members support and care for each other?
2. We will examine the macro level, the organization of our society and its major social institutions as they relate to families.
   - What are the most important characteristics of contemporary American society in which our families are embedded?
   - How are our economic system, government, and class, race, and gender stratification systems organized?
3. We will look at the ways in which the macro level affects the micro level, the daily experience.
   - How does our society shape our family life?
   - What problems do families face in trying to survive within our society?
   - What choices must families make? How are these choices necessitated by social context? How are they restricted by social institutions like the government or the economy?
   - What are the varieties of families in which Michaels, Kimberlys, and others live? How do race/ethnicity, social class, and historical period affect that variation?
4. We will look at the way in which micro-level activities shape the macro-level organization of our society.
   - How do people seeking solutions to problems their families face create change in their society?
   - Since people make changes not only as individuals interacting within their own families, but as organized groups of individuals, what solutions are being offered by the many organizations that are concerned about family issues?

- How do ideas about families fit into the arguments about how our society should be organized? For example, what are the different points of view on various welfare reform bills?
- What should the relationship of the government be to families? Are children best raised by families without government interference or should the government help to provide food and shelter for children?
- Is sexuality legitimate only for married heterosexual couples?
- How can we stop violence?
- Who is responsible for taking care of the sick, the homeless, or the elderly—their families, the government, or both?
- Who should make decisions about abortion?

FAMILIES AS A POLITICAL ISSUE

Is the family a political institution? Not according to the popular notion of family in the U.S. Families are not considered part of the politics of society. Sometimes, furthermore, families are thought of as the opposite of politics—a private space away from the public world (Lasch, 1977). Ernest Burgess, who is considered the father of family sociology, took this position. A basic assumption of his work is that "the family sphere is qualitatively different from the public sphere. The public and private spheres are separate social realms. The family is a psychological relief station from the public world" (Osmond, 1987, p. 113). Others, however, have argued that families are very much part of the "public sphere" including the political debates that take place there, and that, in addition, families are not "psychological relief stations" but often arenas of power struggles (Ferree, 1990).

In order to address the question of whether the family is a political institution we first need to determine what the word politics means. Politics is the expression and organization of power. Jean Lipman-Blumen (1984, p. 6) defines power as "the process whereby individuals or groups gain or maintain the capacity to impose their will upon others, to have their way recurrently, despite implicit or explicit opposition through invoking or threatening punishment, as well as offering or withholding rewards." If these kinds of activities touch upon families or take place within families, then families must be arenas of power and political institutions.

main pt: family is political

In this book I will be arguing that families are political institutions in two ways (Pogrebin, 1983). Families are political both at the macro level and at the micro level. First, at the macro level, families have a political relationship to the rest of society. Second, at the micro level, families include political relationships within their own boundaries. Families are political configurations existing within larger political configurations.

The first type of "politics of family" is evident in the current debates around families and family issues like childcare, education, abortion, sexuality, and welfare rights. These are topics of great concern to the politicians, governmental agencies, and social organizations that we commonly think of as political institutions. What happens to families with respect to these issues is tightly tied to political decisions, the relationships of power, and the battles over power that underpin those decisions. Sometimes the discussions of these issues even go so far as to claim that families do not just touch upon the politics of our society, but that families are at the center of the political world (Harding, 1981; Pogrebin, 1983).

Families: Hot Issue in the U.S.

The decade of the 1980s opened with the White House Conference on Families called by President Jimmy Carter to fulfill one of his campaign promises. The real results of this conference in terms of policy changes were minimal. But the conference paved the way for people from across the political spectrum to come forward with their vision of family, and the differences in these visions were remarkable. Even the question of naming the White House Conference elicited a heated debate. Conservative forces who attended the conference proclaimed, "Recognizing that our nation was founded on a strong traditional family, meaning a married heterosexual couple with or without natural children, it is imperative and we demand that the President immediately correct by Executive Order the name 'White House Conference on Families' and let it be further known in all futures as the 'White House Conference on the Family' " (Diamond, 1983).

Liberals insisted that the name remain "White House Conference on Families" and that the topics of discussion include many different kinds of families like single-parent families, extended families, grandparents raising their grandchildren, gay men and lesbian couples, and many others, in addition to heterosexual married couples and their biological children.

The liberal point of view carried. The name was kept and the content was left broad. The debate between the conservatives and the liberals, however, did not end with the conference and has continued to escalate throughout the 1980s and 1990s.

Susan Harding (1981) argues there are three components for each of these two points of view. The conservatives' position is characterized by three factors: (1) a belief in the importance of maintaining hierarchies; (2) a view of families that collapses the interests of all family members into one whole; and (3) an ideology about family that stresses a narrow and specific morality. The liberal position is characterized by three different factors: (1) concern for equality; (2) a focus on the individual and individual rights and responsibilities; and (3) an ideology about families that is based in reason.

Harding maintains that for conservatives "one's place in the family is defined in relation to the whole, in terms of a family role, and the emphasis is on conformity to standards of right and wrong. Wives and husbands, mothers and fathers, children and parents, naturally have unequal rights and responsibilities in the family and the symbolic authority and prestige of the father are stressed" (Harding, 1981, p. 58).

For liberals, "The emphasis . . . is on the individual, and on the personal well-being, fulfillment, and prosperity of family members. An equitable division of rights and responsibilities is idealized as the best means of expressing individual ability and inclination" (Harding, 1981, p. 58).

Throughout the 1980s and 1990s family issues hit the news as the advocates of these two positions continued their debate. We heard about elections and Supreme Court nominations hinging on the fight over the legalization of abortion. The largest public demonstrations since the 1960s were held in Washington, D.C. advocating pro-choice, "housing now," and gay rights. Pro-life

Sometimes our beliefs about families do not match reality. Nuclear families with young children like the one pictured here comprise a relatively small proportion of all the households in the United States.

advocates were arrested for civil disobedience in dozens of cities. Congress debated the controversial Family Protection Act, which would protect the right of parents to use corporal punishment against children, repeal laws that grant educational equity to both sexes, impose prayer in school, and require that parents be notified before daughters receive counseling or medical care for sexually transmitted disease, contraception, or abortion (Pogrebin, 1983). State and federal legislators fought over childcare bills and welfare reform. Black families reemerged as a central issue in the civil rights movement (Baca Zinn, 1987; Gresham, 1989). Disputes raged over sexuality and homosexuality, touching on everything from health care to rock and roll. Violence against women and child abuse filled the pages of our newspapers. Family has been firmly placed as a key issue in political struggles in the courthouses, the legislatures, and the streets of the contemporary United States. The results of these battles have an impact on every one of our lives.

The Personal Is Political

Families are political in a second way—in the relationships *within* families. These relationships have to do with the relative amounts of power of various family members, the sources of power among family members, and the implementation of power.

If you look back at Lipman-Blumen's definition of power you will see that it can be applied to many of the relationships and activities that exist within families. In your own family certain family members may have more power than others. For example, parents compared to children may have greater ability to offer rewards or to mete out punishment to others in the family. And you undoubtedly have been involved in power struggles within your family over issues like who should do the dishes, who should decide how money will be spent, or which family members can use violence against others in the family.

The politics of family, in both forms, will be a recurring theme in this book. We will be looking at the role family issues play in the broad political struggles within our society. We will also be examining the political relationships within families.

RECENT HISTORY OF THE SOCIOLOGY OF FAMILY

Sociology has never been a uniform discipline. A variety of contending schools of thought within the field has been a characteristic of its history, just as it is today. Critical theorists like C. Wright Mills, symbolic interactionists like Herbert Blumer and Erving Goffman, behavioral sociologists like Ernest Burgess, and structural functionalists like Talcott Parsons were all actively working to shape sociology in the 1940s and 1950s. Some scholars argue that structural functionalism dominated during this period, despite the diversity within the field (Glenn, 1987). In the U.S., the early 1950s was an era of social and political conservatism. The social sciences reflected the public mood and the dominant theoretical perspective within sociology was one that reflected the ideological climate of the times. The framework of structural functionalism adapted itself well to the cultural and political mood of conservatism (Becker, 1979; Giddens, 1977; Gouldner, 1970; Huaco, 1986; Ritzer, 1992).

Structural functionalists examine society as a whole system. They assume that each part of a society fits with the other parts to create a smoothly functioning system. Functionalists believe that society is essentially stable and beneficial for nearly all its members, and thus changes must take place very slowly so that they do not prove too disruptive. Functionalism has been criticized because it emphasizes social order and stability and ignores or justifies problems like inequality.

Talcott Parsons was a key figure in the development and promotion of functionalist theory within sociology. In particular Parsons's work was critical to the development of sociology of family. D. H. Morgan writes about Parsons, "It would not be too much of an exaggeration to state that Parsons represents *the* modern theorist on the family" (Morgan, 1975, p. 25).

According to Parsons, society is best served by families that are functional units in which one male person serves the "instrumental" needs of the family members. By this he meant that men should be the rational, decision-making, money-earning public figures. Parsons thought that the ideal complement to the instrumental husband was an "expressive" wife, a woman who would be the nurturant, domestic, childcare provider. Within this framework, a woman employed outside the home was believed to be disruptive or dysfunctional for her family. In 1955 (pp. 14–15) Parsons wrote: "It seems quite safe in general to say that the adult feminine role has not ceased to be anchored primarily in the internal affairs of the family, as wife, mother and manager of the household, while the role of the adult male is primarily anchored in the occupational world, in his job and through it by his status-giving and income-earning functions for the family." As you can see, Parsons's ideas about families included many questionable assumptions about gender and also ignored the experience of racial ethnic and working-class families. Both men and women were given stereotyped prescriptions for their behavior with little room for individual variation.

The implications for women were particularly troublesome. Another functionalist, Mirra Komarovsky, wrote:

> A social order can function only because the vast majority have somehow adjusted themselves to their place in society and perform the functions expected of them. . . . Even if a parent correctly considers certain conventional attributes of the feminine role to be worthless, he creates risks for the girl in forcing her to stray too far from the accepted mores of her time. . . . At the present historical moment the best adjusted girl is probably one who is intelligent enough to do well in school but not so brilliant as to get all A's . . . capable, but not in areas relatively new to women; able to stand on her own two feet and to earn a living, but not so good a living as to compete with men. (1953, pp. 52–74)

The structural functionalist position on families was also questionable for working-class and racial ethnic families. The assertion that a nuclear family with a male breadwinner and a female housewife was the best type of family ignored the need for wives in working-class and many racial ethnic families to work outside of the home. And it denied the possibility that alternative family forms could provide equally positive environments.

During the 1960s powerful social movements challenged functionalist ideas. The civil rights movement, which grew rapidly in the 1950s and 1960s in the United States, was one that disagreed with the functionalist analysis (Wiley, 1979). Civil rights activists demanded that we acknowledge the diversity within our society in all social activities including family. Many families, especially many African American families, were not organized as Parsons and Komarovsky claimed they should be. Civil rights advocates argued that we needed to understand racial ethnic families in terms of their relationship to the larger society and that we needed to consider the positive aspects of different family forms (Billingsley, 1968; Gresham, 1989; Rainwater & Yancey, 1967; Ladner, 1971; Stack, 1974).

The gay rights movement initiated in the 1970s also reminded us that our ideas about families were too narrow and failed to include the experience of gay men and lesbians. With an estimated 10 percent of the population gay and lesbian, our ideas about intimacy, love, and sexuality needed to be broadened beyond the assumption of heterosexuality for all adults (Altman, 1971, 1982; Brake, 1982b; Rich, 1986; Rubin, 1975; Weeks, 1985).

The women's movement, in particular, focused its attention on families and it too was critical of the functionalist view of families (Benston, 1969; Cronan, 1971; Firestone, 1970; Millett, 1970). Some feminists argued that families like those functionalists portrayed as ideal were really prisons for women. The women's movement took as its central concerns the problems women faced in families in dividing work, expressing their sexuality, and facing violence (Flax, 1982).

Second Wave Feminist Movement

The battle for the right of women to vote that developed in the nineteenth century and ended with the passage of the Nineteenth Amendment on August 26, 1920, was called the First Wave women's movement. The origins of the Second Wave, which emerged in the 1960s, are found in the First Wave and in nineteenth and early twentieth century Marxist writing and socialist movements.

The women's liberation movement in the 1960s and 1970s generated change in nearly all areas of American life, including the study of families.

The immediate predecessors of the Second Wave women's movement included the student movement, the anti-war movement, and the civil rights movement of the 1950s and 1960s. In 1964 women in the Student Non-Violent Coordinating Committee (SNCC), an organization active in the civil rights movement, protested their treatment in SNCC. In 1965 a group of women presented their demands for equality at a convention of the new left and anti-war group Students for a Democratic Society (SDS). The National Organization for Women (NOW) was formed in 1966. In 1967 and 1968 women's liberation groups formed spontaneously in Chicago, Toronto, Detroit, Seattle, and Gainesville (Freeman, 1989). By 1970, "for the first time the potential power of the feminist movement became publicly apparent" (Freeman, 1975, p. 84). And throughout the decade consciousness-raising groups, conferences, protests, speak-outs, and teach-ins abounded. In all of these activities family issues took center stage with debates on abortion, sexuality, housework, childcare, battering, and marriage.

Betty Friedan, one of the earliest feminist authors of this period and the first president of NOW, wrote *The Feminine Mystique* in 1963 about the same white middle-class housewives that Parsons had described. Friedan's depiction, however, was not of the benign, natural, functional family. Instead, she described this state of affairs as "comfortable concentration camps." According to Friedan and other feminists (Firestone, 1970; Mitchell, 1984), Parsons's ideal family was not functional for women because it limited their ability to be independent, active participants in their society.

In response to these movements and sometimes as part of them, family sociology scholars began to question their work in fundamental ways (Morgan, 1975). They asked:

- How do people organize their family lives and what does that organization mean to them?
- Who lives with whom?

- Whom do we call family?
- Are families prisons, havens, or both for women, men, and children?

Revisioning Families

Over the past two decades, sociologists have asked these kinds of questions and have worked to reconceptualize the study of families (Flax, 1993; Hartsock, 1993). Jessie Bernard (1989), a prominent sociologist and a founding member of Sociologists for Women in Society, has named this period of scholarship "The Feminist Enlightenment" because of the important role feminists have played in this revisioning.

This revisioning, however, has not been carried out by feminists alone; a variety of recent family scholars have participated in the effort. Barrie Thorne, whose work has examined several issues related to families, summed up the rethinking of the family that has taken place in the past two decades in five themes (Thorne, 1992).

The first is challenging the myth of the monolithic family. This theme includes both acknowledging and appreciating diversity as it occurs historically or within a particular historical period. The other four are: observing that families are sites of a variety of different activities including production, reproduction, socialization of children, and sex; acknowledging differences among family members within families by gender and age; raising questions about the conceptualization of families as rigid containers that separate us from the rest of society; and concern with the dilemma of creating families that allow for individual freedom and equality, as well as nurturance and sharing of responsibility.

CHALLENGING THE MYTH OF THE MONOLITHIC FAMILY

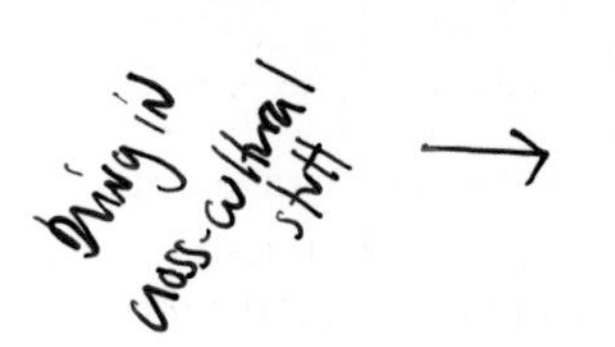

The word monolithic describes something that can be characterized as totally uniform. Frequently we speak of The Family as if all families were the same, as if there really were some way of organizing families that is most prevalent, natural, and beneficial for humans.

The family that was described by functionalist theorists above is one in which a husband and wife and their minor children live together. The father is a full-time worker who earns enough money to allow the mother to be a full-time housewife (at least until the children are in school). This family lives in a single-family home, the parents are not divorced, and the children are natural or adopted. In addition to the problems with characterizing this kind of family as ideal, the type is not descriptive of the experience of most Americans. This kind of family does exist in the United States, but it is a tiny minority.

Only about 10 percent of households in the U.S. are composed of a father who is in the labor force, a mother who is a housewife, and their children. If we were to restrict this category further by counting only households in which the children are the natural or adopted children of the married couple—not their stepchildren—or in which the family lives in a single-family home, the proportion would become even smaller (U.S. Bureau of the Census, 1989a).

The Family as Defined by the Government

Table 1-1 presents data collected by the U.S. Census. One of the questions asked by the census is "With whom do you live?" Answers to this question are grouped into two main categories in Table 1-1: family and nonfamily.

Family is defined in a manner some would find too limiting. According to the government a family is defined as those people who live with people to whom they are related by blood, marriage, or adoption. The category of family is then further divided into two subcategories: married couples and other families. Married couple families consist of husbands and wives and, if they have them, children under the age of eighteen who are the natural or legally adopted children of at least one of the spouses. Other families consist of households that are not married couples but in which household members are related. Most of these kinds of households are single-parent households, but they also would include households in which, for example, three generations live together or a sibling of one of the spouses lives with a married couple.

The second major category, nonfamily, is also subdivided into two types: single people living alone and other nonfamily households. Nonfamily households would include, for example, heterosexual and gay couples living together and college roommates living together. The college roommates probably fit the definition of being nonfamily since their relationship would frequently be based on friendship or even just convenience, but not intimacy. Heterosexual and gay couples, however, might argue that although their relationship does not include a marriage license, it is an intimate one and they should be considered a family.

Appreciating Diversity

The real-life experience of family is so diverse that using the term The Family to suggest a wife, a husband, and their two children is not a useful way to describe accurately the way most people live. The concept of The Family is also problematic because it may imply certain assumptions about how people *ought* to live. If one assumes that there is such a thing as The Family, families that do not fit this mold can be labeled deviant, dysfunctional, or abnormal. This assumption is especially loathsome to scholars concerned with racism. Patricia Hill Collins writes:

> The archetypal white middle class nuclear family divides family life into two oppositional spheres: the "male" sphere of economic providing and the "female" sphere of affective nurturing. . . . Black women's experience and those of other women of color has never fit this model. Rather than trying to explain why Black women's work and family patterns deviate from the alleged norm, a more fruitful approach lies in challenging the very constructs of work and family themselves. (Collins, 1990, p. 47)

All racial ethnic groups in the U.S., including whites, blacks, Latinos, Native Americans, and Asians, have substantial numbers of nonnuclear families. Regardless of race, these nonnuclear families have been subject to criticism. In particular, African American families that have not fit the mold of The Family have been accused of being chaotic, dysfunctional, and broken. In Chapter 5 we will explore the organization of African American families that are not nuclear but also are not chaotic and dysfunctional. We will see the importance of ex-

TABLE 1 • 1
Household Composition, 1990 (in percent)

Family Households	
Married couples	
With children under 18	26.3%
Without children under 18	29.7%
Other families	
Male householder	3.1%
Female householder	11.7%
Nonfamily Households	
Single people	
Men living alone	9.7%
Women living alone	14.9%
Other nonfamily households	4.6%

Source: U.S. Bureau of the Census, 1990.

tended family, shared child raising, and cooperative systems for providing shelter and food for members of the community (Stack, 1974).

Collins (1990) argues that instead of using concepts such as the monolithic family that do not fit the experience of large segments of our society and tend to reinforce prejudice, especially against African Americans, we should research real families. Furthermore, we should keep our eyes open not only to the validity of differences among families but to the potentially positive effects of diversity. "Different" families, compared to The Family, often have more effective and humane ways of organizing social life. In addition, variety in family forms in itself may be a positive characteristic of a society.

The Monolithic Family and the Denial of Historical Change

The concept of The Family sometimes reflects an assumption that one kind of family has dominated all human societies throughout history. Imagining what life was like 1,000 years ago or even 100 years ago is hard to do, especially if we are trying to envision the ways in which people thought or the ways in which they felt about each other or treated each other. We can see the artifacts in museums that show us how people dressed, the vehicles they drove, or what tools they used. But when we try to picture a mother feeding a baby and we speculate about how she felt about the baby, or when we try to picture a woman and man being married and we wonder what was going through their minds, the task becomes more difficult. This is because we tend to imagine people wearing different clothes, using different tools in different dwellings, but thinking, feeling, and acting pretty much the same as we do.

Women have always borne children; and for many centuries, at least, men and women have married each other. But the ways in which motherhood and marriage, as well as other family relationships, have been organized have changed throughout human history. The ways in which mothers and children and husbands and wives, grandparents, aunts and cousins treated each other and thought about each other have varied considerably through all of human history.

One does not have to go back too many years to notice historical change in families. As recently as the turn of the century, for example, children were seen in a very different light as compared to today. Viviana Zelizer describes two legal cases that contrast how children were valued in 1896 compared to 1979:

> In 1896, the parents of a two-year-old child sued the Southern Railroad Company of Georgia for the wrongful death of their son. . . . The court concluded that the child was "of such tender years as to be unable to have any earning capacity, and hence the defendant could not be held liable in damages." In striking contrast, in January 1979, when three-year-old William Kennerly died from a lethal dose of fluoride at a city dental clinic, the New York State Supreme Court jury awarded $750,000 to the boy's parents. (1985, pp. 138–139)

Zelizer explains that before the early part of the twentieth century, children were valued on the basis of their ability to contribute economically. Since the passage of child labor laws at the turn of the century, children have become valued as economically useless but emotionally priceless.

Making the assumption that families have always been the same blinds us to the reality of human history and limits our ability to understand who we are and where we came from. But more importantly, denying historical change means that we cannot see the direction in which we are heading.

If we assume families have always been the same, we must assume that they will not change in the future, or that if they do change, they can only disappear. On the other hand, if we see families as moving through history and constantly changing, we can understand the changes taking place in our own time as a part of that history, *not* as evidence of the collapse of The Family. Scholars are increasingly becoming aware of the need to acknowledge the diversity of family organization both in contemporary American society and throughout human history (Collier, Rosaldo, & Yanagisako, 1992).

OBSERVING THAT FAMILIES ARE THE SITES OF MANY DIFFERENT SOCIAL ACTIVITIES

Many activities take place within families. Juliet Mitchell (1984), a British scholar recognized as a prominent leader in the emergence of the women's movement in the 1960s, wrote that four social structures determine women's fate. Although these four structures are interwoven throughout society, they intersect in many modern Western families and provide examples of the variety of activities that simultaneously take place in families.

Mitchell named one structure production. This refers to the way in which families are economic institutions producing food and shelter, preparing workers to enter the wage system, and purchasing and consuming goods and services.

The second structure Mitchell describes is reproduction. Families also play a central role in the reproduction of the next generation. Women, both supported and constrained by their husbands, the church, the legal system, and physicians, frequently make decisions about fertility within their families. Birth is increasingly in a family context with fathers and siblings attending the birth of new babies. In the U.S. we also have explicit laws designating children's relationship to adults. For example, all children must be under the legal custody of a natural or adoptive parent or guardian.

Families are also sites of the socialization of these children, Mitchell's third structure. Socialization is the process by which humans learn the rules of a society—what is appropriate and what is inappropriate behavior and what one might expect if one behaves in either way. Along with other social institutions like schools, the media, and the courts, families are a key site for the socialization of children in our society.

Finally, Mitchell reminds us that in our society sexuality is part of family organization. Although there is much sexual activity outside of marriage, the primary "legitimate" and often only legal form of sexuality is heterosexual intercourse between husbands and wives. Other expressions of sexuality and sexual relationships are always compared to this standard and are treated unequally and often critically by informal rules and formal laws. For example, gay men and lesbians are harassed and can even be imprisoned in some states for their sexual activities. And the unmarried sexual relationships of young people, heterosexual or homosexual, are publicly denounced in the government's "just say no" campaign in AIDS education materials prepared for the public schools.

Sociologists who approach social phenomena from a macro level have claimed that these different systems within families must be recognized as separate though interlocking and must each be examined both alone and relative to each other. *All* of the structures that make up families must be examined.

If families are complex systems that include many different activities and structures, we cannot understand them if we focus our attention on only one or two of the structures. Think of a family as if it were a human heart. A biologist cannot understand the heart without understanding all of the systems that intersect within a heart: the cardiovascular system, the nervous system, the endocrine system, the muscular system. Likewise, a social scientist cannot understand a family without understanding all of the structures that intersect within a family.

Furthermore, scholars must attend to the relationships among the structures. Once again think about the heart analogy. If a heart does not operate properly, the biologist must look at the ways in which the various systems interact and affect one another in order to find the problem and try to solve it.

Acknowledging the complexity of families and the ways in which various systems interact is critical to problem solving in the study of families as well. Feminists have been concerned with how families operate in ways that perpetuate inequality between women and men and restrict women's opportunities. If we liberate women within one structure but not in others we risk reorganizing inequality while not eliminating it.

For example, many people believed that if women were fully integrated into wage labor, they would be emancipated. But if women's responsibilities for bearing and raising children remain the same, instead of being emancipated women only find that their work load increases. In this case, the structure of production has been changed in the direction of greater equality between women and men, but the structures of reproduction and child socialization have become more problematic since few women receive paid pregnancy leave or childcare from their employers.

All of the activities that Mitchell cited—production, reproduction, socialization, and sexuality—take place both inside and outside of families. In some families all four activities might occur, in others only one or two. Other theorists have focused on other structures not included in Mitchell's framework. For example, emotions such as love or violence might be included as activities that take place in several sites in our society, including some families.

EXAMINING DIVISIONS WITHIN FAMILIES BY GENDER AND GENERATION

We often speak of the happy family, the good family, the violent family, or the strong family. The implication is that families are united and that the experience of all family members is uniform. According to writer Heidi Harmann:

> Such a view assumes the unity of interests among family members: it stresses the role of the family as a unit and tends to downplay conflicts or differences of interest among family members. . . . I offer an alternative concept of the family as a locus of *struggle*. In my view, the family cannot be understood solely or even primarily, as a unit shaped by affect or kinship, but must be seen as a location where production and redistribution take place. As such, it is a location where people with different activities and interest in these processes often come into conflict with one another. (Hartmann, 1981, p. 368)

Hartmann reminds us that families are made of sets of relationships among different members. A family may be a happy place for some family members but not for others. It may be strong because of one or two members. We might call a family violent, but if we look inside that family we could find that the experience of violence is very different for different family members. One person may be a batterer, another a victim of abuse, and another a witness to it—all within the same family.

This differential experience occurs partly because every person is unique. We are all individuals and have different personalities and different experiences that we bring to our family relationships. But in many families there are patterns of differences along the lines of gender and generation. For example, men and women do different tasks in families. Even in the 1990s, household work is usually divided between women and men in gendered ways. We might assume that the increasing proportion of women in the paid labor force and the visibility of the women's movement would have created equality in the way in which housework is divided. Recent studies, however, as we shall see in Chapter 7, show that women still do most of the housework and, when men do housework, they do certain tasks like taking out the garbage and yardwork.

Furthermore, changes in families are experienced in different ways. For example, men and women emerge from divorce with different resources, a topic that will be explored in Chapter 10. Generation, too, is a factor distinguishing family members. For example, violence between parents and children usually victimizes children, as discussed in Chapter 13.

Writers like Heidi Hartmann remind us that when we look inside families, we need to be careful to look at conflict and competition as well as consensus. This does not mean that we will only look at tensions and differences within families, because sometimes families work together or experience themselves as an integrated unit. For example, in Chapter 3 we will observe the way in which African American sharecropping families worked together to protect their members from assault by violent landowners. In this book we will see that families are places where people can be *both* allies and adversaries.

ASKING WHETHER FAMILIES ARE SEPARATE FROM THE REST OF SOCIETY

Sometimes the obvious truth of the many strands of our social roles and relationships is blurred by our belief that families have fixed boundaries. We believe that if we have a good family we will be a good person or we will have a good life. We think of our families as rigid containers that insulate and separate us from the rest of society. But our family relationships and our experience of family life are only one part of our social existence. And our family relationships and our experience of family life are constantly being shaped and reshaped by social relationships and institutions outside of families and vice versa.

If we are to understand families, we need to examine the fluid boundaries of families and the ways in which families and family members interact with other aspects of society (Rapp, Ross, & Bridenthal, 1979).

- Is it true that our families can provide us with a private life unaffected by the rest of society?
- If you were born and raised in your present family, but that family was part of a different society, would your life be the same?
- Would you be the same kind of person as you are now?
- What if you were born into a peasant family in Guatemala, a blue-collar family in Tokyo, or the Trump family in New York City?

These families are all different from yours and from one another because they exist in different societies or in different social classes. The members of these different families would participate in different activities, think different thoughts, and experience different emotions, because not only are they part of a particular family, they are simultaneously part of different societies and different social classes within those societies.

Scholars who are committed to revisioning families have paid attention, in particular, to the ways in which the economy and the government are organized and how these two social institutions interact with families and individual lives (Mitchell, 1986; Gerstel & Gross, 1987). Rosabeth Kanter argues that work, a factor in our economic system, has an important influence on our family life. She states:

> If any one statement can be said to define the most prevalent sociological position on work and family, it is the myth of separate worlds. The myth goes like this: In a modern industrial society, work life and family life constitute two separate and non-overlapping worlds, with their own functions, territories, and behavioral rules. Each operates by its own laws and can be studied independently. . . . a corollary of the myth is the assumed separation of men's and women's domains, with the family woman's place. (Kanter, 1977a, pp. 16, 20)

In the scenario with which we began the chapter, Michael and Kimberly assume that their work will not interfere with their family decisions and vice versa. But what if Kimberly finds that finishing her degree and building her career takes several years, and at age thirty-five she still has not had the two children she wanted? Taking time off to have the babies, and especially taking several years off to stay home with them until they go to kindergarten, will affect her career. The gap in her resume might prevent her from advancing in her field. On the other hand, if she delays childbirth too much longer, she may be physically unable to have children because infertility increases as women age.

CREATING FAMILIES THAT PROVIDE BOTH LOVE AND INDIVIDUAL FREEDOM

The recognition of what is sometimes the oppressive character of women's position in families and of the importance of the work that people do in those families forces us to carefully consider the *complexity* of families and to create goals for reconstructing families and societies in ways that provide for both loving and egalitarian communities.

Michele Hoffnung describes the dilemma between advocating for individual freedom and equality and for societies that are nurturant and collectively run. She writes:

> Mothering is done at home, outside the world of achievement, power and money. . . . It is this aspect of motherhood—its limiting effect on women's public participation at a time when women have won access to the public world—that must inform the next stages of feminist activity for social change. It is not enough for women to be able to do men's work as well as women's; it is necessary to reconsider the value of mothering and to reorder public priorities so that caring for children counts in and adds to the lives of women and men. (Hoffnung, 1989, p. 157)

On the one hand, for example, we want individual women to be able to decide about their own fertility, to choose whether to have children or not without the interference of the government, their parents, or their husbands. On the other hand, we need to recognize that bearing children is a social activity. It necessitates the interaction of at least two people and many people have direct or indirect interest in the birth of a child.

We need to support individual women's right to make their own fertility decisions, as well as to expand opportunities for men to participate more fully in the lives of children (Ehrensaft, 1987). We must also increase the responsibility of our society for children instead of leaving nurturance, or the loving care of children, in the hands of mothers, or mothers and fathers alone.

The Question of Research Methodology

The five themes suggested by Thorne show that sociologists who are committed to revisioning the study of families have asked certain kinds of questions about the social world that were often left unasked by other scholars. They have also challenged the methodologies of social inquiry, the way in which scholars have attempted to answer their questions (Harding, 1987; Reinharz, 1983; Smith, 1974, 1979; Ferree & Hess, 1985).

Social science researchers use a variety of techniques to gather and analyze data, including interviews, observation, case studies, historical documents, experiments, surveys, and content analysis (Reinharz, 1992). When these techniques are used by feminist researchers, however, they are incorporated into a methodology that is characterized by three principles.

The first principle of feminist methodology is the replacement of value-free research with value-committed research (Mies, 1983). Sociologists supporting value-free research argue that they must remain detached and objective about their work. Those advocating value-committed research contend that maintaining a value-free stance is impossible, disrespectful of participants, and impractical.

Researchers who are value-committed maintain that a value-free stance is impossible because people studying people is entirely different from people studying chemicals or cells. People studying people create social relationships between the observer and the observed. For example, when a researcher interviews a participant they are both touched by the experience. The interview is a social interaction that involves listening to each other, trying to understand each other, and incorporating each other's ideas into one's own thinking. This interaction does not allow for neutrality.

Second, value-committed researchers argue that value-free research objectifies the people we study, turning them into objects that are no more important than viruses on a slide. Many researchers have in fact extended this commitment not to objectify people by purposefully incorporating participants into the creation and implementation of research projects so that the lines between researcher and researched become blurred.

Third, value-committed researchers criticize the value-free stance as impractical. If we treat the people we are researching as objects about which we must not express concern, it is unlikely that they will share with us the information about their lives that we seek (Gorelick, 1991).

A second principle of feminist research is to replace the "view from above" with the "view from below" (Mies, 1983). This means that we examine the lives of people who have often been invisible, such as working-class people, racial ethnic people, children, and women.

It also means that we examine their experience from their point of view. In many sociological investigations, the researcher chooses the topics, determines the questions, and explains the findings. In feminist investigations these activities should be done *with* the "subjects."

The third principle of feminist methodology is the integration of research and social action with the intention of changing the status quo (Reinharz, 1992). Mies (1983) contends that feminist studies grew out of the women's movement and that they must stay rooted there if they are to survive.

Feminist researchers are not alone in their adherence to these principles. In particular, researchers who use qualitative methods have objected to modeling social science research after the supposedly value-free work of natural and physical scientists. Scholars whose work is with participatory or collaborative methods have argued for integrating participants into the research process. Action researchers share the goal of challenging the status quo with their research (Cancian, 1993; Reinharz, 1992). What is unique about feminist research is its reliance on the cluster of these three principles and especially its connection to the theoretical themes like those proposed by Thorne (1992).

In this book we will note the ways in which information was gathered and especially the way in which these principles of feminist methodology influenced the sources of data and their interpretation. For example, in the discussion in this chapter of the census data on households, I noted that the definition of family that the government used to compile the tables was inadequate and might be challenged by gay and lesbian rights activists or those concerned with racial ethnic equality. This is an example of the way in which adhering to the theme of paying attention to and respecting diversity might alter the technique of data collection, in this case by reformulating the definition of family.

The Question of Theory

Theory is an essential feature of sociological inquiry. Theory helps us to understand and explain our social experience and it also helps to determine ways in which to resolve the problems we uncover. We cannot solve problems we cannot explain and understand. Nancy Hartsock writes, "Our theory gives us a description of the problems we face, provides an analysis of the forces which maintain social life, defines the problems we should concentrate on, and acts as a set of criteria for evaluating the strategies we develop" (1993, p. 8). Based on her reading of Antonio Gramsci (1971) Hartsock argues that theory not only can serve as a guide to social change, it can itself become a force for change. When people's understanding of social issues is enhanced by theory and their political activities are thereby made more efficient, theory serves as a tool for accelerating the process of social change.

This chapter has focused our attention on the contrast between two theoretical frameworks that are currently important to understanding families, structural functionalism and feminism. Structural functionalism was dominant in the 1950s and although it was challenged in the 1960s it remains prevalent in the work of social scientists (Glenn, 1987). Feminists are among a variety of social theorists who have challenged functionalism.

Feminists are a diverse group of people and in this text we will examine some of the debates that have emerged among feminist sociologists. But Janet Chafetz argues that all feminist theories share three characteristics:

> First, gender comprises a central focus or subject matter of the theory. Feminist theory seeks ultimately to understand the gendered nature of virtually all social relations, institutions, and processes. Second, gender relations are viewed as a problem. By this I mean that feminist theory seeks to understand how gender is related to social inequities, strains, and contradictions. Finally, gender relations are not viewed as either natural or immutable. Rather, the gender-related status quo is viewed as the product of sociocultural and historical forces which have been created, and are constantly re-created by humans, and therefore can potentially be changed by human agency. (1988, p. 5)

In addition to looking at the diversity among feminists, we will be examining the ways in which other theorists have tried to revision our understanding of families. In every chapter several theoretical frameworks will be described and compared. For example, Chapter 4 discusses a symbolic interactionist approach to the response of families to unemployment. In Chapter 6 we will look at what exchange theorists have had to say about the effect of work and family on relationships between husbands and wives. We will examine Marxist analysis of marriage in Chapter 8. In Chapters 11 and 12 we will look at psychoanalytic theory and the relationship between parents and children. And in Chapter 13 we will review social constructionism and its contribution to understanding violence in families. Many other theories will be reviewed as well and in every case the alternative theories will be examined with regard to their strengths, weaknesses, and connection to empirical evidence.

Each chapter in this book will include the four sets of issues proposed by Mills: the micro level, the macro level, the effect of the macro level on the micro level, and the effect of the micro level on the macro level.

To illustrate the micro level, the chapters will open with a vignette like the story of Kimberly and Michael. These stories are based on historical documents, interviews, and other sociological research but they may not be about real people. Their purpose is to capture an image of the ways in which the issues addressed in the chapter might be experienced by individuals in their everyday life. Be careful not to fall into accepting them as the only ways the issues are experienced. Let yourself think about how they might be similar or different for yourself or other people you know. For example, perhaps the story of Kimberly and Michael does not capture your experience at all because you are a single mother, a divorced person, or a gay man. What would your story sound like and how would it fit into the broader picture?

SUMMARY

Analyzing Social Life

- The assumptions people make about what their lives will be like are actually part of a much bigger picture of how their lives are affected and conditioned by society at large.
- It is important to analyze the social phenomena of families in a conscientious and systematic way. One tool that is especially important in analysis is the Sociological Imagination.

The Sociological Imagination: Bridging the Gap Between the Individual and Society

- Sociologist C. Wright Mills argued that the goal of sociologists should be to understand both individual activities and the organization of society, as well as their relationship. He named the ability to bridge the gap between the individual and society the Sociological Imagination.
- The micro level of analysis is like studying society with a microscope. It means studying the everyday life of the social interactions of individual families and individual family members close up.
- The macro level of analysis is like studying society with a telescope. It means studying family life as it is organized through social institutions such as laws, ideologies, divisions of labor, and technology.

Society Is a Human Invention

- Not only are our everyday lives affected by social structure, but people's day to day activities can create change in the social organization of our society.
- We need to examine four sets of issues, according to C. Wright Mills's model, to understand families:

  1. The micro level of individuals and individual families socially interacting.
  2. The macro level of social structure.

3. The ways in which the macro level affects the micro level.
4. The ways in which the micro level affects the macro level.

Families as a Political Issue

- Jean Lipman-Blumen defines power as "the process whereby individuals or groups gain or maintain the capacity to impose their will upon others, to have their way recurrently, despite implicit or explicit opposition through invoking or threatening punishment, as well as offering or withholding rewards."
- Based on this definition families are political in two ways. First, many family issues are political issues within the larger society. Second, each individual family is a political institution itself.
- Throughout the 1980s, family issues were central issues of debate in American politics. The two contending forces in the debate were conservatives and liberals.
- Conservatives are characterized by their focus on hierarchy, holism, and morality.
- Liberals are characterized by their focus on equality, individualism, and reason.
- Although it may sound peculiar to say that families are political institutions, Lipman-Bluman's definition applies when we observe the power plays and jockeying for position apparent in most families.

Recent History of the Sociology of Family

- In the 1950s sociology was a pluralist array of paradigms dominated by the theoretical perspective of structural functionalism, which was strongly identified with Talcott Parsons.
- Structural functionalism asserts that the male-dominated nuclear family with a male breadwinner and a female homemaker is the ideal family form.
- Structural functionalism was challenged in the 1960s by the civil rights movement and then in the 1970s by the gay rights movement and particularly the women's liberation movement.
- Scholars too have challenged our ideas about families.
- This rethinking of the family has been summed up in five themes by Barrie Thorne. The first is challenging the myth of the monolithic family. This theme includes both acknowledging and appreciating diversity as it occurs historically or within a particular historical period. The other four are: observing that families are sites of a variety of different activities including production, reproduction, socialization of children, and sex; acknowledging differences among family members within families by gender and age; raising questions about the conceptualization of families as rigid containers that separate us from the rest of society; and concern with the dilemma of creating families that allow for individual freedom and equality, as well as nurturance and sharing of responsibility.
- Feminist methodology is characterized by three principles: criticism of the supposedly value-free stance of many social scientists; giving voice to those from below; and combining scholarship with social activism.
- This chapter has highlighted the debate between structural functionalists and feminists but there are many other theoretical paradigms that are revisioning the family.

The transition from an agriculturally based economy to an industrialized one created changes in white families that included a decline in fertility and alterations in the expectations and experience of husbands, wives, and children.

2 A HISTORY OF U.S. FAMILIES WITH A FOCUS ON EURO-AMERICANS

Sarah was born in the late 1700s in Massachusetts. When Sarah was a child her mother Ruth told her stories about life in the mid-1700s on the family farm. Ruth spoke of the hard work and the self-sufficiency of the household. The family purchased very little because their own work provided nearly all of the food, clothing, and shelter they needed. Sarah liked to hear Ruth talk about all of the people that lived in the house—Ruth's parents, her six sisters and brothers, apprentices, servants, and frequently a youth who had run into trouble in town and whom the authorities wished to place under the stern guidance of Ruth's father. Ruth's father was the head of the household and led the family in their work, religion, and education. He even chose Ruth's husband for her. When she grew up, Sarah was glad her generation was able to choose their own husbands.

Sarah's mother's and grandmother's lives were controlled by hard physical work. Sarah's life is easier because she lives in Boston with her husband, a bank manager. Many of the items she uses in her household are purchased with his wages. She does not have to weave cloth, carry water, or tend animals. Sarah's job has been to raise her four children and to provide a moral and nurturing environment for them and her husband.

Although her life has been sheltered and privileged compared to the working-class women she sees on the streets of Boston, Sarah has not been entirely happy. She feels that she has been prevented from using all of her talents. When she was a child she was clever with numbers and now she sometimes watches her husband working at home and wishes she could work at the bank.

As Sarah sits in her dining room reminiscing about her memories of her mother and about her own life, there is a knock at the door. A messenger has brought her a letter from her daughter Elaine. Elaine writes, "July 28, 1848. My dearest mother, I am writing to tell you I have decided to break off my engagement to be married. This week I have attended the Seneca Falls Convention and I intend to dedicate my energies to winning the vote for women. We have remained in the shadow of men for too long. Your loving daughter."

- *How much have families changed over the past few centuries?*
- *Would these changes have been the same for Sarah and her kin if she had been from another social class?*
- *How is Sarah's life better than her mother's? How is it more difficult?*
- *How are the differences between Sarah and Ruth's lives related to the development of an urban industrial economy in the nineteenth century?*
- *What kinds of problems do you think Sarah's husband might have because of his role in their family as the only one able to earn money and the person who must always be in charge of the family?*
- *The Seneca Falls Convention of 1848 is recognized as an important event in women's history. What kinds of family issues do you think the women discussed at this meeting? How might they be different from the family issues about which women speak today?*

STUDYING THE SOCIAL HISTORY OF FAMILIES

Until recently, sociologists have been unconcerned with historical change (Goode, 1964). Before 1960 sociologists who were interested in the history of American families had only three or four books from which to choose. Since the 1960s, however, the field of social history and especially the social history of families has burgeoned (Gordon, 1978).

The growth in scholarly attention to the social history of families has resulted from two sources. First, as we noted in Chapter 1, scholars who are revisioning families are concerned with acknowledging the diversity of families, including the ways in which families vary over time.

Second, the growth in the study of the social history of families is related to a change in our thinking about legitimate areas of scholarly interest (Bridenthal & Koonz, 1977; Tilly & Scott, 1978). Until recently, historians tended to focus their attention on dominant figures in society. Their concern was with great men and extraordinary events. Since women and other "lesser" people like peasants, slaves, and workers were less likely to be dominant figures, their stories were not as likely to be examined. The focus of new scholarship in history, however, has shifted to include the everyday life of the common person, which includes family matters.

Chapters 2 and 3 review the family history of people in the United States from the colonial period until the mid-twentieth century. Since race ethnicity has had such a striking effect on the experience of Americans, this history is divided into two chapters, one with an emphasis on Euro-Americans and one with an emphasis on African Americans. This chapter focuses on Euro-Americans although Native Americans, Mexican Americans, and Asian Americans are included.

The central theme of this chapter is that as the Industrial Revolution developed in the U.S., family organization changed with it. As the economy moved through distinct stages, so did families. Along with this general pattern of macro-level changes in the economy causing micro-level changes in the organization of families, however, there are two important variables. First, at each stage important differences existed among families who differed by race ethnicity and social class. Second, people like Elaine in the scenario that opened this chapter organized to try to change both the economy and families; the micro level affected the macro level.

The chapter is organized into three major sections. The first section reviews U.S. history through a number of periods: colonial America, industrialization, the pioneer era, the Great Depression, World War II, and the fifties. Each of these subsections discusses the development of various families as well as the changing relationships between men and women, adults and children. The second section is centered on a theme that traces throughout the history, the problem of patriarchy. The third section describes one of the responses to patriarchy, the organization and development of the women's movement.

STAGES IN THE HISTORY OF EURO-AMERICAN FAMILIES

White middle-class families in the U.S. have moved though three stages from the time the Europeans first began to immigrate to North America in the 1600s until the present. The family form that characterized the first stage, called the Godly Family, lasted from the early 1600s until about the time of the Revolutionary War in 1776.

The Modern Family is the term family historians use to designate the family form like that of Sarah's in the scenario at the beginning of the chapter. This form emerged among white middle-class Americans at the end of the eighteenth century. The Modern Family consists of a breadwinning husband, a housewife, and their dependent children. Table 1-1 in Chapter 1 showed that this family form has declined in the last half of the twentieth century and is no longer the most common one (Coontz, 1988; Degler, 1980; Mintz & Kellogg, 1988). Before its decline, the Modern Family passed through two periods: the Democratic Family, which lasted from about 1780 until the end of the nineteenth century; and the Companionate Family, which lasted from about 1900 until 1970. In the two decades since 1970, family organization has moved into the third stage, the Postmodern Family. Postmodern families will be the focus of our attention in Chapters 4 through 14.

One of the critical factors in the social context of these changes in family organization has been the organization of the economy. An economy is a system by which the goods and services that people need to live are produced and distributed. The macro-level social institution of the economy has changed as industrialization developed through the eighteenth, nineteenth, and twentieth centuries. These macro-level changes have been accompanied by parallel developments at the micro level of family organization.

Changes over time in the economy have shaped the history of the organization of families, as shown in Table 2-1. The first stage, the Godly Family, is associated with a preindustrial, agriculturally based economy. The second stage, the Modern Family, is associated with the development of industrialization. The third stage, the Postmodern Family, is associated with deindustrialization. This chapter focuses on the first two stages. Chapter 4 explains what deindustrialization means and how it relates to recent changes in family organization.

At the same time that macro-level changes in the economy have had an important effect on the organization of families, activity at the micro level among individuals and individual families has also affected the organization of the economy. For example, one of the first goals of the women's movement was to contest the laws that barred married women from owning property.

The Godly Family

Colonists from Germany, Holland, Sweden, Ireland, and other European countries came to settle in the U.S. in the 1600s. Those from England were particularly numerous, especially in what came to be the New England states.

The English colonists who came to live in North America were escaping a world that they did not find to their liking. The 20,000 people who emigrated from England to Massachusetts between 1620 and 1640 had a vision of a new world that included a new kind of family, one that did not exist in England. They wanted to establish a Godly Family that conformed strictly to the teachings of the Bible (Mintz & Kellogg, 1988).

English family life in the sixteenth and seventeenth centuries was unstable because of high mortality rates and a mobile population. High infant and child mortality rates meant families would raise few children to adulthood and high rates of death among adults meant that marriages were short and stepfamilies were common. Many people in England, furthermore, were not allowed to marry. For example, servants, apprentices, and college lecturers were not legally permitted to marry at all. And by social custom young men were not allowed to marry until their fathers died and they received an inheritance with which to support a family. Children born to unmarried women were considered bastards and barred from certain rights granted to "legitimate" children (Mintz & Kellogg, 1988).

TABLE 2 • 1
The History of Euro-American Families in the U.S.

Family Form	Economic System	Time Period
1. Godly Family	Preindustrial agriculture	1620–1780
2. Modern Family	Industrial capitalism	1780–1970
a. Democratic Family		1780–1900
b. Companionate Family		1900–1970
3. Postmodern Family	Postindustrial Capitalism	1970–present

The Godly Family the Puritans established in the colonies was characterized by four factors:

1. The family structure was patriarchal.
2. Families were highly integrated into the community.
3. Each family was a nearly self-sufficient economic unit.
4. All social activities, including education, health care, and welfare, took place within families.

The Godly Family was a patriarchal one. Patriarchy literally means rule by the father. The term, however, is used in a variety of ways. Radical feminists use the term patriarchy to describe a broad system of oppression and control of women by men (Jagger, 1983). Before the emergence of radical feminism in the 1960s, the term patriarchy was used mostly by anthropologists who defined it more narrowly as a system of control by men that was associated with a fairly specific family organization. Gayle Rubin (1975) described this older definition of patriarchy as a family that was dominated by "one old man whose absolute power over wives, herds, children and dependents was an aspect of the institution of fatherhood." Puritan families fit both definitions of patriarchy.

In the religious beliefs of the Puritans, God the father ruled over his children and they modeled their human families after this image. Puritan fathers exercised authority over large households that included their wives, children, and servants. Puritan fathers were the representatives of their families in the social and political affairs of the community; they owned most of the property; and their wives, children, and servants were required by civil and religious law to submit to the patriarch's authority.

In Connecticut, Massachusetts, and New Hampshire statutes called for the death of children who cursed or struck their fathers. Fathers had the legal right and obligation to select a spouse for their children. And even when the children were married adults, fathers could intervene in their affairs.

The role of wives was also determined by the demands of patriarchy. "Puritans believed that a wife should be submissive to his [her husband's] demands and should exhibit toward him an attitude of 'reverence,' by which they meant a proper mixture of fear and awe" (Mintz & Kellogg, 1988, p. 11). Court records show that women who did not obey their husbands were subject to fines and whippings.

In addition to laws that demanded that children and wives submit to the patriarch's authority, fathers also controlled most of the property. Their ownership of the most important means of obtaining food and shelter meant that those who did not own the land had to obey their wishes. During this time, married women were not allowed to own property and single women were not likely to inherit property. In order to survive, women had to be attached to husbands who would provide them with access to food and all of the other materials they needed. Sons could legally own property once they became adults but they rarely were able to accumulate enough capital to do so. Eventually sons were likely to obtain property through inheritance from their fathers but the deeds were almost never passed on until the father died. This meant that adult sons worked for many years under the control of their landowning fathers (Greven, 1978).

Puritans, however, did not believe that families should be isolated patriarchal kingdoms. Families were very much part of the whole community. The integration of family and community is the second characteristic of the Godly Family. Actions that we might consider outside interference in family affairs were accepted as legitimate. For example, in the late 1600s in Massachusetts, the court appointed "tithingmen" who were assigned to watch over ten or twelve households. Their job was to make sure that the internal affairs of the families were harmonious and proper. They sought out and punished parents who did not control their children, husbands who did not support their families, and anyone who fornicated or committed adultery. These tithingmen were devoted to their jobs and successfully identified many people who were not conducting themselves properly. Court documents show that fornication and adultery were by far the most numerous criminal cases (Morgan, 1978). Punishments for not maintaining a proper home included fines, brandings, whippings, and occasionally the death penalty.

When children were found to be unruly they were removed from their home and placed in the household of a patriarch whom the community believed to be a more effective disciplinarian (Mintz & Kellogg, 1988). In the scenario at the beginning of the chapter, Sarah recalls the stories of her grandfather boarding one of these wayward youths.

The third characteristic of the Puritan family was that it was a self-sufficient economic unit. Puritan families were likely to be large. Nine was the average size and many families had as many as fifteen members. This included not only children and parents, but unrelated apprentices and servants and extended kin such as unmarried aunts.

One of the myths of the preindustrial patriarchal family is that it included several generations of kin. The television family "The Waltons," which has four generations living in the same house, is an illustration of this myth. Although some Puritan households were organized like the Waltons', they were rare because of the short life span. Puritan households were large, not because they included several generations of people related by blood, but because they were usually composed of a nuclear family and a number of other unrelated people whose work helped to make the household economically self-sufficient (Modell & Haraven, 1978).

All members of the household participated in the production of nearly all the goods they needed to survive. A few people produced specialized goods, like shoes, furniture, and leather to be sold in distant markets. But like Ruth's family, nearly all white families in colonial America were primarily self-sufficient agricultural units (Mintz & Kellogg, 1988).

The work that took place in these families was enormous. Julie Matthaei (1982) has examined the records of the homemaking tasks of the women during this era. She found that work fell into four categories: mothering, meal provision, clothes provision, and nursing.

During the colonial period, the activity of mothering consisted mostly of pregnancy and childbearing, which occupied much of a woman's life. A woman bore an average of eight children. She would see one in three of her children die before it reached the age of twenty (Matthaei, 1982). Maternal deaths were also common. Childbearing was dangerous work. For example, in Plymouth in the 1600s, records show that 20 percent of women died from causes associated with childbirth.

Providing meals was another onerous activity. A recipe from 1742 for a pork dish begins, "Cut off the head of your pig; then cut the Body asunder; bone it, and cut two Collars of each side; then lay it in Water to take out the blood . . . " (Matthaei, 1982, p. 42).

Women were also expected to provide clothes for their families. Starting from scratch meant growing the cotton and tending the sheep who provided the wool. Thread was made, cloth was woven, and finally clothes were sewn.

The fourth area of domestic work was the provision of medical care. Health care was so closely related to women's other domestic work that cookbooks included, alongside recipes for vegetables, pastry, and wine, ones for medicines, ointments, salves, and sleeping potions.

The final characteristic of Godly Families was that they were the site of social activities that today we expect to find in other institutions like the church or the school. For example, the law required that fathers lead their households in prayer and teach all family members, servants, and apprentices reading and the principles of religion and law, as well as a trade. The family was also responsible for social service tasks we usually associate with outside agencies such as caring for orphans, the ill, and the elderly (Demos, 1970).

Stephen Mintz and Susan Kellogg (1988) assert that the most striking difference between the Godly Family and its present-day counterpart is in the experience of childhood. By the age of seven, children were full participants in the productive work of their families, farming, weaving, gardening, making soap and candles, and all of the other activities of the household.

Children were strictly controlled by their families, especially their fathers. The Puritans believed that children were born sinners and that the task of parents was to drive out that sin through religious study, physical beatings, and psychological pressure. One technique used to ensure that this was taking place was for families to send their adolescent children to live in another household where the adults, unrestrained by parental concern, would be able to administer the proper discipline. Nearly all boys and girls of all classes were placed in other households as students, servants, or apprentices.

The Godly Family lasted for about 150 years until the late 1700s. Rising population and abundant land to the west made it increasingly difficult for fathers to use the distribution of land to their sons as a way to control them. In addition, changes in the American economy opened up new sources of income in the production of goods outside the family. Children who could earn their own money in factories and shops could also defy their parents' wishes about sexual behavior and make their own decisions about when and whom they should marry. Among Euro-Americans, as the industrial economy developed and manufacturing and transportation improved, the Modern Family began to emerge as an alternative to the Godly Family.

Before moving to the next historical period, let us look at a family form that existed in North America at the same time as the Godly Family, the Iroquois Family. The Iroquois Family provides a striking contrast to the Godly Family.

Iroquois Families

When the Puritans arrived in North America to establish their new society and its Godly Family, between ten and twenty million Native Americans inhabited the continent (Amott & Matthaei, 1991). They represented about 300 distinct cultures and 200 languages (Nabakov, 1991). Before their history could be documented many of these peoples were killed by war and disease brought by the Europeans. We know very little about most of them. One group about whom records were kept and saved is the League of the Iroquois.

The League of the Iroquois was made up of six member nations—the Mohawk, Onondaga, Seneca, Oneida, Cayuga, and Tuscaroras—who lived in the area that is now the state of New York. It is estimated that the Seneca numbered about 10,000 when the Europeans began settling in North America (Jensen, 1991).

Judith Brown (1977) and Joan Jensen (1991) are two anthropologists who have investigated the organization of Seneca families. They found that Seneca people lived in extended kin groups that were controlled by older women. These matrons owned the house, the seed, the tools, and the harvest. They supervised the work and distributed the food to the rest of the clan. The land was communally owned by all of the women in common.

The matrons also wielded much social power. In 1390 the Iroquois Confederacy created a constitution called the Great Law of Peace. The constitution stated,

> Women shall be considered the progenitors of the nation. They shall own the land and the soil. Men and women shall follow the status of their mothers. You are what your mother is: the ways in which you see the world and all of the things in it are through your mother's eyes. What you learn from the father comes later and is of a different sort. . . . Clan Mothers and their sisters select the chiefs and remove them from office when they fail the people. . . . Clan Mothers! You gave us life—continue now to place our feet on the right path. (Amott & Matthaei, 1991, p. 36)

Just as the Puritan patriarchs arranged their children's marriages, the Seneca matrons arranged for their children's marriages and lineage was traced through the mother's line. Because the women owned the houses in which the families lived, if a wife wished to divorce her husband she could order him to leave on grounds of sterility, adultery, laziness, cruelty, or bad temper (Evans, 1991). If he wanted to stay, he had to have an aunt or grandmother intervene on his behalf (Morgan, 1965).

The governing body of the Seneca was composed of men. The men, however, were appointed by the matrons who watched the men carefully and quickly deposed any whom they felt was not representing the women's interests. The Seneca matrons could not decide to go to war because that was the business of the male warriors. The women, however, ultimately controlled even decisions of war and peace because they could refuse to provide the warriors with shoes or food (Randle, 1951).

Both of the societies that inhabited what is now the northeastern United States, the Puritan society and the Iroquois Confederacy, were preindustrial. They both had family forms in which power was held by those who controlled the source of livelihood, the land. In the case of the Puritans, the men controlled the land and the family form was patriarchal. Among the Seneca the women controlled the land and the family form was matriarchal.

The Modern Family, Stage 1: The Democratic Family

The Modern Family is divided into two stages. The first stage is called the Democratic Family. This term was coined by Alexis de Tocqueville, a French aristocrat who came to the United States in 1830 and wrote a book about the new society being created in North America and the new family form that was emerging among the white middle class.

The Modern Family differed from the premodern form—the Godly Family—in four ways:

1. Work was split into unpaid domestic work and paid work.
2. Individuals could freely contract marriages based on love.
3. Privacy and the separation of families from the community were promoted for middle-class families.
4. Women had fewer children, but were supposed to spend more time at their "natural" and demanding task of mothering (Stacey, 1990) (see Figure 2-1).

The Split Between Family and Work Among Euro-Americans The rise of industrial capitalism coincided with the rise of the Modern Family (Stacey, 1990). During the early 1800s the operation of families as self-sufficient economic units began to change. In 1776, Watt perfected the invention of the steam engine and the Industrial Revolution rapidly advanced. Family farms began to specialize in crops to sell on the market and the production of food and clothing, soap, candles, and nearly every other commodity moved from the home of the premodern, preindustrial family into the factories of industrial capitalists.

As the jobs moved so did the workers. A new domestic division of labor appeared to replace the old preindustrial form of women, men, and children working together in a common economic enterprise. Among the middle class, husbands were expected to follow the wage-earning jobs. Women were expected to stay home and the occupation of housewife emerged (Mintz & Kellogg, 1988). Women were supposed to use their time to take care of their families, to raise their children, and to maintain the morals of the community.

Among middle-class whites, men and women became divided from one another in their work. As industrialization developed throughout the nineteenth century, men were increasingly brought into the paid labor force where their work was clearly demarcated by time and place. Women's work was unpaid, took place at home, and did not have time constraints. Work and family life came to be viewed as two distinct spheres and women and men were viewed as the appropriate occupants of these two different worlds (Mintz & Kellogg, 1988).

The conceptualization of social life being divided into two spheres still influences popular notions about work and family. Chapter 6 looks at the way in which the idea of a split between work and family is still part of many people's thinking, although the idea has been criticized because it does not accurately depict the experience of all families (Bose, 1987). Many families did not have a breadwinning husband and a housewife in the nineteenth century. This was the case especially for African American families living during the periods of slavery and sharecropping, as we shall see in Chapter 3. In the twentieth century white families with breadwinning husbands and housewives became less and less common as well.

In families where work was divided between a wage-earning man and a housewife, problems occurred. Because housework does not earn money, it is economically invisible and the women who do it appear to be noncontributing members of their society. Chapter 7 describes how this problem remains even today.

In addition, access to goods and services becomes limited to those who can purchase them. Wages become essential to survival. Women who did unpaid housework and did not earn money were dependent on their wage-earning husbands (Stacey, 1990).

Also, the women who were left in these middle-class homes began to have other responsibilities that tied them to unrealistic and stifling roles. As the production of food and shelter moved into the factory, housework became the task of providing the "proper" social environment for children and husbands. This required that women behave in carefully prescribed and restrictive ways. The prescription for proper behavior for wives and mothers has been called the "cult of true womanhood" (Welter, 1978).

The Cult of True Womanhood Between 1820 and 1860 a new definition of womanhood emerged and was widely disseminated in popular magazines and novels, and religious literature and sermons. In the 1700s, womanhood was associated with deviousness, sexual voraciousness, emotional inconstancy, and physical and intellectual inferiority (Mintz & Kellogg, 1988). This image of womanhood changed dramatically in the nineteenth century. Historian Barbara Welter (1978) investigated the image of women in the literature of the early nineteenth century. Welter coined the term "cult of true womanhood" to name the image of women she found.

According to the advice manuals of the nineteenth century, true women were to be judged by how well they exhibited four virtues: piety, purity, submissiveness, and domesticity. If a woman displayed these attributes she was promised happiness and power. If a society was comprised of women with these attributes it was safe from damnation. True women who displayed the four virtues were the pillars of their community (Welter, 1978).

Piety was argued to be the most important attribute and women who were without religion were believed to be restless and unhappy. Welter (1978, p. 314) notes that physicians and scholars "spoke of religion [for women] as a kind of tranquilizer for the many undefined longings which swept even the most pious young girl and about which it was better to pray than to think."

Purity was the second attribute. Women were warned that men were more sensual and prone to sin than women. Women must protect both themselves and these weak men by not giving in. A woman's greatest treasure was her virginity, which she bestowed on her husband on the single most important day of her life—her wedding day.

Books and magazines in the nineteenth century gave practical advice to women to ensure that they remained chaste. In the early 1800s Eliza Farrar recommended, "Sit not with another in a place that is too narrow; read not out of the same book, let not your eagerness to see anything induce you to place your head close to another person's" (Farrar, 1837, p. 293).

The third attribute of true women was submissiveness. Women were supposed to be weak, timid, humble, and dependent. Caroline Gilman (1834, p. 122) advised brides, "Oh young and lovely bride, watch well the first moments when your will conflicts with his to whom God and society have given the control. Reverence his wishes even when you do not his opinion." One advice book went so far as to advise that "if he is abusive, never retort" (Welter, 1978).

The fourth characteristic of true women was domesticity. Women's duty was to provide a domestic life that would keep men home and out of trouble. They were to be skilled in housework, needlework, health care, and flower cultivation.

Capitalist Industrialization and the Working-Class Family In the latter half of the nineteenth century, industrialization intensified. Millions of Europeans immigrated to the United States seeking work in the growing numbers of manufacturing plants. At the beginning of the 1800s about 5,000 immigrants arrived every year. By 1850 that number had risen to 2.8 million per year and between the end of the Civil War and the beginning of World War I, 24 million immigrants arrived.

The Modern Family consisting of a breadwinning husband, a housewife, and their children, which characterized the white middle class, was not an effective form for the working class. Working-class people, especially recent European immigrants who poured into the cities, could not live on the income of one family member.

Sometimes entire families were employed in the factories. In 1828 an advertisement in a Rhode Island newspaper read: "Family Wanted—Ten or Twelve good respectable families consisting of four or five children each, from nine to sixteen years of age, are wanted to work in a cotton mill, in the vicinity of Providence" (Mintz & Kellogg, 1988, p. 94).

The more typical situation, however, was for fathers and children to work for wages and wives to earn money while staying at home (Sacks, 1984). Child labor was extremely common. One study in Massachusetts in 1875 found that 44 percent of the outdoor laborers and mill operators were children (Matthaei, 1982). Although boys were more likely to work than girls, girls were frequently pulled out of school to work so that their brothers could finish their educations.

Young single women during this time were likely to work for wages. In 1890, for example, 70 percent of foreign-born single white women were in the paid labor force (Matthaei, 1982).

Wives who were not employed outside of the home contributed to the family by earning money taking in boarders. Only about 5 percent of today's families take in boarders, but in the 1880s in the working-class towns of

Pennsylvania, nearly one-half of the homes had nonrelative boarders living in them. Women who stayed at home also earned money by making flowers, doing piecework from the clothing factories, or doing laundry.

Decision making about who would work and where was done on the basis of the collective needs of the family as a whole, rather than on the basis of individual preference. If times were hard, working-class children were expected to defer marriage or education to remain with their parents and work (Mintz & Kellogg, 1988).

The Problem of the Family Wage The ideology of the cult of true womanhood and its insistence that women remain in their homes as unpaid housewives did not fit the economic circumstances of the working class. Working-class men were unable to earn sufficient wages to provide for economic dependents. In some cases, as we have seen above, the ideology of the cult of true womanhood was set aside and working-class women found ways to earn wages. Sometimes the ideology was used to support the demand for better wages for working-class men—a family wage.

Samuel Gompers, an early leader in the American labor movement, called for a family wage in 1898. He stated that working men must have "a minimum wage—a living wage—which when expended in an economic manner shall be sufficient to maintain an average sized family in a manner consistent with whatever the contemporary local civilization recognizes as indispensable to physical and mental health, as required by the rational respect of human beings" (Boyle, 1913, p. 73).

Gompers found support for his quest for a family wage among those who were usually his adversaries, like Henry Ford, a major industrialist. Ford strongly supported the family wage and the maintenance of families in which the husband provided for his non–income-earning wife. As a way to allow men to become the sole breadwinners in their families, Ford established the wage of $5 a day for his employees—twice that of workers in other factories at the time. These improvements for working-class men, however, were done at the expense of working-class women. In 1919, Ford fired eighty-two women from their jobs in the Ford plant in Highland Park, Michigan, when management discovered that they were married to men who had jobs.

The family wage was both a boon and a tragedy for women. Women whose husbands were able to earn a family wage like the Ford workers benefited from the higher wages their husbands brought home. At the same time, the policy of a family wage was used to justify pushing women out of jobs and back to their "proper place" as non–wage-earning housewives.

The contradictory nature of the family wage was investigated by Martha May (1982). She argues that the key problem with the family wage was that it was premised on a particular organization of family, one with a male breadwinner and his dependent wife and children. The family wage defined normal men as those who accepted their position as economic provider. The family wage endorsed the cult of true womanhood. It promoted the idea that all women would and should become economically dependent wives and, therefore, could be legitimately excluded from the labor force.

The call for a family wage, therefore, carried a mixed message for working-class people, especially for working-class women. On the one hand the demand

BOX 2 • 1 AMERICA IS FOUNDED ON ITS SELF-RELIANT FAMILIES—OR IS IT?

In 1985 the Gramm-Rudman-Hollings balanced budget amendment was passed, resulting in a number of cutbacks in federal spending for social services. Senator Philip Gramm, a co-author of the bill, is well known for his opposition to "government handouts." In contrast to his beliefs, Senator Gramm's life reveals a long history of government support. Gramm grew up in a family supported by a federal veteran's disability pension. He went to a state-supported college on a grant paid by the federal War Orphans Act and then received a graduate degree funded by a fellowship from the National Defense Education Act. His first job was at Texas A&M University, a federal land grant institution. Senator Gramm is an American success story, but his accomplishments were based in an array of publicly funded support (Broder, 1983; Coontz, 1992).

Senator Gramm's biography shows the contrast between the popular rhetoric that claims self-reliance as a cherished value and longstanding tradition in the U.S. and the real reliance of Americans on collective institutions. This contrast is rooted in a mythical view of our history, especially the legends of the frontier family and the suburban family of the 1950s (Coontz, 1992).

One of the sources of images of the self-reliant pioneer family is the collection of *Little House on the Prairie* books, which depict an isolated family pitted against the elements, on their own, with no help from their neighbors or from their government.

> In reality, prairie farmers and other pioneer families owed their existence to massive federal land grants, government funded military mobilizations that dispossessed hundreds of Native American societies and confiscated half of Mexico. . . . It would be hard to find a Western family today or at any time in the past whose land rights, transportation options, economic existence and even access to water were not dependent on federal funds. (Coontz, 1992, p. 73)

Public funds were used to pay $15 million to purchase Louisiana, to support three years of war with the British to obtain Florida, $150 million to build canals in the Midwest and around the Ohio and Mississippi rivers. The U.S. government paid $9 million for Cherokee land before forcing the Cherokees to march to Oklahoma ($6 million was then deducted from the bill as expenses for the eviction). Public money was used to finance a $97 million war against Mexico, and then to allow the victorious United States to pay $25 million for Texas, California, and parts of New Mexico and Arizona (Coontz, 1992).

At a more local level, pioneer families were not entirely self-sufficient. Frontier life was very much a community affair, sharing work, tools, products, and labor.

Like the legendary frontier family of the nineteenth century, the suburban family of the 1950s is a prevalent image in the myth of self-reliance. The surge in homeownership and other indications of a rising standard of living is believed by many to have been a result of people standing on their own two feet without government interference or community support. In fact, suburban families were the beneficiaries of enormous governmental subsidies. The GI Bill allowed nearly half a generation of young men to obtain a college education. Prior to World War II banks often required a 50 percent down payment on homes. The GI Bill provided for down payments of 5 to 10 percent and guaranteed thirty-year mortgages for fixed rates of 2 or 3 percent. The Veteran's Administration asked for only a dollar down. Fixed rate mortgages, furthermore, freed this generation from inflation—at least in their housing costs—for thirty years (Lee, 1986; Coontz, 1992). Even the construction of the homes was partially federally funded. The development of aluminum clapboards, prefabricated walls and ceilings, and plywood paneling all resulted from government-funded research.

The idyllic families of Beaver Cleaver, Ozzie and Harriet, and the Wonder Years included hard-working people. Their ability to sustain their standard of living, however, was largely based on the "government handouts" Senator Gramm and others decry but which undoubtedly allowed them to succeed.

for a family wage was a radical one that would improve the quality of life of working-class people and free them from some of the oppression of long, grinding hours in the factories. On the other hand, the demand for a family wage was conservative by making working-class women more tightly controlled by and dependent on their husbands. The limited economic opportunities of working-class women were further eroded by the family wage because it legitimated the exclusion of married women from the workplace.

Pioneer Families: Leaving Friends, Facing Hardship The previous discussion of the development of the Democratic Family focused on the eastern part of the United States, which was the first section of the nation to urbanize. In the west during the nineteenth century another social change was taking place, the westward expansion of Euro-Americans. The migration had important effects on the families that made the journey.

Under the Homestead Act of 1862 any citizen or anyone declaring an intention to become a citizen who was over the age of twenty-one could pay a fee of $10 to live on 160 acres of government-owned land. After living on the land for five years and making certain improvements the person could file for final title to the land. From 1840 to 1870 between a quarter and a half million people moved west. It may have been the greatest movement of people in U.S. history (Billington, 1949).

Historians who have studied the diaries and journals of the people who set out on the Overland Trail have found that in nearly every case it was men who decided to take advantage of this opportunity and among married couples, the husband who made the decision for the whole family to go. And only about 25 percent of the women who made the trip said they agreed with that decision. There is no reliable information on how many couples chose to separate when they could not agree to move west (McNall & McNall, 1983). But women's diaries and letters show that among those women who did go many felt sadness and anger at having to leave their families and friends behind.

The close ties among women during the nineteenth century is a topic that Carroll Smith-Rosenberg (1985) has examined extensively. She reviewed thousands of letters written during the 1800s. Collecting these types of data is called content analysis: studying cultural artifacts that were not originally prepared to provide information for a researcher. Smith-Rosenberg examined written records, letters, and diaries. Barbara Welter, who studied the cult of true womanhood, reviewed narrative texts, speeches, and advice manuals for brides and wives. Later in this chapter I will cite Valerie Matsumoto's research on Japanese Americans who were kept in internment camps during World War II. Matsumoto analyzed letters written by the people being held in the camps.

Harriet Martineau (1834), one of the first women sociologists, wrote in the early 1800s that the study of cultural products was a key source of information (Reinharz, 1992). Modern scholars have taken her advice, arguing that especially when studying women, there may be little information in official documents and the researcher must search for these informal sources of information. Shulamit Reinharz warns, however, that one of the problems with such research is that the documents that are left tend to overrepresent more privileged groups of women who were literate and whose cultural products were saved.

Smith-Rosenberg found that the letters written by women to each other reveal deep and intimate relationships. One woman, a twenty-nine-year-old wife and mother, writes to her lifelong friend, "Dear darling Sarah, How I love you and how happy I have been! You are the joy of my life . . . I cannot tell you how much happiness you gave me, nor how constantly it is all in my thoughts . . . My darling how I long for the time when I shall see you . . . your Angelina" (Smith-Rosenberg, 1985, p. 56).

Smith-Rosenberg argues that several factors may have permitted women to form close emotional relationships. First, rigid gender differentiation segregated women and men emotionally from each other. Second, the roles of mothers and daughters overlapped with each other. Third, frequent pregnancies, births, and nursing bound women together. These three factors led women to create supportive networks and intensely emotional and perhaps sexual relationships with each other. Leaving both individual friends and the networks of female support in a community was especially difficult for women whose husbands chose to go west.

In addition to leaving their friends, pioneer women and men faced enormous physical hardship. Anna Howard Shaw wrote,

> When [my father] took up his claim of three hundred and sixty acres of land in the wilderness of northern Michigan, and sent my mother and five young children to live there alone until he could join us eighteen months later, he gave no thought to the manner in which we were to struggle and survive the hardships before us. . . . we were one hundred miles from a railroad, forty miles from the nearest post office, and half a dozen miles from any neighbors . . . and a creek a long distance from our house for our only source of water. (Shaw, 1915, p. 28)

Women's hard labor also contributed significantly to their families' cash income. In some households women's sale of goods like butter and milk amounted to incomes as high as $200 or $300 a year in an economy where a cash income of about $400 per household was typical (McNall & McNall, 1983).

The demands placed on pioneer women, however, ultimately may have helped them to make their case for the right to vote. In 1920 women's suffrage was passed by Congress. But women in the pioneer state of Wyoming, the first state to give women the vote, were granted suffrage in 1869. The fact that a western state was the first to grant women this right has been linked to many factors. Two of these factors that relate to the role of women in families are the recognition of the work accomplished by women pioneers, and the value placed on women because of their relative scarcity in the region (Hymowitz & Weissman, 1978).

Split Households Among Chicanos and Chinese Americans In describing the historical development of the relationship between work and family, our attention thus far has been focused on working-class and middle-class Euro-Americans. The dominant ideology romanticized the split between work and family while making it inaccessible to working-class families because of low wages. For the middle class the split between work and family was an important part of their history and has created problems, especially for women. For the working class the split was less real. In Chapter 3 we will examine the history of African American families and see the blurring of the lines between work and family and women's and men's work.

In this section we will examine two groups of people in the U.S. for whom the split between work and family and men and women was dramatic. Both Chicanos (Mexican Americans) and Chinese Americans were required to divide their households for the men to be allowed to work (Glenn, 1991).

In the late nineteenth and early twentieth centuries Chicano men were recruited to work in mining camps and on railroad gangs which did not provide accommodations for wives and children and sometimes prohibited them from being there (Barerra, 1979). The men were required to leave their families for long periods of time in order to earn a living.

The women who were left behind engaged in subsistence farming and wage labor in urban areas. They had jobs as cooks, maids, and laundry workers in hotels and other public establishments (Camarillo, 1979; Garcia, 1980). In addition to these tasks, Chicanas—Mexican American women—also raised their children and provided health care in communities that often had no running water or public sanitation systems.

After the turn of the century the split between work and family and women's and men's work began to decline as growers increasingly hired entire Chicano families to work as farm laborers in the large fields of the southwest. For many Chicanos the split between work and family disappeared as husbands, wives, and even small children and infants went to work in the fields. While these changes may have improved family life by allowing family members more contact with one another, living conditions were no better. Infant and child mortality was high, wages were low, education was nonexistent, and life expectancy was short (Glenn, 1991).

Among Chinese workers the split was more formalized. A significant number of Chinese immigrants arrived in the western United States during the 1800s. By the end of the century in California, for example, the Chinese constituted 25 percent of the entire work force (Takaki, 1989). Nearly all of these immigrants were men. There are several reasons for this. First, only male workers were offered jobs and when the men were recruited it was with the specification that their wives and children must be left behind. Second, the men's families in China may have encouraged them to leave their wives and children behind, believing that this would increase the probability that the men would remain in contact with their relatives in China and especially that they would continue to send money home. A third factor influencing the number of Chinese women entering the U.S. was immigration law. In 1882 the Chinese Exclusion Act prohibited Chinese women from entering the country even if they had husbands waiting for them. The purpose of the act was to prevent the Chinese from settling permanently after their labor was no longer needed for major projects like building the railways (Glenn, 1991). Fourth, Chinese people were afraid to raise families and make their homes in the U.S. because of racial harassment and attacks against them (Daniels, 1978). "From Los Angeles to Seattle and as far east as Denver and Rock Springs, Wyoming, Chinese were run out of town or beaten and killed by mobs whose members were almost never brought to justice" (Kitano & Daniels, 1988, p. 22). Because of these four factors, very few Chinese women lived in the U.S. during the nineteenth century and as late as 1930 the ratio of Chinese men to women in the U.S. was still 11:2 (Glenn, 1991).

Little is known about the wives and children who were left behind in China. Their numbers, however, were significant since about half of the men who came to work in the U.S. were married. Many of these women never saw their husbands again. Some men were able to visit China twenty or thirty years later. Most of the wives lived with their husband's parents where their behavior, especially their sexual behavior, was carefully monitored. Although the women rarely saw their husbands the marriage contract held a tight rein on their lives (Glenn, 1991).

All of these groups, Chicanos, Chinese Americans, African Americans, working-class Euro-Americans, and middle-class Euro-Americans—were living in the same developing industrial capitalist economy in the United States. But they all experienced a different history of changing relationships between the organization of work and family and between men's work and women's work. Among African Americans the lines were most blurred, as we will see in Chapter 3, and among Chinese Americans the lines were most distinct. Each of these groups also faced problems as a result of the way in which families and work were organized throughout this period. Those problems, however, differed significantly (Dill, 1986).

The Modern Family, Stage II: The Companionate Family

Table 2-1 showed the Modern Family divided into two periods, the Democratic Family, which lasted from about 1780 until 1900, and the Companionate Family, which lasted from about 1900 until 1970. Burgess and Locke (1945) witnessed the transition at the turn of the century and called it the change from institutional marriage to companionship marriage. They argued that the change was characterized by a move from a form of marriage that was controlled by external factors like the law, public opinion, and elaborate ritual to one that was internally directed by "mutual affection, sympathetic understanding, and the comradeship of its members" (Burgess & Locke, 1945, p. vii).

Burgess and Lock's view of this transition implies that the change was relatively uniform and that it occurred not just in terms of ideas but in terms of real behavior. These two assumptions have been questioned by authors who argue that many married couples still lack the characteristics of companionate marriages (Cancian, 1987). Others maintain that the changes have been much stronger at the ideological level than they have in real behavior (LaRossa, 1992). Nearly all scholars agree, however, that ideology and rhetoric about the value of companionate marriage has increased.

The idea of a companionate marriage emerged around the turn of the century. The term companionate marriage, however, was not coined until 1925 in a book by Denver Juvenile Court Judge Ben B. Lindsay and Wainright Evans, *Revolt of Modern Youth.* They used the phrase in their proposal that marriage not be held together by moral duty, but rather by mutual affection, sexual attraction, and equal rights, where husbands and wives were friends and lovers and parents and children were pals (Mintz & Kellogg, 1988). Lindsay and Evans also recommended that disagreements between husbands and wives should be acknowledged as normal, healthy, and resolvable through "communication," another popular catchword at that time (Fass, 1977).

The Companionate Family stressed the importance of sexual gratification in marriage. Husbands and wives were encouraged by the growing numbers of professionals in marriage counseling, sex education, and sex counseling to seek emotional and sexual fulfillment. The emphasis on sex created an interest in contraception (Mintz & Kellogg, 1988), or perhaps increasing access to contraceptives helped fuel an interest in sex. In any event, the rise of the Companionate Marriage paralleled the political struggle to make contraceptives available in the U.S.

The distribution of contraceptives or even information about contraceptives was illegal in the early part of the twentieth century. For example, in 1917 Margaret Sanger, a nurse and the founder of the birth control movement in the U.S., was arrested in New York City for distributing obscene materials when she dispensed diaphragms at a health care clinic (Sanger, 1931). After numerous demonstrations, jailings of movement leaders, and legislative battles, in 1936 the courts ruled that physicians could prescribe birth control for whatever reasons they saw fit (Hymowitz & Weissman, 1978).

The ideal of the Companionate Family also called for changing relationships between parents and children. During this period relationships between children and parents became more intimate and affectionate, and allowed greater freedom for children (Mintz & Kellogg, 1988).

The Twentieth Century and the History of Childhood During the rise of the Companionate Marriage from the end of the nineteenth century until recently, the way people thought about and treated children and the way in which children experienced their lives changed substantially. A key aspect in these changes was the increasing differentiation between children and adults, especially in terms of their responsibility to work. Zelizer (1985) traced the history of childhood in America and discovered some remarkable differences between the economic and sentimental value of children in 1870 compared to 1930.

As we noted in the discussion of working-class families in the nineteenth century, child labor was typical in the late 1800s. In 1900, one child of every six between the ages of ten and fifteen was gainfully employed and this did not count all the children who worked for their parents in sweatshops and farms (Zelizer, 1985).

The contribution of these children to their household income was significant. In Philadelphia in 1880 children contributed from 28 to 46 percent of the total family labor income in two-parent families (Goldin, 1981).

Making children earn a living was seen not only as economically necessary, but morally sound. In 1924 the *Saturday Evening Post* wrote, "The work of the world has to be done; and these children have their share. . . . why should we . . . place the emphasis on . . . prohibitions. . . . We don't want to rear up a generation of non-workers, what we want is workers and more workers" (Zelizer, 1985, p. 67).

The role of paid work in children's lives, however, changed quickly in the late 1920s. By 1930 most children were out of the labor market. They could still work in the newspaper business, motion pictures, and in family business (except

Manufacturing work was not for men only. At the turn of the century, entire families were often employed in the mills and factories.

for mining and manufacturing), and they could work on farms outside of school hours, but the legal and social tide had turned in favor of removing children from the adult world of work (Zelizer, 1985).

As children lost their role as economic assets, they gained sentimental value. Zelizer (1985) traces the changing social value of children from economically useful at the turn of the century to economically useless but emotionally priceless by the 1930s. In her work she examined the process of placing adoptive and foster children as an indicator of children's changing value:

> Nineteenth century foster parents took in useful children expecting them to help out with farm chores and household tasks. In this context, babies were "unmarketable," and hard to place except in foundling asylums or commercial baby farms. . . . The useful child was generally older than ten and a boy. More than three times as many boys as girls were placed. (Zelizer, 1985, pp. 170, 173)

Within this system, mothers of infants had to pay a surrender fee to people who agreed to take their babies. By 1920, however, the situation had reversed and babies could be sold for $100. By 1950, babies were sold—illegally—for as much as $10,000 (Zelizer, 1985).

From 1920 on, the most adoptable child was a girl, preferably a pretty blue-eyed blonde with an attractive personality. Her value, and her brothers' value, was no longer measured by the economic contribution she could make, but by emotion and sentiment.

Families During the Great Depression Families were affected not only by the relatively slow unfolding of economic change over the past few centuries, but also by the turning points, the great upheavals in that history. In Chapter 3 we will look at the way in which the Civil War affected the organization of African American families. In the next two sections of this chapter we will examine two important historical events in the twentieth century, the Great Depression and World War II. In both of these cases the country underwent profound changes. These macro-level shocks reverberated in the micro level of the everyday experience of families and especially in the organization of gender within those families.

In 1929 the stock market crashed, ushering in the Great Depression. Unemployment rose from 3 million that year to 12 and a half million in 1932. The Depression created intense poverty for most U.S. families. It also caused changes in behavior. The marriage rate, birth rate, and divorce rate all fell dramatically (Elder, 1974; Mintz & Kellogg, 1988) (see Figures 2-1, 2-2, and 2-3).

The Depression created other changes in family life. Women, especially wives between twenty and sixty-five years old, increased their participation in the labor force more between 1930 and 1940 than in any previous decade in the twentieth century (Milkman, 1976). Many women sought to compensate for the family's lack of income when their husbands were laid off (Cavan & Ranck, 1938; Komarovsky, 1940).

The economic crisis also had an impact on women's family work. Women attempted to maintain their family's standard of living, despite decreases in their income, by substituting their labor for goods and services they had previously purchased. For example, they increased their activity in tasks like canning food and sewing clothes. Women's responsibility for emotional support was also intensified as unemployment and poverty disrupted people's lives (Milkman, 1976).

World War II, Rosie the Riveter, and Her Family The Depression ended with World War II, which stimulated the economy with the new demands for war materials and ended unemployment with the draft. The boom in industry coupled with the enrollment of huge numbers of men in the armed services created a shortage of workers that could only be made up by hiring women (Milkman, 1991). The women hired at manufacturing plants were called Rosie the Riveters.

According to a popular song during the war:

> Rosie's got a boyfriend Charlie;
> Charlie he's a marine.
> Rosie is protecting Charlie
> Working overtime on the riveting machine.
> (Kessler-Harris, 1982, p. 276)

The government and businesses flooded the media with pitches to encourage women to change their ways and the organization of their families' lives (Milkman, 1991). Women were told that their domestic skills could easily be translated into war work—the new message was that a sewing machine was hardly different from a punch press (Gluck, 1987).

The need for women workers challenged the ideologies that defined good wives and mothers as those who did not work outside the home but who devoted themselves to domesticity. A new image was quickly fashioned. The ideal wartime woman was one who placed her child in a daycare center, found fast, efficient methods to prepare meals and clean house, tied back her hair, and went to work as a riveter, a welder, an electrician, or another kind of production worker.

The entrance of the United States into the war had other effects on family life as well (Elder, 1986). Quick marriages became the rage. "In the month after the surprise attack [on Pearl Harbor] the marriage rate was 60% higher than in the same month the year before" (Mintz & Kellogg, 1988, p. 153). Some men married because they thought they were going to die. Others married to avoid the draft. Some married because husbands received higher wages than single soldiers. The soaring marriage rate was quickly followed by a baby boom.

In 1979, Sherna Gluck (1987) began a research project to find out the way in which these changes affected individual Rosies. She conducted ten oral history interviews with women who had worked in aircraft manufacturing plants during World War II.

An oral history can take many forms. In this case Gluck asked women to talk about their entire lives including the period in which they took their job at the aircraft plant and what had happened to them since that time. Gluck was interested in looking at the change over their lifetimes in the work they did and their ideas about women and work.

She found that the war had been an important shaper of their work and family histories. Although most had been employed before the war, their experience as blue-collar workers in jobs that had previously been reserved for white men changed their thinking about what women should be. Betty Jean Boggs said, "I think it showed me that a young woman could work in different jobs other than say, an office, which you ordinarily expect a woman to be in. It really opened up another viewpoint on life in general" (Gluck, 1987, p. 103).

Other women used the opportunity to make changes in their family life. Marie Baker said she was unhappy with her marriage before the war started and left her husband when she heard of the openings for women in the aircraft plant. She explained, "I needed a job because I was going to be very independent. I wasn't going to ask for any alimony or anything. I was just going to take care of myself" (Gluck, 1987, p. 221).

Tina Miller, a black woman, pointed out that the new opportunities in the job market brought about by the war caused changes in her family life. But more significant, in her opinion, were the changes that took place in race discrimination during the war. She explained, "The war made me live better, it really did. My sister always said that Hitler was the one that got us out of the white folks' kitchen" (Gluck, 1987, p. 24).

Gluck's research is an example of feminist research (Reinharz, 1992). According to Gluck, although oral history is not unique to feminist scholars, it is particularly well suited to their interests. "[W]omen's oral history is a feminist encounter because it creates new material about women, validates women's experience, enhances communication among women, discovers women's roots and develops a previously denied sense of continuity" (Reinharz, 1992, p. 126).

Japanese American Families During the War

The United States entered World War II on December 7, 1941, when the Japanese bombed Pearl Harbor, Hawaii. On December 8 the financial resources of Japanese Americans and Japanese nationals who lived in the U.S. were frozen and the F.B.I. began to arrest leaders in the Japanese American community. Within two months President Roosevelt signed Executive Order 9066, which suspended all civil rights of Japanese Americans and authorized the removal of 120,000 Japanese and Japanese Americans—60 percent were American citizens—from their homes. The exclusion order was posted in neighborhoods where Japanese Americans lived. It read: "Instructions to all persons of Japanese ancestry: All Japanese persons, both alien and non-alien [American citizens] will be evacuated from the above designated area by 12:00 Tuesday" (Nakano, 1990, p. 135). They were incarcerated in concentration camps for an average of two to three years. No formal charges were ever brought against them and they had no opportunity for a trial of any kind (Nagata, 1991).

After being forced from their homes, many of the evacuees were taken to internment camps in California. Others were first sent to sixteen temporary compounds, called "assembly centers" by the army. These were structures like horse stalls at race tracks, fairgrounds, and livestock exhibition halls, in addition to tar paper-covered barracks which were built for the camps. After one to six months in the "assembly centers" they were loaded onto trains and taken to desolate areas in the deserts and semideserts of Arizona, Utah, Colorado, Wyoming, and Idaho and the swampy lowlands of Arkansas (Matsumoto, 1990).

People were given one week to prepare for the evacuation and could bring to the camps only as much clothes and personal belongings as each person could carry. They were forced to either abandon their homes, businesses, and personal possessions or sell them within a few days at a fraction of their worth. By November, 1942 the involuntary evacuation was complete. The camps were not closed until the end of 1945.

At the camps, people were housed in military barracks that had been subdivided into cramped apartments measuring about 20 by 20 feet, sometimes furnished with steel army cots and sometimes with gunny sacks filled with straw to serve as beds (Takaki, 1989). Families were housed together but single people were placed with smaller families so that on average eight people lived in each unit. There was little privacy. The walls to the units in the barracks did not reach the ceilings and sounds traveled throughout the barracks. In addition, eating, showers, and toilets were all communal (Matsumoto, 1990). One person remembered,

> There was constant buzzing—conversations, talk. Then, as the evening wore on, during the still of the night, things would get quiet, except for the occasional coughing, snoring, giggles. Then someone would get up to go to the bathroom. It was like a family of three thousand people camped out in a barn. (Takaki, 1989, p. 394)

Valerie Matsumoto (1990) studied life in the internment camps by examining letters that people wrote while living there. One letter written by Shizuko

During World War II Japanese Americans were forced to abandon their homes and businesses and live in isolated camps. The hardship and organization of life in the camps caused disruption and change in their families.

Horiuchi in 1942 read, "The life here cannot be expressed. Sometimes we are resigned to it, but when we see the barbed wire fences and the sentry tower with floodlights, it gives us a feeling of being prisoners in a 'concentration camp' " (Matsumoto, 1990, p. 373).

Every able-bodied person in the camps was required to work. The pay was the same for young and old, women and men. Most people received $16 a month; professionals like physicians and teachers earned only slightly more, $19.

Internment in the camps had a traumatic affect on the adults and children forced to live there. In addition, the internment of their parents and grandparents continued to adversely affect children and grandchildren who were born after the camps closed. Like victims of the Holocaust, many people held in the camps found it difficult to express their feelings about their imprisonment, even years after their release. Children born after the war ended observed that the experience had been painful and traumatic but were unable to obtain enough concrete information to understand their parent's reaction. Research on these children shows that the experience had a negative effect on their lives and altered their perception of their own place in U.S. society. These third-generation children whose parents had been imprisoned developed a sense of foreboding and secrecy about the internment. Compared to third-generation Japanese Americans whose parents had not been imprisoned they also felt more vulnerable, more fearful that a future internment might occur, and more likely to prefer Japanese Americans over whites for friends (Nagata, 1991).

In addition to these negative effects, some researchers have argued that there may have been some positive outcomes as well. Before the war, parents, and especially fathers, in Japanese American families had much authority over their children. Marriages were arranged and decisions that were made by fathers were usually strictly followed by the children and wives. Internment changed those relationships (Pleck, 1991). The paid employment of all family members and the nearly equal wages earned combined with the harsh living conditions and lack of privacy made it difficult to maintain rigid boundaries around families and the strong authority of fathers over their families. Although the internment in the camps was a horrific event, the results of the uprooting meant that children and wives gained more control over their lives (Matsumoto, 1990).

The Fifties When the war ended a new era in U.S. history began and a new era in family organization also emerged. This period was called the Fifties although it actually lasted from about 1947 to 1963. After the war the government and businesses began a propaganda campaign to reverse the images of working women and ideal families that had been popularized during the war. Women were asked to give up their jobs to the returning soldiers and many women who did not voluntarily quit were fired. Women were pushed back into low-wage "pink-collar" jobs. Working wives and mothers were no longer idolized, as in the song "Rosie the Riveter," but instead were condemned for shirking their "proper" duties.

In a postwar article Agnes Meyer wrote in *Atlantic Monthly,* "Women have many careers but only one vocation—motherhood. . . . It is for woman as mother, actual or vicarious, to restore security in our insecure world" (Meyer, 1950, p. 32).

At the same time that ideologies about gender and family were being altered the economy was experiencing an enormous boom. In 1960 President Eisenhower reviewed the decade's success in his State of the Union address. He noted the 20 percent increase in real wages, the 15 million new housing units built, and the increasing numbers of people who owned houses, cars, and refrigerators (Mintz & Kellogg, 1988).

The combination of the conscious effort to push women out of the "men's" jobs they had held during the war and a booming economy created the conditions for the strange anomaly known as the Fifties. In an overview of several demographic trends—births, marriages, and divorces—in the twentieth century the 1950s appear as a brief reversal.

Figure 2-1 shows how the longstanding decline in the birth rate stopped in the 1950s and births rose sharply. A relatively small generation of parents gave birth to a large generation of babies (Cherlin, 1992a). Between 1940 and 1957 the fertility rate climbed 50 percent (Pleck, 1991). The increase in births was so remarkable it was given a special name—the postwar baby boom—and people born during the period are called baby boomers.

Figure 2-2 shows that the marriage rate during the 1950s was also unusual relative to the rest of the century. The marriage rate declined to a low in the 1930s and then steadily climbed during the war. Right after the war the rate shot up and remained high through the 1960s when it again began to decline. The 1950s had the highest rate of marriage of any generation on record: 96 percent of the women and 94 percent of the men were married. Even more remarkable were the record low ages at which women (20.2 years on average)

FIGURE 2 • 1
Annual Total Fertility Rate of Euro-American Women, 1800–1989

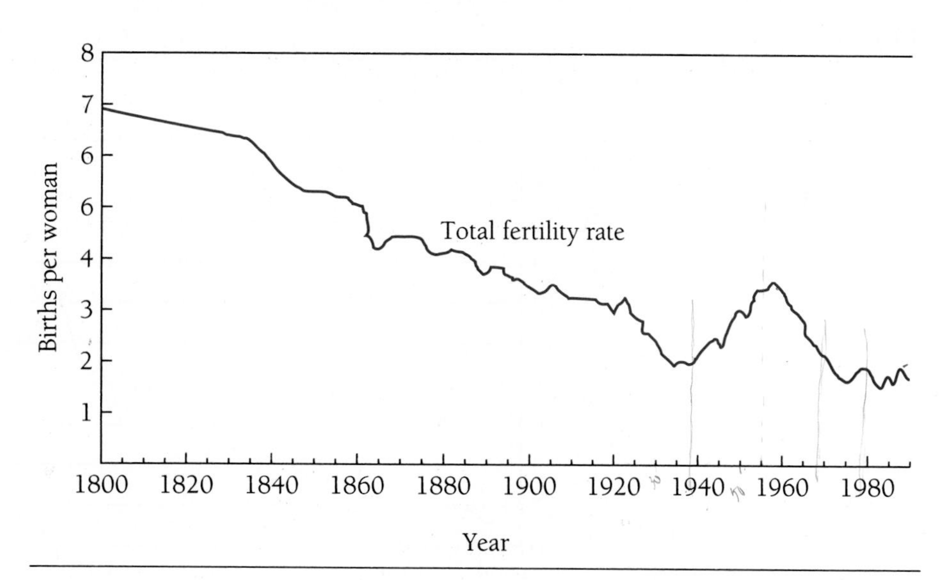

Sources: Thornton and Freedman, 1983, p. 13; Lamanna and Riedmann, 1991, p. 352; *Statistical Abstract of the U.S., 1992,* p. 67.

FIGURE 2 • 2
Rate of First Marriage for Women, 1921–1984 (in 3-year averages)

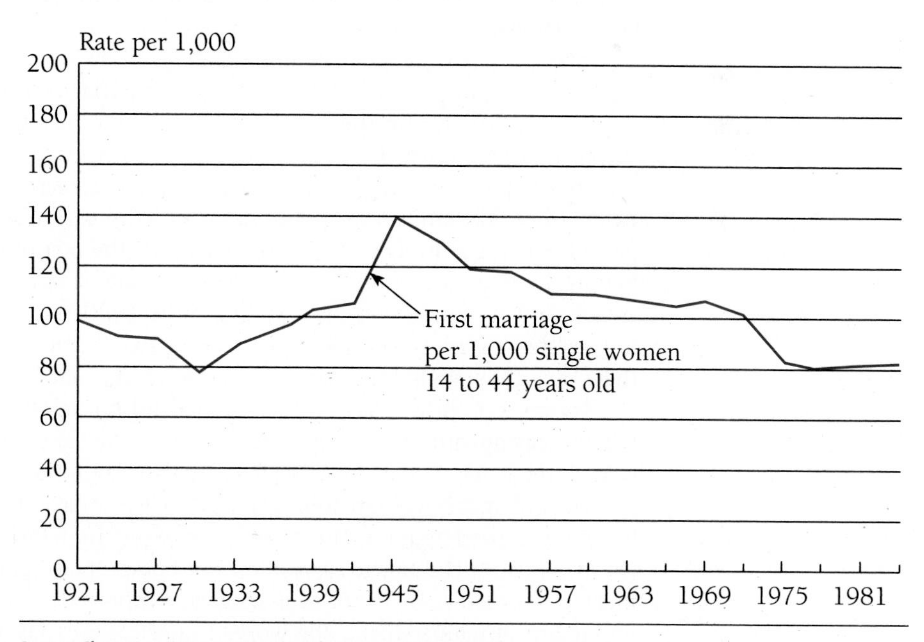

Source: Changes in American Family Life; Current Population Reports, Special Studies, Series P-23, No. 163; 1989d. U.S. Dept. of Commerce, Bureau of the Census, p. 7.

FIGURE 2 • 3
Annual Divorce Rate, United States, for 1920–1988

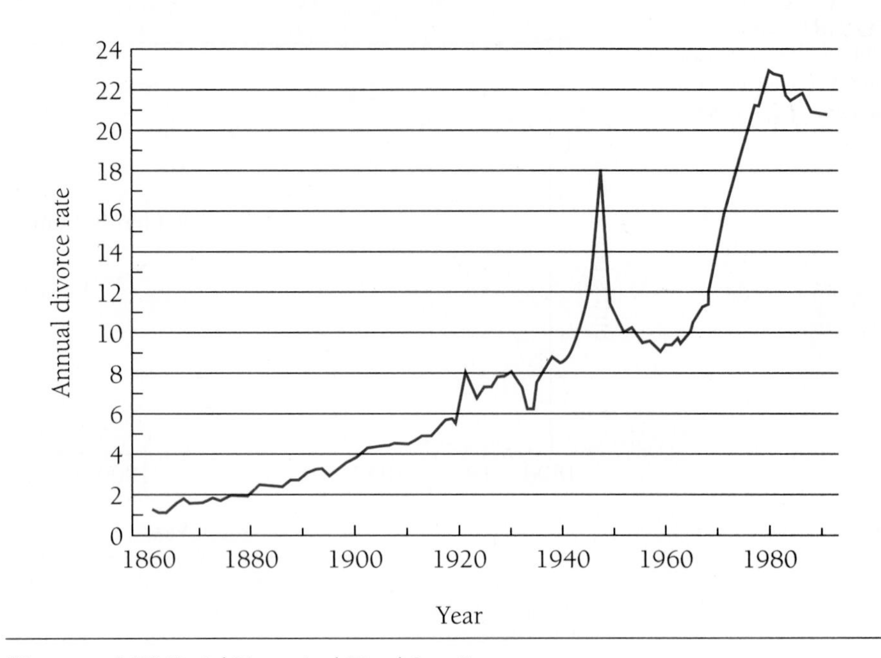

(Divorces per 1,000 Married Women Aged 15 and Over; For 1860–1920: Divorces per 1,000 Existing Marriages)

Source: Cherlin, 1992a, p. 21.

and men (22.6 years on average) married in the 1950s (Levitan, Belous, & Gallo, 1988).

The divorce rate peaked immediately following the war, perhaps making official the changes that had taken place in marriages during the war. But Figure 2-3 shows that during the decade of the 1950s the divorce rate was low and stayed that way until the 1960s when it began to rise rapidly into the 1980s.

In Chapters 9 and 10 we will examine marriage and divorce rates more extensively. The important point about this discussion is that the Fifties were a period that was out of sync with the trends of the rest of the century. Sometimes people look at the marriage and divorce rates today, compare them to the 1950s, and conclude that the 1970s and the 1980s are somehow unusual, showing families in crisis. A look at the overall sweep throughout the century shows that "it is the 1950s that stand out as more unusual" (Cherlin, 1982, p. 137).

Figure 2-4 shows that one trend that did not decline during the 1950s was the increasing numbers of women entering the labor force. Although women were forced out of the "men's jobs" of the war industry, many remained employed and more women joined them. White married women increased their labor force participation rate from 17 percent in 1950 to 30 percent in 1960 (Evans, 1989). The image of women not being "real" employees, however, permeated the media. In 1956, *Look* magazine reported, "No longer a psychological immigrant to man's world, she works rather casually, as a third of the U.S. labor force, and less toward a big career than as a way of filling a hope chest or buying a new home freezer. She gracefully concedes the top job rungs to men" (*Look*, 1956, p. 35).

FIGURE 2 • 4
Labor Force Participation Rates of African American and European American Women, 1900–1989

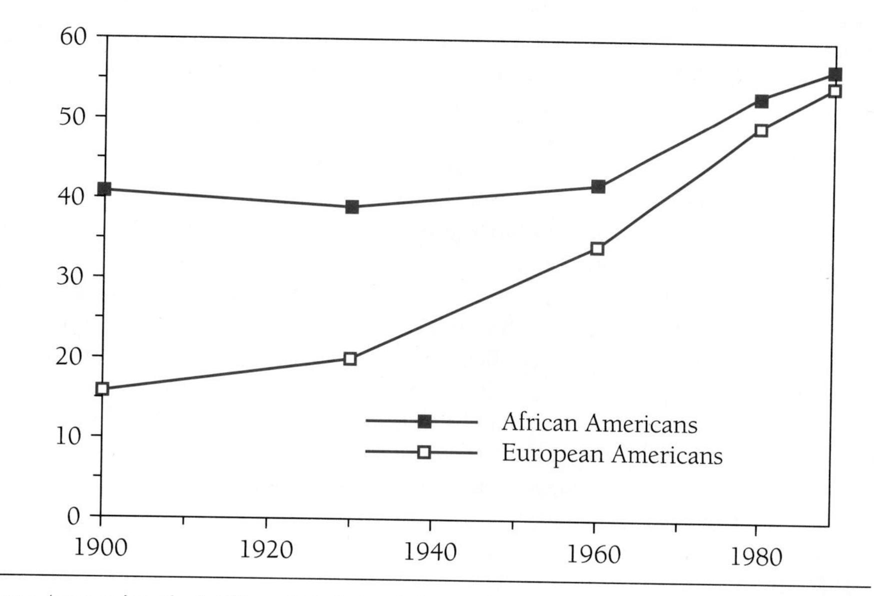

Sources: Amott and Matthaei, 1991, p. 166; *Statistical Abstract of the United States, 1992,* p. 381.

THE EVOLUTION OF PATRIARCHY

This overview of the history of families in the U.S. shows great changes over time in several features of the organization of families. It also shows diversity within each historical period. One issue that reveals a combination of both change and constancy is the relationship between women and men within families. The three women of three generations mentioned in the opening scenario—Ruth, Sarah, and Elaine—each had different relationships with the men in their lives, indicating historical change. On the other hand, there is an element of constancy in the inequality between women and men, the greater power of men over women, and the struggle of women to make relationships more equal. The term that has sometimes been used to describe the inequality between women and men is patriarchy.

In the discussion of the Godly Family at the beginning of this chapter I used the term patriarchy to describe the organization of families during this period. The Puritan household was a patriarchal one according to the definition offered by Gayle Rubin. Puritan fathers had legal, moral, and social authority over the members of their households, including their wives, children, servants, apprentices, and boarders. This classical form of patriarchy declined with the development of industrialization and the emergence of the more isolated nuclear family, especially among the white middle class.

Patriarchy is used in a second way, however. Contemporary feminists use the term to describe a broad range of family and nonfamily relationships that include the domination and oppression of women by men. Zillah Eisenstein offers a broad definition of patriarchy: "a sexual system of power in which the male possesses superior power and economic privilege" (Eisenstein, 1984a, p. 17).

According to this broad definition, patriarchy can exist in any social institution, including but not exclusively families. A Senate that has only a handful of women members and that does not speak for women's interests is patriarchal. An economic system that is gender segregated and in which women are likely to be low-paid employees working for better paid and more powerful male bosses is patriarchal.

Within the broad definition, patriarchy can also exist regardless of specific legal rights women may have. Twentieth-century women compared to Puritan women have greater legal rights. They do not have to submit to the authority of their husbands as much as Puritan women did; they have the vote and can own property. But if men compared to women still have more power and economic privilege, then patriarchy exists (Walby, 1986).

In this chapter we have seen the near elimination of patriarchy in the narrow classical sense. According to the broader definition, however, patriarchy appears to have varied significantly over time, but it is still very much with us.

THE WOMEN'S MOVEMENT

Verta Taylor (1983) asserts that social movements are responses to social problems. Social movements can ebb and flow but until the social problem is resolved the movement does not entirely disappear. She writes:

> All social movements originate in some contradiction or conflict in the larger social order. When a social movement persists throughout modern history, as the American feminist movement has, we can only assume that the injustices, grievances, and oppression on which it is based are deeply rooted in society. (Taylor, 1983, p. 434)

One response to the problem of patriarchy was the development of the women's movement. During the nineteenth century, working-class women needed a social movement that called for both changes in the economy and changes in the family. They needed more than a family wage; they needed an economic system that allowed both men and women to earn decent wages. They also needed a social movement that allowed women to remain single if they chose or allowed them to be their husband's equal, not tied to ideologies of the cult of true womanhood and not dependent on him for their survival.

Middle-class women also were faced with problems as a result of the development of industrialization and the Modern Family. Although the economy provided middle-class women with a comfortable life, the cult of true womanhood and their role in their families did not. Like Sarah in the opening scenario, many middle-class women were unhappy with the limitations of their role as "true women." A feminist movement developed in response to these and other problems.

The first published works of feminist protest are dated in the early 1600s, but it was not until the middle of the 1800s that the women's movement became fully focused and organized in the U.S. (Lengermann & Niebrugge-Brantley, 1992). July 19, 1848, marked a particularly important event in this development. On that day 300 people including Susan B. Anthony, Elizabeth

Cady Stanton, Frederick Douglass, and Lucretia Mott convened a Women's Rights Convention in Seneca Falls, New York. At that meeting they approved a Declaration of Sentiments and twelve resolutions (Hole & Levine, 1971).

The resolutions passed by the Seneca Falls Convention covered a wide range of reforms including woman's suffrage and the right of women to work and to keep their wages. The convention even advocated changes in dress, calling on women to abandon their cinched waist, multilayered petticoats and replace them with loose-fitting bloomers.

Some of the proposals made at Seneca Falls and subsequent feminist gatherings also directly addressed issues related to family. Women demanded the right to divorce and to retain custody of their children after a divorce. They also sought the right of a married woman to retain her own property and to be safe from marital rape (Stanton, 1889).

The women's movement continued to press for change throughout the nineteenth and twentieth centuries. Their work addressed a wide range of topics including slavery, suffrage, and equal rights but issues related to women's role in families were always prominent. The Second Wave of the women's movement in the 1960s also emerged in response to many issues related to families.

THE MICRO-MACRO CONNECTION

This chapter shows how families have changed over time in the U.S. One of the major aspects of the macro level of society, the economy and changes in its organization, profoundly affected the micro level, the organization of families and the everyday experience of family life. For example, the patriarchal Puritan family could not survive the development of industrial capitalism. The growth of jobs that were controlled by factory owners rather than fathers undermined the hold that the patriarchs had over their families. In Chapter 4 we will continue to examine some of the relationships between the economy and families by looking at changes in today's economic system and considering some of the effects on families that might result.

The relatively constant evolution of the macro level of economic conditions is marked periodically by major shifts at critical turning points. These turning points are related to (1) changes in technology—for example, the Industrial Revolution; (2) the economy—for example, the Great Depression, and (3) war—for example, World War II. Shifts at the macro level of society alter the everyday experience of families at the micro level.

Although the economy has been identified as an aspect of the macro level of society that has been especially important in shaping family life, other social institutions that make up the macro level of society have had an effect as well. Political institutions affect families. For example, governments that decide to go to war or that place families in internment camps create changes in gender relationships and divisions of labor within families. Government proclamations that offer huge tracts of land, like the Homestead Act, change family relationships.

Popular media and religious institutions are another feature of the macro level of society that can have an effect on families. For example, we observed the activities of these two aspects of the macro organization of our society in the creation of the cult of true womanhood.

The macro level, in addition to being complex and dynamic, also may affect various members of society differently. For example, the families of Ford workers faced different issues from their middle-class counterparts and Chinese Americans and Mexican Americans faced still different problems. Although all of these groups—white middle class, white working class, Chinese American, and Mexican Americans—lived within the same broad social context, their experience of that context was remarkably dissimilar.

The macro level also affects members of families differently. The case of the family wage is an especially good example. The battle for the family wage created different problems for husbands compared to wives in blue-collar families. In another example, the decision to move west implied different losses and gains for husbands compared with wives. The macro context includes not just an economic system or a political system that creates the same social context for everyone; the macro level also includes systems of gender and racial ethnic inequality that affect families and family members. Chapter 5 will continue this discussion by focusing our attention on issues of diversity among contemporary U.S. families.

In sum, the macro level affects the micro level. The economy, the government, social class, and race ethnicity influenced and sometimes determined the organization of families and the experience of people within those families. History also shows us the ways in which individual people attempted to shape their society, the effect of the micro on the macro. The birth control movement, the labor movement, the women's movement, and the reformers who fought for child labor laws all represent individuals organizing to address issues related to family life. These social movements were composed of people working together to both revise and reformulate the macro organization of society.

SUMMARY

Studying the Social History of Families

- History is important to our understanding of families but scholars have only recently turned their attention to issues of the family.
- A central theme in the history of families is the importance of parallel changes in the economy.

Stages in the History of Euro-American Families

- American family history can be divided into three periods related to changes in the economic context: the Godly Family was associated with the preindustrial system, the Modern Family with the rise of industrial capitalism, and the Postmodern Family with deindustrialization.
- The Godly Family (1620–1780) was characterized by: patriarchal organization, integration with the community, economic self-sufficiency, and inclusion of all social activities.
- Coexisting with the Godly Family of the European immigrants was the Iroquois Family, which was extended and female dominated.

The Modern Family, Stage I: The Democratic Family

- The Modern Family (1780–1970) emerged with the development of industrial capitalism. This period is divided into two sections: the Democratic Family and the Companionate Family.
- The Modern Family differed from the premodern form in four ways: work was split into paid and unpaid; marriage was to be freely contracted based on love; privacy of the family in the community was promoted; and women had fewer children but were supposed to spend more time on their "natural" task of motherhood.
- The rise of industrial capitalism and the Modern Family created a split between paid work in the growing factories and mills and the unpaid domestic work carried on at home.
- For most middle-class whites this resulted in men going to work for wages, and women staying home to do housework.
- The creation of the occupation of housewife in the middle-class family was accompanied by numerous books and sermons telling women how to play their new role. Barbara Welter called the new image of woman the cult of true womanhood.
- According to the advice manuals women were to be pious, pure, submissive, and domestic.
- In working-class families wives were likely to be part of the labor force along with their husbands and children. If they stayed home they earned wages by taking in piecework or boarders.
- The inconsistency between the ideal of women staying home with the inability of working-class men to earn a wage that provided for a wife and family created the demand for a family wage.
- The family wage was both a boon and a tragedy for working-class women. It provided more income for husbands but it was also used as a way to force women out of jobs.
- Millions of families left the East during the westward expansion. The decision to leave was usually made by men. Women suffered much from the break with their friends and the physical hardship.
- Research on this experience by Smith-Rosenberg is an example of content analysis.
- Chinese American and Mexican American families were required to split their households from their work in an extreme manner. The men traveled thousands of miles to other countries or regions to work while wives and children remained at home performing other work.

The Modern Family, Stage II: The Companionate Family

- The Companionate Family, the second stage of the Modern Family, emerged in the early years of the twentieth century.
- The Companionate Family ideal was characterized by a concern for the sexual relationship between husbands and wives, as well as support for equality between spouses.
- The development of the Companionate Family included changes in the role of children. In the late nineteenth and early twentieth centuries

children were valued as workers and played a significant role in the financial support of their families.

- This image of children changed, however, and as the twentieth century progressed children were more and more likely to be valued as objects of affection rather than as wage earners.
- The Depression caused a fall in marriage, divorce, and birth rates.
- Families faced with unemployment frequently sent women into the labor force to try to maintain the family's income and increased their unpaid domestic work to save money.
- World War II brought many women into the paid labor force to produce arms and to replace men who joined the armed services. Ideas about what good mothers and wives should be like began to incorporate the image of the worker.
- Sherna Gluck studied "Rosies" through oral histories.
- Japanese American families were held in internment camps during the war. Extreme hardship and the employment of all family members greatly altered relationships within families and in particular diminished the authority of fathers.
- The 1950s were an anomaly in the twentieth century, with high rates of marriage and fertility and low rates of divorce.
- One trend that continued through the century, including the 1950s, was the ever-increasing number of women in the paid labor force.

The Evolution of Patriarchy

- Patriarchy can be defined in two ways. The patriarchal family in which the father has legal, moral, and social authority has disappeared.
- The second definition of patriarchy is that it is a system in which men have more power and greater access to resources. This kind of patriarchy has been maintained throughout the past few centuries although it exists in different forms in different historical periods and in different social groups.

The Women's Movement

- Women who faced inequality in society have organized in the last two centuries to address the problems associated with patriarchy, which are often but not always related to issues connected to families.

The Micro-Macro Connection

- History shows the way in which macro forces affect families' micro-level experience of everyday life.
- The macro level of society includes the economy, the government, religious institutions, and the media.

- The macro level also includes systems of inequality of race ethnicity, social class, and gender, which create a diversity in the experience of families and family members.
- The micro level has also altered the macro organization of society as people have worked in social movements such as the labor movement, the birth control movement, and the women's movement to try to reform family organization and to address issues related to families.

During slavery cotton production dominated the American economy, and African American men, women, and children were forced to work in the fields to keep the industry booming.

3 A HISTORY OF U.S. FAMILIES WITH A FOCUS ON AFRICAN AMERICANS

The year is 1840. Seventeen-year-old John is a slave on a large South Carolina cotton plantation owned by a man named Franklin. When John was ten, the slaveowner on the plantation where he was born sold him and his father and brothers. John was unable to keep track of his brothers and father, knowing only that a man named Brady from Mississippi bought them. His mother stayed at the old plantation and John was separated from all of them to go to work at Franklin's.

When John was younger, his mother used to walk to his plantation occasionally to see him. She was permitted to make the twelve-mile walk between dusk and dawn and if she came home after dawn she was beaten. John's mother died when he was thirteen, but he was not allowed to attend her funeral because she died during harvest and he had too much work to do.

John has a woman friend, Mary, whom he cares about and would like to marry, but Franklin, his owner, plans to sell her because she has already suffered a miscarriage and may be unable to bear any more children. Franklin has been talking about making John sleep with another slave woman, Caroline, because she is robust and he believes that John and Caroline would produce strong children.

John hears about a woman named Harriet Tubman who helps slaves escape. He decides to convince Mary that they should both go with her on the Underground Railroad up to Canada.

- *What kinds of family life were possible for African Americans under slavery?*
- *How did the ideas and experience of family for whites differ from that of African Americans during slavery in the U.S.?*
- *What kinds of choices did slaves have to make in their decisions about family issues like marriage and children?*
- *How does the extreme case of slavery help us to see the importance of the effect of social context on families?*
- *What happened after slavery? How did the new system of sharecropping, and later industrialization, affect family organization among African Americans?*

DIVERSITY IN AMERICAN FAMILIES

The United States is a nation of diversity. Many groups of people came to live here from Europe, Asia, Africa, and the rest of the world. In addition, before European contact there were millions of Native Americans living in dozens of different nations throughout the hemisphere. This means that the cultural roots and the history of Americans are varied and complex.

In addition, many groups have been exploited or oppressed in a variety of ways. Many Native American groups were completely wiped out and others were forced to march across the country to be "resettled" on the lands of other Native American peoples. As we saw in Chapter 2, Japanese who lived in the U.S. during World War II were forced out of their homes and communities and interred in camps. Asians, Puerto Ricans, and Chicanos have been and continue to be discriminated against and exploited as cheap labor in fields and sweatshops across the country.

These various groups have their own histories of family organization, which are different from that of white Americans. Family sociologists, therefore, are presented with the dilemma of needing to pay close attention to diversity, but finding it impossible to cover all of the possibilities in one volume such as this text.

TABLE 3 • 1
Racial Ethnic Populations as a Proportion of Total U.S. Population, 1800 to 1990

	1800	1860	1900	1940	1970	1990
Whites	68.6	85.6	87.9	89.8	83.3%	75.3%
Blacks	31.3	14.1	11.6	9.8	10.9	11.9
Hispanics	N.A.	N.A.	N.A.	N.A.	4.5	9.0
Asians/Other	N.A.	0.2	0.5	0.5	1.3	3.8

Source: U.S. Bureau of Census. 1975. *Historical Statistics of the United States, Colonial Times to 1970.* Washington, DC: U.S. Government Printing Office, p. 14; Hacker, 1992, p. 15.

In our examination of the historical development of families in this book, we focused in Chapter 2 on Euro-Americans during the colonization of the U.S., the westward expansion, and the period of immigration and urbanization. In that chapter we also explored a few examples of Japanese, Chinese, Mexican American, and Native American families. In this chapter we will turn our attention to African Americans during the eighteenth, nineteenth, and twentieth centuries.

Among all the possible groups that could be examined African American families have been chosen for special attention for several reasons. First, there is a large body of literature on African Americans and African American families. Less research has been done on other groups of people, for example, Cherokees or Vietnamese Americans.

Second, African Americans are the largest racial ethnic group in the U.S., making up 11.9 percent of the total population in 1990. As Table 3-1 shows, whites account for 75.3 percent of the total population and Hispanics and all other groups make up the remaining 12.8 percent. The proportion of African Americans in the total population was even more pronounced in the nineteenth century. In 1860 African Americans made up 14 percent of the total U.S. population and 36.4 percent of the population in the southern states (U.S. Bureau of the Census, 1975). In 1800, ten years after the first census, African Americans comprised almost one-third of those counted.

Third, African Americans are the only group of people that were legally enslaved for a long period of time (more than 200 years) in the U.S. They are the only group that was legally segregated in a large region (the entire South) for a long period of time (from 1877, the end of Reconstruction, until the passage of the Civil Rights bill in 1964). Finally, African Americans are the only group of people for which a civil war was fought at least in part over the question of whether to end their enslavement.

In a sense, then, African Americans are a special case and when we study their family history we need to keep that in mind. On the other hand, examining the changes over time of the organization of African American families and their relationship to the changing social context should help us to develop principles that might be tested in the cases of the many other peoples of the United States.

Throughout the chapter we will see the way in which the macro level of social organization affected the micro level of everyday life for African American people. The key element in the macro level that we explore in this chapter is the political economy. A political economy is the manner in which a society organizes its political and economic institutions. Political institutions refer to the organization and distribution of power and decision making. Economic institutions are those that produce and distribute the goods that people use to survive.

Slavery, for example, was a particular type of political economy that had an enormous effect on the family lives of slaves. Our review of the macro level of political economy and its effect on families will show how the political economy changed and in turn created changes at the micro level. In addition, this history will show the ways in which African Americans working in social networks at the micro level fought back in attempts to alter both their family organization and the larger, oppressive political economy.

This chapter is organized into two major sections. The first section reviews the history of African Americans and is divided into a number of subsections that cover slavery, sharecropping, and the industrialization of the South. I discuss the relationships between African American men and women and adults and children in each of these periods and compare the organization of African American families to that of middle-class whites. The second section focuses on the central role played by issues related to families in the struggle to create equality and to end the oppression of African American people. In Chapter 5 we will pick up this history again and look at recent developments in African American family history during the last half of the twentieth century.

FOUR PERIODS IN AFRICAN AMERICAN HISTORY

The history of African Americans can be traced through four historical periods, each creating a different social context for families. The four historical periods are: (1) slavery, (2) sharecropping, (3) industrialization, and (4) the New South (Scott, 1988). This chapter will examine the first three of these periods.

Slavery

During the period of slavery, which lasted from colonial times until the end of the Civil War, the experience of most African Americans was tied to a particular region in the United States—the South. "At the end of the nineteenth century, nine out of ten Afro-Americans lived in the South" (J. Jones, 1985, p. 80). Although the Southern states were part of the United States, they maintained a political and economic system that differed dramatically from that of the North. Until the end of the nineteenth century the Southern states had a separate and unique political economy—the American slave system.

The political economy of slavery was characterized by segregation between whites and African Americans, the concentration of power and wealth in the hands of very few whites, and the ownership of nearly all African Americans as a form of property. The dominant economic activity in the region was the production of agricultural products, especially cotton, for export to the North and to Europe. Huge plantations covered the South, each using the labor of hundreds of slaves. The plantation economy was enormously wealthy and the production of the slave laborers created a large proportion of the total economic output of the entire nation. In the mid-1800s 5.4 million bales of cotton a year were produced by Southern plantations, accounting for 75 percent of the total of U.S. exports (Keller, 1983).

About 90 percent of African Americans were slaves who had no civil or legal rights but were considered the property of the slave owners. The other 10 percent of the African American population during this time were not slaves. They were called freedmen but their rights were curtailed as well. Most freedmen lived in urban areas working in skilled trades.

Most whites were not slaveholders and were not part of the plantation economy. About 25 percent of Southern whites owned one or more slaves and about 15 percent owned large numbers on enormous plantations. The large slaveholders owned huge plots of land where the soil was richest. The rest of the whites had little power in political institutions and were not part of the dominant economic activity. They worked as subsistence farmers in less productive agricultural areas.

Early in the development of the slave system in the U.S. the Northern states of New York, New Jersey, Delaware, Pennsylvania, Massachusetts, Connecticut, and Rhode Island had legally sanctioned slavery. In fact, in New York City in 1741 twenty-three armed slaves burned down a slaveowner's house. The incident developed into an insurrection by hundreds of African American and poor white people, 200 of whom were eventually arrested. Thirty-one African Americans and four whites were subsequently executed (Zinn, 1980).

The number of people who were enslaved in the Northern states, however, was relatively small and the system disappeared there in the eighteenth century. In the South, however, slavery was legally established in Virginia in 1640 and lasted in the states that seceded from the union—Florida, Virginia, North Carolina, South Carolina, Georgia, Alabama, Mississippi, Louisiana, Arkansas, Texas, and Tennessee—as well as in the slave states that did not secede—Delaware, Kentucky, Maryland, and Missouri—until the passage of the Emancipation Proclamation in 1863.

Since the Civil War, the Southern political economy has remained somewhat different from that of other regions in the United States because segregation and discrimination were legal until the 1950s. A system of legal segregation is called de jure segregation. Those black people who moved north or west also lived segregated lives in separate neighborhoods, held different jobs, and attended separate schools but in those areas the system of segregation usually was not stated in the laws. This is called de facto segregation because although it is not specified in the laws it nevertheless exists. Throughout the United States African Americans have experienced a history that was different from but highly interrelated with that of other racial ethnic groups. Family organization for African Americans, not surprisingly, has been different from that of whites and other groups of Americans.

Family Life Under Slavery Historians are still debating whether the nuclear family was the most common family form among slaves. A nuclear family is one in which a legally married husband and wife and their children live together. Slavery interfered with this kind of family organization. Slaves could be bought and sold, seized in payment of a master's debts, and inherited. African American families under slavery, therefore, lived in constant fear of separation.

Even for women and men who lived together as husband and wife, legal marriage was not possible. Since slaves were property they had no legal right to make a contract, including a marriage contract. The slaveowner had complete legal authority over his slaves and it was the slaveowner who decided who could live with whom.

Neither was monogamy legal for slaves. Monogamy is the practice of marrying only one person at a time and remaining sexually exclusive with that person. Slave women were constantly sexually abused by white men and they had no legal right to resist that abuse. "To oppose the rape of black women in effect meant opposing slavery. A black woman's body was not considered her own. Control over her body was passed from white person to white person along with a bill of sale" (Hymowitz & Weissman, 1978, p. 51).

In 1855 a nineteen-year-old slave woman was hanged because she defended herself against a rapist. Celia, who had no last name, had been repeatedly raped by her owner since she was fourteen years old. She had borne two children as a result of the rapes and had begged the rapist's two "legitimate" adult daughters to help her to stop him but they refused. In desperation one night she killed the master. The jury did not accept her argument that the murder had been in self-defense and she was sentenced to be executed (McLaurin, 1992).

Slave parents also had no legal rights in regard to their children. Children could be sold away from their mother at any age: "The young of slaves stand on the same footing as any other animal" (Davis, 1981, p. 7). In a famous statement Sojourner Truth, an African American woman active in the abolitionist movement and the women's movement, describes her experience as a parent under slavery: "I have borne thirteen chillun and seen 'em mos' all sold off into slavery and when I cried out with a mother's grief, none but Jesus heard . . . " (Loewenberg & Bogin, 1976, p. 235).

The threat of separation of parents and children was evident in the play of slave children. In the 1930s, former slaves were interviewed about their lives under slavery. David Wiggins (1985) studied the play of slave children by examining these statements. He found that one of the games the slave children played was called "Auction." One of the children would be the auctioneer and conduct a simulated slave sale.

Like all social systems, slavery was not uniform in its expression. The laws forbade slaves to marry and slave masters had the final say in the organization of the family lives of their slaves. Nevertheless, some slaveholders were more lenient and allowed some small "privileges," acknowledging slave marriages. More importantly, these issues were arenas of constant struggle. In some cases slaves were able to bend the rules in order to attempt to maintain relationships of parenthood and monogamous marriage.

Herbert Gutman (1976) is the most prominent of the scholars who argue that despite the legal sanctions against nuclear families, the slaves themselves maintained complex, organized systems of rules, expectations, and emotional relationships among husbands and wives and parents and children. Slaves had their own rituals, for example, that displayed their marriages to their community. One of these rituals was called jumping the broom. The bride and groom jumped over a broom handle as a symbol of their commitment to each other.

Gutman examined historical documents that showed how slaves maintained strong family ties even when it meant risking severe punishment. In the scenario at the beginning of this chapter John tells how his mother used to walk thirteen miles to visit him at night. This story is based on the true story of Frederick Douglass, who was a slave, and who wrote about his mother traveling thirteen miles to visit him after he was sold to another plantation.

Slave men who were married to women who lived on different plantations were allowed to visit their wives, whom they called "abroadwives," but the journey was dangerous:

> The men were given passes to visit their wives on Wednesday and Saturday nights and all day Sunday. In their journey to the other plantation the black men had to deal with "Patterollers" who were patrols of whites whose task was to see that no slaves travelled about at night without the proper credentials. The Patterollers were notorious for their brutality. (Hymowitz & Weissman, 1978)

If slaves visited other plantations without carrying the proper documents they were likely to be beaten or killed. In 1712 the state of South Carolina passed "An Act for better ordering and governing of Negroes and slaves. If a slave was caught away from his or her plantation without a 'ticket' from the master's plantation, 'it is hereby declared lawful for any white person to beat, maim, or assault, and if such negro or slave cannot otherwise be taken, to kill him'" ("An Act . . . ," 1992).

Others tried to find their kin by running away. Franklin (1988) cites numerous advertisements referring to runaway slaves whom slaveowners believed to be fleeing to make contact with their families. One ad describes a fourteen-year-old who was thought to have fled to Atlanta where his mother had been sent. Another ad was for a mother who had escaped to find her children at another plantation. Husbands sought wives and wives husbands.

On plantations, slaves also preferred to organize their lives to emphasize the importance of nuclear family units. For example, sons were frequently named after their fathers and both sons and daughters were named after blood relatives (Gutman, 1976). Slave women preferred cooking for their own families to taking their meals in a communal kitchen even when it meant additional work. In some cases where plantation owners tried to establish communal kitchens, slaves refused them (Hymowitz & Weissman, 1978). Thus, the ability to maintain control over family life and especially the right to establish nuclear families was a point of constant struggle between slaves and slaveowners.

Family Organization and Black Women's Role in a Community of Slaves
Angela Davis is an African American scholar and activist who has examined the organization of families under slavery and the relationship between African American family organization and African American women's political activity. Davis's (1981) point of view differs somewhat from Gutman's. Like W. E. B. Du Bois (1969), one of the earliest scholars of African American culture, Davis argues that slaves may have aspired to live in nuclear families and may have gone to great lengths to maintain contact with their families, but slavery ultimately disrupted the nuclear family. Davis is quick to point out, however, that the difficulty of maintaining nuclear families did *not* mean that slaves had no families.

Gutman's work tends to emphasize those people who were successful in establishing and maintaining nuclear family organization. Davis's work tends to emphasize the alternative kinds of families that were established within the slave community. Both of these scholars' research indicates that even within a system as harsh and restrictive as slavery, a variety of families existed.

Davis believes her work shows that the dominant family organization among slaves was an extended family that consisted of both kin (people related by blood) and fictive kin (people who are not related by blood but are close personal friends). She argues that the intensity of the labor demanded of slaves and the disruption of marriages and relationships between parents and children meant that people relied on a wider circle of social contacts than a nuclear family. This wider circle included both relatives and others in the community.

When Wiggins (1985) examined the narratives of former slaves talking about the games slave children played, he found evidence of the cohesiveness of the entire community of slaves and the lack of boundaries between groups within the community. A theme that ran through all of the narratives was the absence of any games that required the elimination of players; all players stayed in the game. He argues that this was a reaction to the possibility that members of a slave community might be sold suddenly. Keeping all players in the game shows the value placed on the maintenance of community that characterized the culture of African Americans under slavery.

One of the features of the extended community-based family was a sharing of responsibility for food and shelter. In a community made up of nuclear families, each family is an economic unit that divides the labor among its members. In the community of slaves, Davis argues, much of the domestic work of supporting the community was done communally by slave women. For example, "most infants and toddlers spent the day in a children's house, where an elderly slave was in charge" (Hymowitz & Weissman, 1978, p. 47).

Davis points out that the only work that was done by slaves that contributed to the survival of the slaves themselves—as opposed to contributing only to the lives of the slaveowners—was the domestic work in the community that fed and provided clothes, health care, and shelter for its members. All of the other work done by slaves, in the fields and in the slaveowner's house, was done for the white plantation owners and served no useful purpose for the slave community.

James Curry, a former slave, described the work of his mother, a house servant who labored both for the plantation owner and for the other slaves:

> My mother's labor was very hard. She would go to the [plantation owner's] house in the morning, take her pail upon her head, and go away to the cow-pen, and milk fourteen cows. She then put on the bread for the [white] family breakfast, and got the cream ready for churning, and set a little child to churn it, she having the care of from ten to fifteen children, whose mothers worked in the field. After clearing away the family breakfast, she got breakfast for the slaves . . . which was taken at twelve o'clock. In the meantime, she had beds to make, rooms to sweep and clean. Then she cooked the [white] family dinner. . . . Then the slaves' dinner was to be ready at from eight to nine o'clock in the evening. . . . At night she had the cows to milk again. . . . This was her work day to day. Then in the course of the week, she had the washing and ironing to do for the master's family . . . and for her husband, seven children, and herself. . . . She would not get through to go to her log cabin until nine or ten o'clock at night. She would then be so tired that she could scarcely stand; but she would find one boy with his knee

out, and another with his elbow out, a patch wanting here and a stitch there, and she would sit down by her lightwood fire, and sew and sleep alternately, often till the light began to streak in the east; and then lying down, she would catch a nap and hasten to the toil of the day. (Blassingame, 1977, pp. 132–133)

Davis (1981) maintains that the critical contribution of women doing this survival labor within the slave community enhanced their value to the community. Women were essential to the community's survival. Davis also argues that the recognition of women as critical members of the community allowed them to emerge as important political leaders.

Davis found that African American women took leadership in many community activities. For example, African American women played an active role in the Underground Railroad and other forms of insurrection against the slave system. Women were likely to initiate and implement acts of sabotage, arson, and assassination as part of an ongoing battle against slavery. Sometimes domestic work itself was openly political, for example, when women fed runaway slaves, thus aiding them in their journey (Aptheker, 1943).

This kind of resistance was so common that a law was passed in South Carolina requiring that plantation owners regularly search slave quarters for runaways and arms. The act read: "Be it further enacted by the authority aforesaid, that every master, mistress or overseer of a family of this Province, shall cause all his negro houses to be searched diligently and effectually, on every fourteen days, for fugitive and runaway slaves, guns, swords, clubs and any other mischievious weapons" ("An Act . . . ," 1992, p. 260).

The political activity of African American women was not practiced without African American men. Rather, just as the oppression and exploitation of black women was equal to that of black men, there was an equality in their struggle against the system (Davis, 1981).

Davis provides much evidence that the African American family was community based as opposed to nuclear and that it included many people, kin and nonkin, working together. Why did this particular type of family occur in the slave community? Davis contends that the extended character of the families in the slave community resulted from the difficulties of establishing nuclear families (see also Du Bois, 1969). Others (Hill et al., 1989; Sudarkasa, 1988) take a different position on this question. They argue that the organization of African American families during slavery and up to the present has been influenced by the cultural roots of West African societies.

Hill et al. (1989) describe West African families as differing from nuclear families in three dimensions. First, the African concept of family includes many people who may not actually live in the same household. Second, extended kin (blood relatives like grandparents and cousins) are thought of and treated as family instead of limiting family to nuclear members only. Third, Africans include as family many people who may not be related by blood. These "unrelated" people may be incorporated into a family through informal adoption and foster care.

These characteristics were ones that Davis found typical of families in the slave community. She argues that these families were created in response to the experience of slavery. Hill et al. (1989) and Sudarkasa (1988) maintain that these families were part of the culture of black Americans because their ancestors came from West African cultures.

Comparing White Middle-Class Women to African American Women in the Nineteenth Century "Third wave feminism looks critically at the tendency of work done in the 1960's and 1970's to use a generalized, monolithic concept of 'woman' as a generic category in stratification and focuses instead on the factual and theoretical implications of differences among women" (Lengermann & Niebrugge-Brantley, 1992, p. 480; see also Dill, 1983; Hooks, 1984; and Chow, 1987). Jacqueline Jones (1985) is an example of a Third Wave feminist. Her work allows us to make a comparison between black women and white women living in the United States during the nineteenth century.

We noted in the last chapter that during the nineteenth century, white women in middle-class and upper-class families were identified with domesticity. Jones (1985) also found that in addition to the work they did for the white plantation owner, black women also performed much domestic work in the slave community. The division of labor in both Northern middle-class white families and in the Southern slave community assigned women to cooking, childcare, and sewing.

As we described in the last chapter, according to the cult of true womanhood, white women were supposed to make their family work the center of their lives. Their identification with domesticity diminished white women's importance in the community, since their domestic work kept them away from activities in the economic and political world that were the seats of power. As a result, women's work was invisible and their attempts to have any impact on politics were discouraged or even banned.

We have seen that African American women were doing the same work as middle-class white women but in a very different context and with a very different effect. Black women's domestic activities performed in service to the men and children of the slave community *enhanced* their importance to that community and their ability to serve the political struggle against slavery. At the same time that the African American community allowed, encouraged, and even demanded that African American slave women run an Underground Railroad and burn down plantations, the white community would not even allow white women to speak before a crowd (Dill, 1986).

This contrast between white women and African American women shows that the assignment of women to certain kinds of family work—cooking, childcare, sewing, etc.—does not necessarily give women power nor diminish their power. It is the combination of the division of work in families and the relationship of families to the dominant political structure that creates opportunities for women to emerge as leaders or to be victims.

Reproduction Slave owners attempted to treat women and men slaves in the same way, working them all to their maximum ability, and treating them all with the same ruthlessness, regardless of their sex. One problem with this strategy, however, was the need to replenish the numbers of slaves. Laws passed in 1807 made it illegal to import slaves from other nations, so the population of slaves could only be increased by allowing slave women to bear children. The physical demands of pregnancy, childbirth, and nursing an infant were not easily combined with starvation, brutality, and the exhausting and difficult work on the plantations.

The "solution" that the plantation owners developed was to divide the work between different groups of slave women. Therefore, women in the border states of the slave system were forced to bear large numbers of children and fertility levels in the upper South neared human capacity. Records show that slave women in Kentucky had high birth rates, bearing as many as twelve or thirteen children. Records also indicate that some slaveowners offered the women freedom or let them stay in their marriages if they bore these large numbers of children (White, 1985). Lacking the large, lucrative cotton plantations of the Deep South, slaveowners in the border areas turned the production of slaves into their dominant industry.

At the same time, the huge agricultural operations of the Deep South, which were highly productive, literally worked people to death. Women in the lower South were worked so hard that they were often unable to bear children. Slave women in Mississippi, for example, had low birth rates because of their large number of miscarriages, especially during the cotton boom years of 1830–1860 (J. Jones, 1985).

In nuclear families, the task of reproduction is divided between the husband and wife in the family. The wife bears the children and frequently cares for the young infants. Under the slave system, the task of reproduction was divided among *women* based on where they lived and what role they played within the system as a whole.

The Question of Separate Spheres The division of labor was different for slaves and middle-class whites. In Chapter 2 we noted that in the nuclear families of middle-class whites in the nineteenth century, work was divided into two "spheres," paid work in the labor market for men and unpaid domestic work for women. Work and family were separate and men's work and women's work was different. In Chinese American and Mexican American families work and family and men's work and women's work were also separate. Under the slave system, both men and women labored at the same exhausting jobs for no pay. The lines between work and family and between women's work and men's work were blurred in the African American family under slavery (Dill, 1986; Matthaei, 1982).

Based on the experience of white middle-class women a conceptualization of a split between work and family has been developed. This conceptualization does not fit the experience of African Americans (Brewer, 1988). Rather than attempting to explain why African American families "deviate" from this "norm," Collins (1990) suggests that we reconstruct our thinking about families based on our discovery of these kinds of diversity. The different relationship between work and family and men's and women's work, in particular, helps us to see how useless definitions of "the family" are. *Family* has many meanings depending on the social context in which a family exists.

The discovery of the contrasts in the experience of "separate spheres" among Euro-Americans, African Americans, Chinese Americans, and Mexican

Americans teaches us a second lesson: that diversity is not just between whites and nonwhites. Diversity occurs across many lines and we should not speak of people of color as if they were a monolithic group.

Sharecropping

After the slave system was abolished by the Civil War, a new economic system called sharecropping was established in the South and a different form of family developed for African Americans who lived there. The sharecropping system guaranteed that ex-slaves would continue to be available to work the land after the Civil War. Jay Mandle (1978) defines sharecropping as a

> crop lien system in which sharecroppers pledged their unplanted crops to landowners or local merchants (frequently at high interest rates). The arrangements of sharecropping varied, but typically sharecroppers received the use of a house, tools, farm animals, and a plot of land in exchange for a portion of the crop they produced. The rest of the crop was due to the landowner. Regardless of how good a year it was, the landowner demanded his due. If there was drought or insect infestation, the sharecropper still owed the landowner for the tools, farm animals, and the value of the crop that would have been available in a good year. This resulted in sharecroppers being in constant debt to the landowners and, therefore, unable to leave the plantation. (Mandle, 1978, p. 18)

Although the Civil War made it illegal for plantation masters to own slaves, for the most part the old slaveholders still owned the land. African American people who had lived in the region as slaves before the war now lived as sharecroppers on farms that were owned by the landholders after the war. In other ways sharecropping was similar to slavery. Both slaves and sharecroppers worked the land for the plantation owner, had few rights, were frequently subjected to barbarous treatment at the hands of the landowner, and lived hard, impoverished lives.

In addition, sharecroppers had little opportunity to move away from the plantations. Sharecroppers were obligated to give the landowners a specific amount of salable produce regardless of the actual production of the farm. If the crops did not come in as well as the landowner stipulated they should, the sharecroppers still owed the same amount. Sharecroppers, therefore, tended to sink further and further into debt. Debtors who wished to leave the plantations were in some cases forced to remain, creating a debt peonage system backed by the local sheriff. Although the sharecroppers were no longer slaves, in a sense they were enslaved by the sharecropping system.

In one way, however, sharecropping was quite different from slavery: in the organization of African American families. Sharecroppers retained their ties to an extended community network, as they had under slavery. Nuclear families, however, became the dominant form of family life under the sharecropping system. At the end of the Civil War, observers described masses of black couples coming to the Freedmen's Bureau offices to legalize their marriages (Giddings, 1984). And by "1870, 80 percent of black households in the Cotton Belt included a male head and his wife" (J. Jones, 1985, p. 62).

The macro level of Southern society, the political economy, changed from a system of slavery to a system of sharecropping. In some ways this transition was

Work for African Americans was much the same after the transition from slavery to sharecropping. Families, however, changed in important ways as African Americans sought to allow women to reduce their work in the fields so that they could take care of their families.

not as dramatic as many had hoped when they fought for the abolition of slavery. The transition, however, was significant enough to alter the organization of most families from an extended community-based form to one that was nuclear and more male dominated.

The basic unit of labor under sharecropping was the nuclear family. Under slavery the basic unit was the individual slave, who was often kept separated from his or her nuclear family and could be sold away from kin at any time. As we described above, this made it difficult and frequently impossible for people to maintain nuclear families. In contrast, nuclear families organized as economic teams with men as their head were the core of the sharecropping system.

Men sharecroppers were the ones who were in direct contact with the landholders, the creditors, and the people in the market who purchased the crops. Women were tied into the sharecropping system through their husbands. Wives and children worked in the fields under the direction of the father.

When a landholder allowed a sharecropper to live on his land, he assumed that the sharecropper's entire family would work together on the land. If the landholders suspected that all family members were not working the land they would arm themselves and ride through sharecropping communities to make sure that women and children were doing their share by working in the fields.

Who Shall Control Women's and Children's Labor? Jacqueline Jones (1985, p. 45) says that one of the great fears among whites after the Civil War was that "black people's desire for family autonomy, as exemplified by the 'evil of female loaferism'—the preference among wives and mothers to eschew wage work in favor of attending to their own households—threatened to subvert the free labor [nonslavery] experiment." The Boston cotton brokers, for example, claimed the disastrous cotton crop of 1867–1868 resulted from the decision of "growing numbers of Negro women to devote their time to their homes and children" (Giddings, 1984, p. 63).

African American women in sharecropping households were not loafers. They did, however, attempt to control their own labor by considering the needs of their families. The women tried to divide their labor between housework and fieldwork to allow them enough time for their families. Often they were not successful. "In 1870 more than four out of ten black married women listed jobs, almost all as field laborers. By contrast, fully 98.4 percent of white wives told the census taker they were 'keeping house' and had no gainful occupation" (J. Jones, 1985, p. 63).

Field work was not the only work required of African American women. In the 1930s, the Commission on Race Relations interviewed African American people about their lives. A stockyard worker in Mississippi explained how women in his community were required to work for whites. He said:

> Men and women had to work in the fields. A woman was not permitted to remain at home if she felt like it. If she was found at home, some of the white people would come and ask why she was not in the field and tell her she had better get to the field or else abide by the consequences. After the summer crops were all in, any of the white people could send for any Negro woman to come and do the family washing at 75 cents to $1.00 a day. If she sent word she could not come she had to send an excuse why she could not come. They were never allowed to stay at home as long as they were able to go. Had to take whatever they paid you for your work. (J. Jones, 1985, p. 157)

From the point of view of the ex-slaves, freedom meant the opportunity to create autonomous independent families. Nearly every family that had an able-bodied man aspired to a division of labor that allowed African American women to take care of their own families, and during Reconstruction large numbers of women dropped out of the wage labor system and limited their fieldwork under sharecropping. "The institution of slavery had posed a constant threat to the stability of family relationships; after emancipation these relationships became solidified, though the sanctity of family life continued to come under pressure from the larger white society" (J. Jones, 1985, p. 58).

Women were pressured to work in the fields and the homes of rich white people, which limited the time they had for their own families. They were also limited by the circumstances of poverty and instability that characterized the sharecroppers' lives. The women, nevertheless, tried to maintain "homelikeness" in the constantly moving households of the sharecroppers.

BOX 3 • 1 DID THE ABOLISHMENT OF SLAVERY END THE INVOLUNTARY SERVITUDE OF AFRICAN AMERICAN CHILDREN BY WHITE PLANTATION OWNERS?

Just as the ratification of the Thirteenth Amendment did not guarantee the right of African American families to make choices about the activities of wives and mothers, it also did not guarantee that families could make choices about their children. Rebecca Scott (1985) found that in North Carolina after the Civil War, apprenticeships were used as a way to reenslave African American children. People under the age of twenty-one could be legally indentured to employers.

For a brief period following the Civil War, from 1865 until 1877, the former Confederate states were placed under martial law and the Union Army attempted to make sure that the Southern political leaders abided by the agreements made at the end of the war. This period in American history is called Reconstruction. One of the institutions that was established during Reconstruction was the Freedmen's Bureau, which was supposed to try to rebuild the South and to aid the four million former slaves. The Bureau received numerous complaints about children who were seized from their parents and taken to work on former slaveowners' plantations.

Children could be legally bound if they were "base-born" (their parents were not legally married) or if their parents "did not employ them in some honest industrious occupation." An estimated ten thousand children were bound in Maryland, despite the objections of their parents (Giddings, 1984, p. 57). "Often no money was specified to be given at the end of the term, children were bound without parental consent, no trades were specified, or children were bound beyond the legal age" (Scott, 1985, p. 196).

The former slaveowners fought hard to reestablish their system. For example,

> William Cole [a former slaveowner] of Rockingham [North Carolina] wrote to the Freedmen's Bureau in 1865, "I understand you are authorized to bind out to the former owners of slaves all those under the age of twenty-one years, under certain conditions," . . . explaining that about thirty of his former slaves were under twenty-one and that he would like them all bound to him. (Scott, 1985, p. 194)

African American people fought equally hard to contest the reestablishment of slavery. One ex-slave woman named Huley Tilor wrote to the North Carolina Freedmen's Bureau about her former master:

> Dear Sir if you please to do a Good Favor for me if Pleas i have been to Mr. Tilor about my children and he will not let me have them an he say he will Beat me to deth if i cross his plantasion A Gain an so i dont now what to do About it an i wish if you pleas that you wood rite Mr. Joseph tilor a letter an let him Give me my childrin Mr. cory i want my children if you Pleas to get them for me i have bin after them an he says i shall not have them and he will not Pay me nor my children ether for thar last year work and now want let me have my childrin nother an i will close huley tilor. (Scott, 1985, p. 198)

> Sharecropping and tenant families frequently moved at the end of the year—in some cases hounded off a plantation by an unscrupulous employer, in other cases determined to seek out a better contract or sympathetic kinfolk down the road. . . . They tacked brightly colored magazine pictures or calendars on the walls of bare, drafty cabins; fashioned embroidered curtains out of fertilizer sacks; and planted flowers in the front yard. (Jones, 1989, p. 34)

Gender Equality and Changes in Families Under Sharecropping The historical documents show that the African American family was a central arena of the struggle between the old slavemasters and the ex-slaves. Former slaveowners

BOX 3 • 2 PARENTS AND CHILDREN IN NATIVE AMERICAN FAMILIES

Throughout the periods of slavery and sharecropping African American parents were prevented from exercising authority over their children and in many instances from even being able to live with them or spend time with them. Parents and children challenged this separation by running away to be with one another and, during Reconstruction, calling on the government to assist them in keeping their families together. During the late nineteenth and early twentieth centuries a similar struggle was being waged by Native Americans.

The Revolutionary War is often thought of as one that was fought between two opposing forces seeking to control North America, the British and the Americans. A third party, Native Americans, played an equally important role. Most Native Americans fought on the side of the British and when the war ended and the British returned home, Native Americans stayed to continue their effort to claim the right to live in their ancestral lands. Their struggle, however, was unsuccessful. The European Americans pushed westward, seeking resources and land to expand the cotton plantations (Zinn, 1980).

In 1820 there were 120,000 Native Americans living east of the Mississippi River but by 1844 fewer than 30,000 were left. This change was brought about by a government policy euphemistically called Indian Removal. Some Native American groups were forced to cede land to the U.S. government and white settlers. Those who refused were killed or run off by troops and their property was destroyed or confiscated. Those who took refuge in the surrounding area were eventually rounded up by the army and forced to march west to states like Arkansas. Eventually they were driven even further west to what became the reservations of Oklahoma and other western states. Dale VanEvery (1976) wrote of one of the marches out of Arkansas in the mid-1800s, "By midwinter, the interminable, stumbling procession of more than 15,000 Creeks stretched from border to border across Arkansas. . . . The passage of the exiles could be distinguished from afar by the howling of trailing wolf packs and the circling flocks of buzzards" (VanEvery, 1976, p. 142).

Once the Native Americans were removed to the isolated reservations, around 1880, the federal government sought "to obliterate the cultural heritage of Native Americans and to replace it with the values of Anglo-American society" (Trennert, 1990, p. 224). The key strategy for accomplishing this was to remove children from their tribal homes and place them in boarding

wanted to continue to dominate black people by controlling their work and their family organization. Newly emancipated people wanted to control their own labor and that of their family members.

Susan Archer Mann (1986) argues that there was another struggle being waged, the one *within* sharecropper families, between husbands and wives. The development of the nuclear male-headed household (at least as an ideal) marks a change that was not entirely positive for women. Certainly women were greatly relieved by the abolishment of slavery and within the sharecropping system they chose to contribute as much of their labor as possible to their families. But the equality between women and men that was present under slavery was diminished when the sharecropping system developed, placing wives under the control of their husbands (Janiewski, 1983). Paula Giddings (1984, p. 61) states that "following the Civil War, men attempted to vindicate their manhood largely through asserting their authority over women."

The absorption of African American women into more nuclear households also may have pulled them away from their extrafamilial connections in the larger community, like those Davis described as characteristic of the slave com-

BOX 3 • 2 Continued

schools, which were located a long distance from where their families lived. Parents were required to enroll their children for a minimum of three years and the Bureau of Indian Affairs (BIA) paid for transportation only for one round trip. During the summer most students could not afford to return home and were therefore placed in local homes to work for fifty cents to one dollar a week (Littlefield, 1989). Alice Littlefield (1989) reviewed student files and found that a number of parents requested that their children be sent home because of illness or economic hardship or because they believed the schools were not treating their children properly. In every case the school superintendent denied the request. Children usually did not see their parents for several years.

At the turn of the century, 17,000 Native American children were sent away to boarding schools while only 5,000 attended federal day schools (Szasz, 1985). The schools were run as paramilitary institutions with rigid discipline, beatings, highly regimented daily activities, students in uniforms, marching in formations, and rising to 5:00 a.m. drills. There was little emphasis on academics. Boys were likely to be trained in a trade and girls in domestic work. Each hour of academic work was matched by three hours of industrial work, most of it unpaid (Szasz, 1985). Children were not allowed to speak their native language and at some schools were punished if they were caught doing so. In addition, although the schools were government run and therefore nondenominational, students were required to participate in service organizations sponsored by Christian churches (Trennert, 1979). In addition, children were malnourished, and crowded conditions led to widespread tuberculosis and trachoma (Szasz, 1985).

Official government policy in regard to Native Americans in the late eighteenth and early nineteenth centuries sought to separate parents from children, to diminish the authority of parents over their children, and to prevent parents from passing on their own cultural heritage and values to their children. The U.S. government believed that if children were removed from their families and their communities they would be better assimilated as useful workers in the dominant white society. Children resisted by frequently running away and breaking rules, especially by pilfering food (Littlefield, 1989). Parents resisted by hiding their children and not allowing government officials to send them to boarding schools (Amott & Matthaei, 1991).

munities. Throughout the period of sharecropping and industrialization African American women continued to try to keep their commitment to the larger community, especially in their efforts to develop, support, and teach in schools for black children.

Jacqueline Jones (1985, p. 99) reminds us that our concern for the loss of equality of black women in gender relations must be tempered by the acknowledgment that under sharecropping, "black working women in the South had a more equal relationship [than northern middle-class white women] with their husbands in the sense that the two partners were not separated by extremes of economic power or political rights; black men and women lacked both."

She also argues that the impact of the larger political and economic context must be remembered. The maintenance of more isolated nuclear families was not for the purpose of empowering black men. The "seclusion" of women in nuclear families where men were the head of the households and the ones in direct contact with the white landowners and merchants was a strategy to protect women from a real threat. One ex-slave vowed "to support his family by his own efforts; never to allow his wife and daughters to be thrown in contact with

Southern white men in their homes" (Lerner, 1973, p. 292). When the Union troops left and Reconstruction ended a backlash occurred and tremendous brutality was directed against all African American people. One of the legacies of slavery was the sexual assault of African American women by white men. One way to try to prevent this was to make African American husbands the social and economic link between the family and the white landowners. To minimize the contact between white men and African American women was an attempt to hold back the constant threat of sexual violence.

Sharecropping and the formation of male-headed nuclear families for African Americans who lived under the system was, therefore, a result of several intersecting factors. African American families were contested terrain. Who would control these families, white landowners, African American men, or African American families themselves? And where would African American women fit in?

Industrialization

Agriculture remained the core industry in the South, with cotton as the most important crop, until after World War II. But throughout the twentieth century the introduction of machinery displaced more and more sharecroppers. Rural Southerners whose jobs were taken over by farm equipment were forced to move North and into the urban areas in the South.

This period in history is known as the Great Migration. Until the turn of the century, nearly all African Americans lived in the Southern states. Between 1910 and 1930, one million African American people (10 percent of the total black population) moved from the South to the North. And even larger numbers moved from the rural South to Southern cities during that period (Marks, 1985).

The move was not just a geographic one, it was also economic, marking a transition in the political economy of the South. This transition was not just one of black workers moving from one locale to another, but of workers moving from agricultural work to urban manufacturing and service jobs. The transition occurred simultaneously with the transition in the political economy of the South from one based on farming to one based on industry and service.

The move had different results for men and women. African American men and white people moved from agricultural work into manufacturing jobs; African American women moved from agricultural work into domestic work. "Less than 3% of all black working women were engaged in manufacturing in 1900 compared with 21% of foreign-born and 38% of native-born white working women" (J. Jones, 1985, p. 166).

It is difficult to say why African American women entered domestic work rather than manufacturing. The pay in manufacturing work was slightly higher than for domestic work and factory work often meant somewhat less degrading supervision by whites compared to what the domestic worker had to endure. For these reasons black women undoubtedly would have preferred industrial jobs to domestic service. They probably did not have that choice. There is some evidence that black women were forced into domestic work by racially discriminatory laws that barred them from factory jobs.

Even African American women who did manage to secure work in the factories had a difficult time. "Black women who fled from the degradation of domestic service only to find themselves in the hot, humid tobacco stemmers' room paid a high price in terms of their general health and well-being" (J. Jones, 1985, p. 135).

African American women remained concentrated in domestic work for another five decades. Until the middle 1960s the largest single occupation of African American women was domestic work. "By 1950, 60 percent of all African American working women (compared to 16 percent of all white working women) were concentrated in institutional and private household service jobs" (J. Jones, 1985, p. 235).

African American families changed as women went to work in the homes of white people and spent long hours away from their own families. Some employers insisted that their maids "live in," which meant that domestic workers were able to see their own families only every other weekend. Once again an African American woman's time with her own family was a battleground between herself and her white employer.

A similar battle was being fought in California during this same period. Japanese American women who had also immigrated from rural agricultural economies to the growing U.S. urban centers found themselves limited in their occupational choices. Large numbers became live-in domestic employees for whites. They also would have preferred factory or shop employment but like African Americans were barred from those jobs by race and gender discrimination.

Being on call twenty-four hours a day by an employer was intolerable for African American and Japanese American women and both fought against live-in domestic service. African Americans compared to Japanese Americans resisted and defied their employers more openly (Glenn, 1990), although both groups used a number of resistance strategies.

Bonnie Thornton Dill interviewed domestic workers who had been employed in the first half of the twentieth century. She found they used several techniques to gain control of their work, including direct confrontation, threatening to quit, chicanery, and quitting (Dill, 1988). Although they were not all successful, many were able to work out some degree of autonomy and gain some respect on their jobs. Ultimately they were successful in changing domestic work from live in to day work which meant they could go to their own home at night (Dill, 1988; Glenn, 1990).

In Chapter 7 we will continue this investigation of domestic work by African American women, as well as other racial ethnic women. The important points to be remembered here are that private domestic work was overwhelmingly the largest occupation for African American women in the early part of the twentieth century and that work involved conflicts between maids and their employers over a maid's right to spend time with her own family.

AFRICAN AMERICAN FAMILIES IN THE STRUGGLE FOR EQUALITY

Although each of these historical periods shows different political economies and different family forms, one theme runs through them all: the centrality of family in the battle waged by African Americans for equality. Jacqueline Jones (1985, p. 4) states, "Throughout American history, the black family has been the focus of a struggle between black women and the whites who sought to profit from their labor."

During the slave period, the goal of slaveowners was to use their slaves to make as much profit as possible. Slaves had very different goals. They fought to maintain their family ties, to keep their community alive, and to topple the slave system. Central issues of contention between slaveholder and slave were related to family organization, such as the right of slaves to choose a spouse, to live with their spouses and children, and to be free from sexual assault by plantation owners. Despite enormous difficulty, slaves often kept track of their kin and created their own rituals recognizing marriage and parenthood, regardless of the slaveowners' laws.

Slaves also organized insurrections and other forms of rebellion such as running away from the plantations. In the opening scenario John mentions that he has decided to go north with Harriet Tubman on the Underground Railroad. The Underground Railroad was a large network of black and white people who smuggled thousands of slaves out of the South before the Civil War.

Harriet Tubman was the "conductor" on the Underground Railroad. She was born into slavery and after escaping she returned to the South, first to bring her husband and other family members to freedom and then to make nineteen more trips carrying more than 300 slaves to freedom. But Harriet Tubman was not alone. Over 3200 people ran the Underground Railroad, transporting 2500 slaves a year to freedom between 1830 and 1860 (Zinn, 1980).

During sharecropping the battle over African American families continued. Landowners physically threatened women who did not work in the fields and kidnapped children to be field workers. Sharecropper families fought back by defying the landowners and often the sheriff in attempts to run away or at least to maintain some control over their lives. Sharecropper women tried to spend as much time as they could on their family work and parents sought to retain their rights to protect their children and make decisions about when and where they would work. Sharecropper men attempted to "seclude" women and children away from whites to protect them from sexual assault.

In the first half of the twentieth century the Great Migration and industrialization pulled Southerners from the rural areas into the cities. African American women who were pushed into wage labor as maids fought with their white mistresses over the time they could spend with their own families and over control of their work. Employers ignored maids' family connections and obligations, demanding that they be on call twenty-four hours a day. Maids had to defy their employers to claim even a small part of their lives for their families and themselves. Eventually the maids' protest spilled into the streets in the Montgomery Bus Boycott, which I discuss below.

During Reconstruction, an attempt was made to integrate African Americans into Southern society. African American men were allowed to vote and many African Americans were appointed or elected to government positions. For example, in South Carolina in the years immediately following the Civil War, fifty members of the state legislature were black and only thirteen were white. But as the old plantation owners regained political power, the situation deteriorated. In 1883 the Civil Rights Act of 1875 was declared unconstitutional. In 1896 the final blow occurred with the Supreme Court decision in the case of *Plessy v. Ferguson*.

From the end of the 19th century until the middle of the 20th century, Jim Crow laws created a legal apartheid in southern United States.

Homer Plessy, a black man, rode a segregated train through Louisiana and was required to sit in a car reserved for African Americans. He sued the railroad, claiming that his constitutional rights had been violated. The Supreme Court ruled in favor of the railroad, making legal the doctrine of "separate but equal" or what were called Jim Crow laws. Much like the apartheid system in South Africa, the doctrine of "separate but equal" meant that all public facilities, in the South, including restaurants, drinking fountains, public restrooms, schools, blood banks, and public transportation, could be segregated.

Black people who rode the bus were required to pay at the front of the bus, then get off, walk to the back door, and get on in the back of the bus. They were allowed to sit only in the seats in the back of the bus even if there were empty seats in the front section "for whites only." If a white person got on and there were no "white seats" left a black person was required to give up his or her seat. One evening in 1955 Rosa Parks, a black woman, refused to give up her seat. She was arrested and placed in jail. Her action sparked the Montgomery Bus Boycott, which was to become a major event in the history of the civil rights movement. Many of the riders on the city buses were African American maids who rode to work at white homes across town. The maids and other African Americans boycotted the buses for many months and eventually were successful in changing the Jim Crow laws affecting public transportation.

In Chapter 5 when we examine diversity among contemporary American families, we will continue the discussion of the history of African Americans and the relationship between changes in the macro level of social context and changes in the micro level of families' everyday experiences. More than half of all African Americans now live outside the South and most of them live in urban areas. As we noted in Figure 2-4 in the last chapter, black women have continued to increase their numbers in the paid labor force. Along with these changes in the context of African American families, the organization of those families has also changed: Table 3-2 shows, for example, that a growing number of African American households are now headed by single women. And the organization of the African American family has continued to be controversial, with the debate over the Moynihan Report in the 1960s and the existence of an "underclass" in the 1980s and 1990s. These issues will be examined in Chapter 5.

STRUCTURATION THEORY AND THE IMPORTANCE OF AGENCY

Chapter 1 discussed C. Wright Mills's notion of the Sociological Imagination and the importance of acknowledging both the way in which the larger social context affects people's everyday experiences and interactions, and the way in which day-to-day social activities help to shape the larger social context. The interplay between the micro and macro levels of society have continued to be of interest to social theorists. Anthony Giddens has worked to develop sociological theory around the question of the relationship between the micro and macro levels of society. He has been especially committed to drawing our attention to the micro-level social activities that influence the macro level. He refers to this activity as agency (Ritzer, 1992). Giddens writes, "Agency concerns events of which an individual is a perpetrator. . . . Whatever happened would not have happened if that individual had not intervened" (Giddens, 1984, p. 9).

George Ritzer argues that Giddens was influenced by Marxism and that his work has grown from the dictum, "Men make history, but they do not make it just as they please; they do not make it under circumstances chosen by themselves, but under circumstances directly encountered, given, and transmitted from the past" (Marx, 1869/1963, p. 15). Giddens's concept of agency fits the first clause of the sentence, "Men make history." Giddens attempts to integrate agency with his notion of social structure, which he divides into four clusters: symbolic orders, political institutions, economic institutions, and the law (Ritzer, 1992). These clusters create those circumstances described in the last half of the quotation, circumstances not chosen but present nevertheless.

This chapter's review of African American history provides evidence of the importance of structure in the description of systems like slavery, sharecropping, and industrialization, which set the conditions under which African Americans lived. It also shows evidence of agency in the efforts of African Americans and others to shape their history. Giddens's structuration theory helps us to see these two levels and, most importantly, the constant interplay between them.

THE MICRO-MACRO CONNECTION

We see from this history that the macro level of social organization makes an enormous difference to family organization and family life. The division of labor between women and men, the political relationship between women and men, and even the reproduction of children are greatly affected by the kind of political economy in which those relationships are embedded.

The startling and alien quality of the slave system provides powerful evidence of the importance of understanding the macro level of society, the social

TABLE 3 • 2
Recent Changes in Black Family Organization, 1970, 1980, 1990

Type of Family	1970	1980	1990
Married couples	68.3	55.5	50.2
Single male–headed	3.7	4.1	6.0
Single female–headed	28.0	40.3	43.8

Source: U.S. Bureau of the Census, 1991. *Household and Family Characteristics, 1989, 1990.* P-20, no. 447. GPO.

context, in order to understand the organization of families and the everyday experiences of family members. We would never make the mistake of saying that John and Mary, from the story at the beginning of the chapter, *chose* the family life they lived without acknowledging the limitations placed on those "choices" by the slave system. Although our own society is not as grotesque as the slave society, it is equally powerful in its effect on individual lives. When we try to understand our own families, we commonly assume our social context to be something natural or inevitable and we believe that we make free and individual choices. This history of African American families helps us to see that our social surroundings profoundly affect and sometimes even determine what our lives will be like.

Once again we observe that society and families have not remained the same throughout history. The macro level of society is not static but changes sometimes slowly and sometimes dramatically. The history of African Americans shows both the steady evolution and dramatic turning points of the last two centuries. It is not unusual to hear people in contemporary American society speak of "the downfall of the family," as if families had remained unchanged for centuries and have only recently deviated from some fixed form. But the Johns and Marys of the slave period led very different lives from the Johns and Marys of the sharecropping period or the Great Migration. This history, like that in Chapter 2, reminds us that change is a constant factor and that families have always been in transition.

We also see how the historical change was experienced differently by African Americans compared to white Americans. This suggests that when we speak of families we must be careful to remember that families vary by race, class, region, nation, and by many other factors. These variations form stratification systems that are part of the macro level of society and which create significant differences in the micro-level experiences of members of various strata. In the next two chapters we will develop this discussion of stratification and the way in which it affects family life.

The two chapters on history show the ties between gender, family, and political economy. Social contexts are made of many factors and even systems of factors. Social structure itself is complex. The history of a society is really made up of many histories of many social structures.

In this history of African American society we can see that systems—family, gender, race, and political economy—make up part of the macro level of social structure and that these systems are interrelated. As families vary according to their social context, so do gender relations and family organization. The relative power of women and men, the tasks assigned to women and men, and the importance attached to those tasks all differ according to the time, place, and racial ethnic group in which they occur.

This chapter has described the day-to-day experience of African Americans at the micro level. It has also described the social organization of slavery, sharecropping, migration, and industrialization at the macro level. And it has shown how this macro organization affected the everyday life of African American people.

The relationship between macro- and micro-level forces is not one way. This chapter also shows the effect of the micro on the macro, of individuals working together to change the system, whether in slave insurrections, in the struggle over black families, or in the fight for equality for African Americans and for women.

SUMMARY

Diversity in American Families

- The scenario about John and Mary shows the importance of race ethnicity to the families of people living in a system of intense racial oppression.
- Families in the United States show tremendous diversity. African Americans were chosen as the focus of this chapter because of several unique characteristics: information is available, African Americans are the largest group, they are the only ones who were legally enslaved for centuries, and a Civil War was fought over their enslavement.

Four Periods in African American History

- African American history can be traced through four periods: slavery, sharecropping, industrialization, and the New South. This chapter looks at the first three of these.
- Slavery lasted from the early 1600s until the end of the Civil War and was practiced primarily in the South.
- Under slavery men and women were not allowed to marry and parents had no rights in regard to their children.
- Slaves were sometimes able to maintain ties with spouses and children by traveling to other plantations. But these journeys were difficult and dangerous.
- Herbert Gutman's work on African American families under slavery emphasizes the way in which slaves attempted to maintain nuclear families.
- Angela Davis argues that most nuclear families were disrupted by the slave system and that slaves developed an alternative family system that consisted of extended kin and nonkin networks within the community.
- Davis further asserts that women's central role in providing for the survival of the community through domestic work allowed African American women to emerge as political leaders in the fight against slavery.
- Davis argues that domestic work was a basis of power for African American women in slave communities, in contrast to Northern middle-class and upper-class white women of the same historical period who were kept powerless partly by being tied to domestic roles.
- Slaveowners wanted to work men and women equally for the maximum profit possible. But after the importation of slaves was outlawed, they also needed women to bear children to perpetuate the system.
- Slaveowners devised a division of labor between "breeding" slave women in the border states and "nonbreeding" women in the Deep South.

- Middle-class white women were separated from middle-class white men, doing different work and staying at home while men went into public.
- This was not the case for African American women, for whom lines between work and family and men's and women's work were blurred, and the concept of separate spheres was entirely inappropriate.
- Sharecropping developed after the Civil War and the abolishment of slavery. Like slavery, sharecropping tied people to the land owned by rich whites who were frequently former slaveowners.
- Sharecropping was very different from slavery in terms of family organization. The dominant form of family under sharecropping was the male-dominated nuclear family.
- The sharecropping period was characterized by a struggle between African American families and white landowners over the labor of African American women and children.
- The establishment of nuclear families during sharecropping was a blessing to African American women coming out of the slave system but it also increased the control that African American men had over women.
- Industrialization brought further changes, driving people out of rural agricultural jobs in the South and into urban jobs in the North and the Southern cities.
- African American women went into domestic work because they were banned from manufacturing jobs by discriminatory hiring practices.
- Domestic work was low paid. It required long hours away from one's own family.

African American Families and the Struggle for Equality

- A recurrent theme throughout this history is the centrality of family in the struggle for equality for African Americans, from the underground railroad to the Montgomery Bus Boycott.

Structuration Theory and the Importance of Agency

- African American history presents an especially good example of the way in which social structure simultaneously shapes the lives of individuals while those individuals create and change social structures.
- Giddings refers to this mutual relationship as *structuration theory*. He calls the social activities of people influencing social structure *agency*.

The Micro-Macro Connection

- This history depicts the impact of the macro level of social structure on individual lives and families.
- Second, the history also shows how social structure itself is constantly changing.
- Third, the history shows how social context can affect different sectors of society (in this case white and black Americans) differently.
- Fourth, the history shows that social context is complex and consists of many interrelated systems, including family, gender, and political economy.
- Fifth, it shows the constant efforts of various individuals and groups of individuals to try to change social structure—the effect of micro activities on macro organization.

The Great U-Turn marked a major shift in the U.S. economy, causing unemployment, poverty, and homelessness. Some families were forced onto the road to try to find work.

4 FAMILIES AND THE ECONOMIC SYSTEM

Jackie grew up in a blue-collar family in Flint, Michigan. Her husband Al was a riveter at the GM plant until the company laid him off. They have a five-year-old son named AJ. Jackie has worked on and off in a small parts plant and as a waitress and a beautician. After Al's unemployment benefits were cut off he decided to go south to find work. Some of his buddies said there were a lot of construction jobs in Arizona, so they had a big yard sale, then took all the clothes and household goods their car could hold and moved to Tucson. When they got there Al found a job but it only paid minimum wage and was not steady—nothing like what he used to make in his union job in Michigan. On his pay in Arizona they could not save enough to pay the deposit on an apartment so they lived in the car. Jackie found part-time work at the

Circle K but eventually lost her job because she had a hard time looking decent, since she had no place to get cleaned up before she went to work. Finally they decided to apply for welfare but when they inquired at social services they discovered that Arizona doesn't provide grants to families that have two able-bodied adults. They also discovered that if protective services found that AJ was living in a car with them, they would take him away and place him in foster care. Now Jackie is afraid she is pregnant but she has no insurance and cannot afford a doctor or an abortion. Standing in line at the outdoor soup kitchen she begins to feel sick. She sits down and cries.

- *Jackie and Al's story sounds melodramatic and unreal. How extraordinary is their experience?*
- *Are all American families facing economic hardship in these times?*
- *What are all the circumstances that bring Jackie to the final scene?*
- *What factors affect the economy, causing it to go through cycles of boom and bust?*
- *What is the solution to Jackie's family's economic crisis?*

THE CRITICAL LINK BETWEEN FAMILIES AND THE ECONOMY

Chapter 1 outlined five themes that have been central to a revisioning of families (Thorne, 1992). One of the themes was the importance of the link between families and their economic context.

In Chapters 2 and 3 we examined the history of the United States from the colonial period to the middle of the twentieth century. During that period there were a number of important changes in economic organization and there were critical changes in family organization as well.

This chapter brings us up to date by looking at what is happening right now in our economy and to American families living through these times. The chapter is divided into four major sections. The first is a brief discussion of the theoretical roots of the importance given to the link between families and the economic context. The second section examines the economic context of contemporary U.S. families. The third reviews the responses to economic hardship at the micro level by describing coping strategies and political organizing efforts. The fourth section proposes an analytical framework for understanding the current economic crisis and how it might be resolved.

The most basic activity in which humans participate is providing themselves with the necessities of physical survival. Like other animals, humans must find food and shelter to keep themselves alive and able to produce the next generation. Sociologists call the organization of the production and distribution of the necessities of life the economic system.

In order to understand our societies and our lives we need to take into consideration the way in which our economic system is organized. As we have seen in the previous two chapters and will observe throughout this book, economics is not the only important factor in our social surroundings but it is an essential one.

Because economic activities are so critical to our existence, they affect all other aspects of our lives. Not surprisingly, when significant changes take place in the production and distribution of goods and services, dramatic changes in other parts of our society occur as well. Both new problems and new possibilities emerge. In the historical review of the preceding two chapters we noted the way in which changes in the economy such as the development of industrialization or the transition from slavery to sharecropping had important effects on the lives of the people living during those transitions.

This chapter begins by describing the problematic character of our current economic situation. At the end of the chapter we will consider some of the potentially favorable possibilities that may emerge from the difficult changes we are witnessing.

Theoretical Assumptions

Dialectical materialism is the underlying logic of Marxist theory (Chafetz, 1988). One aspect of this approach is materialism. Materialism refers to a philosophical assumption that the material realities of human life shape the ways in which people think and structure their society and their individual lives (Chafetz, 1988). Materialists maintain that "specifically the manner in which people satisfy their needs by acting on the environment (which is their labor) structures their thought systems (ideologies, religions), family relations, systems of social inequality, political and educational systems and so on" (Chafetz, 1988, p. 7).

Not all sociologists are materialists. For example, Max Weber was not a materialist; he was an idealist. For him, ideologies, especially religious ideologies, formed the basis from which social and economic structures grow (Chafetz, 1988). An idealist like Weber would approach the problem of contemporary families quite differently from a Marxist materialist. An idealist would emphasize the kinds of ideas that dominate in our society and the ways in which those ideas shape our families.

The organization and content of this chapter illustrate a materialist point of view. The way in which we organize our material lives, the production and distribution of the necessities of physical survival, creates the framework in which to understand contemporary families. Based on the materialist premise that we need to understand our economic situation in order to understand what is happening to our families, the next section will describe the current plight of the U.S. economy.

The Contemporary Economic Scene in the U.S: The Great U-Turn

The United States emerged from World War II ready to move into a leadership position in the world economy (Berberoglu, 1992). Unlike Japan, the Soviet Union, and European nations, the United States suffered little damage to its industrial base during the war. Buildings and factories were left standing, and mines, roads, airports, harbors, and the energy distribution and communication systems were relatively untouched.

At the same time, the damage in other nations created a huge market for the goods that Americans could produce. American steel, autos, tractors, food, clothes, building materials, and chemicals were needed by consumers and businesses all over the world. This combination of demand for goods and little competition for markets created an enormous boom in the U.S. economy (Tanzer, 1974). Throughout the 1950s and 1960s Americans grew to expect that each year would bring a higher standard of living and a great number of people found their expectations realized.

But in the middle of the 1970s this economic growth in the U.S. reversed (Berberoglu, 1992). The wages of American workers began to decline. Then in the 1980s, because of the decline in wages, the tax base began to dry up and the federal, state, and local governments entered a period of fiscal crisis (O'Connor, 1973).

The federal government responded by raising taxes for all but the very wealthy. Ninety-five percent of the American population saw their taxes increase, especially those at the bottom of the socioeconomic scale. "The tax rate for poverty level families increased from 1.8% of their income in 1979 to 10.8% in 1986" (Feagin & Feagin, 1990, p. 59). And from 1987 to 1988 low-income families saw their taxes rise again, nearly 20 percent.

The very wealthy, however, saw their taxes cut during this period. For example, "the wealthiest 1% of the population had an average tax cut of $44,750 in 1988" (Feagin & Feagin, 1990, p. 59). In addition to benefiting from these cuts in personal income tax, the wealthy benefited from tax cuts on the corporations they owned:

> A large proportion of the largest companies paid little or no tax in at least one year since the mid-1970's. . . . Forty large corporations that paid no taxes in 1986 . . . [and] at least sixteen large corporations, whose combined profits totaled almost $10 billion in 1987, not only paid no taxes, but together received more than $1 billion in tax refunds from the federal government in that year. Topping this list were General Motors, with profits of $2.4 billion and a refund of $742.2 million and IBM with profits of $2.9 billion and a refund of $123.5 million. (Feagin & Feagin, 1990, pp. 60–61)

Corporations have paid a smaller and smaller proportion of taxes since World War II. In the 1940s corporate taxes accounted for 34 percent of all the taxes collected by the federal government. By 1960 that proportion had fallen to 20.2 percent, by 1970 to 16.9 percent, by 1980 to 12.4 percent, and by 1989 to 10.97 percent (Berberoglu, 1992).

The thinking behind this protection and enhancement of the incomes of those at the top of the economic system was that their good fortune would eventually "trickle down" to the rest of us. The strategy did not work. The economic crisis coupled with enormous tax cuts for the very wealthy and for corporations resulted in a sharp increase in the national debt that could not be solved by taxing everyone else more. In 1970 the national debt was close to $0. From 1980 to 1990 the national debt increased from about $900 billion to $3.2 trillion (Berberoglu, 1992).

In response to the fiscal crisis, the government began to cut back on programs to supplement the wages of working people. As we move into the 1990s wages are lower, for most of us taxes are higher, and the trend appears to be toward greater declines and continued government cutbacks at the moment when we need help the most (Moore, Sink, & Hoban-Moore, 1988; Block et al., 1987).

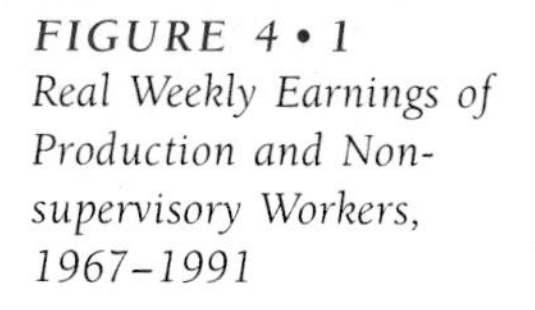
FIGURE 4 • 1
Real Weekly Earnings of Production and Non-supervisory Workers, 1967–1991

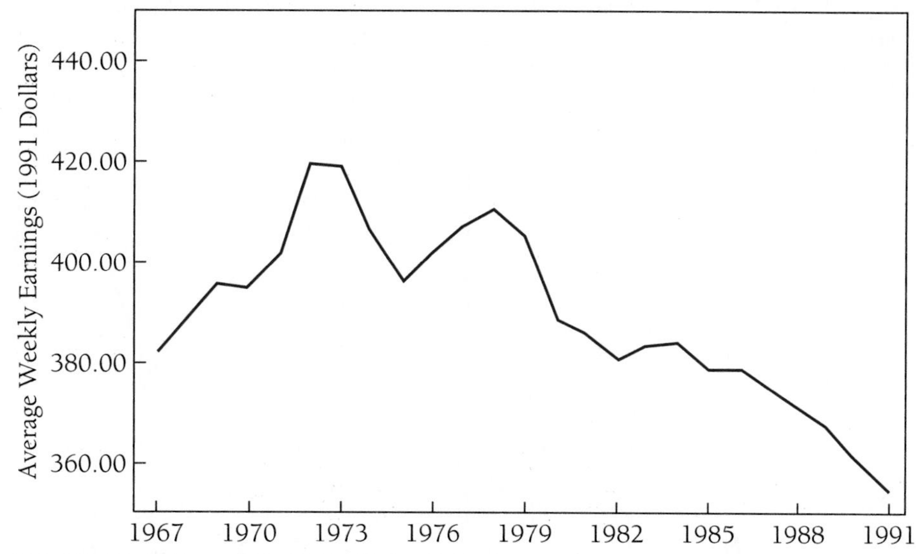

Source: Mishel & Bernstein, 1992, p. 136.

This reversal in the direction of the U.S. economy has been called the "Great U-Turn" by economists Bennett Harrison and Barry Bluestone (1988). Figure 4-1 illustrates the Great U-Turn, showing the increase in wages in the late 1960s until the mid-1970s and the decrease since then. The figure shows that by 1991 average wages for production and nonsupervisory workers—about 80 percent of the workforce—had declined to below the level of the early 1960s.

In Figure 4-1 the numbers have been adjusted for inflation. This means that the real (unadjusted) numbers before 1991 were lower. But the *buying power* for all of the years is comparable when the numbers are adjusted.

For example, in 1960 a worker may have brought home $100 (unadjusted) per week but in that same year he or she could buy a new car for only $2500. One hundred dollars sounds low to us but the buying power of that $100 was greater than it is today because of inflation. Since the price of cars, and everything else, has risen over the past decades it is more realistic to compare buying power than to compare absolute numbers. When the numbers are adjusted the buying power can be compared.

Falling Wages

In discussing the decline in wages there are three categories of workers to consider. In the first category are workers who have lost their jobs. Figure 4-2 shows that unemployment rates fluctuate but have inched consistently upward and are especially high among African Americans. Those who lose their jobs often find new jobs but at lower wages.

Al, the riveter from Flint in the scenario at the beginning of this chapter, is an example. Because he was willing to make an enormous effort—moving his family across the country—he was able to find another job. Like many laid-off workers in this category, however, he found that reemployment did not bring him his previous level of pay and his previous standard of living.

FIGURE 4 • 2
Unemployment Rate, 1965–1991 (percent of labor force)

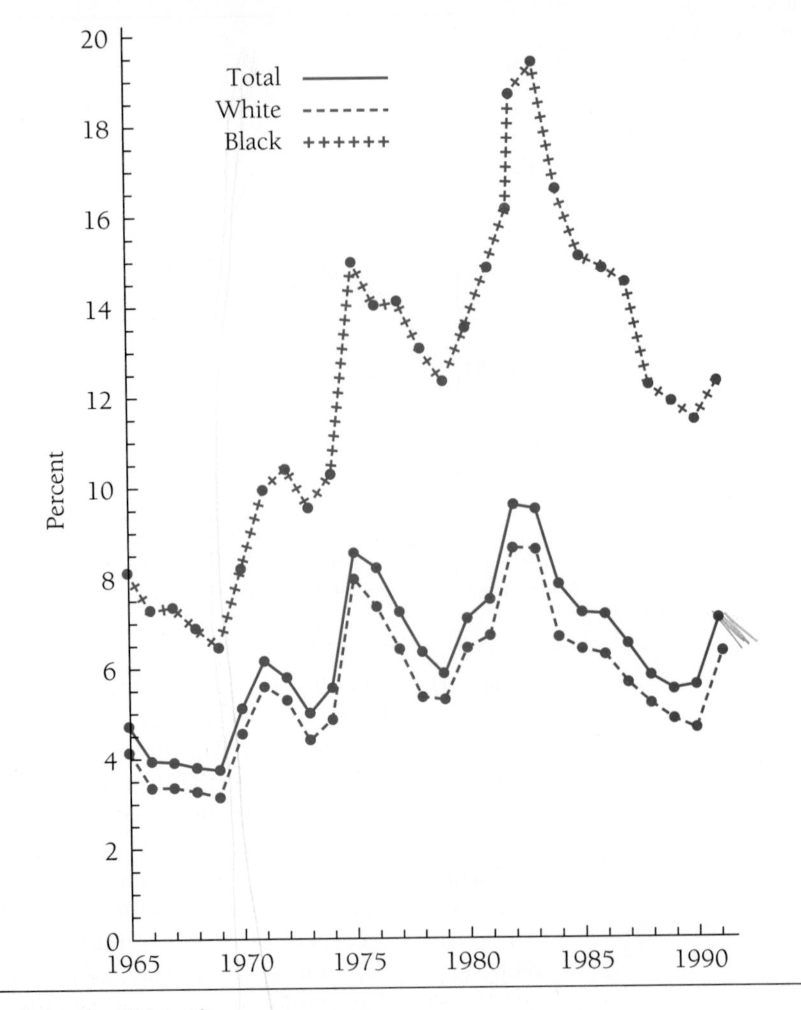

Source: Berberoglu, 1992, p. 59.

Other workers have retained their jobs but their wages have not kept pace with inflation. Figure 4-3 shows the increase in inflation, which has been especially severe since the 1970s. In some cases workers have received increases but their raises were not as great as the inflation rate. Others have received no increases. For example, the wages for many state jobs have been frozen at the same time that prices have continued to rise. Other workers have been forced to accept concessions in which their pay has actually been reduced for the same job (Sheak, 1990).

A third category of workers are those who are new entrants to the labor market and who have taken newly created jobs. Table 4-1 shows the Bureau of Census' projection of where new jobs will be added to the economy. Most of these categories are not highly paid occupations. As a result, on the average, new jobs are likely to pay the minimum wage. Between 1979 and 1984 more than half of the eight million new jobs created in the U.S. paid less than $7,000 per year (Harrison & Bluestone, 1988). The jobs that were being removed from

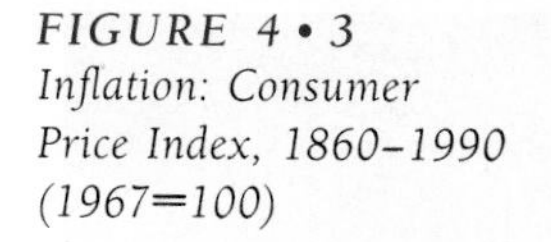

FIGURE 4 • 3
Inflation: Consumer Price Index, 1860–1990 (1967=100)

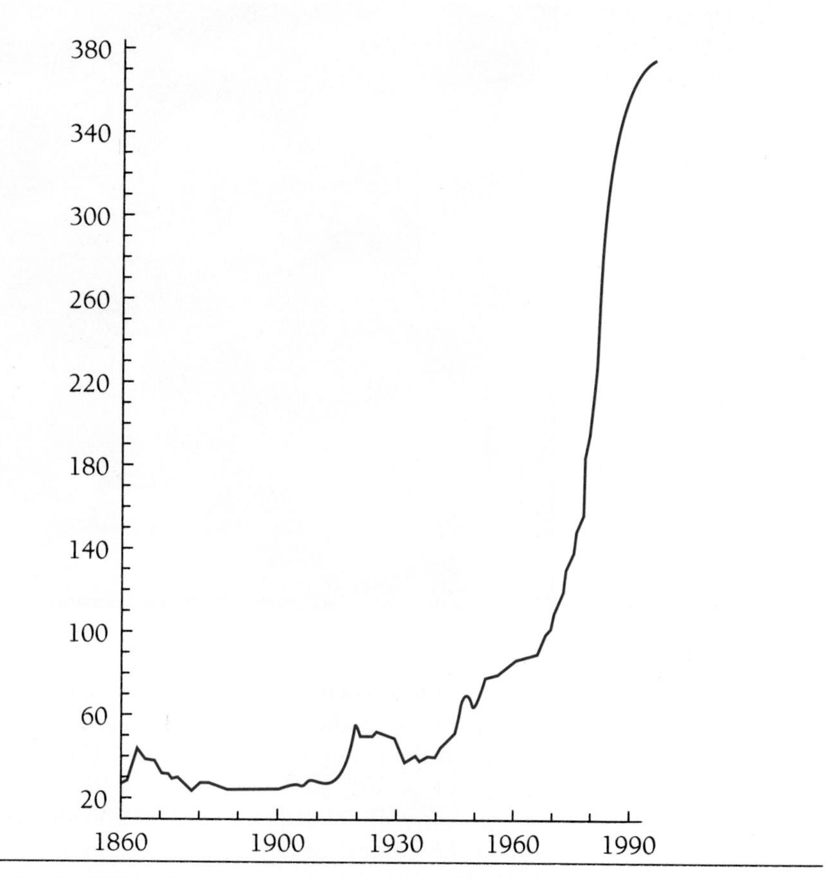

Source: Berberoglu, 1992, p. 60.

TABLE 4 • 1
Ten Largest-Growing Jobs 1988–2000

Occupation	Number of New Jobs
Retail sales	730,000
Registered nurse	613,000
Janitor/maid	555,000
Waiter/waitress	551,000
General manager	479,000
Office clerk	455,000
Secretary	385,000
Truck driver	378,000
Nursing aide/orderly	369,000
Receptionist	331,000

Source: Gilbert & Kahl, 1993, p. 81.

the labor market were those that were better paid. Between 1981 and 1987 "the industries where employment contracted paid $10,404, or 47% more in annual compensation than the expanding industries" (Mishel & Bernstein, 1992, p. 177).

Young people have an especially difficult time making it today because nearly all of the new jobs that were added to the economy in the 1980s and 1990s were low skill and low wage.

Sometimes when we see people like Jackie and Al living on the streets we wonder what they did to become homeless. These statistics show that structural changes in the economy, not individual choices, are an important factor in the creation of poverty and homelessness in U.S. families. Macro-level structural changes like plant closings, layoffs, and inadequate wages in remaining jobs are beyond the control of any individual and they are events about which people like Jackie and Al have little choice.

Economic Decline and Poverty

Falling incomes and growing unemployment have taken their toll. The U.S. Bureau of the Census reports that poverty rose sharply in the late 1970s and has remained high through the 1980s and early 1990s. Poverty was defined by the government in 1991 as income less than $13,359 for a family of four. Figure 4-4 shows how the percentage of the total population who are poor declined from the late 1950s to the mid-1970s when it rose sharply. Since then the proportion has declined some but began inching up again in 1990. In 1991 the absolute number of poor people in the U.S. was 33.6 million (Gilbert & Kahl, 1993).

Economic Decline and Growing Inequality

Table 4-2 gives us a closer look at the decline in wages since the 1970s. It shows that those families with children and those families with young heads of household have seen the largest decrease in income. As in Figure 4-1, the average incomes have been adjusted for inflation.

Table 4-3 shows that poverty rates for black and Hispanic families were higher than for whites (Winnick, 1988). In 1990 10.7 percent of the white population was poor, 31.9 percent of the black population was poor, 28.1 percent of the Hispanic population was poor, and 12.2 percent of the Asian or Pacific Island population was poor (Mishel & Bernstein, 1992).

FIGURE 4 • 4
Percent of total population who are poor, 1959–1990

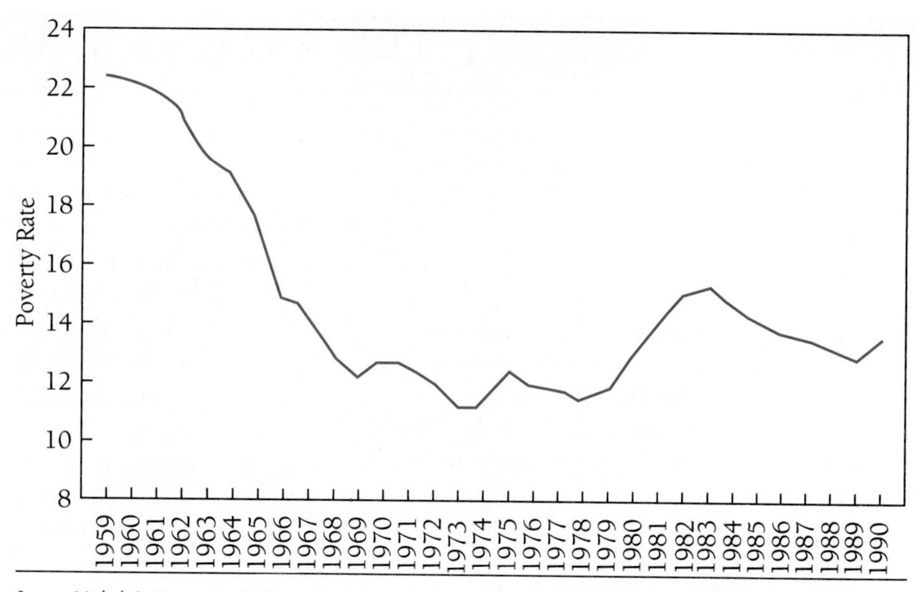

Source: Mishel & Bernstein, 1992, p. 272.

TABLE 4 • 2
Summary of Changes in the Median Annual Earnings of Family Heads, 1973–1986

	1973	1986	Change 1973-1986
Family heads—all ages	$20,970	$15,912	−24.1%
Family heads under age 30	$19,243	$13,500	−29.8%
Family heads under age 30 with children	$19,736	$12,000	−39.2%

Source: Children's Defense Fund, 1988a, p. 5.

TABLE 4 • 3
Poverty by Race/Ethnicity

Year	Poverty Rates		
	White	Black	Hispanic
1967	11.0%	39.3%	NA
1973	8.4	31.4	21.9%
1979	9.0	31.0	21.8
1989	10.0	30.7	26.2
1990	10.7	31.9	28.1

Source: Mishel & Bernstein, 1992, p. 286.

Increased Debt

One way in which families have sought to counter their declining incomes is by increasing their debt. As a result more people are in debt and the average debt is larger than ever before. "In 1990 at least 65% of U.S. households were in debt, either with a mortgage or a consumer loan, and 55% have zero or negative net financial worth; they owe more than they own in financial assets" (Pollin, 1990, p. 9).

BOX 4 • 1 TEN KEY FINDINGS OF THE CHILDREN'S DEFENSE FUND IN 1988

One: an economic disaster has affected America's young families, especially those with children.

Two: Poverty among children in young families has skyrocketed.

Three: The growing economic plight of young families has been caused by sweeping changes in the American economy that have reduced the earnings of young workers and undermined their ability to marry and form families.

Four: Young Black and Hispanic families have suffered particularly severe earnings and income losses.

Five: Education still pays, but a high school diploma is no longer an adequate defense against poverty for young families.

Six: While young female-headed families are by far at the greatest risk of poverty, young married couple families also have suffered, avoiding large income losses only by having both parents work.

Seven: Inequality of income has grown substantially among young families.

Eight: The youngest families find it increasingly difficult to obtain an adequate income.

Nine: Home ownership is now beyond the reach of most young families.

Ten: Young adults are least likely to have health insurance or access to the health care they need as they start their families.

Source: Children's Defense Fund, 1988a, pp. vii–ix.

The total amount of money that U.S. consumers owe has grown steadily over the past four decades. In 1950 total consumer credit was $23 billion; in 1960, $60 billion. Between 1970 and 1980 it increased from $132 billion to $349 billion, and by 1989 it had risen to $729 billion (Berberoglu, 1992, p. 73). The trend is in the direction of continued increase.

Table 4-4 shows the dramatic rise in household debt relative to annual income between 1949 and 1991. The table indicates that the ratio between what people earn in a year and what they owe grew from about 30 percent in 1949 to almost 87 percent in 1991. On the average, we now owe almost as much as we earn per year. Not surprisingly, this debt load has been accompanied by an increase of delinquencies on mortgage loans, mortgage foreclosures, and bankruptcy rates (Pollin, 1990).

Housing

The cost of housing has risen faster than wages in the past two decades. This has made it difficult for most families and impossible for some to provide themselves with shelter (Dolbeare, 1983).

> In the 1950's, two thirds of all U.S. families could have afforded the average new house without spending more than 25% of their income, by 1970 that proportion had dropped to one half, and by the early 1980's, only one in ten could afford the average new dwelling without devoting more than 25% of their income on housing. (Parker, 1991, p. 173).

The shaded columns in Figure 4-5 show the increase from 1970 to 1980 in the median sale price of new homes (50 percent of the sale prices fell above and

TABLE 4 • 4
Household Debt Burden, 1949–1991

Year	Debt as Percent of Personal Income
1949	30.3%
1967	61.7
1973	61.5
1979	66.6
1989	82.4
1991	86.7

Source: Mishel & Bernstein, 1992, p. 266.

FIGURE 4 • 5
Percentage of U.S. Families Able to Afford Median-Priced New Houses, 1970–1980

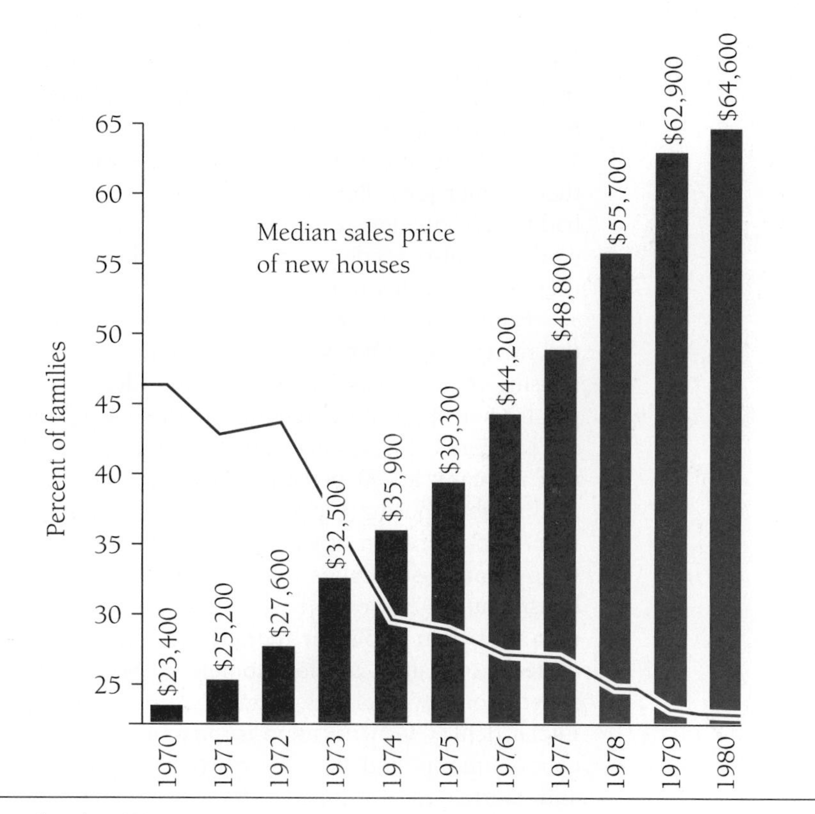

Source: Berberoglu, 1992, p. 75.

50 percent fell below the median. The median is not the mean or average, which is the sum of the prices of all the houses divided by the number of houses). The figure shows that the median price of new houses rose from $23,400 in 1970 to $64,600 in 1980.

Figure 4-5 also shows a descending line, indicating the percentage of families that could afford a median-priced house by paying 25 percent of their income on the mortgage. The figure shows that this percentage of families fell from 46 percent in 1970 to 22 percent in 1980. Since 1980 the trends have continued in the same direction. Housing costs have continued to rise while the percentage of families able to afford the housing has continued to decline.

Robert Parker explains the specifics of the economics of housing:

> In 1975, the average new single-family home cost $44,600. Assuming a standard 8.75% loan, a 20% down payment, and housing costs at 25% of income, a family needed to make a down payment of nearly $9000 and would have a monthly payment of $280. In contrast, the average price of a new home in 1988 was around $120,000. Assuming a 30 year mortgage and an interest rate of 10%, a home purchaser would need an annual household income of about $50,000 and an additional $30,000 to cover the down payment and closing costs. (Parker, 1991, p. 173)

Renters have faced the same problems. Between 1970 and 1988 median rent increased 192 percent while renters' income increased only 97 percent (Gilderbloom & Appelbaum, 1988). Al and Jackie in the scenario at the beginning of the chapter were confronted with this problem when they arrived in Tucson. Their income declined because the only jobs available paid less than their former jobs. But because of inflation the cost of rent was more than they had paid in previous years. In addition, because Jackie and Al had moved to find work they also needed to be able to make an initial payment that included both their first month's rent and a deposit.

Families have used three strategies to try to maintain shelter in the face of the rising costs. One strategy is to pay a larger proportion of their income for housing. Economists believe that the ideal maximum amount a family should pay for housing is 25 percent of their household income. Paying over 30 percent is considered excessive (Timmer & Eitzen, 1992). Yet researchers have found that as many as 70 percent of poor families pay 50 percent or more of their income for housing (Parker, 1991). Of course, this means that they have less money to spend on other necessities like food, fuel, and health care.

A second strategy is to try to find cheaper housing. The number of low-rent housing units, however, has declined. Between 1980 and 1988 2.5 million low-rent (less than $250 per month) units were lost and were not replaced (Committee on Health Care for Homeless People, 1988). The loss of these units resulted from two related trends: gentrification and redevelopment (Timmer & Eitzen, 1992). Gentrification is the practice of converting low-income housing to condominiums and upscale apartments and then charging much higher rents and mortgages (Kasinitz, 1984). Redevelopment is the replacement of low-income housing with office buildings, stores, and luxury apartments in an effort to revitalize the inner city. Both gentrification and redevelopment are a boon to urban developers and real estate interests because they increase the value of the property and generate higher income for the property owners. The two trends of gentrification and redevelopment have dealt a serious blow, however, to low- and middle-income people seeking shelter.

Furthermore, at the same time that the number of low-rent dwellings were declining the number of people seeking such shelter was growing. In 1970 7.3 million people needed shelter that cost less than $250 a month and 9.7 million housing units were available. In 1985 the number of people seeking low-rent apartments shot up to 11.3 million while the number of available units declined to 7.9 million (Timmer & Eitzen, 1992).

Third, families have looked to the government for assistance in helping them meet their housing costs. During the 1980s, however, housing assistance programs were dismantled by the Reagan and Bush administrations (Timmer & Eitzen, 1992). Between 1980 and 1987 federal support for subsidized housing was reduced by 60 percent and most of the remaining funds reflect subsidy commitments undertaken before 1980. Federal support for development of new low-income housing has disappeared (U.S. Congress, House Commitee on Ways and Means, 1987). Between 1981 and 1989 federal support for subsidized housing dropped from $32.2 billion to $6 billion. The Department of Housing and Urban Development (HUD) authorized the construction of 183,000 subsidized units in 1980 but only 20,000 in 1989 (Appelbaum, 1989).

Government subsidization of housing for the rich increased during this period. "The largest federal housing program is the tax deduction for mortgage interest payments and property taxes" (Hope & Young, 1986). In 1986 these tax deductions for homeowners totaled $42.4 billion. More than 95 percent of the benefits went to upper-income households (Swanstrom, 1992).

In the 1980s, Cushing Dolbeare compared these kinds of housing benefits received by the rich to those received by poorer households. She found that the average amount of federal housing assistance was a little more than $10 per month for households earning less than $10,000 per year, but those households earning $50,000 and above received in excess of $155 a month. Dolbeare's study showed that 25 percent of all federal housing assistance went to the 7 percent of households with incomes of $50,000 or more, while roughly half that much went to the 25 percent of households earning less than $10,000 per year. In all, her report identified $33 billion in housing subsidies for households earning more than $50,000 (Parker, 1991).

Growing Homelessness

The most life-threatening result of the Great U-Turn is homelessness. During the 1980s, millions of people were pushed out of their homes and found themselves living on the streets.

The government has had a hard time estimating the number of homeless people. In 1983, the U.S. Department of Health and Human Services said there were 2 million homeless. A year later another branch of the government, the U.S. Department of Housing and Urban Development, told Congress that there were only 350,000 homeless in the nation.

The government has been criticized for reducing its original estimate (Appelbaum & Dreier, 1992). The smaller estimate has been questioned because it is based on a faulty survey of the numbers of people who were housed in shelters. For example, the estimate given for New York City was 12,000 while the actual number of people in shelters on the night of the count was 16,000. Second, the estimate includes counts from only a few selected cities. Most importantly, it did not count homeless people who did not stay in shelters (Kozol, 1988). For example, Jackie and Al would not have been counted as homeless because they lived in their car.

The first problem encountered when trying to count homeless people is defining the phenomenon. To overcome this problem, in 1987 Congress attempted to define homelessness more precisely:

> 1) an individual who lacks a fixed regular and adequate nighttime residence;
>
> 2) an individual who has a primary nighttime residence that is:
>
> A) a supervised or publicly operated shelter designed to provide temporary living accommodations (including welfare motels, congregate shelters, and transitional housing for the mentally ill);
>
> B) an institution that provides a temporary residence for individuals intended to be institutionalized; or
>
> C) a public or private place not designed for, or ordinarily used as, a regular sleeping accommodation for human beings. [This would include, for example, people who live in the subways in New York, in tents in the desert in Arizona, in caves in the Midwest, in a box in Washington, D.C., in abandoned buildings and under bridges in Atlanta, and in cars in suburban shopping malls.] (Committee on Health Care for Homeless People, 1988, p. 2)

Using this broader definition of homelessness the Alliance Housing Council (1988) estimated that in 1988 2 million people were homeless for at least one night in the past year, an additional 4 million were at extreme risk of becoming homeless, and the numbers were continuing to grow (Committee on Health Care for Homeless People, 1988). Phillip Clay (1987), a professor at MIT, conducted a study funded by Congress that concludes that nearly 19 million people will be homeless in the U.S. by 2003.

The Rich Get Richer

On the average, wages have declined for working Americans in the past two decades. Does that mean that everyone has seen a decrease in their income? Not necessarily. Although most people have experienced downward mobility, some people, particularly those at the top, have seen their fortunes increase. Their taxes are lower, the government is providing them with housing subsidies, and Tables 4-5 and 4-6 show that their income and wealth has also increased.

When sociologists talk about financial gaps among different kinds of households, they distinguish between income and wealth. Income is the amount of money earned by an individual from wages and salary or the unearned money derived from investments. Table 4-5 compares the difference in trends for income among different sectors of our society, the very rich, the rich, and everyone else. Table 4-5 shows that the Great U-Turn describes the change in income for most but not all Americans. The vast majority of people in the U.S. became poorer since the mid-1970s. At the very top of the socioeconomic system, however, households made large gains in income during this period.

Wealth refers to the assets an individual owns. These assets might include money, houses, stocks, buildings, bonds, land, or other items. Table 4-6 shows the different trends in the ownership of wealth. Like income, wealth has increased in a few households at the top of the socioeconomic system while it has decreased for the majority of people at the bottom of the system.

TABLE 4 • 5
Mean Income of Households, 1977–1990, by Fifths and Top Percentiles (in 1990 Dollars)

	1977	1990	% change
Population fifths			
Poorest	$8,738	$7,424	−15
Second	$21,253	$19,428	−9
Third	$32,780	$30,942	−6
Fourth	$44,082	$44,850	+2
Richest	$82,455	$104,248	+26
Top 5 percent	$141,705	$206,162	+46
Top 1 percent	$294,874	$548,969	+86

Source: Gilbert & Kahl, 1993, p. 105.

TABLE 4 • 6
Net Wealth Held by Different Groups, 1963 and 1983 (percent)

Wealth group	1963	1983	% change
Super-rich (upper 1/2%)	25.4	35.1	+38
Very rich (2nd 1/2%)	7.4	6.7	−9
Rich (90–99th%)	32.3	29.9	−7
Everyone else (90% of total)	34.9	28.2	−19

Source: Kloby, 1991, p. 45.

Families on one side of this line face increasing difficulties in staying afloat financially or even surviving. Families on the other side of the line are becoming more wealthy. And the gap between the two groups is growing.

ECONOMIC DECLINE AND U.S. FAMILIES

The statistics we have seen in this chapter show a dramatic decline in the economic fortunes of most people in the United States. One of the important results of this decline has been the creation of intense poverty. The crisis at the macro level of the economy has created a crisis among many families who have found themselves facing problems like homelessness, unemployment, impoverishment, and downward mobility. In this section we will examine the effect of changes in the economy and the Great U-Turn on families.

Homelessness

When the problem of homelessness first captured the public's attention, it appeared that nearly all of the new homeless were single men. As the decade of the 1980s progressed it was revealed that a large and growing proportion of the homeless are families. In 1986, the U.S. Conference of Mayors estimated that in the 25 cities they represented, families comprised about 28 percent of the homeless population (Committee on Health Care for Homeless People, 1988).

In 1988, for example, there were 28,000 people in emergency shelters in New York City. About 10,000 were single people and about 18,000 were parents and children. About 75 percent of the families were headed by single parents (Barak, 1992). The average homeless family included a mother and two or three

BOX 4 • 2 DO ALL PEOPLE IN THE U.S. HAVE THE RIGHT TO BURY THEIR DEAD AND VISIT THEIR FAMILY MEMBERS' GRAVESITES?

Kozol (1988), while examining homelessness in New York City, looked into the question of what happens when a homeless person dies. He found that funeral and burial costs amount to about $3500 in New York City. Until 1986 the government provided up to $250 for these expenses and since then has raised the amount to $900. Most homeless people who die in New York City, therefore, are buried in Potter's Field, a pauper's grave located on Hart Island.

The unembalmed bodies are taken to the grave by the truckload in rough wooden boxes, which are buried in trenches twenty to thirty at a time. This forty-five-acre site is strictly a place of disposal, not remembrance. There are no grave markers and after thirty years the graves are bulldozed under to make room for more trenches. "Between 1981 and 1984, nearly half the children who died in New York City before their second year of life were buried at Potter's Field. Almost a third of all persons buried at Potter's Field during those years were infants" (Kozol, 1988, p. 192).

Hart Island is also a prison facility and inmates perform the burials. Because it is a prison site, no one—including family members of the deceased, even parents—is allowed to attend the burial or visit the area.

Many choices we assume should be made by families are taken from homeless families: the choice of where their children will live and with whom, the choice of who will share their close physical surroundings, and in the case of death, the choice of where their children (and other kin) will be buried and whether they will be able to attend the funeral or visit the gravesite.

children. On average the parent was twenty-seven years old and the child six (Bassuk et al., 1986; Kozol, 1988).

Homeless parents typically are isolated and have few if any social supports. In one survey (Bassuk et al., 1988) homeless mothers were asked to name three persons on whom they could depend for support: 43 percent named no one or only one other person and 25 percent named one of their children as their main source of support.

Researchers (McChesney, 1986) have found that mothers have used a number of varied and creative means to stave off homelessness, such as living with friends. Often, however, homeless people have been unable to call on one possible source of support, their own parents or siblings, for three reasons: (1) their parents were dead; (2) their parents or siblings lived too far away; or (3) they were estranged from their families.

Shelters: An Inadequate Solution Despite the fact that families make up a large and growing proportion of homeless people, most shelters are designed for single individuals. Homeless families, therefore, must frequently break up—a mother sent to the women's shelter, a father sent to the men's shelter, and children placed in the custody of child welfare authorities (Committee on Health Care for Homeless People, 1988).

The problem of the government separating children from their parents is one that appears in a number of studies of homeless people. The dilemma confronted by Jackie and Al in the opening vignette—wanting to ask for assistance but fearing that the Department of Social Services would place AJ in foster care if they found he was homeless—is one that faces many homeless parents.

In addition, the government financially punishes families that stay together. In the state of New York, for example, an Aid to Families with Dependent Children (AFDC) child that stays with his or her parent receives $262 per month for food, clothes, and rent. If that same child leaves his or her natural parent and goes to a foster home the grant is increased to $631 per month.

In those shelters that do allow parents to stay with their children, families have other problems. A common practice is to require people to leave the shelter early in the morning and not return until evening. Parents must find a safe spot protected from the weather for their children during the day. Second, the shelters themselves are not safe. They are usually large rooms full of closely spaced cots, which allow no isolation of children from people who may have any number of physical, psychological, and social problems. Third, shelters are located away from schools and parents often lack transportation, making school attendance erratic at best (Traveler's Aid Program and Child Welfare League, 1987).

Some homeless families who receive AFDC have been housed in welfare hotels as an alternative to staying in shelters. Twenty-eight states have elected to use this form of emergency assistance (U.S. Congress, 1987b).

Welfare hotels, however, also have serious shortcomings. First, they do not have facilities for food storage or preparation. Cooking is not allowed in the rooms. This means that more expensive cooked food must be purchased. It also makes it difficult to provide nutritious meals for children. For example, milk, fresh fruits and vegetables, meat, cheese, and infant formula cannot be kept without refrigeration (Committee on Health Care for Homeless People, 1988). When Jonathan Kozol (1988) interviewed people in the world's largest welfare hotel, the Hotel Martinique in Manhattan, a continuing complaint was hunger.

Second, although welfare hotels provide more privacy than shelters, families still must share accommodations with prostitutes and drug users. For example, homeless people have been assigned by the District of Columbia to stay in the Annex, a rooming house for prostitutes (Kozol, 1988).

Third, welfare hotels are extremely costly. Massachusetts spent $16,000 per year per family for a room in a welfare hotel (Gallagher, 1986). This amount would provide a spacious apartment in a good neighborhood. By law, the money is not allowed to be spent that way; it must be used to rent temporary space in a welfare hotel.

Blue-Collar Layoffs

Researchers who have examined the effect of the economic crisis and specifically unemployment on blue-collar workers have found that it results in three problem areas. The first is economic difficulty. Perucci and Targ (1988) found that 72 percent of their sample of families who had a member laid off after a plant closing in the Midwest reported that they had serious economic problems nine months after the closing. In their sample, 56 percent had been reemployed, 44 percent full-time and 12 percent part-time. Hourly pay averaged $5.73 per hour compared to the $7.35 they had received on the average at their old job.

The second result is social/psychological difficulty. Louis Ferman and Mary Blehar (1983) interviewed blue-collar men who became unemployed in the recessions in the mid- and late 1970s. The researchers were struck with the disillusionment and the resulting psychological anguish that developed as a result

During the 1980s manufacturing workers were faced with declining numbers of jobs because of many factors, including automation. Those people who were able to keep their jobs were asked to take concessions of pay cuts. Steelworkers were among the many who fought to keep the plants open and to hold the line on losses in pay and benefits.

of unemployment. Most men started off optimistically thinking they would soon find another job and return to work, but as the weeks went by the real suffering began. The layoff itself, therefore, was not as traumatic as the subsequent period. Ferman writes:

> We began by thinking that the actual episode of job loss was the big trauma, but we were dead wrong. What we're finding is that job loss in many cases is only mildly traumatic compared to what follows—searching for new jobs, dashing of hopes that the old employer will call again, being rebuffed by new prospective employers. These are the events that try the patience and sanity of most workers. (Ferman & Blehar, 1983, p. 590)

The third problem area is marital relationships (Voydanoff & Donnelley, 1988). Perucci and Targ (1988) asked married workers nine months after their plants closed whether unemployment had affected their relationship with their spouse. 51 percent said their marriage had not been affected at all, 33 percent said it had been affected negatively by the layoff, and 16 percent reported increased happiness. In the section below on coping we will discuss the reasons why these differences exist among families. But before moving to a discussion of how families cope with the problems, let us look at the effect of the Great U-Turn on one more group, the middle class.

Downward Mobility in the Middle Class

Blue-collar workers were hardest hit by the layoffs of the U-Turn. Table 4-7 shows, however, that nearly one in six of the displaced workers was a manager or professional and more than one in five was in a technical, sales, or administrative support position (Newman, 1988).

TABLE 4 • 7
Displaced Workers by Occupation of Lost Job

	Percentage of total who lost jobs
Managerial and professional specialty	15%
Technical, sales, and administrative support	22%
Service occupations	5%
Precision production, craft, and repair	20%
Operators, fabricators, and laborers	36%
Farming, forestry, and fisheries	2%

Persons over 20 years old, with tenure of 3 or more years, who lost their jobs between 1/81 and 1/86 because of plant closings or moves, slack work, or the abolition of their position or shifts. n = 5,130,000

Source: Newman, 1988, p. 25.

Layoffs of managers and professionals created long-term unemployment for many. Twenty-seven percent of managers and professionals who were displaced remained unemployed for six months or more (Nulty, 1987).

This long-term unemployment creates economic difficulties. Katherine Newman (1988) argues that perhaps even more importantly it destroys a person's self-esteem. Job loss and downward mobility causes social disorientation and feelings of failure and loss of control.

The personal crisis of unemployed individuals spills over into their families. Newman studied the families of managers who had been laid off and found:

> Every aspect of the family's life becomes strangely different. . . . The ultimate victim of downward mobility is not just the breadwinner's occupational identity, nor even the checkbook and savings account, but the whole family's sense of normalcy. Nothing seems the same. (Newman, 1988, p. 60)

One of the daughters of a man who lost his business told Newman of the catastrophic change brought by her family's economic collapse. She said, "We went from one day in which we owned a business that was worth probably four or five million dollars in assets and woke up the next day to find that we were personally probably a half a million dollars in debt. Creditors called the house and started to send threatening notices" (Newman, 1988, p. 96).

The families that Newman interviewed initially responded to the economic crisis by trying to cover up their downward mobility. They tried to make it appear that nothing had changed. They created stories to account for their situation and they held on to houses and cars to perpetuate their previous lifestyle. Eventually, the pretense could no longer be maintained and with the loss of the job and material possessions they also lost their friends.

Friendships among these managers' families were based on participating in certain kinds of activities, such as dinner parties, dances at the country club, attending shows and symphonies, and dining at expensive restaurants. When their income declined and they were unable to sustain these luxuries, their friends began to disappear.

These social losses were accompanied by psychological distress, particularly self-blame. At first others pointed the finger of blame at the unemployed person.

One man explained how his wife was quizzed by others about whether he was trying hard enough to find work. He said,

> People say to her, "With all the companies on Long Island, your husband can't find a job? Is he really trying? Maybe he likes not working." This really hurts her and it hurts me. People don't understand that you can send out 150 letters to headhunters and get 10 replies. Maybe one or two will turn into something, but there are hundreds of qualified people going after each job. (Newman, 1988, p. 5)

Eventually the blame turns to self-blame. One manager describes it:

> I'm beginning to wonder about my abilities to run an association, to manage and motivate people. Having been demoted has to make you think. I have to accept my firing. The people involved in it are people I respect for the most part. So I can't blame them for doing what they think is right. I have to say where have I gone wrong? (Newman, 1988, p. 78)

Newman (1988) also found that the self-blame spilled into the next generation. Children of the downwardly mobile families worried that failure was genetic and that they would become like their fathers—"losers," "friendless," "unable to find a job," with "a life that has fallen apart."

RESPONDING TO ECONOMIC CRISIS

Thus far, this chapter has focused on the way in which the macro organization of society affects the micro level. We have looked at the way in which massive structural changes in the economy, the Great U-Turn, have affected families and individuals within families. We have observed that the effects are substantial and that they differ by age, class, race, ethnicity, and occupation.

In this section we will shift our focus to look at the way in which individuals and families are responding to the Great U-Turn. First we will examine the ways in which families have attempted to cope with economic crisis.

Unemployment, as we have seen, can be devastating. Some families respond by falling apart and others seem to adapt to the situation. What are the differences between these two groups?

Patricia Voydanoff (1984) argues that there are several mediating factors in a person's response to a layoff. These mediating factors can make the difference in a family's ability to cope with the event. Mediating factors can be divided into two types: family definition of the event and family resources.

Family definition has three components. The first is the suddenness of the layoff. If the family interprets the layoff as something they had anticipated, they are better able to cope with it (Hansen & Johnson, 1979). Second, family definition includes assessment of responsibility. When the worker is blamed for the job loss, coping is more difficult (Cobb & Kasl, 1977). Third is a sense of failure. Even if the worker is not seen as having caused the job loss, if he or she is perceived to have failed his or her responsibility to the family as a breadwinner, difficulty with coping is intensified (Nye, 1974).

The factor of family resources also has three components. First is financial resources. Savings, home ownership, and other financial resources buffer the response to the job loss (Voydanoff, 1963). Second is family-system characteristics. Families that are characterized by cohesion and adaptability are better at coping (McCubbin et al., 1980). Third is social support. Family support mediates the effects of the layoff for the unemployed person and further social support from extended kin and the community contributes to the stability of the whole family (Cobb & Kasl, 1977). This includes both material support like loans and food as well as emotional support.

Voydanoff (1984) concludes that family coping strategies are multifaceted combinations of all of these factors. Families that cope most successfully with unemployment develop effective coping methods such as placing more members in the labor market to make up for the lost income, managing finances more efficiently, strengthening internal and external family support, and redefining the situation as an opportunity for positive changes as well as a hardship.

Defining Unemployment and the Theory of Symbolic Interactionism

Voydanoff's (1984) assessment of the coping behavior of families hit by unemployment fits into the theoretical framework of symbolic interactionism. George Herbert Mead was one of the first to develop the theory of symbolic interactionism. Mead was a philosopher who was strongly influenced by the philosophy of pragmatism and his view of symbolic interactionism was based on pragmatist assumptions (Ritzer, 1992).

Pragmatism is characterized by three assumptions. First, reality does not exist independent of those who are observing and experiencing it. Reality "is actively created as we act in and toward the world" (Hewitt, 1984, p. 8). Second, people base their understanding of the world on what has proven useful to them. Third, "people define the social and physical 'objects' they encounter in the world according to their use for them" (Ritzer, 1992, p. 327).

Each of these factors fits into the description Voydanoff gives of the behavior of families of the unemployed people she interviewed. The experience of unemployment is not an event that exists outside of their interpretation of the event. The entire family actively works to construct an understanding of the event based on their assessment of why the layoff occurred and especially the culpability of the individual who was laid off.

Voydanoff (1984) notes that external sources of support are important to the ability of families to cope. But more important is the families' knowledge of resources and the ways in which they use them to create ways to survive unemployment.

According to symbolic interactionists,

> In the process of social interaction, people symbolically communicate meanings to the others involved. The others interpret those symbols and orient their responding action on the basis of their interpretation. In other words, in social interaction actors engage in a process of mutual influence. (Ritzer, 1992, p. 352)

Voydanoff's observations show how families as networks of interacting individuals actively create an understanding and a response to the layoff of a family member. All families do not accomplish this in the same manner. The event has different meanings for different individuals. Some families learn how to cope, but in other cases the layoff is perceived and experienced in a distressing and disruptive manner.

Social Victims or Social Critics

Voydanoff's work shows the ways in which some families adapt to unemployment. Do some workers move beyond coping with unemployment to become social critics of the society in which unemployment has become so prevalent?

It may appear that this is not often the case. Many accounts in the popular media of workers who have lost jobs portray unemployed workers as passive victims—baffled, depressed, cynical, and resigned (Frisch & Watts, 1980).

Frisch and Watts (1980) were curious about the validity of this image of workers they saw in the *New York Times*, so they obtained copies of the complete interviews with workers who were quoted in the newspaper in articles about the economic crisis and compared the interviews with the edited quotes that were actually published. They found that when journalists edited the interviews for inclusion in their articles, they selected quotes that were emotive, exclamatory, ungrounded, arbitrary, and individualistic. The published quotes only reflected part of the perception of the workers they interviewed. The journalists tended to omit quotes from workers who were reflective, informed, and offered a political analysis.

Ramsey Liem (1988) interviewed unemployed people in Boston to see if the workers' comments reflected helplessness and victimization, as portrayed in the media, or if they expressed informed social criticism. He found that they did both.

Liem offers as an example of victimization the response of one man explaining what unemployment means. The man says: "It means being without power to control the things you normally like to have control over in your life like happiness, for me anyway. I lost confidence. . . . I'm actually scared to go on a job interview" (Liem, 1988, p. 143).

Workers also talk about their unemployment in ways that do not blame themselves or internalize the problem but articulate an understanding of the inequities of the system as a source of their difficulties. For example, another man told Liem: "Help [in finding a job] wasn't available because in my opinion the corporate system from government on down is not really concerned with the plight of the unemployed person be they white collar or blue collar" (Liem, 1988, p. 144).

Liem argues, in addition, that workers express their social criticism in their refusal to give up their desire to find meaningful work and decent pay. Finally, Liem's research provides evidence that workers were knowledgeable about the industry in which they had been employed and able to explain the conflict of interest between themselves and management that touched upon the layoffs. For example, one assembly line worker who had been laid off said,

> Polaroid goofed a lot of things. They overproduced and overspeculated. Some guy making $100,000 a year made a film cartridge for $27. They had to sell it by government standards for $10 and also the chemicals leaked into the camera. Suddenly making of the camera ceased and the people making that camera were higher in seniority than us so they had to get rid of the excess and we were the excess baggage. (Liem, 1988, p. 147)

National Unemployment Network

Voydanoff's work shows unemployed people and their families adapting. Liem shows them criticizing. A third response exists as well. Some people have moved beyond thinking about the sources of their problems to acting collectively in a manner that can change the system (Deitch, 1984).

In 1979 approximately 90,000 steel workers were employed in the Pittsburgh area. By 1983 their numbers had declined to 41,000. Furthermore, 2.7 jobs were dependent on every job in steel so that a decline in employment in steel created an even greater decline in job opportunities in other kinds of work (Carnegie Mellon, 1983).

Two types of grassroots community responses emerged from the economic shocks of the soaring unemployment in the region. One type of response focused attention on stopping plant closings. The other attempted to modify the impact of the job loss on those who had been laid off (Deitch, 1984).

Thomas Bonomo (1987) has studied one large organization that worked on both of these fronts, the Mon Valley Unemployed Committee (MVUC). The MVUC, in the heart of the steel-producing area in Pennsylvania, established the National Unemployment Network. The MVUC had two goals: stop plant closings and create new unemployed committees throughout the country. In July, 1983 they presented their program for action in their newsletter.

> 1. Secure a decent living standard for all:
>   —Jobs or Unemployment Compensation Until Jobs are found.
>   —Save our Homes
>   —Food: Full Use of Surplus, Increase in Programs
>   —Guaranteed Utilities/Moratorium on Utility Shutoffs
> 2. Build a unified force of all workers, employed and unemployed around our common interest: Plant closing, unemployment compensation, job retraining, and other vital issues.
> 3. Change the government's budget priorities—Money for Jobs not War—Fund Social Services that Help People, Not Kill them.
> 4. Fight the disproportionate effect of unemployment on women, minorities and handicapped workers in all our efforts. Discrimination only serves to divide us and we must fight to build a unified movement. We must support and strengthen affirmative action programs. (Bonomo, 1987, p. 18)

In 1984 the MVUC boasted an operating budget of $160,000, 1000 members, and 15,000 to 18,000 sympathizers—no small operation. Bonomo (1987) concluded, however, that although this was a significant start much work remained to be done. For the MVUC to have been truly effective, it needed to link its efforts with other groups besides laid-off industrial workers, expand nationally, and improve the materials and methods used for publicity.

By 1993, MVUC had pretty much disbanded. The deepening recession and increasing difficulty to survive as unemployment benefits dried up and social supports were cut by the government made it difficult to sustain the organization (Bonomo, 1993).

MVUC's impact was short-lived. Nevertheless it provides an example of how victims of economic change can organize to identify their problems and create ways to address them. The next section describes the activities of another group even more devastated by the Great U-Turn but able to continue its efforts through the 1980s and into the 1990s.

Homeless But Not Helpless

In 1980 the homeless movement began to emerge. At first much of the activity was organized by advocates: people who were not homeless themselves but were concerned about homelessness and worked through churches, governmental agencies and other community organizations. Initially most of these organizations worked to increase the charity available in the form of soup kitchens and shelters.

The second wave of organizations were more likely to include or even be led by homeless people themselves. Their work has been more political and they have demanded housing and jobs rather than shelters and soup kitchens. One of these organizations is the National Union of the Homeless, founded in Philadelphia in 1983 by Chris Sprowal, a hospital worker who was laid off and found himself homeless (Barak, 1992). The primary tactic of the union has been to pull down the boards from abandoned housing, especially public housing units. Homeless people have moved into these buildings, many of which are fully equipped with utilities, water, and heat, and then have faced arrest for trespass (Barak, 1992).

In October of 1989, 250,000 people from all over the country, including both homeless people and advocates, went to Washington, D.C. to protest their plight. In the winters of 1990, 1991, and 1992 protests continued at the local level in several cities, including Los Angeles, Chicago, Detroit, New York, and Philadelphia.

MVUC and the National Union of Homeless are two of several organizations that have sprung up or reemerged in response to the problems created by the Great U-Turn. In addition to these activist efforts, scholars have also attempted to analyze the problems and their causes and possible solutions. In the next section we will review some of their ideas.

WHAT IS BEHIND THE GREAT U-TURN?

This chapter has presented much evidence that we are living through difficult economic times and that those difficulties are having serious consequences for most American families. What underlies these problems?

One way of trying to answer that question is to step back for a minute and look at the current era within the broad sweep of human history. If we could see a picture of the last several thousand years of human history, we would notice that at two times in that history, revolutionary changes have taken place in the organization of the production and distribution of the necessities of life. The first was the Neolithic Agricultural Revolution about 8000 years ago. The second was the Industrial Revolution, which began about two hundred years ago (Eitzen & Baca Zinn, 1989).

Important changes are now occurring in our society. Is a third revolution in the making?

If a revolution in the economy is occurring it would certainly have an enormous impact on the rest of society and it might cause the kinds of unsettling problems we have reviewed in this chapter. Before trying to answer the question of whether a revolutionary transition is taking place now, let us look at how transformation occurred in the two previous revolutions so that we have some sense of what these transitions look like.

The First Revolution

During the Neolithic Agricultural Revolution, humans began to control their environment by settling down, planting food to harvest, and domesticating animals to eat. This was very different from the life they had led earlier: following the herds and gathering the food that was naturally available. Throughout the agricultural period, people developed better and better tools. They invented hoes, shovels, and picks and they invented methods of making the metals to produce these tools. They devised ways to harness the energy of animals to use as vehicles, and to plow and turn mills. Production increased enormously and most importantly, humans had greater control of that production than they had before the Agricultural Revolution (Childe, 1948; Beals, Hoijer, & Beals, 1977).

When I defined economics at the beginning of this chapter I said that it included two tasks: production and distribution. The previous paragraph shows the changes that occurred in production during the Neolithic Agricultural Revolution. Distribution also changed.

Before the Agricultural Revolution, distribution was relatively equal for two reasons. First, the numbers of people living together were so small that each group needed every individual to survive in order to maintain the survival of the group. Distribution had to be done in a way that kept as many people as possible alive. Second, since people had so little control over the production of food and shelter, there was little left over to create inequality in distribution.

After the Agricultural Revolution, people were increasingly able to accumulate surpluses beyond subsistence. Distribution became less equal as some powerful people accumulated these surpluses, creating a rich and powerful elite.

The effect of the Agricultural Revolution on human existence was overwhelming. Some of the features of our society that we think of as natural developed during this period. For example, nations were established, money was invented, and so was the concept of government, calendars to keep track of seasons, writing, and numbering systems. The Neolithic Agricultural Revolution lasted for about 10,000 years from roughly 8,000 B.C. to the late 1700s.

The Second Revolution

The Industrial Revolution created another turning point in human history: Machines were added to human and animal energy to produce the necessities of life. Once again, production increased enormously. Although machines enhanced human energy, they could not run on their own. The Industrial Revolution is marked by the pressure to put as many people as possible into manufacturing for as many hours as possible. In Chapter 2 we observed the way in which industrial capitalism developed in the U.S. and how factories employed entire families, men, women, and children. People were driven off the land into the cities and the factories (Beals, Hoijer, & Beals, 1977).

Once again the changes in production were paralleled by changes in distribution. Wages were increasingly used as a means to distribute the goods and services produced by the offices and factories. Prior to the Industrial Revolution, people used what they produced or traded what they produced for the materials they needed. Workers who were employed in factories could not use what they produced. Instead wages were paid to workers who then could purchase what they needed from those who owned the factories.

The effect of the Industrial Revolution was felt in every part of society. Cities, banks, factories, world wars, international economies, and as we saw in Chapter 2, relatively isolated mobile nuclear families, at least among the white middle class, emerged during this period. The Industrial Revolution began in the late 1700s and began to shift to a postindustrial period in the 1970s.

Postindustrial Society: A Third Revolution?

We may now be at another monumental turning point in the economic, and thus social, organization of humanity (B. Jones, 1985). Eitzen and Baca Zinn (1989) argue that four forces have converged in the past two decades to create such a turning point. The first force is the expansion of industrial capitalism throughout the world, resulting in increased competition for U.S. business. The second force is the increased investment by U.S. businesses in other countries, which moves jobs from the U.S. to places like Mexico and Central America. The third force is the shift in the economy from manufacturing to information and services, which leaves many workers unskilled to perform the jobs that are being created.

All of these are important but it is the fourth force that may be the most significant: the development of microelectronics, computers, and robotics. The microchip and its application in robotics is creating the possibility of the production of the necessities of life *without* human labor. Figure 4-6 shows the dramatic effect this has had on manufacturing, where production has expanded enormously while the number of workers involved in that production has declined. Microelectronics may be ushering in a new era that does not yet have a name, but it might be called an Electronics Revolution.

Unlike industrial production, electronic production can be done with less and less human energy and ultimately, perhaps, with none at all. The key difference between this production system and an industrial one is that industrial production enhances human energy while electronics/robotics replaces human energy (B. Jones, 1985). Machines can build everything we need to stay alive. Computers can guide their production. Computers were invented by humans and initially involved a lot of input from humans. But computers increasingly can learn from other computers, can program themselves and other computers, and can make decisions.

There is debate over whether the trend we see in increasing automation will actually continue to develop and will continue to eliminate jobs (Block, 1984). For example, some have argued that services or the need for assembling and transmitting information will simultaneously produce more jobs to make up for those lost in manufacturing (Dassbach, 1986). One could argue, however, that the same technologies that have eliminated so many manufacturing jobs are already having a significant effect on reducing the need for human labor services and information processing (Leontief & Duchin, 1986). For example, the use of the word processor has eliminated the jobs of many secretaries who used to type multiple letters and keep elaborate filing systems of written materials in cabinets. Electronic mail has eliminated postal jobs through the use of computers and electronics in communications. Automatic teller machines have replaced

FIGURE 4 • 6
Factory Jobs Disappear. Thanks to automation and computers, manufacturers have increased output by more than 400% since 1947 with only a 17% increase in their work force.

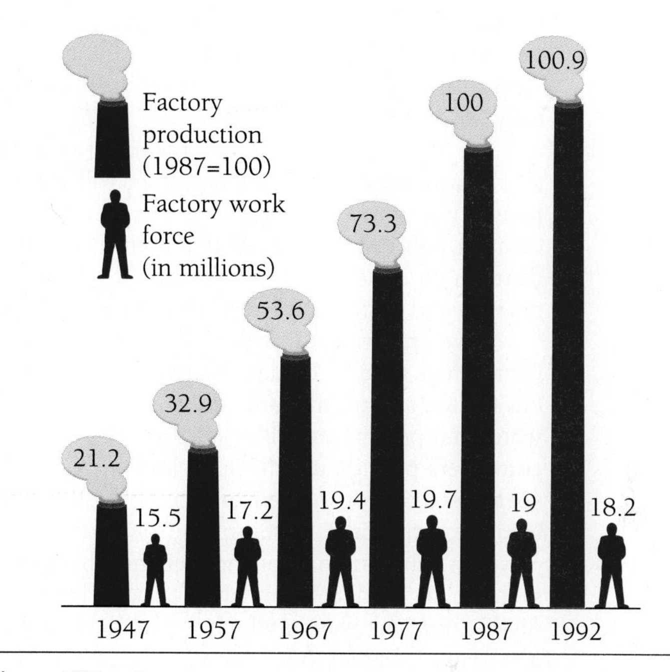

Source: Ullmann, 1993, p. 1c.

bank tellers. Computers have replaced library clerks (Machung, 1984). It is difficult to say for sure whether electronics will continue to be applied in ways that eliminate jobs but thus far the impact of automation has been significant.

If we look at the total hours of paid employment per capita, it appears that this figure has declined sharply across the twentieth century in the major Western capitalist nations. In short, technological changes have produced an expanded flow of goods and services with a declining amount of work per man, woman, and child (Block, 1988, p. 195; see also Kuznets, 1971).

Like changes in each of the two previous revolutions, the changes in production in the third revolution will demand another form of distribution. In the discussion of the Industrial Revolution I noted that wages were increasingly used during that period as the sole means to distribute the goods and services being produced. In an electronically based system of production people are not working as many hours (if at all) to produce. Therefore, they cannot earn wages to allow them access to the goods and services (Block, 1988). This factor plays a key role in the problems described in this chapter (Oxford Analytica, 1986). Workers' wages have decreased because of rising unemployment and those who have held onto their jobs have seen a decrease in the buying power of their take-home pay. Consequently fewer and fewer people are able to purchase the necessities of life for themselves and their families.

Barry Jones (1985) suggests that we might resolve some of the problems of unemployment, poverty, and homelessness that are being created by the Electronics Revolution by inventing a citizen wage. A citizen wage system would distribute the goods and services to those who needed them regardless of the number of hours they worked in production (see also Block, 1988). This is not a new idea. What is new about Jones's argument is the link he draws between the utopian notion of distribution based on need and the real potential of electronically based production.

The questions of whether the Electronics Revolution will develop further and what kind of distribution system will emerge from it are ones that will demand enormous efforts by large numbers of people in thinking, planning, and creating political change. At this point it is difficult to predict either the kind of distribution system that will work best or the further effects the changes in production and distribution will have on every aspect of our society.

If we are living through a revolution and moving toward a society in which for the first time in human history the necessities of life can be produced without (or with a minimal amount of) human energy, we can look forward to some wonderful possibilities (Block, 1988). We could see a world in which robots and computers provide us with food and shelter and where people would be able to spend all or most of their energy doing those things that only humans can do: creating, exploring, learning, and nurturing.

Ironically, in spite of the fact that this new electronically based production system can produce more than enough for everyone, because of our system of distribution, the way in which we are currently experiencing this revolution is as a system of scarcity. Our economy seems to be running in reverse as more and more of us are unable to find subsistence. As robots and computers eliminate jobs and deskill others, we see layoffs, unemployment, lower wages, poverty and homelessness, and all of the resultant social and psychological problems of families trying to cope with those problems. Harley Shaiken writes: "the technology provides choices: what is selected depends on who does the choosing and with what purpose in mind" (1984, p. 2).

In Chapter 3, we could see that the problems that slaves faced in trying to maintain a decent life for themselves and their families could not be resolved without abolishing slavery. Today we cannot imagine anyone wanting to preserve the slavery system. The maintenance of an economic and political system that caused such human misery seems unthinkable. During the period of slavery the micro experience was so severely constrained and dehumanized by the macro organization of slave society, there was no choice but to make radical change.

What about the problems of poverty and homelessness we face today? Can we continue to maintain the status quo? What changes must we make? How far must we go to create a society that will allow people to live and raise their children in decency?

Chapter 1 reviewed the abstract idea of "the macro level," and Chapter 3 described a social structure that existed long ago and that we now agree was completely untenable. In the current chapter, the macro level becomes more personal because we are speaking of our own times. We are no longer considering a society that lives only in history books. We are assessing our own society and our own history. We must be the ones to decide what should stay and what must go. Will we need to be as radical as the Abolitionists were in their day?

THE MICRO-MACRO CONNECTION

The focus of the last three chapters has been on the relationship between the economy and families. Chapters 2 and 3 covered the historical period from the colonial era to the 1960s. This chapter brings us up to date by reviewing the past three decades. We see that throughout history and even today economic context shapes the organization of families. While this relationship holds all the time, it becomes most apparent when a crisis occurs in the economy. Since the mid-1970s a crisis has occurred in the American economy which has caused a Great U-Turn, resulting in serious problems for most American families.

The relationship between families and the economy, however, is a two-way street. The macro level of social organization has affected the micro level; the economic crisis has tossed and battered families. But in return the micro level has actively reacted. Families have responded individually by offering psychological support for their members and attempting to shuffle resources by going into debt, reducing the quality of their lives, seeking support from the government, and finally by surviving in the streets.

Some people have also tried to address these problems by changing the social structure. People organizing at the micro level have fought for changes at the macro level. Some have developed new ways of looking at the issues that are critical of the social structure. Others have formed organizations to make demands for stopping plant closings and aiding families whose breadwinners have lost their jobs and who have lost their homes.

In the last part of this chapter we saw that beyond the macro structure of our contemporary society is an even larger social context, the whole sweep of history. Our own society is a part of that sweep, and may in fact be at a particularly important turning point. The profound changes in the macro level of the economy are creating profound changes in other areas of our society as well. These social changes are so large and new that in these early stages it is difficult to imagine their final shape.

SUMMARY

The Critical Link Between Families and the Economy

- The story of Jackie, Al, and AJ, a homeless family, depicts the grim experience of economic difficulties an increasing number of families face today.
- The previous two chapters looked at this link in the eighteenth, nineteenth, and early twentieth centuries. This chapter brings us up to date on the relationship between family and economy.
- Because economics is crucial to our survival, changes in economic organization have important effects on other aspects of our lives, including family.
- The philophical assumption that underlies the argument that economics is a critical determinant in human society is called materialism and is identified with Marxist theory.
- From the conclusion of World War II to the mid-1970s the U.S. economy expanded and provided an increasingly better standard of living for most Americans.

- Since the mid-1970s the economic fortunes of most people in the U.S. have declined. This turnaround is called the Great U-Turn.
- In addition to a decrease in wages, all but the wealthy have also been required to pay a greater proportion of their income in taxes, making take-home checks even smaller.
- Workers who have seen their income decline fall into three categories: those who kept their jobs but whose wages did not keep up with inflation; those who lost their jobs and were unable to find comparable work; and those who entered the workforce in new jobs that were likely to pay a low wage.
- On the average, all workers saw their wages sink but the decline was greater for some groups than others.
- Falling incomes have caused more people to carry a heavier debt load.
- Lower incomes coupled with rising housing costs have made it difficult for families to afford shelter.
- Despite greater need for assistance, during the 1980s and early 1990s the federal government cut back on housing support for low- and middle-income families.
- In contrast, housing subsidies for the wealthy have remained substantial.
- Not all Americans have suffered because of the Great U-Turn. Those who have remained at the top of the socioeconomic system have seen enormous gains in their income and wealth in the past two decades.

Economic Decline and U.S. Families

- Families are an increasingly large proportion of the homeless.
- Low-income families that were affected by the Great U-Turn have been pushed out of their homes and have become homeless.
- Estimating the number of homeless is a subject of much controversy.
- Shelters are usually not designed for families and may result in children being separated from parents and husbands from wives.
- Welfare hotels have been established in some cities to provide shelter for families but they are also inadequate.
- Rising unemployment has affected blue-collar families in three ways: greater economic problems, social and psychological difficulties, and strains in marital relationships.
- Managers and professionals have also suffered significant unemployment rates.
- Middle-class people who have lost their jobs because of the Great U-Turn suffer financially but perhaps more significantly they suffer social and psychological losses.

Responding to Economic Crisis

- Some families are able to cope with unemployment while others are not. Success in coping depends on the way in which families define the problem and the kinds of resources they have.
- Voydanoff's assessment of the coping strategies of families of the unemployed is an example of the theory of symbolic interactionism. Symbolic interactionism is identified with George Herbert Mead, a philosopher whose work rested on the philosophical assumptions of pragmatism.

- The media often portray laid-off workers as victims. Interviews with laid-off workers often reveal another image, one in which workers express informed social criticism that blames management and social structure for their unemployment rather than blaming themselves.
- Some people have moved beyond criticism to organizing a collective political response to high unemployment.
- In Pennsylvania, for example, the Mon Valley Unemployed Committee organized a large active group to call for reforms to stop plant closings and assist laid-off workers.
- Another example of the micro level affecting the macro level is the numbers of homeless and their advocates who have organized to end homelessness.

What Is Behind the Great U-Turn?

- Human society has moved through two revolutions: the Agricultural Revolution and the Industrial Revolution.
- These revolutions are characterized by changes in both production and distribution and these economic changes have been paralleled by huge changes in social life.
- Some scholars believe that we are currently moving into a third transition that could be called the Electronics Revolution.
- The Electronics Revolution is characterized by the replacement of human labor with computers and robots. This automation creates unemployment and greater competition for jobs, driving down wages, which results in poverty.
- Automation, however, ultimately could create a world of plenty with time to develop the abilities that only humans have, to nurture, create, explore, and learn. Such a change may demand radical social changes in our thinking and in our organization of society.

The Micro-Macro Connection

- This chapter has shown the way in which the macro system of the economic system has had a profound effect on contemporary U.S. families and the individuals in those families.
- In response to the problems the economy has created people have attempted to cope with the system or to challenge it by demanding changes in the social structure.
- The chapter also shows that the macro system exists in its own social context, the sweep of human history. Our society is part of that sweep and may even be at an important turning point.

Families became a central issue in the debate around what caused the 1992 LA rebellions and what should be done to prevent this kind of social upheaval. Conservative political spokespersons blamed the organization of families on the outbreak, arguing that improper family life led to inappropriate behavior. Other more radical politicians contended that families were victims of the same social problems—poverty and injustice—that were the root cause of the uprisings.

5 FAMILIES AND THE ORGANIZATION OF RACE, CLASS, AND GENDER

April, 1992. Four white police officers were acquitted of charges of beating black motorist Rodney King although millions had witnessed the attack on video. A rebellion ripped through Los Angeles leaving 38 dead, 8000 arrested, and billions of dollars in property damage.

May, 1992. Vice President Dan Quayle announced that the riots were rooted in a lack of proper family organization among the citizens of South Central Los Angeles. He said, "I believe the lawless social anarchy which we saw is directly related to the breakdown of family structure, personal responsibility and social order in too many areas in our society. . . . For those concerned about children growing up in poverty, we should know this: Marriage is probably the best anti-poverty program of all" (New York Times, *1992, p. A20). Quayle then called for social sanctions against unmarried women who bear children, stating, "Ultimately marriage is a moral issue that requires cultural consensus and the use of social sanctions" (Rosenthal, 1992, p. A20).*

Congresswoman Maxine Waters, who represents the South Central district where most of the L.A. rebellions occurred, spoke in the same week. She presented an alternative point of view, saying the problem was not families, but the economy. At a rally held in Washington, D.C. in response to the rebellion, Waters "marched onto the stage booming 'jobs, jobs, jobs' " and then pointed out that many of the looters were people with no previous criminal records. Some were women who wanted shoes and bread for their children (Newman, 1992, p. A20).

According to Waters, upheavals like the one in Los Angeles will not be stopped by focusing on changing families or by punishing women who do not marry. She said, "Listen up, America. We want to talk to you today. Our children are hurting, our mothers are tired and our young men are angry" (Newman, 1992, p. A18). Waters said the people of South Central were calling for justice and the crowd agreed. When she finished her speech they broke into a chant, "No justice, no peace."

These events illustrate a debate that places family at the center of the question of what causes social inequality and social unrest. From each point of view family is important, either as cause or effect. Quayle, a conservative spokesperson, says that improper families are the cause of poverty and unrest. Waters says that the economy is the cause of poverty and unrest and that families are victims of the unjust and failing economic system.

- *Were you surprised that family played such an important role in the analysis of the Los Angeles rebellion?*
- *Is this the first time that the family has been central in the debate over the causes and solutions to the problem of inequality in the United States?*
- *What is the relationship between family organization and inequality by race, class, and gender?*
- *Do different classes and races of people have different families? Do some have "better" families?*
- *What about gender? Is gender inequality expressed in the same way in families of different classes and racial ethnic groups?*
- *What exactly are race, class, and gender?*

SYSTEMS OF STRATIFICATION

In the last chapter we looked at the broad social context of contemporary American families and noted that the economic crisis was a critical factor in the declining financial situation for nearly all families. We also noted, however, that the effect of the economic crisis was not the same for all families. Some suffered more than others from the downturn and a few families at the top of the socioeconomic system were actually thriving.

BOX 5 • 1 WHO SHOULD BE CONCERNED ABOUT GENDER, RACE, AND CLASS?

Michael Kimmel tells a story in his book *Men's Lives* (Kimmel & Messner, 1989) about overhearing a conversation between a black woman and a white woman. The white woman was trying to convince the black woman that their similarities as women were greater than their racial differences. The black woman asked the white woman, "When you look in the mirror, what do you see?" The white woman replied, "I see a woman." The black woman responded, "That is precisely what I mean. When I look in the mirror I see a black woman. For me race is visible every day because I am not privileged in this culture."

Kimmel was startled when he heard this exchange. He thought to himself as he overheard the women talking, "When I look in the mirror I see a human being, universally generalizable. The generic person" (Kimmel & Messner, 1990, p. 57). His race, gender, and class were invisible to him. He had not thought about how they affected his life.

This experience led Kimmel to begin thinking about gender, race, and class and why they are so visible to some but not to others. He concluded that "There is a sociological explanation for this blind spot in our thinking: the mechanisms that afford us privilege are very often invisible to us. What makes us marginal (unempowered, oppressed) are the mechanisms that we understand because those are the ones that are most painful in our daily lives" (Kimmel & Messner, 1990, p. 57).

The people who are at the bottom of the stratification systems of race, class, and gender may be the first to notice those systems and they may be the most persistent in drawing our attention to their importance. But race is not an issue of concern only for racially oppressed people. We are all defined and influenced by our race. Class inequality is not just a question for poor people. Social class affects everyone's life. And gender is not just a problem for women. Every discussion of gender has significance for men's experience as well. We all need to see how these systems shape every individual's life.

The situation of families within society and, therefore, their response to social forces varies because all families are not the same. This issue of diversity will be the focus of this chapter.

Sociologists use the word stratification to describe systems of social inequality among different groups of people. Stratification is organized around many factors across human societies. For example, in some societies like those of Ireland or Israel religion is a critical factor. In an aristocratic system like that of premodern England, bloodline is the most important issue. In the U.S., three systems of stratification, race, class, and gender, are most important.

People in our society are classified as members of various categories according to their gender, race, and class. Each of these social markers has an important effect on individuals. Gender, race, and class affect a person's access to jobs, housing, or education.

Family experience is also affected by these systems of stratification. Our chances of marrying, divorcing, bearing children, our health and even cause of death are related to where we are in the stratification systems of gender, race, and class.

This chapter is divided into four major sections. The first examines the stratification systems of social class, race, and gender. The second section looks at the way in which each of these systems affects family organization. The third section briefly sums up the observation, from this review of the diverse character of contemporary families, that attempting to develop one definition of the family may be a useless and impossible exercise. The fourth section looks at the contemporary political debate over efforts to define "the family" and the various political groups that are attempting to promote their vision of good families.

What Are Class, Race, and Gender?

Social Class Not all societies have social classes. For most of human history people lived in hunting/gathering societies that were not stratified by social class. Contemporary U.S. society, however, is a capitalist society and social class is a critical feature of its organization. Our class system is a form of stratification that is characterized by four factors:

1. Class is economically based and is measured by economic variables such as occupation, income, and ownership of wealth (Giddens, 1991).
2. Class is a social relationship. A society with one social class would be classless because class rests on the assumption of inequality. In order to have inequality the system has to have at least two strata. Furthermore, social class is not just a category but a relationship because class assumes one has a relationship of inequality with other people in the society. For example, in our society working-class people have less money than members of the owning class, which puts them into two different economic categories. But working-class people also have a social relationship with the owning class because working-class people work for the people who own the factories and offices (Ollman, 1976; Wright, 1985).
3. Class system boundaries are fluid. Unlike other systems of inequality, like slavery, for example, people can move in and out of the class into which they were born. Boundaries between classes are not clear-cut. Social interaction, even marriage, occurs between members of different social classes (Giddens, 1991).
4. Class positions are in some part achieved. Although it is difficult to be born into a working-class family and later become a member of the owning class, for example, it is not impossible (Giddens, 1991).

Contemporary American society can be divided into three social classes: an upper class, a middle class, and a working class. The upper class could also be called the owning class as I referred to it in the definition of class above. This small elite owns most of the wealth in the United States. The wealthiest 10 percent of the population owns 90 percent of corporate stocks and 95 percent of bonds (U.S. Bureau of the Census, 1991).

The middle class includes those people who are self-employed such as small business owners and farmers. It also includes managers and professionals. The income range of this group is large. A middle-class person could be a self-employed physician with a lucrative practice or a farmer who owns a small peach orchard and barely keeps ahead of expenses. Professionals range from wealthy attorneys to low-wage school teachers.

The definition of social class above states that social class is a relationship and explains how this relationship is structured for owning-class and working-class people. Middle-class people frequently have jobs that literally put them in the middle, between working-class and owning-class people. For example, a manager of a plant is a middle-class person who may have a relatively high income but little real control over the business. A manager of a plant who does not own the plant must obey the wishes of the owner. In that sense the manager is like a working-class person in his/her relationship with the owner. On the other hand, a manager does control the workers and in that sense is like the owning class (Wright, 1985; 1990).

The working class are those people who are not professionals or managers and who earn wages by working for someone else. Large groups of workers like factory workers, restaurant workers, and clerical workers fall into this category. Within the working class there is also a range of incomes from relatively well-paid skilled blue-collar workers in union shops to the cashiers at fast-food restaurants and day-labor construction workers.

Race Nearly every time we fill out a form, we are asked to identify our race. Most of us respond by checking one of the boxes. Table 5-1 shows which box people in the U.S. checked in the 1990 census. Race is a widely recognized and critically important *social* factor in shaping our lives. What most people don't recognize is that people cannot be separated into biologically different races (Lewontin, Rose, & Kamin, 1984).

For a long time, scholars attempted to develop categories of races into which all humans could fit. Some came up with only a few categories, some came up with a few dozen. Around 1940 biologists were forced to abandon the concept of race among humans when they discovered that the amount of genetic variation from one individual to the next is enormously greater than any variation between "racial" groups. Any attempt to divide people into a few races on the basis of genetics, or even some more superficial measure, was impossible.

> For example, the Kikuyu of East Africa differ from the Japanese in gene frequencies, but they also differ from their neighbors, the Masai, and, although the extent of the differences might be less in one case than in the other, it is only a matter of degree. This means that the *social* and *historical* definitions of race that put the two East African tribes in the same "race" but put the Japanese in a different "race" were biologically arbitrary. . . . In practice, "racial" categories are established that correspond to major skin color groups and all the borderline cases are distributed among these or made into new races according to the whim of the scientist. (Lewontin, Rose, & Kamin, 1984, pp. 120, 126).

And even if we attempt to distinguish among "races" on the basis of something more superficial than genes like physical appearance we run into barriers. For example, we often refer to black people as a race that can be designated by dark skin and dark curly hair. However, the original inhabitants of Australia have very dark skin but their hair is wavy and sometimes blonde (Giddens, 1991). Where would we classify these people? Would we classify them in the same category as the person from Ohio with dark skin and wavy light brown hair whose mother is African American and whose father is Euro-American?

Dividing people into *biologically* distinct races is impossible for three reasons: (1) there are so many varieties of groups of people; (2) there is so much variety within each group; and (3) there are so many children whose ancestors came from different groups.

TABLE 5 • 1
Racial Ethnic Distribution in the U.S., 1990

White	75.6%
Black	11.7
American Indian, Eskimo and Aleut	.7
Asian or Pacific Islander	2.8
Other	.1
Hispanic	9.0
n = 248,709,873	

Source: Bureau of Census, 1990, Summary Population and Housing Characteristics, p. 59.

Nevertheless, people continue to check a box on the forms because race is an important *social* category. Although it is entirely "made up" we learn to overlook the irrationality of the system and to create ways in which to classify ourselves and everyone else into a few "races." These classifications then have enormous repercussions for the people so classified.

In addition, we may identify with a culture we believe is related to race. These cultural distinctions are called ethnicity. For example, an African American may identify him/herself as a member of the black community and may study African American history or become interested in African American music. A Native American may attempt to learn about the languages and religions of his/her "race."

Ethnicity can be based on distinctions by religion, nationality, or race. Jews, Greeks, and Irish are ethnic groups. These categories are related to religion and nationality. Ethnic groups based on the social category of race are called racial ethnic groups. When I use the term race in this book, I refer to socially defined, not biologically determined, racial ethnic groups.

One of the results of the fact that race is not a biologically valid category but still a critical social factor is the problem of naming. In this book, for example, I use both the terms African American and black. In the 1960s the Black Power movement called for the replacement of the term Negro with the term black. In the 1980s black people began to call for another change to African American. They argued that the name black did not fully capture the importance of culture and historical roots to the African American community. Some authors use one word, others use another. Government documents usually use the word black.

Finding an appropriate name for the Latino population in the U.S. is even more difficult. The government uses the term Hispanic but that is not often used by authors who are Latino or who write about Latinos because Hispanic is an English word. The word Latino is more acceptable because it is Spanish. Latino, however, is not completely accurate because the Latino population is made up of many different cultures and nationalities. For example, Latinos include Cubans, Cuban Americans, Puerto Ricans, Mexicans, Mexican Americans, Chicanos, and Central and South Americans who come from many different countries and cultures. This same problem emerges in the use of the term Asian, which can include people whose ancestry is Japanese, Chinese, Korean, Vietnamese, Filipino, and many others from Asian nations. Although they may share an ancestry from a particular region of the world, there are also significant differences in their history, culture, language, and social organization. In this book I use an eclectic mix of terms based on the use by the scholars whose work we are discussing.

Gender Most humans can be divided into one of two categories, male or female. This biological category is called sex and it is determined by physical factors like genes and sex organs. All known societies also create expectations about these two categories of people. The constructs a society creates about what females and males should be like—women (or girls) and men (or boys)—are called gender.

These social constructs vary widely from one society to another. When Margaret Mead (1935) studied three societies in New Guinea she found that each had an entirely different way of defining masculinity and femininity. Among the Arapesh both women and men were gentle and passive. Their neighbors the Mundugumor expected both men and women to be aggressive and violent. In the third group, the Tchambuli, the women were domineering and wore no adornment, while the men were gossipy, nurturant, artistic, and spent much of their time with the children.

Gender also may vary within a society (Rogers, 1978). For example, in our society, masculinity might be defined in one way among pro football players and another way among surgeons.

In some societies gender is clearly distinguished from sex. For example, among some Native Americans in western North America and the plains, females may act "like men." They may take on the gender of masculinity in their group, even though they are biological females. These females may dress, speak, work, and in every way behave as men are expected to do. They may even marry another female and treat her as a wife (Blackwood, 1984).

Gender and sex are collapsed into one category in the contemporary United States. In our society, many people believe—mistakenly—that biological sex determines gender. For example, people may believe that because one is male one is more interested in making money and less interested in caring for children. They may also believe that if they see a woman who is skilled at taking care of children it is because she is a female and not because she has been socialized to care for children or because she has chosen to learn how to care for children.

Whereas sex, for the most part, is a given, gender is not. Gender is socially defined. It comes from people being exposed to certain ideas and expressions of gender and from their being required or choosing to participate in the categories of masculine and feminine.

Both sex and gender have an important effect on families. Even if men and women were absolutely socially identical in our society, their experience of parenting would often differ because of their sex difference. Only females can be pregnant, give birth, and breastfeed. Bearing and raising children, therefore, includes activities that make the experience different for biological mothers than for fathers. But in addition, society creates ideas and activities for parenting that differ based on gender and are entirely unrelated to biology.

Gender has an important effect on nearly every facet of families' experiences. And families play an important role in the maintenance of gender.

SOCIAL CLASS, RACE, AND GENDER AND FAMILY LIFE

The stratification systems of class, race, and gender constitute a major feature of the macro level of social organization in our society. They exist beyond the control of any individual and are so pervasive they sometimes become invisible. But they weave in and out of our lives, sometimes overlapping, and sometimes contradicting each other, but always defining and shaping our lives and our relationships with others.

In this section we will examine the way in which class, race, and gender create different experiences within families. The emphasis will be on the effect of the macro organization of our society, which includes these three systems of stratification, on the micro level of society, the everyday experience of families. We will also observe the way in which families respond to the macro system by helping to preserve inequality, attempting to survive in spite of inequality, and creating ways in which to resist inequality and thereby alter the institutions of inequality.

The section investigates different social classes and racial ethnic groups. Gender is also covered as the section discusses how women and men relate to each other in families in all social classes and racial ethnic groups.

Upper-Class Families

Life in an upper-class family is not often open to scrutiny by the public or by researchers. As a result less is known about the private life of the members of this social class than of others. Rich people, however, know a lot about each other. Their preoccupation with maintaining boundaries between themselves and others has been noted by a number of scholars who have studied the elite (Domhoff, 1970; Eitzen, 1985; Mills, 1956). Families are a key way in which "membership" is identified. Being from a "good" family is essential and sometimes even overrides financial status. For example, when one of the "best" families lose their fortune, family members may be still counted as upper class at least for a time because of their ancestry (Bedard, 1992).

Georg Simmel (1907–1978) wrote that "Aristocrats would get to know each other better in an evening than the middle class would in a month." He meant that wealthy people identify themselves by membership and background, while middle-class people identify themselves by individual achievement. Therefore a person who knows the meaning of various memberships and connections among the upper class can draw a complete picture of a person. An essential piece of information in determining "membership" is family lineage.

Families play a critical role in keeping an individual in the upper class:

> The most important single predictor of a son's occupational status is his father's occupational status. . . . A man born into the top 5% of family income had a 63% chance of earning over $25,000 a year in 1976 (being in the top 17.8% of family income). But a man born into the bottom 10% of family income had only a 1% chance of attaining this level. (Braun, 1991)

Women in elite families play a special role in maintaining boundaries. "Women serve as gatekeepers of many of the institutions of the very rich. They launch children, serve as board members at private schools, run clubs, and facilitate marriage pools through events like debuts and charity balls" (Rapp, 1982).

Families also help to maintain an individual's social standing among the wealthy class by teaching family members how to maintain their class position. For example, one rule upper-class children learn is not to "spend down capital" (Millman, 1991). This means that they should use only the interest, not the principal, of an inherited estate. The wealth that has been accumulated may have taken generations to acquire and is thought of as belonging to the family line, not to individuals.

Tax laws reinforce the idea that wealth belongs to all generations of a family rather than to individuals. Inheritance taxes can be reduced if the inheritance of an estate skips generations. When the inheritance is claimed only every other generation, taxes must be paid only every other generation. For example, if a wealthy person wills his or her estate to grandchildren rather than to children, one tax is paid rather than two (Millman, 1991). This increases the motivation to teach children to live on the interest and not to touch the principal and that the family fortune should be shared only within a small circle of kin.

Volunteer work is an especially important activity in the production and maintenance of social status (Daniels, 1988). Susan Ostrander (1984) interviewed thirty-six upper-class women about their activities "to uphold the power and privilege of their class in the social order of things" (Ostrander, 1984, p. 3).

Marriage was one issue about which they spoke. One woman explained, "A compatible marriage first and foremost is a marriage within one's class" (Ostrander, 1984, p. 86). The women talked about debuts as critical events to ensure their children met the proper prospective mates. Social clubs were also cited as places to keep themselves away from those the women referred to as "anybodies."

Athletic games and activities were also mentioned as important. The women believed that these activities enhanced the ability of their children to stay in their class. They spoke of the lessons of "discipline, confidence, competition and a sense of control" (Ostrander, 1984, p. 94).

A good education in a prestigious upper-class school was also a goal because of both the academic training and social networks it afforded their children. The women spent much time planning and orchestrating all of these activities.

Upper-class families are largely responsible for maintaining their own position within the stratification system. They pass wealth down within families. They teach their children how to maintain their position and they bring their children into the social institutions such as elite schools and clubs that further reinforce their membership in the class. Women play a special role in maintaining the class and especially the boundaries around the class.

Along with the maintenance of individual families within the class or the maintenance of the class itself is the maintenance of the system of inequality. In a system of finite resources where some have control over a large proportion of those resources, others have control over less. As we saw in Chapter 4 finite resources are unequally distributed in the U.S. Today the net worth of the top 1 percent of the population is greater than the net worth of the bottom 90 percent

(Forbes, 1992). Families are essential to the constant work of retaining those resources and creating relationships of difference and inequality between themselves and other classes. "The family as an institution ensures the continuity of the have-nots as well as entrenching the power and privilege of the haves" (Morgan, 1985, p. 214).

Middle-Class Families

Four factors characterize middle-class families: (1) geographical mobility resulting in residence away from kin; (2) replacement of kin with other institutions for economic support; (3) reliance on friendship rather than kinship for affective support and exchange; and (4) investment of resources lineally (Rapp, 1982).

In order to maintain their income, middle-class families may have to move around. For example, people in middle-class occupations are frequently asked to move when their company needs them to work at another site. Middle-class professionals may find that to get a raise or further their career they must take a job with another company in another state.

These moves remove them from extended family ties and when economic help is needed middle-class people may rely on nonfamily sources. For example, a middle-class family that needs money for a down payment on a house would go to a bank for a loan. Both upper-class and working-class families might be more likely to seek assistance from their kin.

Middle-class families may also replace kin with friends in seeking emotional and social support. In the discussion of working-class families that follows we will see how working-class people convert friends into kin in order to facilitate sharing material goods (Stack, 1974). Rayna Rapp (1982) argues that middle-class people do just the opposite. She states that middle-class people refrain from sharing with extended kin and maintain friendships that do not include sharing resources. In this way middle-class families are better able to accumulate material wealth rather than dispersing it. Middle-class families stress upward mobility based on not sharing what they have accumulated (Millman, 1991).

The wealth that each relatively independent middle-class household is able to accumulate is invested lineally—between parents and children—rather than laterally among extended family and close friends, as is the case in working-class households. Rapp (1982) cites the investment of the middle class in education for its children and in extravagant wedding gifts as examples of sharing lineally.

Geographic Mobility One of the sources of the independence/isolation of middle-class families mentioned above is the geographic mobility that accompanies their occupations. While a large proportion of families are dual career, most researchers have looked at this issue as it exists in families where the husband needs to move. These moves may enhance the career of the husband for whom the move is being made, but they create hardship for wives and children. "Very few women do not suffer some losses as the result of a family move. These may include giving up friends, community and sense of self-worth and identity, close contact with relatives and often, a job or career possibility" (Gaylord, 1984).

Gaylord (1984) cites advice distributed to wives of husbands whose companies transfer them:

> Acceptance and harmony are two basic ingredients a wise woman preserves for her family regardless of the physical changes her husband's job may require. Your attitude and sense of "sportsmanship" set the pace for the family. . . . [When he announces the transfer] let him know you are happy, able and willing to be his partner in this new step. When you have done this, you have passed the ultimate test of wivesmanship with flying colors. (Upson, 1974, p. 15)

Wives may pass the test with flying colors but research (Weissman & Paykel, 1972) indicates they sometimes internalize the stress of the move, resulting in feelings of loneliness, marital friction, difficulties with children, career frustrations, and identity confusion. Children also report emotional difficulties with moving, especially those between the ages of three and five, and fourteen and sixteen (Seidenberg, 1973).

Gaylord (1984) concludes that corporations that relocate their employees need to take an active role in providing emotional support for these families and even offer short-term group therapy for those who have severe problems.

Upper-Middle-Class Families and Corporate Wives The middle class can be divided into two tiers, the upper middle class and the lower middle class. Whereas wives in the upper class work to maintain the boundaries of their class, wives in the upper middle class work to help keep their husbands in the occupations that place them in the class (Kanter, 1986). These women are called corporate wives although they may be married to professionals like physicians or professors as well as corporate executives.

Upper-class women are not likely to be in the paid labor force. Women in the upper middle class, however, may have their own professions. Regardless of their own work histories, they work to support their husbands' careers by playing an adjunct role, a support role, and a double-duty role (Fowlkes, 1987).

The adjunct role helps the professional man turn out the highest quality work possible. For example, the physician's wife may help him to run his business by decorating his office and managing his staff. The professor's wife may work as a research assistant, provide clerical support, and negotiate social interactions with her husband's colleagues.

The support role helps the professional man to sustain his drive. In this role the wife actively provides encouragement and emotional support and passively accepts the central role of his career in their lives.

The double-duty role provides time for the professional man to devote to his career. The wife maintains all other aspects of their lives, taking as little time as possible from him. She runs the house and raises the children in ways that make few demands on her husband (Fowlkes, 1987).

Rosabeth Moss Kanter (1977b) was one of the first scholars to investigate the life of corporate wives. She discovered that wives of executives were not only part of their husbands' careers but part of the corporation. The women's activity in support of their husbands was essential and men were often hired because of the talents of their wives. The men's jobs required two people and although the company did not formally hire the wives, a woman's talent and experience were essential parts of her husband's resume and his on-the-job performance.

Kanter (1986) identified four kinds of tasks that corporate wives accomplished in support of their husbands and the corporations for which they worked. First, a wife directly substituted for work that could have been done by a paid employee. She might type and keep accounts, for example. Second, wives provided indirect support with services that could have been purchased, such as hostessing dinners with potential clients. Third, a good corporate wife provided consulting services for her husband, for example, by discussing business decisions with him. Fourth, she provided emotional support, sending him to work in a proper state of mind and keeping him satisfied with his work.

The years a corporate wife puts into the company is a real career. This career does not proceed randomly but is likely to follow a predictable progression along with that of her husband. Kanter (1986) divided the corporate wife's career into three phases.

The first phase is the technical phase in which she provides primarily technical assistance and personal support to her husband but has little direct contact herself with the corporation. Kanter observed that in families at this stage, the central problem was handling the exclusion/inclusion dilemma.

The wife is excluded from recognition as a part of the corporate team during this phase. Her husband is busy learning about his job and about techniques and language to which she is not privy. Her life, however, is dominated by her husband's work. She must figure out how to fit in even though she has limited information about the actual tasks of the job.

The second phase is the managerial phase. The wife becomes more involved socially with other members of the corporation, entertaining her husband's colleagues and their wives.

In this phase the woman begins to be recognized as part of the company and her behavior is openly evaluated in terms of its contribution to the business. The central problem in this phase is deciding how to interact with the family friends who are also her husband's colleagues. The time constraints of the family's life mean that these two kinds of relationships are frequently intermeshed. Their only friends are the people with whom her husband works. But these relationships cannot be open and honest if the men are also in competition on the job.

The third phase is the institutional phase in which the husband reaches the highest levels of the company and his wife must maintain social ties not just among employees of the corporatation but between the corporation and the larger community. In this third phase, the line between the upper levels of the middle class and the lower levels of the upper class begin to blur and the task of the corporate wife begins to look very much like that of the wives of the elite.

This phase is characterized by the loss of distinction between public and private life. Wives and children become the evidence of a man's humanity. Private life must be made to seem truly private while open to public exposure to ensure that the husband appears as a complete person and a good "family man."

Rosabeth Moss Kanter's (1986) research is an example of a feminist ethnography (Reinharz, 1992). Her research took place over several years and involved the collection of many kinds of data including interviews, documents, and field observations. Ethnography is an established research methodology. The factors that make Kanter's work feminist are (1) her attention to gender as a basic feature of all social life and (2) her success at rendering visible an aspect of women's lives that had been invisible (DiIorio, 1982).

The Black Middle Class Black middle-class families are similar to white ones in the focus of their lives on home and family (Bedard, 1992). Charles Willie's (1983) research on black middle-class families shows them to be achievement oriented, upwardly mobile, immersed in work, and with little time for leisure. Education, hard work, and thrift are perceived to be the means to achievement.

There are also some interesting differences between black middle-class and working-class families and white middle-class and working-class families. Attitudes about education are one example. Middle-class and working-class black families place an enormous amount of emphasis on education for their children because they perceive education as the road to success and as a way to overcome racial discrimination (Wilkinson, 1984). Lower-middle-class black families prioritize education and encourage their daughters to choose education over marriage (Higgenbotham, 1981).

Willie's (1985) work showed that in contrast working-class white families are more ambivalent and sometimes even negative about education for their children. They "worry that highly educated children will no longer honor family customs and maintain cohesion with their relatives" (Anderson, 1988, p. 177).

A second race difference is the perception by black middle-class families of cultivating community responsibility:

> Middle-class black parents insist that their children get a good education not only to escape possible deprivations but to serve as symbols of achievement for the family as well as for the race. . . . Each generation is expected to stand on the shoulders of the past generation and to do more. All achievement by members in black middle-class families is for the purpose of group advancement as well as individual enhancement. (Willie, 1988, p. 183)

In contrast, Willie (1985) found that white families emphasize freedom, autonomy, and individualism. The negative feature of this emphasis is that individualism can shatter family solidarity and can lead individuals to display narcissistic attitudes and hedonistic behavior (Willie, 1988).

In the black middle-class family "Individual fulfillment is seen as self-centered activity and therefore is less valued. What counts in the black middle class is how the family is faring" (Willie, 1988, p. 184). The down side of the emphasis on solidarity is that it stifles experimentation. Risk taking is discouraged and individuals may hesitate to try more experimental and creative activities.

Willie (1988) concludes that blacks and whites can learn from each other on this question. "Too much creativity has been stifled in middle-class blacks who have been trained to put family needs above personal needs. And too many individuals have drifted aimlessly in middle-class white families who have been taught to put individual freedom before collective concern" (Willie, 1988, p. 184).

The third difference concerns the question of gender equality. A number of studies have shown a greater level of equality between husbands and wives in black families compared to those in white families (Morgan, 1985; Middleton & Putney, 1960; Willie, 1983, 1985, 1988; TenHouton, 1970; Mack, 1978). As you should recall from Figure 3-4 in Chapter 3, black women are more likely to have been in the labor force than white women. Egalitarian ideologies are stronger among blacks than whites and black men are more likely to share in housework and childcare than white men (Anderson, 1988). Willie (1988) asserts that gender equality is a worthy goal and that black families have been pioneers in this effort. Therefore, he concludes, "the egalitarian family form is a major contribution by blacks to American society (Willie, 1988, p. 186).

Working-Class Families

Working-Class White Families White working class families are characterized by three factors: (1) the ideological commitment to marry for love, not money; (2) the importance of extended kin and other networks to economic and emotional survival; and (3) the appearance of separation of work and family. Within each of these factors is a contrast between what people believe and what they really experience (Rapp, 1982).

Working-class couples marry for love. Person after person in Lillian Rubin's (1976) interviews of blue-collar couples said they had married for love and that love provided a way to escape from the difficulties of their parents' home. One young woman recalled, "We just knew right away that we were in love. We met at a school dance, and that was it. I knew who he was before. He was real popular; everybody liked him. I was so excited when he asked me to dance, I just melted" (Rubin, 1976, p. 52).

In contrast, upper-class couples recognize their marriages as a way to preserve their class identity (Millman, 1991). Upper-class couples may marry for love but they are conscious that love should only occur between themselves and others of their class. Middle-class people may also marry for love, but as we saw in the discussion above of middle-class families, the overriding task of middle-class families is also an economic one, to enhance the earning power of the breadwinner.

Working-class people are also affected by the economic realities of their lives. Working-class families must operate as economic units. The economic tasks of families are less a part of their dreams about marriage than they are a part of the reality of their married life. "The economic realities that so quickly confronted the young working-class couples of this study ricocheted through the marriage dominating every aspect of experience, coloring every facet of their early adjustment. The women finding their dreams disappointed felt somehow that their men had betrayed the promise implicit in their union" (Rubin, 1976, p. 75).

The second characteristic of working-class families is the reliance on extended kin and others "to bridge the gap between what a household's resources really are and what a family's position is supposed to be" (Rapp, 1982, p. 175). Rapp (1982) says that working-class families are normatively nuclear. By this she means that they believe that independent autonomous families are the best form and they believe that for the most part their families are independent and autonomous.

Observations of their real behavior, however, reveal much sharing of baby sitting, meals, and small amounts of money especially among extended kin (Rubin, 1976; Stacey, 1990). Sometimes these extended kin relationships became problematic and half of the women Rubin (1976) interviewed said that the struggle over who comes first, your wife or your mother, was a source of contention between themselves and their husbands. For example, one woman told Rubin, "He used to stop off there [at his mother's house] on his way home from work and that used to make me furious. On top of that they eat supper earlier than we do, so a lot of times, he'd eat with them. Then he'd come home and I'd have a nice meal fixed, and he'd say he wasn't hungry. Boy did that make me mad" (Rubin, 1976, p. 88).

The third characteristic of working-class families is the appearance that work and family are completely separate. Blue-collar jobs do not include bringing work home and one's occupation does not carry over into one's identity in the way a middle-class professional's might. But work and family are not entirely separate in the working class, where work affects family life and family affects the workplace. In Chapter 6 we will examine some of the specific ways in which this interaction takes place.

Working-Class African American Families: The Moynihan Report and Its Historical Context In Chapter 3 we reviewed black history from the days of slavery up to the middle of the twentieth century. Throughout that time, black families were a focus of the struggle of African Americans for equality. During slavery, African American people fought plantation owners and the slave system for the right to marry and live with their spouses and children. During the sharecropping period, black families struggled for the right for wives and mothers to devote time to their families instead of working for whites. As industrialization developed, African American women moved from the farms to the cities to take jobs as domestics. Here they challenged their employers for the right to work shorter hours to spend time with their husbands and children.

In the last half of the twentieth century, African American families have continued to be a volatile political issue. Some have blamed African American families for a myriad of urban problems, as we saw in the opening scenario. Advocates of African American families have fought back, expressing an alternative point of view. They argue that black families have been scapegoated and are not to blame for poverty and civil unrest. Furthermore, they argue, black families have been the victims of poverty and inequality resulting from the real structural sources of these problems.

One important event in this recent history was the publication of a report entitled "The Negro Family: A Case for National Action" (Moynihan, 1965), commonly called the Moynihan Report. Before examining the Moynihan Report, let us look at its social context.

The 1950s and 1960s were an important period in American history because of one of the most significant social movements in the twentieth century, the civil rights movement, which protested the unequal treatment of African Americans in the U.S. Civil rights activists argued that socially powerful institutions like the legal system, government, schools, businesses, and landlords had created poverty and injustice in the black community. In 1965 a Labor Department document entitled "The Negro Family: The Case for National Action" appeared with an alternative point of view. This document came to be known as the Moynihan Report after its author, Daniel Patrick Moynihan, the senator from New York.

The Moynihan Report blamed the dilapidated housing, poverty, unemployment, and inferior education that faced African Americans on the organization

In the 1950s the civil rights movement called attention to the poverty, unemployment, and poor housing and education faced by African Americans. It blamed these problems on discrimination and an unjust government. Moynihan presented an alternative analysis in 1965, placing the blame on what he argued was the faulty family organization of African Americans.

of black families. Where the civil rights movement saw these same problems and found their cause in the racism in the most powerful sectors of society, Moynihan blamed the victims.

Moynihan argued that black families were disorganized and female dominated. He maintained that black men were humiliated and emasculated by domineering black women. According to Moynihan the only hope for saving the black family and therefore the community was to reestablish black men as the rightful heads of their families (Giddings, 1984). Moynihan (1965, quoted in Gresham, 1989) wrote, "Ours is a society which presumes male leadership in private and public affairs. . . . a subculture such as that of the Negro American, in which this is not the pattern, is placed at a distinct disadvantage" (Gresham, 1989, p. 118). In order to overcome this "disadvantage," the Moynihan Report advised "that jobs had primacy and the government should not rest until every able-bodied Negro man was working even if it meant that some women's jobs had to be redesigned to enable men to fulfill them" (Giddings, 1984, p. 328).

The Moynihan Report also suggested that if black men were to take their rightful place as head of the family and community they would need to bolster their skills in behaving in a properly masculine manner. Moynihan suggested they join the army. He wrote, "There is another special quality about military service for Negro men: it is an utterly masculine world. Given the strains of the disorganized and matrifocal family life in which so many Negro youth come of age, the Armed Forces are a dramatic and desperately needed change: a world away from women, a world run by strong men of unquestioned authority" (Moynihan, 1965, p. 42).

Moynihan reframed the debate around civil rights so that the opposing sides were no longer African Americans versus an unrepresentative government or poor people versus the power structure. New lines were drawn by the Moynihan Report between black men and women over who would have access to scarce jobs and who would "dominate" in families.

Several scholars and the African American community in general reacted critically to the Moynihan Report. People like Joyce Ladner (1971), Andrew Billingsley (1968), and William Ryan (1971) led the debate against Moynihan's assertions (Giddings, 1984; Rainwater & Yancey, 1967).

One of the most controversial features of the report concerned the so-called "black matriarchy." The term matriarchy means rule by the mother. At the core of Moynihan's argument was the characterization of African American women as dominant authoritarian figures—matriarchs.

Robert Staples (1981) actively attacked this idea, calling black matriarchy a myth. He asked: If black women are so dominant and powerful, why do we not see great numbers of black women in Congress, and why do we continue to see black women earning less than white men and women and black men?

Staples argued, furthermore, that when we see black women actively working to ensure that their children are fed and when we see black women fighting shoulder to shoulder with black men for integration, education, and civil rights, we should be proud, not critical. Staples commented, "While white women have entered the history books for making flags and engaging in social work, black women have participated in the total black liberation struggle" (Staples, 1981, p. 32).

Carol Stack (1974), an anthropologist, decided to systematically investigate Moynihan's thesis by doing fieldwork in a low-income black neighborhood she called the Flats. Her work became one of the most influential alternative views of poor black families (Katz, 1989).

The Flats Were African American families in the Flats disorganized matriarchies? This was the question with which Stack began her research. After two years of observing and interviewing the residents of the Flats, Stack (1974) concluded that the families there were neither nuclear nor male dominated. Nor were they disintegrating, nonexistent, or matriarchal. Instead, Stack found families that were complex organized networks characterized by five factors: kin and nonkin membership, swapping, shared child raising, fluid physical boundaries, and domestic authority of women.

Networks were comprised of both kin and nonkin—parents, siblings, cousins, aunts, uncles, and grandparents, as well as nonkin who became "like family" because of their extended interaction and support of network members.

BOX 5 • 2 MATRIARCHIES AND PATRIARCHIES IN RACIAL ETHNIC FAMILIES

At the same time that the Moynihan Report was criticizing African American families for being matriarchal and proposing that African American men be encouraged to take a more dominant role in their families, another racial ethnic group—Chicanos—was being criticized for being patriarchal (Baca Zinn & Eitzen, 1990). In 1966 Arthur Rubel and Celia Heller both published books on the Mexican American community that were highly critical of Chicano families. Moynihan, Rubel, and Heller used the same family pathology argument but they made the argument for opposite characteristics of racial ethnic families.

Like Moynihan, Rubel and Heller argued that the reason Mexican Americans were not making sufficient progress in the U.S. and remained impoverished was because of their faulty family structure. The difference in the two arguments is that Moynihan claimed that the flaw in African American family was their matriarchal character. Rubel and Heller blamed the poverty of Mexican Americans on their patriarchal families.

Mexican American families supposedly were characterized by an intense segregation between men and women and the nearly total dominance of Chicanas by Chicanos within their families. As a result, children were not being raised in a sufficiently modern manner so that they could successfully compete in contemporary American society. In order to lift themselves out of poverty Chicano families must adapt more modern nonpatriarchal values and ways of organizing their families.

There are three problems with this argument. First, it assumes that Chicano families are more patriarchal than white families. Studies of parenting in Chicano families indicate that they are more democratic and egalitarian than the old sterotypes suggest (Baca Zinn, 1989b). Furthermore, comparisons of white men, African American men, and Chicanos conclude that there are few differences in their expression of "machismo" (Senour & Warren, 1976).

Second, it asserts that the ideals of masculinity in Chicano families are a result of the culture of the community rather than a response to the oppression of Chicano men by a racist social structure. Baca Zinn (1989) points out that there has been no research conducted on the impact of discrimination on the way in which men in oppressed groups express gender. Ramos (1979, p. 61) speculates that what has been called "machismo" may be a "way of feeling capable in a world that makes it difficult for Chicanos to demonstrate their capabilities."

Third, it assumes that family structure rather than discrimination and an inadequate economic system should be blamed for causing poverty. This assumption places both the Moynihan Report and Rubel's and Heller's work within the framework of the "culture of poverty" approach. This approach is discussed in the section entitled "Theoretical Models: The Culture of Poverty."

After living in the Flats for two years and sharing rides and childcare, even Carol Stack began to be integrated into the network as a member of the family and began to be called sister by one of the women in the Flats.

The stereotypical middle-class white family is bound together through blood or legal relationships of marriage and adoption. In the Flats, people recognized these ties. More importantly, however, familial networks in the Flats were also bound together by social relationships based on "swapping."

Swapping Swapping refers to the borrowing and trading of resources, possessions, and services. In times of need a member of the network could rely on other members for money, food, clothes, a ride, or childcare. In return the member was obligated to share what he or she had with those in need. Because resources were scarce, people in the Flats constantly redistributed them in order to survive.

Stack describes an example of a swapping network. The description illustrates the many different kinds of resources that are swapped and the complex system that keeps those resources moving in an efficient and fair manner:

> Cecil (35) lives in the Flats with his mother Willie Mae, his oldest sister and her two children, and his younger brother. Cecil's younger sister Lily lives with their mother's sister Bessie. Bessie has three children and Lily has two. Cecil and his mother have part-time jobs in a cafe and Lily's children are on aid. In July of 1970 Cecil and his mother had just put together enough money to cover their rent. Lily paid her utilities, but she did not have enough money to buy food stamps for herself and her children. Cecil and Willie Mae knew that after they paid their rent they would not have any money for food for the family. They helped Lily by buying her food stamps, and then the two households shared meals together until Willie Mae was paid two weeks later. A week later Lily received her second ADC check and Bessie got some spending money from her boyfriends. They gave some of this money to Cecil and Willie Mae to pay their rent, and gave Willie Mae money to cover her insurance and pay a small sum on a living room suite at the local furniture store. Willie Mae reciprocated later on by buying dresses for Bessie and Lily's daughters and by caring for all the children when Bessie got a temporary job. (Stack, 1974, p. 37)

Bloodmothers and Other Mothers Child keeping is a special form of swapping in the Flats and other black communities (Collins, 1990). Poverty makes it difficult for parents to care for children alone. In addition, a value placed on community responsibility is historically rooted in the culture of West Africa and the slave community of the South. Sharing child care in the black community is common, with various adults in addition to the parents sharing or entirely taking over the responsibility for raising a child.

Sometimes child keeping may be shared among parents and other adults for a short time. In other cases it may be for an extended period of years. Sometimes the child lives with one adult at a time. In other cases the child is literally shared, staying in one residence one night and another the next or eating with one adult and sleeping with another.

Children do not see this as being without a real parent but rather as having a number of real parents. Adults, likewise, do not treat their children differently depending on whether they are their natural children or network children (Collins, 1990).

The importance of sharing child raising in the black community is shown by the magnitude of informal adoptions among African Americans. Of the 1 million black children living in families without their parents, 10 percent are in formal foster care homes. Extended family networks have successfully found informal adoptive homes for 900,000 children, a much greater proportion than in the white community (Hill, 1977). Most black children are "adopted" by their grandparents or aunts and uncles. "Two-thirds of black children under 18 living with relatives today are grandchildren of the heads of household, while one-fifth are primarily nieces and nephews" (Hill et al., 1989, p. 196).

Extended Network Families in Racial Ethnic Communities Child sharing among an extended network family is not unique to African American communities. John Red Horse (1980) describes this kind of family organization in some Native American societies. He explains, "An Indian family, therefore, is an active kinship system inclusive of parents, children, aunts, uncles, cousins and grandparents . . . and is accompanied by the incorporation of significant nonkin [who] become family members" (Red Horse, 1980, p. 463).

Red Horse notes that sharing in the Native American community is sometimes informal as it is for African Americans in the Flats but also may be formally marked by naming rituals. In naming ceremonies, which may occur immediately after birth or later in a child's life, the child is given a name and an adult is chosen as the namesake. After the ceremony, the adult is responsible for the child and is obligated to set a good example and to help care for the child or to take over childcare completely if the parent is unable to care for the child.

In the Chicano community a similar system of shared childraising occurs, called *compadrazgo* (Dill, 1986). Many parents designate nonkin, *compadres,* as godparents (*padrinos* and *madrinas*). Godparents celebrate holidays and important rites of passage like first communion and marriage with their godchildren. They are also relied upon for economic and social support in times of need and to substitute in case of the death of a parent (Comarillo, 1979).

Asian American families, especially those that are recent immigrants to the U.S., also rely on a network of kin and nonkin. John Matsuoka (1990) explains that among Vietnamese and other Southeast Asian immigrants in the U.S., a quickly expanding population, extended family includes not only those who are currently alive but ancestors and families of the future. Children are taught that their primary duty is to their family lineage. The dominant American ideology that emphasizes the individual and his or her place in a nuclear family has been problematic for Asian immigrants who believe that one's connections are much broader (Kitano & Daniels, 1988). Asian families illustrate the way in which child sharing not only implies a broad range of people who are responsible for children but also a range of people to whom children are obligated.

Bell Hooks (1984) argues that these kinds of child raising arrangements represent more than cultural differences between racial ethnic families and white middle-class nuclear families; they are revolutionary. She writes,

> This form of parenting is revolutionary in this society because it takes place in opposition to the idea that parents, especially mothers, should be the only childrearers. . . . This kind of shared responsibility for child care can happen only in small community settings where people know and trust one another. It cannot happen in those settings if parents regard children as their "property," their possessions. (Hooks, 1984, p. 144)

Child sharing challenges our ideas about children. It asks us to reconsider whether children are only appendages of their parents or if they are integrated members of society (Collins, 1990). Child sharing also demands community organization that is communal and does not isolate individuals or small groups of individuals—nuclear families—from one another.

Household and Family The domestic networks that comprise the families in the Flats are often spread over several addresses. On the other hand, people who are not nuclear family members may "double up" within a household.

Chicanos rely on extended kin networks of madrinas *and* padrinos *to provide children with social and material guidance and support.*

Where people sleep and eat and where they contribute money for the rent or spend their time is not necessarily concentrated in one physical location. The physical boundaries of families in the Flats are fluid. They range over several addresses, they change, and they overlap.

In middle-class nuclear families, in contrast, households and families tend to be the same. Nuclear family members live in one single family home and other people do not live with them. A person who assumes that nuclear families are the only possible way in which to organize a family might look at the families in the Flats and conclude that no family existed. A more careful examination, however, reveals that a family form does exist, although it is quite different from that of the middle-class nuclear family.

Women's Domestic Authority This description of life in the Flats indicates that Moynihan's (1965) portrayal of the black community as one in which families were disrupted or chaotic is false. The families in the Flats were quite different from the stereotypical middle-class white family. But they were highly organized and provided a source of survival in an impoverished community.

Moynihan (1965) also proposed that black families were matriarchal. Stack (1974) investigated this issue as well and concluded that women in the Flats were not matriarchal.

In a matriarchal society, power over households and the community as a whole is controlled by older women. In Chapter 2, for example, we observed a matriarchal society in the Iroquois Confederacy during the colonial period of U.S. history. The Flats was not matriarchal because women were not powerful in the community. Power in the Flats was wielded by landlords, employers, and especially the government through the welfare office.

Stack found that women in the Flats also did not have matriarchal relationships with the men in their network families. Women had more authority relative to men than women in white middle-class, male-dominated nuclear families. But decisions in the Flats tended to be made by groups of people that included both women and men in the network.

Immigrant Mexican Families Julia Rodriguez (1988; see also Zavella, 1987) found similar patterns in the communities of Mexican immigrants she studied in Southern California to those Stack found in the Flats. She studied women who came north both to follow their husbands who were seeking work and to find jobs themselves. Their emigration from Mexico depended on their ability to obtain support from kin and nonkin in Mexico who could help them to obtain documents, pay for travel, and arrange for childcare. In addition, some had to find childcare for children they left temporarily in Mexico while they moved to the U.S.

Once they arrived in the U.S. the women quickly worked to become familiar with their new communities and establish new networks to exchange goods, as the women had in the Flats. The women also needed to establish information networks because of their special needs as new immigrants or undocumented workers to find employment, housing, health care, and schools in a new environment.

Like the African American people in the Flats, new Mexican immigrants have created a family organization that helps them to survive and to be able to move long distances to resettle.

Contrary to the negative stereotype sometimes portrayed of poor racial ethnic families, these examples show complex, well-organized families. Most of the families, however, are not isolated, male-dominated nuclear families. This challenges classical definitions of family and suggests that how to define "the family" may be an impossible and irrelevant question.

WHAT IS A FAMILY?

In the 1940s George Murdock offered a definition of family: "A social group characterized by common residence, economic cooperation, and reproduction. It includes adults of both sexes at least two of whom maintain a socially approved sexual relationship and one or more children, own or adopted, of the sexually cohabiting adults" (Murdock, 1949, p. 1).

Murdock's definition is widely cited but it has also been roundly criticized (Gerstel & Gross, 1987). The way in which people define family and certainly the way in which they live their family lives frequently varies on every one of the components suggested by Murdock.

Work like that of Stack (1974) helps us to see the variable character of family. She concluded her studies by offering a definition of family that fit the families in the Flats:

> Ultimately I define "family" as the smallest, organized, durable network of kin and non-kin who interact daily, providing domestic needs of children and assuring their survival. The family network is diffused over several kin-based households. . . . An arbitary imposition of widely accepted definitions of the family, the nuclear family or the matrifocal family blocks the way to understanding how people in the Flats describe and order the world in which they live. (Stack, 1974, p. 31)

TABLE 5 • 2
Children Age 18 and Younger Below the Poverty Level, 1970–1990

	1970	1980	1990
White	10.5%	13.4%	15.1%
Black	41.5%	42.1%	44.2%
Hispanic	N.A.	33.0%	39.7%

Source: U.S. Bureau of the Census, Current Population Reports, Series P-60, no. 175, 1990c, p. 456.

In the scenario at the beginning of the chapter, Vice President Quayle's definition of families is very much like Murdock's. He argues that based on this definition, families in poor black neighborhoods have been destroyed or are nonexistent. Stack (1974) asserts that families are varied and that our definitions need to be equally diverse. How would you define family? Does your own family fit this definition? Can we define "the family"? Is it necessary to do so?

RENEWED INTEREST IN POOR FAMILIES

The debate over poor black families has not abated. The 1980s brought forth a renewed interest in poor families. And although most poor families are white, the stereotype of the poor family is of one that is headed by a black single woman. In this section we will look at the history of poor families in the 1980s and 1990s and the way in which they have been portrayed in political debates like that in the opening scenario of this chapter.

Most poor families in the U.S. are white, but as we observed in Chapter 4, African Americans, Latinos, and Native Americans, especially those with children, are disproportionately likely to be poor. Forty-four percent of African American children live below the poverty level; 15 percent of white children live below the poverty level. The poverty level of Latino children is 40 percent (U.S. Bureau of the Census, 1990c). Table 5-2 shows how these numbers have grown in the past twenty years.

Poor families are likely to be single female–headed households. Almost 45 percent of single female-headed households with children were poor and women and children make up the majority of the poor. Of all the poor people in the U.S., 27 percent are children, 49 percent are women, and 24 percent are men (U.S. Bureau of the Census, 1990c). Diana Pearce (1978) coined the term "the feminization of poverty" to describe the trend toward larger numbers of single women and children in poor households. Pearce (1978) attributed the feminization of poverty to two factors: the responsibility of women for children and discrimination in the labor market against women. Single women are responsible for more mouths to feed than single men are but when women enter the labor force they are paid less than men. The result is more poor single women and more poor children.

The "feminization of poverty" is compounded by race because race discrimination further limits access to housing, education, jobs, and good pay. Among female-headed households with young children, 50 percent of those headed by whites are poor, 69 percent of those headed by blacks are poor, and 72 percent

TABLE 5 • 3
Percent of Families with Children Aged 6 and Under Who Are Poor, 1990

	% in Category Who Are Poor
White	
Married Couple	8.6%
Single Mother	49.9%
Black	
Married Couple	20.2%
Single Mother	69.1%
Hispanic	
Married Couple	29.6%
Single Mother	72.0%

Source: U.S. Bureau of Census, *Statistical Abstract of the U.S.*, 1992, p. 457.

of those headed by Latinos are poor. Table 5-3 shows racial ethnic contrasts in poverty among married couple and single mother families with young children.

During the 1980s the proportion of single women heading households and the prevalence of poverty in these households increased. A series of reports and media images emerged attempting to explain these phenomena. One of the popular media's presentations on single mothers that was seen by a large section of the population was a documentary by journalist Bill Moyers called "The Vanishing Black Family: Crisis in Black America," which aired on January 25, 1986. Patricia Collins (1989) refers to the video as the film version of the Moynihan Report.

Moyers's video presented a description and analysis of single female–headed households that was colorful and entertaining but also distorted and startlingly racist and sexist (Harrison, 1987). The video was distorted because it focused on black families, making it appear as if poor and single female–headed households are mostly African American. This is incorrect.

The statistics above show that black and Hispanic families are more likely than white families to be poor. Black families, however, make up only about 12 percent of the total population and Hispanics 9 percent. The largest proportion of poor families and of single-parent families are white. Table 5-4 shows the proportion of poor children who are white, black, Latino, and Asian.

The video is sexist because once again, as in the Moynihan Report, the message is that black women have too much power over black men and that the male-dominated nuclear family is the best family organization and the only solution to the problems faced in low-income black communities.

The image presented in the Moyers video of single female–headed black families was also distorted because it was a pathological one. Young black parents were portrayed as immoral, ignorant, irresponsible, and inarticulate. The women were getting pregnant, having babies, and living on welfare checks and the men were strutting, smiling, and abandoning the women and children.

These images are of course insulting to African Americans and the question that emerges is, why are these images promoted? Some authors have suggested that the film was politically motivated. One critic argues: "Whenever the system is in crisis (or shows signs of becoming transformed); whenever blacks get restless (or show strength); whenever whites in significant numbers show signs of

TABLE 5 • 4
Poor Children in the U.S. by Racial Ethnic Group

White	41%
Black	35
Latino	21
Asian or Pacific Islander	3

Source: Children's Defense Fund, 1988b.

coming together with blacks to confront mutual problems (or enemies), the trick is to shift the focus from the real struggle for political and economic empowerment to black 'crime,' degeneracy, pathology and—the 'deterioration' of the black family'' (Gresham, 1989).

The Moynihan Report came out as an alternative to the message carried by civil rights leaders during a period of social unrest over the question of equality for African Americans. Civil rights leaders blamed the system and the powerful people who worked to maintain it. Moynihan blamed the victims.

Moyers's video came out during a period of economic crisis following the Great U-Turn. People like Congresswoman Maxine Waters claim that the problem is the economy and the ineffective leadership of those who wish to maintain the status quo. The speech by Vice President Quayle was in direct response to one of the most destructive social upheavals in the U.S. in recent history, the Los Angeles rebellion. Quayle argues that the victims are to blame.

Are poor and black single female–headed households pathological? Are they the source of the many social problems we see in our society? Or are they, as Gresham (1989) suggests, victims who are being scapegoated?

The debate around this question, which gained much attention in the 1960s and reemerged in the 1980s, continues to be a focus in the 1990s. The debate brings together the three systems of inequality we reviewed at the beginning of the chapter—gender, race, and class— and the relationship of these stratification systems to families. Those people who are at the bottom of each of these systems of inequality—poor black women and their families—are at the core of the debate.

Theoretical Models: The Culture of Poverty

The debate over the causes and effects of the problems faced by poor families and their solutions involves two opposing models for understanding the relationship between family structure and poverty. One model is cultural and one is structural (Baca Zinn, 1989a).

The cultural deficiency model claims that poor people have a different and inferior culture compared to that of middle-class and upper-class people. Baca Zinn (1989a) argues that currently the deficiency model actually has three variations: culture as villain, family as villain, and welfare as villain. The central thesis of all of these variations of the cultural deficiency model is that bad culture causes poverty. Within this framework, poverty is an effect, not a cause.

The Moynihan Report in the 1960s, the Moyers video in the 1980s, and the comments by the vice president are all illustrations of the cultural deficiency model. They maintain that certain individuals within our society have adopted values and behaviors that make them unable to succeed economically and socially and these bad values and behaviors are passed on within families.

The University of Michigan's Panel Study of Income Dynamics (PSID) has collected data that challenge the cultural deficiency model (Baca Zinn, 1989a). First, the PSID found that the poor in one year are not necessarily the poor in the next year. People move in and out of poverty status. This means that poor people are not a deviant subgroup that are always poor.

Also, the panel discovered that the values of welfare-dependent families and non-welfare-dependent families were not significantly different. These two groups share similar ideas regarding their desire to be good parents, their concern for the future, and their motivation to achieve.

A second variation of the culture of poverty argument is that the lack of a two-parent family is what causes and sustains poverty. According to this point of view, unmarried mothers and divorced single parents create a deficit in their children's lives that leads to poverty (Baca Zinn, 1989a).

Hill and Ponza's (1983) work based on data from the PSID shows that being in a two-parent family does not necessarily keep people out of poverty. African American children in two-parent families had lower incomes over the decade between 1970 and 1980 than white children in single-parent households during that same time period.

Mary Jo Bane (1986) has looked at the importance of divorce in creating poverty. She found that among whites there was a transition from nonpoor to poor after divorce. Among African Americans, however, poverty both preceded and followed divorce.

In both of these studies poverty is not seen as a problem of "deviant" single-parent families. Poverty, especially for African Americans, is a common experience regardless of family structure. The data do not support the family deficit argument that certain types of family organization, single-parent families, *cause* poverty.

The third variation of the cultural deficit model blames welfare for poverty. This argument contends that welfare makes people lazy and causes poor women to have babies and poor men to abandon these women and their children (Murray, 1984).

Research comparing different welfare benefits in different states, however, shows no systematic variation in family structure (Baca Zinn, 1989a). Furthermore, increases and decreases in welfare spending over time do not produce predictable changes in family organization (Ellwood & Summers, 1986; Danziger & Gottschalk, 1985). For example, from 1960 to 1972 welfare spending increased and from 1972 to 1984 it declined. During both of these periods family composition increasingly included more female-headed households. If welfare spending causes single-parent households to increase there should have been a reversal of this trend after 1972.

Despite its popularity in the media and among conservative politicians the cultural deficit theory is not well supported by systematic research. An alternative model for understanding the links among poverty and family organization is the social structural model.

Social Structural Model

The second model Baca Zinn (1989a) describes is a social structural model that emphasizes the trends we reviewed in Chapter 4. The transition to an electronically based economy, the accompanying decline of the industrial manufacturing sector, and the shift of jobs from the central cities of the midwest and northeast to the suburbs, the south, and to other countries have created unemployment and a decline in wages for those left with jobs. From the perspective of the social structural model, massive changes in the structure of the American economy have created poverty (Baca Zinn, 1989a).

In particular, racial ethnic people have had their source of livelihood ripped from them. Many jobs have been eliminated entirely by automation and others have been sent far away. Richard Hill and Cynthia Negrey (1989) found that 50 percent of the African American men who worked in manufacturing lost their jobs between 1979 and 1984.

Even those jobs that stayed in the region, however, were inaccessible because they were outside the neighborhoods of working-class African Americans and beyond public transportation routes. John Kasarda (1985) found that one-half of the African American population in Philadelphia and Boston did not own cars. In New York two-thirds of African American and Latino households lacked personal transportation.

Shifts in the economy eliminate jobs or remove them to inaccessible sites. Unemployment then affects family organization. For example, the lack of jobs inhibits marriage. William Wilson and Kathryn Neckerman's (1986) research shows that high unemployment is a major factor in explaining the low number of marriages among young African American men.

The social structural model asserts that economic decline causes poverty, which creates changes in family structure. The decline in the number of jobs created by major shifts in the economy causes significantly greater unemployment among blacks and Latinos, which in turn causes fewer marriages and greater numbers of single-parent households. Compared to the cultural model the structural model has gained much less attention in the popular media but it appears to be better supported by the research evidence.

Do Families Contain the Seeds of Resistance?

Thus far our discussion in this chapter has emphasized the way in which the macro level of social structure, particularly social stratification, affects the micro level of family organization and the everyday experience of people living in families. We have also observed how families have attempted to cope with problems of poverty and inadequate housing by pooling their resources and restructuring their families. These responses are examples of the way in which the micro level attempts to respond to the macro level by attempting to hold back the most devastating effects of inequality and oppression.

People at the micro level have also created ways in which to try to change the macro level. Sometimes these activities are unconscious, like those described by Hooks (1984), and sometimes they are organized efforts to demand changes in the macro organization of our society. In her description of child raising in low-income households above, Hooks (1984) suggests that the family system that developed in the Flats is a revolutionary one. By that she means that it challenges our ideas about what the relationship between children and adults should be and it implies radical changes in the organization of society beyond families.

Other writers have suggested that poor families can be revolutionary in a more general sense (Glenn, 1991). Mina Caulfield (1974) argues that families can provide a base of operation for people attempting to challenge their governments. Families provide a vehicle of resistance in three ways. First, families aid the physical survival of their members. Second, families preserve and teach the culture of the community, thereby keeping alive not only individuals but also a point of view. Families can teach a history and culture, and an alternative way of thinking and acting, all of which incorporate an ideology that questions the current organization of power. For example, in Chapter 12 we will look at specific proposals being made by scholars to socialize children by teaching them alternative, critical points of view. Bem (1983) tells us to teach our children to be feminists. Peters (1988) tells us how to teach our children to recognize and challenge racism.

Third, families create or strengthen the organizations that struggle against the powerful (Naples, 1992). For example, during the labor organizing drives in the U.S. at the turn of the century workers had difficulty creating organizations that crossed lines among those who were unionized and nonunionized, skilled and unskilled, men and women. One of the ways in which these groups of workers were brought together was through their family connections. Unskilled young women meat packers were able to unite with skilled older men butchers because the women were the daughters and nieces of the men (Lembcke, 1991; Benenson, 1985).

The Mothers of East Los Angeles

The Mothers of East Los Angeles (MELA) is a contemporary example of a conscious organized effort by people working together at the micro level to affect the organization of the macro level of our society. It also is an example of women using their family networks and family roles as the basis of political action.

MELA is a group of about 400 Mexican American women who initially came together in the mid-1980s to protest the state's proposal to build a prison in their neighborhood. Since then they have fought the building of a toxic waste incinerator. They have elected representatives, lobbied the state legislature, and defeated a proposal to build an oil pipeline through their community. Their goal is to bring new schools to their neighborhood instead of prisons, and safe work sites instead of hazardous industries (Pardo, 1990).

The profile of a typical member of MELA is of a low-income, high school–educated, highly religious (Catholic), middle-aged (forty to sixty) Mexican American woman. Some scholars have identified these characteristics as "retardants" to political activism. Pardo's (1990) work shows, however, that the women in MELA have transformed their social identity from "retardant" to activist in the politics of East L.A.

Pardo (1990) describes how the members of MELA have transformed their oppressed social position into a basis of empowerment. She cites five areas of this transformation. The first of these is the transformation of gender-based roles, responsibilities and networks into a political resource (1990, p. 2). For example, women used experiences and friends they had met in their childrens' school activities. Second was the transformation of "invisible" women into a focal point (1990, p. 3). Many women literally expanded their discussion and activity from the neighborhood to the state capitol. Third was the transformation of ethnic and class identity into a basis of community unity (1990, p. 3). The women took pride in who they were and contrasted their grassroots style to the high paid lobbyists they confronted. Fourth was the transformation or redefinition of "mother" to include militancy and political responsibility (1990, p. 4). Fifth was the transformation of "unspoken sentiments of individual women into collective voices of community protest" (1990, p. 5). This final transformation included personal change when they needed to take strong public positions to win their battles.

Pardo concludes that the women in MELA defy the common perception of who will become a political activist. The problems they face within their community have forced them to reconstruct a definition of themselves as militant fighters. Their activity reminds us to keep an open mind when we look for avenues for change and the people who might lead us there. This chapter has presented much evidence of the oppressiveness of the stratification systems of class, race, and gender. The Mothers of East L.A. assure us that it will take more than that to prevent people from seeking justice.

THE MICRO-MACRO CONNECTION

A stratification system with three major components—gender, race, and class—creates an important part of the macro level of our society. Each of these systems of stratification has an effect on the organization of families. Social class creates certain tasks and opportunities that impinge on families. Upper-class families are called upon to maintain the system. Middle-class families work to keep themselves financially afloat. Working-class and poor families must invent creative ways to organize their families to survive.

The system of race stratification also shapes families. Gender differentiates the experience of women and men within families of all social classes and racial ethnic groups.

In addition, the three systems work together to create an especially vulnerable family type that finds itself at the bottom of the three systems—poor, black, single female–headed households. The question of whether these kinds of families are the victims or the cause of urgent problems of poverty, inequality, and civil unrest is a focus of debate among political activists on every side.

In previous chapters I have emphasized the importance of the way in which the micro level affects the macro level. As evidence of this we have seen examples of political movements that have been critical of the maintenance of the status quo. In this chapter I introduced the idea that there are also political activists who seek to maintain the present organization of the macro level, including its systems of stratification.

Upper-class women, for example, work to reproduce the social class system and the boundaries between themselves and other classes. People like former Vice President Quayle, Senator Daniel Moynihan, and journalist Bill Moyers take the position that it is not the system that should be criticized and changed but rather the individuals who are unsuccessful within that system.

At the other end of the spectrum, scholars like Carol Stack and Maxine Baca Zinn write about alternative critical views. Their work at the micro level challenges the organization of the social structure. Other people have also called for changes in the macro organization of our society. Congresswoman Maxine Waters says that we need to restructure the United States to eliminate poverty and inequality and until we do there will be no peace. The Mothers of East L.A. continue their efforts to build a political organization that will address the problems of inadequate schools, jobs, a clean environment, and housing in their community.

The macro level has an important effect on our families and our individual lives. Part of that effect is to differentiate families by race, class, and gender. The micro level in turn fights back, but not in only one way. People who hold different places in the stratification system have different interests, different ways of resisting, and different goals in their efforts to shape the macro level of society.

SUMMARY

Systems of Stratification

- The differing opinions expressed in the scenario at the beginning of the chapter illustrate two perspectives on the way in which our society is socially stratified.
- Stratification systems are structures of inequality that can be organized around any variable.
- In our society gender, race, and class are the three key stratification systems and they all have important effects on families.
- Our class system is characterized by four factors: it is economically based; it is a social relationship, not just a category; boundaries between classes are fluid; and positions within the system are in some part achieved.
- Contemporary American society can be divided into three social classes: an upper class, a middle class, and a working class.
- Biological races do not exist among humans, but race is an important social category.
- Ethnicity refers to cultural distinctions based on religion, nationality, or the social category of race.
- Most people can be assigned to one of two categories by biological sex, but gender is more variable.
- Gender is a social category that describes the kinds of behaviors we expect and observe within our society.
- Both sex and gender have important effects on families.

Social Class, Race, and Gender and Family Life

- Upper-class families protect their privacy from others outside their class and provide an important means of entry into the class.
- Women in elite families play a special role as gatekeepers to the class.
- Families provide a mechanism by which individuals stay in the upper class. The upper class maintains its boundaries as well as maintaining the class stratification system itself.
- Four factors characterize middle-class families: geographic mobility; replacement of kin with other institutions for economic support; reliance on

friendship for affective support and exchange; and investment of resources lineally.

- Geographic mobility, a common experience in the middle class, has been especially hard on the mental health of wives and children of men who are required by their jobs to move.
- Upper middle class wives play three roles for their husbands: an adjunct role, a support role, and a double-duty role.
- Wives who are married to executives are sometimes called corporate wives. Their support is essential to their husbands' career. The wives work for both their husbands and the corporation.
- Black middle-class families are similar to white ones but they differ in three ways: their greater emphasis on education; their prioritizing of family needs over individual fulfillment; and their greater expression of gender equality.
- Working-class white families can be characterized by three factors: ideological commitment to marrying for love and not money; reliance on extended kin for support; and the appearance that jobs and families are separate.
- The importance of family as a political issue in the African American community has continued into the twentieth century. Civil rights activists blamed the political and economic system for poverty and inequality. Daniel Moynihan blamed these problems on the organization of black families.
- Central to Moynihan's critique of black families was the assertion that women were too dominant and men did not hold the "normal" position of power in African American families.
- Several scholars critiqued Moynihan's position. Carol Stack entered an African American community to study the organization of family and produced an alternative view of families.
- Carol Stack did research in the Flats and found the families there were characterized by five factors: kin and nonkin membership; swapping; shared child raising; fluid physical boundaries; and domestic authority of women.
- Swapping refers to the borrowing and trading of resources among members of the family networks.
- Children in the Flats and also in many Native American, Chicano, and Southeast Asian communities were raised by other members of the community as well as their biological parents.
- Bell Hooks claims this communal child rearing is revolutionary because it challenges our ideas about the relationships between children and adults and because it demands cooperative community organization.
- The physical boundaries of families in the Flats were fluid, with people living, eating, sleeping, and paying rent at several addresses.
- Women in the Flats had domestic authority but were not matriarchal.
- The kind of family networks Stack found in the black community are also found in the immigrant Mexican community of Southern California.

What Is a Family?

- Murdock's definition of "the family," which he claims is universally applicable, is similar to the stereotypical middle-class white family.
- Stack's definition of family is quite different. Although it fits the families she studied in the Flats, families are so variable that no single definition will do for all of them.

Renewed Interest in Poor Families

- In the 1980s the debate over black families reemerged with concern about poor and single female–headed families.
- Most poor families are white but the media image of poor families is that they are African American.
- Poor families are likely to be headed by single mothers, so the debate about poor families must consider the problem of gender and the feminization of poverty.
- Moyers's video provides an example of the image of black families that was being projected by the media in the 1980s. Like the image in the Moynihan Report, this image was a negative one. The video has been criticized for being distorted and some have suggested that its distortion was politically motivated.
- Baca Zinn asserts that two competing models to explain poverty exist: the culture of poverty model and the social structural model.
- The culture of poverty model claims poor people have an inferior culture and an aspect of that inferior culture is their deficient families. Baca Zinn uses research to show that this model is not supported by the evidence.
- The structural model asserts that changes in the economy have created poverty, which in turn shapes families.
- Families can provide a vehicle for resistance for people trying to change the system.
- The Mothers of East L.A. are a political group that has worked on issues of concern in their community since the mid-1980s.
- The women of MELA defy the common perception of who will become a political activist. They have taken the social factors that make them oppressed members of our society and transformed them into avenues to empowerment and social change.

The Micro-Macro Connection

- Stratification systems of gender, race, and class form an important part of the macro system that affects the organization of families.
- Scholars and activists have challenged the maintenance of the status quo and the blaming of families for problems they believe to be structurally based.
- Other people are interested in maintaining the status quo. They feel that the inappropriate organization of families underlies the unrest that challenges the system.

Like many parents, this Latino father depends on a childcare center to help take care of his son while he works.

6 WORK AND FAMILY

Cindy and Jordan are seniors in college and have been good friends throughout their years at the university. Both will be graduating with degrees in accounting at the end of this semester. They sit in the student union discussing where they will be in five years.

Cindy says, "I might go back to school for an MBA later, but for now I want to get out and make some money." Jordan nods his head in agreement and says, "Yes but what about a family? You'll probably be married with a kid by then, too."

"No," says Cindy, "I want to get established first, with a good position and a house and car before I settle down. It's too hard to do everything and I want to use my degree. What about you and Allie?"

"We're all set to get married next fall. I don't think it will be that hard. We want to have kids too but we plan to share. That way we can both establish our careers and raise a family," Jordan replies. "Allie can work during the day and I can work during the evening at the bank. It's just a matter of scheduling and commitment. Having a wife and kids at home will help me to keep my life more balanced and not just job, job, job."

"What if the boss wants you to entertain clients from out of town, or travel to a conference, or just work late, or what if the baby gets sick, then what?" says Cindy, "No, it's not for me, I've made my choice. Women have to make choices these days and thank God they can."

Cindy and Jordan are at a similar juncture in their lives, finishing college, making plans for their career, and thinking about where family will fit in. Despite their similarity they have come to different conclusions. Cindy is convinced that she must make a choice between work and family because they would interfere too much with each other. Jordan believes he can integrate the two activities. He also thinks that family pressures will be good for his personal well-being, keeping him from being a workaholic.

- *What kinds of assumptions are Cindy and Jordan making about the organization of work and family and the intersection of the two?*
- *Are their opinions based on a perception of work and family as two separate spheres or do they see work and family holistically?*
- *How does gender affect their thinking?*
- *Are Cindy and Jordan limited in their thinking about what they can do about their social environment? Are there demands they could make on their employers or the government that might alter their possible choices?*

MYTH OF SEPARATE SPHERES

In Chapter 2 we examined the emergence of a split between work and family among middle-class whites during the development of industrialization. Many productive activities that took place within families in preindustrial society were moved to factories as society industrialized. Work that took place outside of families was separated spatially and temporally from work and other activities within some families.

Along with these geographic changes, ideas began to change as well and new ideas emerged about what families are like or should be like. The new ideas separated work and family into two different spheres. The new ideology also separated women from men, placing men in the public work sphere and women in the domestic family sphere.

During the nineteenth and early twentieth centuries this familial ideology was problematic for most people. It was problematic for upper-class and middle-class white women who often felt isolated and stifled in their domestic role. African American and working-class families found the ideology problematic because it was held up as an ideal but was often impossible for them to attain. The idea of men in the public sphere earning money and women in the domestic sphere taking care of the home did not accurately depict their lives. Work and family were not separate spheres for African Americans who as slaves and sharecroppers were all forced to work constantly.

Among white working-class families the idea of separate spheres was also largely mythical because nearly all members of working-class families worked for wages, including wives who stayed home and did piecework or took in boarders to earn money. In these families women were not separated from the work sphere, although their paid work and unpaid family work sometimes existed in the same physical space, the home.

As the twentieth century has progressed the idea of separate spheres has become increasingly mythical not only for racial/ethnic people and working-class whites but for the white middle class as well. The majority of women in the U.S. are now in the labor force—57.5 percent in 1991 (Ries & Stone, 1992). Even women who are wives and mothers in families are likely also to be in the paid labor force (Moen, 1992). Furthermore, as we will discuss in this chapter, work shapes and penetrates families and families influence and overlap work. Unlike the myth of separate spheres, work and family are inseparable in real lives (Kanter, 1977a; Kelly, 1979; Bose, 1987; Chow & Berheide, 1988).

This chapter is divided into four major sections. The first is an examination of the problematic conceptualization of work and family as separate spheres. In the second section we will look at the ways in which work affects family and in the third, how family affects work. Finally, we will consider the demands for changes that individuals are making to try to alleviate the difficult tension between work and family.

The Study of Work and Family Is Distorted by the Concept of Separate Spheres

One of the results of the conceptualization of work and family as two separate spheres, one for women and one for men, is that the *study* of work has been segregated into two models, a "job model" and a "gender model" (Feldberg & Glenn, 1979). When men workers are studied a "job model" is used to explain their behavior. The job model focuses on working conditions, opportunities, and problems in trying to understand men's behavior at work and their expectations about their jobs. This model assumes that men in the paid labor force are in their proper sphere and that their experience and obligation as members of families is unrelated to their work experience and largely irrelevant to their lives.

When women workers are studied a "gender model" is used. This model ignores working conditions or job characteristics and instead focuses on personal characteristics of the worker and her family circumstances to explain her behavior and expectations at work. The gender model, furthermore, assumes that women put their family first in making decisions about work. This model assumes that women in the paid labor force are out of their "true" sphere. Women's experience at work, therefore, is determined or overshadowed by family.

These different models result in distorted interpretations of men's and women's lives. To illustrate this issue, Roslyn Feldberg and Evelyn Glenn (1979) surveyed Robert Blauner's (1964) book, *Alienation and Freedom,* a classic study of textile workers. They argue that in his research Blauner used the two different models—gender and job—to interpret women's and men's work behavior.

Blauner reported that women workers compared to men were more closely supervised and did more physically demanding, machine-paced work. Women workers also complained more about their work than the men did. The women said they had to work too fast, there was too much pressure at work, and they became tired.

When Blauner (1964) discussed why the women complained he seemed to forget what he had found about the difficult and oppressive conditions of the women's work. Using a gender model to explain the women's complaints, Blauner said that they complained because compared to men, they had less physical stamina and greater home responsibilities as housewives and mothers. Blauner's explanation ignored the job-related explanation provided by his own data that the work the women did was more oppressive because it was more closely supervised and more physically difficult; it required constant motion and was machine-paced. Instead he relied on an explanation that emphasized gender and the connection between women and families.

Another example of gender bias in Blauner's (1964) work was revealed when Feldberg and Glenn (1979) reviewed his interpretation of the responses of men and women workers to their jobs. Blauner found that in the textile mills both women and men worked under difficult conditions, yet they expressed relatively little dissatisfaction with their jobs and little aspiration to find better ones. For the men, Blauner explained that their lack of dissatisfaction and aspiration resulted from their low levels of education and the scarcity of alternatives in the small one-industry towns in which they lived. For the women, Blauner switched to the gender model and explained that the women were not dissatisfied because "work does not have the central importance and meaning in their lives that it does for men, since their most important roles are those of wife and mother" (Blauner, 1964, p. 87).

Feldberg and Glenn (1979) concluded that both the job model and the gender model are partial explanations. When they are used alone they distort and obscure rather than help us to understand. What is needed is an integrated model that takes into consideration both work and family and the links between them for both women and men.

Young Women Speak of Work and Family in Their Future Figuring out how work and family fit into our lives is not a problem just for scholars, of course. Everyone has ideas about how these two activities fit together and how they would like to experience them. And almost everyone has trouble determining what the choices are, what choices they should make, and especially how they will be able to attain those choices. Ruth Sidel (1990) inteviewed young women about their expectations for the future. Much of their responses included consideration of their future activities as employees and as wives and mothers. Based on their responses, Sidel grouped them into one of three categories: New American Dreamers, Neotraditionalists, and Outsiders.

The New American Dreamers plan to have it all. They optimistically predict that they will be successful, affluent, and independent. They will enter professions that have been dominated by men. They will also be extraordinary wives and mothers with lots of children and a big house. But first they plan to establish their careers.

The New American Dreamers speak of having careers as models, judges, and physicians. They believe that all they must do is establish their goals and work hard.

One sixteen-year-old New American Dreamer told Sidel,

> It's your life. You have to live it yourself. You must decide what you want in high school, plan your college education, and from there you can basically get what you want. If you work hard enough, you will get there. You must be in control of your life and then somehow it will all work out. (Sidel, 1990, p. 25)

The young women whom Sidel (1990) placed in the category of New American Dreamer put work before family both in time and priority. However, they also believed they would eventually be successful in both areas. The New American Dreamers had contemplated the ways in which work and family might affect each other in their lives, especially the ways in which work might interfere with their ability to excel as wives and mothers, or the way in which family might interfere with their ability to reach the top at work.

In the scenario at the beginning of the chapter, Cindy represents a New American Dreamer. She wants to be successful in a business career and she believes she must choose between work and family. Her choice is to establish herself as an accountant and then consider marriage and child bearing.

The second category Sidel called Neotraditionalists. These women had considered the interaction between work and family. They too wished to have interesting and well-paid work but they named family activities as their top priority. One woman, a senior in college, explained:

> I want to be smart. I want to be somebody. I want to make money. I want to be a successful lawyer but my personal life comes first. I want to be a lovely wife, do my husband's shirts, take Chinese cooking lessons and have two children. (Sidel, 1990, p. 37)

In the scenario at the beginning of the chapter, Jordan shares the point of view of the Neotraditionalist women. He wants to have both a career and a family. He has chosen to place priority on marriage and children. (Because Sidel interviewed only women we have no basis for comparing young men to young women on the question of work and family. This question would be an interesting one to pursue.)

The third group of young women Sidel interviewed she named Outsiders. They felt they had little chance of success in any arena, including work and family, and they could not imagine negotiating the two. Linda, age seventeen, expresses the position: "I don't plan. I don't look to the future. I can't plan 'cause my plans never work out. They never go through" (Sidel, 1990, p. 67).

Where do you or the young women and men you know fit into these three models? How do the personal lives of the three types mesh with the social reality in which they live? In the next section, we will look at the reality of work and family and their intersection in order to understand where these young women's dreams fit and to determine what might need to change to allow them to successfully negotiate the choices.

Women in the Paid Labor Force

What is work? Nona Glazer (1987) suggests that when we speak of the contemporary United States, we define work as "those activities which produce goods and/or services and/or provide for the circulation of goods and services which are directly or indirectly for capitalism" (Glazer, 1987, p. 240).

Much of Glazer's research has been on unpaid work such as housework. In the next chapter we will examine the unpaid work that takes place in families. In this chapter we will focus on paid work and its intersection with families.

Women are still largely identified with unpaid domestic activities and, in fact, women do an enormous amount of unpaid housework. One of the most significant changes in American society in this century, however, has been the steady increase of women in the paid labor force.

Table 6-1 shows that while men still are more likely than women to be in the paid labor force, the majority of women are working for wages outside of their homes. Although the numbers vary some by race, within the categories of white, black, and Hispanic the majority of women are earning money.

Figure 6-1 shows that the numbers of women in the labor force have grown steadily thoughout the last half of the twentieth century. The increases have been especially dramatic for women with children (Burris, 1991). For example, in 1950 only about 12 percent of women with children under six years old were in the labor force. By 1991 that proportion had jumped to 67 percent.

Why Have Women Increasingly Entered the Labor Force? Explanations for why women have entered into the labor force in the twentieth century can be divided into three groups: supply side, demand side, and social structural (Strober, 1988). Supply-siders (Becker, 1965; Mincer, 1962) argue that as the century progressed women's work at home decreased in its value. Falling birth rates and labor-saving devices for housework made unpaid domestic labor less important for the family economy. The supply of women available for paid work outside of the home thus increased, putting pressure on the labor market to open up jobs for women.

The second explanation focuses on the demand for women in the labor market. Economists like Karen Oppenheimer (1969) assert that jobs that have been identified as female occupations such as secretary, nurse, and waitress have increased. The demand for women to fill these positions has created a pull on women to enter the labor force.

The third explanation emphasizes the broader social structure (Vickery, 1979). Improved birth control and increased life span for women are two aspects of social structure that have been identified as important factors in relation to the increasing numbers of women in the work force. In addition, improved technology in factories has made it cheaper and easier to purchase goods and services than it is to produce them at home. For example, automated factories can produce clothes more cheaply and efficiently than individual seamstresses can. These social structural factors that make up the social context of women's work at home and in the labor force have created the impetus for bringing women into paid work. This third explanation is similar to the supply side argument. The difference between the two is that supply-siders emphasize the changes in families—fewer children, less housework. Those who focus on social

TABLE 6 • 1
Labor Force Participation Rates, by Race and Sex, 1990

	All	Whites	Blacks	Hispanics
Men	76.1%	76.9%	70.1%	81.2%
Women	57.5%	57.5%	57.8%	53.0%

Source: Ries & Stone, 1992, p. 308.

FIGURE 6 • 1
Women in the Labor Force by Presence of Children, 1947–1991

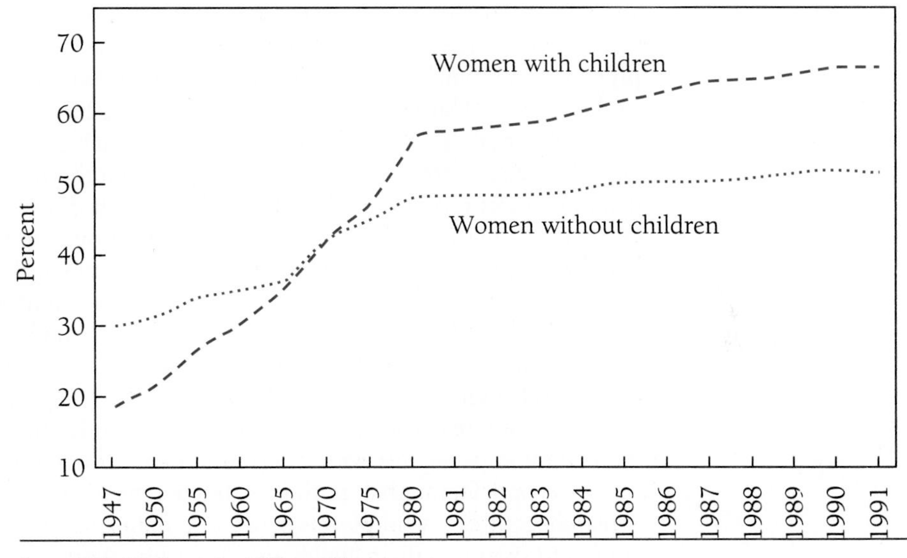

Source: Mishel & Bernstein, 1992, p. 408.

structure, in contrast, emphasize the social factors outside of families that influence these changes.

Myra Strober (1988) argues that all three of these explanations provide a piece of the whole picture. And she adds that the recent decline in real income has given an additional boost to women to enter into the labor market. Haya Stier (1991) notes, however, that despite the pervasiveness and strength of these forces the effect has not been uniform for all women. Her research on Asian American women indicates that although recent immigrants have all of these pressures to place them in the labor force, they may still be impeded from doing so because of factors having to do with their immigrant status, most importantly their language skills.

Supply factors, demand factors, other structural factors, the economic climate, and racial ethnic diversity are all macro-level features that play a role in creating the impetus for women to enter the paid labor force in increasing numbers. Let us now move to the micro level to see how individual women make the choice to work outside the home.

How Do Women Choose to Work? Most women work because of economic need. But women work for social rewards as well. Most employed women would not leave their paid employment even if they did not need the money: "56% of full-time homemakers say that they would choose to have a career if they had to do it all over again, and only 21% of working mothers would leave their current job and stay at home with the children" (Scarr, Phillips, & McCartney, 1992, p. 415).

Kathleen Gerson (1987) decided to investigate more thoroughly the question of how women choose to enter the paid labor force. She interviewed sixty women to ask them how they choose between work and family and found their choice was largely the result of two influences: (1) early childhood experience, and (2) constraints and opportunities they encountered as adults.

Twenty percent followed a traditional model in which as girls they were prepared emotionally and practically to be housewives and remained committed to that choice. Thirty percent were also socialized as children to become housewives and mothers, but as adults experienced rising aspirations to enter the labor force and increasing ambivalence about motherhood. Another 20 percent had hoped even as children to enter the paid labor force and they retained that desire into adulthood. The last 30 percent had grown up wanting to have careers but experienced falling employment aspirations when they became adults.

Four factors tended to push women into the domestic path. First, women were influenced by their experience within their families. A stable relationship with a male partner reinforced the idea among the women that marriage provided a safe, secure place that allowed wives to stay home. A stable marriage also frequently resulted in family decisions that prioritized the husband's career, thereby diminishing the wife's opportunities in the paid labor force.

A second factor that pushed women toward a domestic choice was their experience in the workplace. Upward mobility at work and a rewarding job tended to diminish their likelihood of quitting work.

Third was the ability of the husband to provide an adequate income. Low wages for husbands in the labor market caused an economic squeeze in the family, which pushed wives into the paid labor force.

Fourth, women considered the social rewards and costs of domesticity. Those who had neighbors and friends who devalued being full-time housewives and mothers were more likely to choose to seek paid work outside the home.

Gerson's (1987) research illustrates the interaction of work and family. It shows how girls' family experiences combine with both their family and work experiences as they become women. The combination of factors throughout their lives creates pushes and pulls into and out of the paid labor force. The tensions between work and family as well as within them provide the impetus for the decision to enter the paid labor force or not.

Chicanas and a Fifth Way of Choosing to Work for Wages Patricia Zavella (1987) interviewed Chicanas who worked in canneries in California about the tensions between work and family in their lives and how they made the decision to get a job in the cannery. The women she interviewed had been encouraged and socialized as girls to become full-time housewives and mothers. A typical pattern was to work outside the home in the early days of their marriage and quit when their first child was born. Then when it became clear that their husbands' paycheck could not adequately support the household, the women returned to work in the cannery. Choosing to work in the cannery, however, was not perceived as an alternative to their responsibilities as wives and mothers but as an extension of that role.

Gerson presents the two roles, employee and wife/mother, as two opposing possibilities. The women in Zavella's (1987) research saw them as contiguous. Gerson describes the women she interviewed as having been socialized as children to be either domestic or career oriented. She describes their experience as adults as pushing them into or away from the domestic choice and she develops a four-part model to summarize the possible factors.

The cannery workers fit into a fifth category in which the women seek outside employment, not because they wish to move away from their domestic role, but because they wish to be better wives and mothers. The Chicanas argue that they chose to go to work because of their obligation to their families. When Zavella asked one woman why she had sought work in the cannery, for example, she said, "I did it for my family. We needed the money, why else?" (Zavella, 1987, p. 88). Another woman who was asked this question responded by motioning to her child, who was carrying a large doll, and said, "That's why I work, for my daughter, so I can give her those things" (Zavella, 1987, p. 134).

This position is different from that of the women in Gerson's work. The women in both Gerson's and Zavella's research were moved to take a job outside the home if their husbands' wages were inadequate. The difference between the two groups was in their perception of this choice. The women in Gerson's research chose to work for wages as an alternative to staying home. Gerson describes their decision as choosing between two separate options. The women in Zavella's inteviews, in contrast, insist that work and family are not two alternatives but that working for wages is a way of fulfilling one's role in family.

Third Wave Feminism The term Third Wave feminism, as I described in Chapter 3, refers to recent work in feminist theory that focuses its attention on differences among women. Much of the most significant work in Third Wave feminism is done by racial ethnic scholars (Lengermann & Niebrugge-Brantley, 1992).

This discussion of the contrast between the work of Gerson and Zavella shows the way in which theories based on the experience of one group of women may not sufficiently capture the experience of another group. Zavella, who is a Third Wave feminist, offers insights from her research on Chicanas that make two contributions. First, her work reminds us once more that Woman is not a monolith. All women are not the same and race ethnicity makes a significant difference in our lives. Second, Zavella's work enriches the work of scholars like Gerson because it broadens the analysis, making it not just more inclusive but also a more powerful theory.

HOW DOES WORK INFLUENCE FAMILIES?

Chow and Berheide (1988) assert that the relationship between work and families can be looked at in three ways: as separate spheres, as spillover from work to family, and as interactive. We have already discussed the separate spheres model and the problems associated with it. The second model is one in which the connections between work and family are noted but in which attention is paid almost exclusively to the way in which work spills over into families. Within this spillover model work is seen as more fundamental and causal, and families are seen as secondary results of work organization and experience. The third model is an interactive one in which the mutual dependence of families is emphasized. The interactive model is the one that organizes this chapter. Here we will review the way in which jobs shape and influence families and then move to issues associated with the way in which families shape the occupational experiences of workers and sometimes the organization of the workplace.

Rosabeth Moss Kanter (1977a) was one of the first scholars to examine the effect of work on family. She cites five ways in which the organization of paid work affects families:

1. Absorptiveness of an occupation—the extent to which a job draws in an employee and his or her family members.
2. Time and timing—the amount of time a job requires and when the time is demanded.
3. Income (or lack of income, for example, if a breadwinner becomes unemployed) provided by the job.
4. World view—the values and ideas promoted at work that influence the employee in his or her behavior at home.
5. Emotional climate—the social-psychological influences at work that affect employees and their relationships in families.

Absorption

Different occupations demand different amounts of commitment by workers. For example, a receptionist in a physician's office may commit only the amount of time when she is on the job to her career. When she leaves the office for the weekend she can stop thinking about her job and stop identifying herself by her occupation. The physician in that office, in contrast, rarely leaves the role of her occupation. She is called Doctor whether she is at work or not. She probably brings work home and may be on call most of the time. Her occupation is part of her identity.

The contrast in the absorption of the workers in these two jobs will influence their families. The receptionist's family can remain unconcerned with her job and they will rarely be called upon to participate in her work. The physician's family frequently will be identified as part of the profession. They may be asked to participate in community events as a way for the doctor to maintain a proper image or they may be asked to assist the physician in building the practice by helping to decorate the office and advertise the business.

Time and Timing

Another way in which jobs differ in their impact on families is by the demand on time placed on the employee (Eckenrode & Gore, 1990). Time can be divided into three categories: (1) the number of hours of work employees perform per day or week, (2) when the work is done during the day or week, and (3) long-term time commitments. If a profession demands long hours away from family members, less time is available for participation in family activities. For example, a nurse who frequently works double shifts at the hospital will have

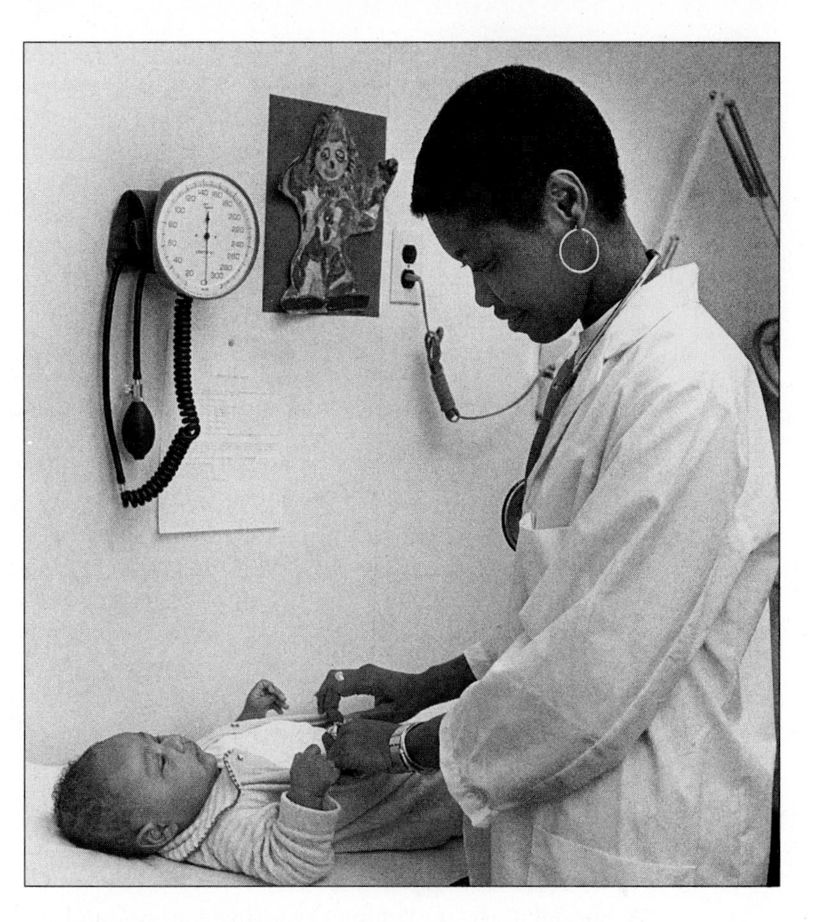

The profession of medicine is so demanding that it presents an especially strong example of the tensions employed parents feel in trying to be both good employees and good parents. Many women physicians have attempted to negotiate the two jobs by practicing in settings where they are not on call all of the time.

little time for doing anything else other than sleeping when she gets off work. A teacher, in contrast, may have more time to spend with his family in the afternoons and in the summer.

The amount of time is only one facet of the issue of time. Equally important is what time of the day work is scheduled. Families must organize their activities around the work schedules of employed family members. In addition, certain work schedules are out of sync with "normal" scheduling and, therefore, interfere with participation in particular kinds of activities. For example, a waiter who works the weekend night shift will rarely have time to spend with his wife and friends if that is the time slot they use for their social activities (Lein, et al., 1974).

The importance of timing is illustrated by an experiment with workers in an aircraft parts plant (Meyersohn, 1963). Workers at the plant were allowed to schedule their hours so that every month they would have a Monday off without reducing their total hours on the job for the month. At first the workers were enthusiastic about the plan. After less than a year, however, they asked to have the plan discontinued. They complained that their Mondays were boring, lonely, and contributed little to family or leisure activities because other family members were at work or school during the day on Mondays.

The third way in which time at work can affect family life is in the long-term demands of a job. For example, some jobs may demand large amounts of time for relatively short periods. The manager of a grocery store might find that when the store first opens he must work long hours training new personnel. After a few months, when the staff is in place, he is able to cut back his hours. A college professor, however, may need to spend large amounts of time on research and publishing for several years in order to receive tenure (Steiner, 1972). Both of these people would find that they had removed themselves from family activities and had caused other family members to readjust to the time required of the job. The professor, however, might find that the new patterns had become too entrenched for her to reintegrate herself after she is granted tenure.

Conflicts Between Work and Family Life: Gender and Time When the Department of Labor asked workers, "How much do your job and family life interfere with each other?" more than one-third said they experienced either moderate or severe work–family conflicts. There were no significant differences between women and men respondents. Women were as likely as men to say that work and family interfered with each other (Pleck, Staines, & Lang, 1980).

When asked what kinds of conflicts they experienced, however, gender was a factor. "Men more often reported excessive work time, at least in part because they worked more hours than women" (Pleck, Staines, & Lang, 1980, p. 30). For women the most often cited problem was schedule conflicts. It was not the amount of time at work that concerned women, but rather when the work had to be done.

The problems workers had with work–family conflicts broadly affected their lives. Both men and women who said they had work–family conflicts reported significantly less satisfaction with their job, family, and life in general.

A Special Case of Timing: Working Parents and Their Children Can employed women be good mothers? Research that has examined the children of employed mothers shows that "the school achievement, IQ test scores, and emotional and social development of working mothers' children are every bit as good as that of children of mothers who do not work" (Scarr, 1984, p. 25).

Some research has shown that children of mothers in the paid labor force are better off than children of mothers who are not. Ramey, Bryant, and Suarez (1985) argue that especially in inadequate family environments, when employed mothers place their children in high-quality daycare, the children have a better opportunity for social and intellectual development then they would have received at home. Daughters of employed mothers seem to be especially likely to be advantaged (Hoffman, 1984).

> Daughters of working mothers are more independent, achievement oriented, and socially competent with more flexible sex-role perceptions, better family relations and higher self esteem. Positive effects on cognitive and social development for daughters often extend into adulthood. (Mischel & Fuhr, 1988, p. 207)

The picture of the employed mother, however, is not entirely positive and there are important additional factors to be considered (Mischel & Fuhr, 1988). One is the mother's satisfaction with her role. Farel (1980) compared children whose mothers stayed home and mothers who were employed. But he also subdivided these groups according to those who were pleased with their role and those who were not. He found that children of mothers who stayed at home and who believed that working outside the home would be detrimental for their children, scored highest on several measures of social and cognitive development. Children of mothers who stayed at home but wished they had a paid job, however, scored lowest.

Fathers' involvement with children is also an important variable. Some researchers (Etaugh, 1980) have found that mothers' employment can have a negative effect on boys' cognitive development and academic achievement, but equally crucial in the equation is paternal involvement. Greater involvement by fathers seems to be particularly beneficial for young sons (Mischel & Fuhr, 1988).

Teenage Workers and Their Families The discussion above suggests that when workers spend time on the job it may weaken their family ties. Research that has been done on teenage girls who work supports this argument. Girls who work report a decrease in feelings of closeness to their parents (Steinberg, et al., 1982).

The experience for boys, however, is different. Boys who work report increased feelings of closeness to their parents (Steinberg, et al., 1982). Greenberger (1987) explains this difference by calling our attention to differences in the relationships between daughters and their parents and sons and their parents. Girls generally report much greater closeness and self-disclosure to parents (especially mothers) than same-age boys (Douvan & Adelson, 1966; Kandel & Lesser, 1972).

Greenberger argues that work helps both girls and boys grow up and begin to take on adult roles. But because they begin with different kinds of relationships with their parents, their maturation is expressed in different ways. For girls, working increases their feelings of autonomy and self-reliance and lessens their dependence on their parents. Work helps them to move away from their parents.

For boys the opposite occurs. Their experience at work increases their identification with their parents, who are also workers, and brings them closer, especially to their fathers.

This research shows that work has an important effect on an individual's experience in a family. But work does not occur in isolation from other factors. Here the equally salient factor of gender combines with work to affect boys and girls in different ways.

Income

The third way in which work may affect families is through the amount of money brought into the family by the employment. This economic issue was addressed in Chapter 4 where we discussed the current economic crisis and its effects on families.

Wages can also have an impact on the internal dynamics of families, either disrupting or creating and reinforcing gender inequality. Women still tend to earn less than their husbands do. On the average wives earn about 29 percent of the total family income. For black families this number is significantly higher. Black wives contribute almost 50 percent to their family income (U.S. Bureau of the Census, 1983). Although this amount is significant and indispensable, it is still less than their husbands' income. The inequity in wages reinforces gender inequality both inside and outside the family. Women who earn less than their husbands do are more economically dependent on their husbands than vice versa. When a husband makes more money than his wife the family must treat the husband's job as more important because it contributes more to the total income (Gerstel & Gross, 1987). Unequal wages between women and men can inhibit women's ability to attain equality within their families:

> Marital power is higher for women who work full time than those who work part time or not at all, and it is greatest among women with the most prestigious occupations, women who are most committed to their work, and those whose salaries exceed their husbands'. Working women have more say especially in financial decisions. This tendency for employment to enhance women's power is strongest among lower and working class couples. (Moore & Sawhill, 1984, p. 158)

In Phillip Blumstein's and Pepper Schwartz's (1990a) interviews, one wife, Leanna, explained this connection succinctly. Leanna, who works in a delicatessen, earns less than her husband Gordon, who is a police officer. She says:

> Gordon still has the last word on everything. We get annoyed with each other over that, but when I start to push back, he reminds me just who supports me and the children . . . he gives me the final line which is something like, "If you're so smart, why don't you earn more money?" or how dumb I am 'cause if I had to go out and support myself I'd be a big fizzle. (Blumstein & Schwartz, 1990a, p. 125)

Earning any wages, however, helps to move husbands and wives in the direction of greater equality. Although Leanna is unable to get as much acknowledgement of her contribution as she would like, research indicates that her situation would probably be worse if she were a full-time housewife. Women who earn wages exercise greater power in their families than women who are not employed (Blumstein & Schwartz, 1990a; Ybarra, 1977, 1982a, 1982b; Baca Zinn, 1980).

Exchange Theory In order to understand the process by which having both husband and wife in the paid labor force enhances gender equality, scholars have used a theoretical framework called exchange theory. George Homans (1958) was the major figure in the development of this theory in the 1950s (Ritzer, 1992).

Exchange theorists argue that people enter into social relationships in which they exchange rewards with each other. This theory assumes that people need rewards and work to acquire rewards to enhance their self-interest. When two individuals interact they attempt to negotiate an exchange of rewards, each trying to gain as much for him or herself as possible through the exchange.

For example, in the case of a husband and wife, both seek from each other such rewards as status, love, control over family finances, and decision-making power. When the woman is employed she brings to the negotiating table more chips with which to bargain because she can contribute more to the total resources available within the family.

Wives or husbands who are not in the paid labor force are not entirely without power. But money is a particularly useful reward because it can be exchanged for many other rewards. Some exchange theorists (Bushnell & Burgess, 1969) have argued further that money is an especially good asset to bring to an exchange because it is nearly impossible to satiate the desire for money.

Robert Blood and Donald Wolfe (1960) are important figures in the study of power in marriage. They rely on exchange theory in their description of marital power. They write,

> The sources of power in so intimate a relationship as marriage must be sought in the comparative resources which the husband and wife bring to the marriage rather than in brute force. A resource may be defined as anything that one partner may make available to the other, helping the latter satisfy his needs or attain his goals. . . . The partner who may provide or withhold resources is in a strategic position. . . . Hence power accrues spontaneously to the partner who has the greater resources at his [or her] disposal. (Blood & Wolfe, 1960, p. 12)

A Feminist Critique of Exchange Theory Some feminists (Hartsock, 1983; England, 1989; Glenn, 1987) have been critical of exchange theories like that expressed by Blood and Wolfe (1960) because exchange theorists often assume that women and men are the same; they ignore gender. Critics of exchange theory argue that husbands and wives come to the "bargaining table" not just as two individuals but as gendered people. The wife has learned to behave in certain "feminine" ways and is expected by her husband, herself, and most everyone else to behave in those ways. Likewise, the husband is expected to be "masculine" (Rosen, 1987).

The negotiating options available to the two are, therefore, different and are restricted by their gender. Paula England (1989) believes that one especially important gender difference is that in our society and much of the Western world to be masculine is to be independent and selfish, "a separative self." To be feminine is to be concerned for others, "an emotionally connected self" (England, 1989, p. 15).

As support for this distinction, England cites the work of Carol Gilligan (1982). In examining the moral reasoning of women and men, Gilligan found that when women decide what is moral they base their beliefs on an ethic of caring and responsibility, which flows from their feeling of connection between themselves and others. Men, in contrast, base their moral decisions on "an ethic of principled noncoercion, which presumes and seeks to honor the other's separateness" (England, 1989, p. 16).

Exchange theory assumes that all individuals operate from their own self-interest. Gilligan's work suggests that while this may be true for many men it is less likely to be true for women. Furthermore, women who behave in this way are stepping out of the expectations for a properly feminine person.

For example, imagine a husband and wife trying to decide how to spend the money they have received from their income tax refund. His salary is larger than hers and he wins the argument by claiming he should, therefore, decide how the money is spent. Although she has lost the argument, they both have played their gender roles appropriately. He is masculine because he makes more money and his selfish, independent reasoning has won the argument. She is feminine because she makes less money and has given in to him, and also has been able to do something to make him happy.

But what if she has a larger salary and therefore wins the argument about how to spend the money? She has won the argument but neither spouse has behaved in an "appropriate" manner. He not only lost the argument (which is unmasculine in itself) but has been openly exposed as earning less money. She has won the argument by behaving in a selfish, adversarial, "unfeminine" manner.

These imbalances mean that although exchanges may take place between husbands and wives they are not exchanges between equals. Furthermore, given our culture's definitions of femininity and masculinity, the goal of serving one's self-interest fits men more comfortably than it does women (Hochschild, 1989).

Blumstein and Schwartz (1983) conducted a study that illustrates the importance of gender in exchanges among intimate couples. They interviewed four types of cohabitating couples: married couples, unmarried heterosexuals, gay men, and lesbians. In their research they found that differences in earning caused the largest power differences in couples of gay men. Heterosexuals, married or not, fell in the middle and power differences among lesbians were not related to their earnings at all.

England (1989) says that this shows that money and power are linked when the negotiating team includes men and the links are strongest when the exchange is between two men. Exchange theory is valid for explaining men's behavior because the social construction of masculinity idealizes men who behave as separate selves seeking individual rewards. But for explaining relationships between two women, exchange theory has little validity. And in mixed couples its validity is diminished because women make up half of these couples and they are likely to behave in an emotionally connected way.

World View

According to Kanter (1977a) the fourth way in which work affects family is by bringing a world view from the workplace to the family. World view is an aspect of culture. Culture can be defined as the values a given group holds, the norms they follow, and the material goods they create (Giddens, 1991). Culture is usually associated with nations and societies, but occupations can also be said to have distinct cultures (Hughes, 1958).

Occupations can create ways in which to think about the world, including oneself. They can create rules about appropriate behavior on the job or off. And they can create material goods to represent these ideas.

People who are socialized at work into their occupational culture may bring their beliefs and ways of doing things into their families. One way to distinguish among occupations is to divide them into two types: entrepreneurial and bureaucratic. These two occupational culture types might have very different ways of raising children (Kanter, 1977a).

An entrepreneurial father who is a car salesman works in a job that demands independence from and competition with his co-workers. As a parent he would be more likely to encourage his children to develop strong self-control, independence, and an active, manipulative approach to the world. A bureaucratic mother who works in a welfare office must interact with her co-workers in a cooperative manner in order to provide all the necessary services to her cli-

BOX 6 • 1 WHAT HAPPENS WHEN HUSBANDS AND WIVES TRY TO SHOW THEIR LOVE BY GIVING GIFTS, BUT THE GIFTS DO NOT FIT GENDER STEREOTYPES?

The discussion of husband–wife interactions and exchange theory shows that exchanges between men and women in adversarial or competitive situations falter because of gender inequality. Hochschild (1991) notes that even when husbands and wives are trying to cooperate, their efforts may be unsatisfying. She refers to these unsuccessful efforts as "mis-givings."

Hochschild argues that an essential component of intimate relationships is "an economy of gratitude" (1991, p. 499). Men and women give each other gifts of love, time, support, and material things. But for these things to be experienced as a gift both parties must recognize them as such. A gift is something that is extra, "beyond what we normally expect" (Hochschild, 1991, p. 499). A gift is also something that the receiver wants and feels grateful for.

A "mis-giving" is a gift that is not perceived by the giver and receiver in the same manner. Hochschild cites the case of one couple she interviewed to illustrate a "mis-giving."

Peter and Nina were a married couple. Nina was a successful entrepreneur and able to contribute a hefty salary to her household. She was glad to do this because it enabled her husband to work as a bookstore owner, a job he loved but which was not well paid. Hochschild says that "Nina offered Peter her large salary as a gift" (1991, p. 506). Although Peter was grateful he could not fully accept the gift. He felt ashamed because she earned so much more than he did. Instead of making Nina and Peter feel good, as the exchange of a gift should do, it made them feel uncomfortable. Gender interfered with the exchange and created a "mis-giving."

ents. She would encourage her children to be accommodating and to seek direction from others. In these two examples, the context of the parents' work causes the employee to value certain attitudes and to develop certain ways of behaving. He or she then brings home these ideas to the family and uses them in raising the children.

Another example of the way in which work shapes workers is the forging of gender identities at work. Usually the family is cited as the central gender socializer, but work can also create gender as well (Westwood, 1984). In our society "it is paid work that confers adulthood and therefore it is on the shop floor that girls become women" (Pascall, 1986, p. 53). But the rules about what women workers should be like are oppressive and gendered. Women workers are more closely monitored and more highly supervised (Wolf & Fligstein, 1979). Stafford and Duncan (1978) found that women workers were allowed one and one-half times less rest time than men. Women workers are also paid less. To be a woman worker is to be a person who is overworked, dependent, subordinate, and unworthy of equal pay (Westwood, 1984). This formulation of gender for women at work can be brought home to family interactions, reinforcing the idea that wives, mothers, and daughters should be dependent, subordinate, and unworthy.

David Halle's (1987) work suggests that gender messages can be brought home about men too. He interviewed men in factory jobs. Thirty percent of the men he interviewed said that their wives complained that their social status was too low because they had blue-collar jobs. The roles men played at work as powerless and subordinate people were "unmasculine." When they came home this "unmasculine" identity came with them and their wives were critical of them.

Emotional Climate

Emotional climate is the fifth way, Kanter (1977a) argues, that work affects families. Job satisfaction, tensions, and pressures at work are three areas of emotional climate that can affect an employee and his or her family (Eckenrode & Gore, 1990). If a job is stressful and unsatisfying, an employee may use his or her family to compensate for the emotional rewards that do not come at work. Or employees may displace anger and frustration at home (Blau & Duncan, 1967; Hoffman & Manis, 1979; McKinlay, 1964). Families of police officers (Norlicht, 1979) and military personnel (Myers, 1979) have been studied in this regard and provide evidence that these highly stressful occupations may be related to high incidences of wife and child abuse.

Emotional demands at work may also affect workers' interactions with family. Kanter (1977b) argues that people in high-interaction occupations may develop interaction fatigue and withdraw at home. The corporate wives she interviewed told her that they noticed when their husbands were training others and needed to be highly involved with the trainees; the husbands were distant and insensitive when they came home, as though they had been burned out at work.

David Halle (1987) argues that tensions in blue-collar jobs may create socializing patterns that compete with the time men have to spend with their families. He interviewed working-class men and found that friendships at work were essential to making their boring and stressful jobs tolerable. This camaraderie frequently spilled over after work and competed with wives' claims to their husbands' time. About half the men he interviewed went straight home from work, but the other half spent a good amount of time with their male friends after work. One might assume that when emotional needs are created by job pressures, people bring those needs home to be met by their families. Halle's interviews show how work can create emotional needs that can sometimes be met by social relationships on the job. But in solving that problem, husbands may create another set of problems because their friendships with work buddies pull them away from their wives, illustrating another way in which the conditions of work touch employees' interactions with their families.

Work, Family, and the Concept of Dialectics

"Most analyses of work and family in the modern American context have settled into a comfortable economic determinism" (Kanter, 1977a, p. 53). Economic determinism is the belief that the way in which the economy is organized, including the organization of work and the workplace, determines the way in which all other factors in society, including family, are organized. Chow and Berheide (1988) refer to this as the spillover model of work and family. In Chapter 4 I introduced the Marxist argument that, based on a materialist outlook, economics is an essential aspect of human society. Sometimes people identify economic determinism with Marxism. Marxists, however, are not economic determinists. Within a Marxist framework economics is a necessary but not sufficient element in social organization.

The underlying logic of Marxist theory is a particular kind of materialism called dialectical materialism (Chafetz, 1988):

> The dialectical method of anlaysis does not see a simple, one-way, cause-and-effect relationship among the various parts of the social world. For the dialectical thinker, social influences never simply flow in one direction as they do for cause-and-effect thinkers. To the dialectician one factor may have an effect on another, but it is just as likely that the latter will have a simultaneous effect on the former. (Ritzer, 1992, p. 44)

We have examined the ways in which work shapes family but that is only half the story of the relationship. Families also influence work.

THE INFLUENCE OF FAMILIES ON WORK

First, families place workers into the labor force by making demands on family members for economic support (Duncan, Featherman & Duncan, 1972). For example, not surprisingly, wives whose husbands have below-average incomes have the highest rates of labor force participation (Ryscavage, 1979).

Families also socialize new workers for the labor force, teaching them values and attitudes about the meaning of work and appropriate behavior on the job. For example, families teach their children communication skills and how to take advantage of education opportunities that will allow them to obtain training for certain types of work (Smith, 1987). The work that parents do to shape their children into acceptable members of society will be further discussed in Chapter 12 on parenting and Chapter 13 on children.

Families also introduce workers to specific information about job choices (Mortimer & Kumka, 1982; Spenner, 1981). Fathers are especially influential in their sons' vocational choices, particularly when the father is in a prestigious occupation himself and he has a close relationship with his son (Hetherington, 1979; Hoffman, 1961).

Bringing the Family to Work

In assessing the effect of work on family I described the way in which work culture can be brought home to influence family life. This effect can also operate in reverse: the culture of family can be brought to work by employees. The activities and values that people express at home can be carried over to their work and become important influences in the relationships among workers and between workers and their bosses.

Harry Braverman (1974) reviewed the history of labor relations between workers and management under capitalism and concluded that the history had been one of struggle over technology. Company owners seek to develop technology that removes control from workers and places it in the hands of management. Skilled workers are able to demand higher wages and they are able to control the way in which production is carried out. For example, a skilled craftsman in a glass factory tells the bosses what materials are needed, how many glasses can be blown in an hour, and how many assistants he needs. In addition, because his skill is relatively scarce, he can also negotiate good pay for himself.

Chicanas bring their family culture to work with them where they celebrate important family transitions like weddings and births with their co-workers and create familial relationships by sharing information and supporting one another on the job. In addition to making the work environment more tolerable, this kind of familial activity can also create networks and opportunities for political activity such as union organizing.

On the other hand, if management can develop automation to use unskilled or semiskilled workers whose task is reduced to attending the machines as they make the glasses, management can pay those workers less. Furthermore, management, not workers, can determine the rate of production and the number of workers necessary.

This struggle over the conditions of work has been central to the history of industrialization and it continues to take place. When women workers engage in the struggle with management they sometimes bring with them to the workplace certain skills, values, and social connections to use in the struggle that are related to their roles as wives and mothers.

Two "feminine" strategies that Louise Lamphere (1985) and Patricia Zavella (1987) discovered in their interviews and observations of women workers are: "1) the organization of informal activities often focusing on the female life cycles such as birthday celebrations, baby and wedding showers, potlucks and retirement parties and; 2) the use of workers' common identities as women, wives and mothers in interworker communication" (Lamphere, 1985, pp. 520–521). These activities, which are closely associated with women's role in families and might be described as bringing the family to work, provide a humanization of the workplace and a vehicle of support for coping with inhumane working conditions (Zavella, 1987).

In addition, sometimes these activities can create or enhance a culture of resistance on the shop floor. The activities and associations create friendships, trust, and a shared understanding that can then become the basis for collective action (Costello, 1985; Sacks, 1984; Zavella, 1987; Benson, 1978).

The women come to know each other well. They keep track of who is getting married, whose daughter is having a baby, and whose husband has died. They develop ways in which to communicate across lines of race ethnicity, age, and job category. Women workers who socially interact with one another on the job can easily move from talking about family affairs to sharing ideas about the conditions of work, interpretations of events that differ from management's, and the possibility for change.

Zavella (1987) observed indirect examples of resistance. She noted the way in which the networks based on "feminine" activities served as vehicles of information distribution about how to get around rules and how to bypass steps in production. Other researchers, especially those who have focused on Chicanas (Ruiz, 1990; Coyle, Hershatter, & Honig, 1980), have found that the familial culture the women create at work has directly supported militant actions at work. For example, Lamphere (1985) found that those connections allowed or at least enhanced the ability of women in the apparel plant she studied to stage a sucessful wildcat strike.

Managing the Contradiction

In Chapter 1 I suggested that people are not just passive products of their environments. C. Wright Mills (1959) argued that social context is a human creation; people build and rebuild the societies in which they live. This theme has been illustrated in the other chapters in the discussions of social movements that have formed with the intention of changing public practice in order to solve personal problems.

Sometimes people who are faced with difficult or impossible tasks, instead of trying to change the social structure, create ways in which to cope with the social structure, as we saw in Chapter 4 in the discussion of families coping with unemployment. In the case of the conflict between work and family, women physicians are an example of a group that has managed to solve or at least ameliorate the contradiction between work and family by specializing in areas where there is less conflict.

Unprecedented numbers of women have entered the medical profession since 1970 (Bourne & Wikler, 1982). "In 1970 there were 22,000 active women physicians comprising 7.1% of the population. By 1986, 79,600 women physicians comprised 15.3% of the profession" (Sidel, 1990, p. 171). And this proportion is likely to grow. In 1987, 36.5 percent of the first-year medical students were women (Sidel, 1990).

The profession of physician has long been identified as a particularly masculine occupation (Hughes, 1945). The women who enter the field as students and then as professionals sometimes have found it difficult to negotiate the masculine character of the field.

Medicine is a "greedy" institution for both men and women, making enormous demands on their time and attention. It appears, however, that women find it particularly problematic (Lorber, 1984). Linda Grant and her colleagues

(1987) looked at the differences between women and men medical students as they progressed through their training. When asked how much time they believed they would be spending on work and family, women compared to men predicted they would be spending more hours on family and men compared to women predicted they would be spending more hours on their practice. As the years went by at medical school both women and men increased the number of hours they believed they would spend on their practice and decreased the hours they would spend on their families. Grant and colleagues (1987) conclude that the entrance of women into the medical profession has not changed the profession. Medical school training appears to be turning women into men and both women and men into physicians who are more focused on their practice and away from their families.

One of the characteristics of the medical profession that illustrates its "maleness" is that the "organization of work in medicine, the sequence of a medical career, and key professional norms are predicated on a conventional male role in the family and on the male biological clock" (Bourne & Wikler, 1982, p. 113). The time period in which the physician is expected to be most intensely involved in training and building a career is the same time as the favored childbearing years for women (Hochschild, 1971).

Women who enter the medical profession know they will have to play the roles of doctor, wife, and mother, and social worker (to their professional husbands). They fear the logistical and personal difficulties that playing all of these roles may create (Bourne & Wikler, 1982). The women medical students that Patricia Bourne and Norma Wikler interviewed, however, defined these difficulties as personal ones and put the burden on themselves to work it out. One student told them:

> When I was thinking of going into medicine, I told myself, "Well you're not going to have any children because you're not going to have any time. If you want to be a good mother, you're going to have to have all your time for the formative years." But now I feel differently. However, that hasn't changed my ideas about children, but I do feel it can be done. I think it requires more energy and a lot of planning and a lot of help from the outside. (Bourne & Wikler, 1982, p. 116)

Women physicians are encouraged to enter specific kinds of settings in which to practice their profession as a way to juggle their work and family. For example, one medical school dean suggested: "Let me give you an example—physicians in student health services in colleges and universities. It's an ideal situation for women physicians" (Bourne & Wikler, 1982, p. 120).

In another interview a male physician suggested that women work in a health maintenance organization:

> Medicine is a taxing field, around the clock. It's harder to accept the kinds of responsibilities that are necessary, if she has in mind having a family, if we were operating on the entrepreneurial model. Now there's a change toward shared responsibility. It's easier for women to fit in. At Kaiser, for instance, I know of two women who are splitting a position. (Bourne & Wikler, 1982, p. 120)

BOX 6 • 2 IS A "MOMMY TRACK" THE SOLUTION TO JUGGLING WORK AND FAMILY?

Felice Schwartz (1989) proposes that corporations distinguish between women who are "career primary" and those who are "career and family." Those women who are "career primary" are those who will not interrupt their careers for children and can, therefore, remain on the fast track with the men. Management should identify these women and then invest more in them, for example, by providing them with training and experience and by expecting more from them.

The women who are "career and family" are those who do want to have time to raise children. They should be put on the Mommy Track. The Mommy Track employees could then be allowed to stay in middle management and not be under pressure to excel and compete for promotions.

The Mommy Track may help women by allowing them time to be wives and mothers, but it diminishes them as professionals, reducing their pay and their prestige. For example, in the profession of medicine we have observed that women physicians often have different career patterns, at least in part because of their desire to marry and raise children. Women physicians also earn less. In 1987, for example, women physicians earned on the average only a little over half of what men physicians did (Sidel, 1990).

Second, the track may be partly chosen by women, but sometimes women may be pushed into the track because of the expectations of others. In the quotes in the discussion of women physicians we see this advice being given by a medical school dean and a male colleague.

Third, the Mommy Track can legitimate discrimination against women. For example, if you were an employer and had to choose whether to hire a woman on the Mommy Track or a man, who would you choose?

Fourth, Glass (1992) notes that higher level jobs, compared to middle and especially lower level jobs, actually give workers more flexibility and authority to create solutions to problems that stem from job–family conflicts. For example, a higher level salaried employee compared to an hourly employee has more control over when she comes to or leaves work. Because the salaried higher level employee is paid for accomplishing tasks rather than putting in a certain number of hours she could, for example, drive her child to school in the morning even if it meant she were a few minutes late. An hourly employee would not have this option or at least would be docked pay for her tardiness.

And finally, the premise of Schwartz's argument for the Mommy Track is based on the assumption that employing women in management is more costly than employing men. There are no published data on this question (Ehrenreich & English, 1989).

Women physicians appear to be taking this advice, especially if they are mothers. The presence of children is the key factor in the reduced number of hours that women physicians practice (Grant, Sampson, & Lai Rong, 1990). Women compared to men physicians tend to specialize in different areas, work fewer hours per week, and see fewer patients (Sidel, 1990). "Women are far more likely to choose specialties with regular hours, such as dermatology or pathology. . . . They are far less likely to choose surgery, which is often thought of as the most rigid and hierarchical of the medical specialties and which is known for its years of infamously arduous training" (Sidel, 1990, p. 171).

This kind of placement of women into positions that facilitate their ability to maintain their work in families, by working in settings that are less difficult but also less prestigious and less lucrative, has recently been dubbed the Mommy Track (Schwartz, 1989) (see Box 6-2).

For some women, like those in medicine described above, the difficulty of negotiating work and family can cause them to develop coping mechanisms. The difficulty of balancing work and family can also encourage employees to make demands for changes to allow them to both work and enjoy a family life (Hyde & Essex, 1991; Zigler & Frank, 1988; Ferber & O'Farrell, 1991).

These demands fall into two categories: alter the organization of work and provide help with family demands, especially childcare. One type of alteration in the organization of work that has recently gained attention is flextime.

Flextime refers to flexible work schedules that allow workers to balance the specific demands of their work and family life (Rothman & Marks, 1989). An innovation that began in the 1970s, flextime allows workers to vary their arrival and departure times each day and the total number of hours worked per week. One variation, the compressed work week, allows workers to work full-time in three workdays. Another version of flextime allows workers to work from their homes through computer hookups. Figure 6-2 shows the proportion of full-time workers who have flexible work schedules.

Although flextime would seem to address some of the conflicts of work and family, it often is not an ideal solution (Negrey, 1990). First, flextime is flexible for both employers and employees. Workers may be asked to work a different schedule every week, which interferes with their ability to plan other activities.

Second, flextime can mean that the number of hours is flexible as well as the schedule (Glueck, 1979; Swart, 1979). If workers on flextime wish to work as many hours as possible they must always be on call in case their employer needs someone to work extra hours. These workers have time away from work but they are really never off the job since they must constantly be awaiting schedule changes.

Third, if the number of hours is flexible, then pay also varies. Workers with varying number of hours may earn less then they need, first because they are working fewer total hours, but also because they earn less per hour. Table 6-2 shows the percentages of women workers and men workers who are part-time. "Part-time workers make up less than one-sixth of the labor force but account for two-thirds of those receiving minimum and sub-minimum wages. In 1988 part-time workers had a median wage of $4.68 per hour, compared with $7.70 for full-time workers" (Conway, 1990, p. 205). In addition, workers who are part-time usually do not receive the same benefits as those who work full-time (Hartmann, 1991). Figure 6-3 shows the discrepancy in health benefits for those who work part-time compared to those who work full-time.

Flextime has been hailed as a solution to work–family conflicts, especially for women. About two-thirds of the part-time and temporary workers in the U.S. are women (Hartmann, 1991). Flextime has in fact been shown to reduce worker tardiness and absenteeism and has increased productivity (Sullivan, 1984).

In a speech presented to Congress, Heidi Hartmann (1991), the director of the Institute for Women's Policy Research, outlined a number of problems with flextime from the point of view of the employee. She concluded that in order to benefit employees, flextime needs to be accompanied by reliable and sufficient wages and benefits and the decision to reduce hours needs to be made voluntar-

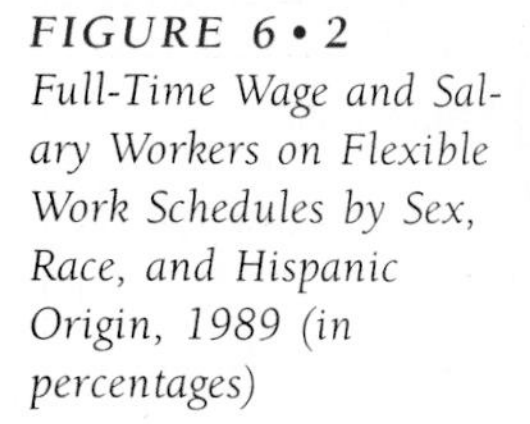

FIGURE 6 • 2
Full-Time Wage and Salary Workers on Flexible Work Schedules by Sex, Race, and Hispanic Origin, 1989 (in percentages)

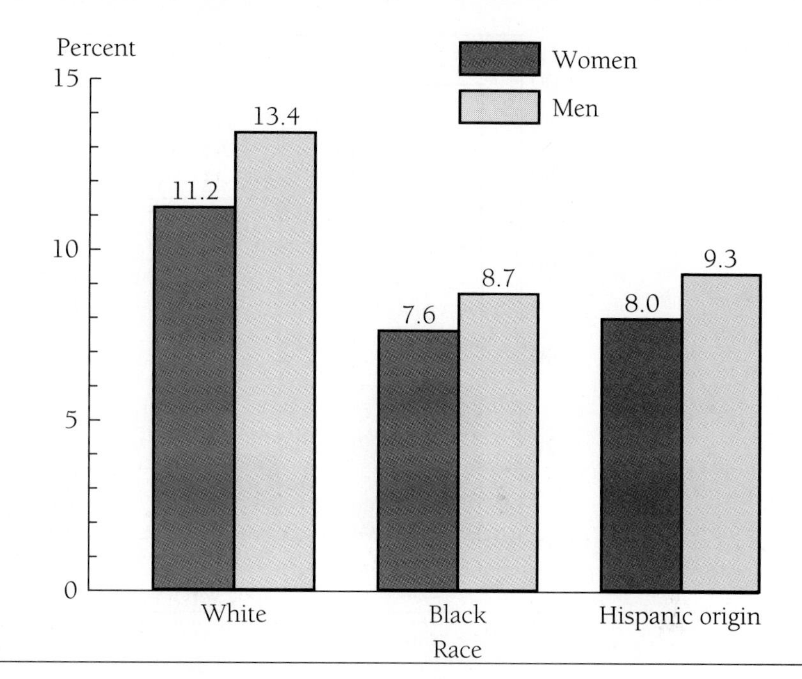

Source: Ries & Stone, 1992, p. 382.

TABLE 6 • 2
Part-Time Workers as a Percentage of All Workers, by Sex and Race, 1990

Part-Time Workers	White	Black
Women	26.3	17.9
Men	9.8	11.1

Source: Ries & Stone, 1992, p. 312.

ily by the employee (Negrey, 1990). A large proportion of part-time workers are not in these positions because they have chosen to be. For example, "in 1988, the labor force included an average of 14 million voluntary and 5 million involuntary part-time workers" (Conway, 1990).

Family Leave

A second approach to alleviating tensions between work and family is to provide more support for employees, such as parental and family leave. In 1985 Pat Schroeder (D—Colorado) and William Clay (D—Missouri) introduced a bill in Congress that would require companies with more than twenty-five employees to provide up to eighteen weeks of unpaid leave for workers who needed the time to care for newborn, newly adopted, or seriously ill children, or dependent adults. The bill required that an employee who took parental leave would not lose pay level or seniority (Stetson, 1991). Table 6-3 shows the proportion of companies that already offer parental leave. Unlike seventy-five other countries, including all other advanced industrial societies, the United States does not mandate job protection for pregnancy and childbirth (Moen, 1989). In many of these other countries employees not only have the right to parental leave, they

FIGURE 6 • 3
Partially or Fully Paid Medical Benefits Offered to Employees, by Full- or Part-Time Status, 1991

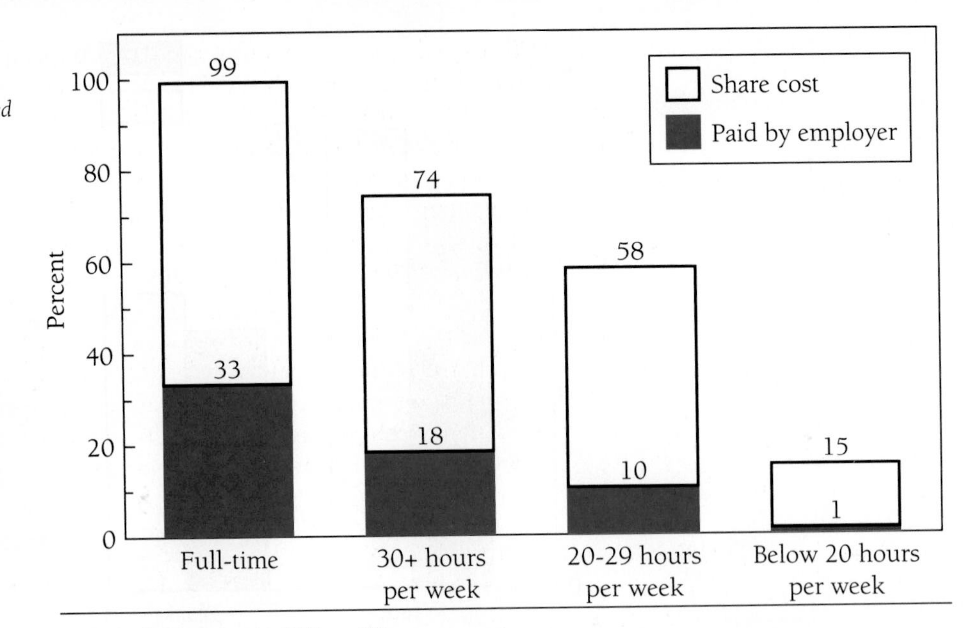

Source: Mishel & Bernstein, 1992, p. 236.

TABLE 6 • 3
Availability of Parental Leave to Full-Time Employees of Medium- and Large-Size Firms by Occupation, 1989 (in percentages)

Occupation	Percent with Paid Leave Available		Percent with Unpaid Leave Available	
	Maternity	Paternity	Maternity	Paternity
Professional and administrative	4.0	2.0	39.0	20.0
Technical and clerical	2.0	1.0	37.0	17.0
Production and service	3.0	1.0	35.0	17.0
All employees	3.0	1.0	37.0	18.0

Source: Ries & Stone, 1992, p. 385.

are guaranteed an income while they are off as well. In Japan, for example, women receive twelve weeks off from their jobs at 60 percent of their regular pay when they give birth (Googins, 1991).

The proposals considered in the United States have not included pay—just the guarantee that their jobs would remain open for them when they returned. Nevertheless conservative business interests like the Chamber of Commerce opposed the bill, claiming that it would be too expensive, especially for smaller companies, and that it would pave the way for mandated paid leaves. They also argued it would lead to too much government interference in business and would jeopardize American competitiveness in the international market (Stetson, 1991).

The bill was altered to apply only to companies employing more than fifty people and to allow up to ten weeks of unpaid leave for family work, including care for young or sick children and older people, and was renamed the Family and Medical Leave Bill. It was finally passed by both houses of Congress in 1990. A number of Republicans lobbied President Bush to support the act, arguing that it would help workers, especially women, cope with the demands of work and family. President Bush, however, vetoed the legislation a few months later (Stetson, 1991). The president said, "Parents not government should make the important decisions about health, child care and education . . . I believe in personal responsibility" (Goodman, 1992, p. 4c). The bill again passed the House and Senate in 1992 and was eventually signed by President Clinton in 1993.

The passage of the bill is significant because it means that people will not lose their jobs if they take time off work to care for family members (Kamerman, 1991). Many workers, however, will be unable to make this choice because they cannot survive without their paycheck. The problem of economic access to family leave will need to be further explored (Bookman, 1991).

THE MICRO-MACRO CONNECTION

In Chapter 1 I described the task of sociology as one in which connections are drawn between the individual micro experience and the broader macro level of social context. In this chapter we have been more specific, focusing our attention on the particular features of the social context, work and family. Furthermore, this chapter has emphasized the dialectical relationship between these two.

Dialectics is a philosophical term that describes the way in which various social phenomena interact with one another. In this chapter we have observed the dialectical relationship between work and family: work influencing and shaping families and families pushing back to affect and change work. Where previous chapters have shown the complexity of the macro level, this chapter has shown how features of the macro level can contradict one another. This chapter reveals contradictions among complex factors. The macro level is not only multifaceted, the "parts" do not necessarily fit easily together.

The relationship between work and family, furthermore, is only one level of interaction. A second dialectical relationship exists between individuals and their social context. Both of these large macro-level social structures—work and family—affect individuals. Individuals are called on to figure out ways to blend the two in their own lives. This balancing act confronts most of us and demands that we create ways to accomplish it.

In the opening scenario, we observed two methods individuals might use to accommodate the demands of work and family. Cindy has decided to put marriage and children on the back burner while she emphasizes work in her life. Jordan believes he, in partnership with his wife, will be able to combine both activities. The women physicians in the discussion of the Mommy Track provide a third alternative. They will deemphasize their work in order to allow themselves an opportunity to marry and raise children.

The last two sections of the chapter suggest another approach. People working for policy reform like flextime and parental leave are demanding that employers change workplace organization to make it easier for employees to manage the balancing act. The deliberations of Congress on flextime and family leave reflect the frustration at the micro level of society where individuals feel they need structural change. And we see individuals organizing at the micro level to restructure the macro social context by influencing the government.

SUMMARY

Myth of Separate Spheres

- The opening scenario shows how work and family intersect in problematic ways for many people in the contemporary United States.
- Industrialization during the nineteenth century created a temporal and spatial separation between work and family for middle-class Americans. Ideology about families also emerged during this time, reinforcing the idea that work and family should be separate and that women and men should each occupy one of these separate spheres.
- The separation was not evident in working-class and African American people's lives but the ideology persisted. As the twentieth century unfolded the idea of separate spheres has become mythical for most middle-class families as well.
- Despite the fact that work and family are not separate spheres, the myth has persisted in our ideas about family and has influenced scholarly work.
- Researchers have tended to explain men's behavior by looking at their working conditions, and women's behavior by looking at their family experience.
- The problem of combining work and family is an important one for young women. Their expectations about how to achieve that combination divide them into three categories: New American Dreamers, Neotraditionalists, and Outsiders.
- More than half of the women in the U.S. are now in the paid labor force and their numbers have been steadily increasing since the turn of the century. The increase is especially remarkable among women with children.
- Three different explanations have been offered for why women have entered the labor force: supply side, demand side, and social structural.
- Women work for both economic and social rewards.
- Both childhood socialization and adult experience influence women in their opinions about working outside of the home.
- Four factors pushed women into a domestic path: a stable marriage, few rewards on the job, adequate household income without their paycheck, and friends who were full-time housewives.
- Chicanas who chose to work in a cannery in California were following a fifth model: work was an extension of their domestic role.
- Zavella's work on Chicana cannery workers is an example of Third Wave feminism, which focuses its attention on differences among women, especially by race and class.

How Does Work Influence Families?

- Work affects family and family affects work.
- Rosabeth Moss Kanter, one of the influential scholars in the study of work and family, divides the effects of work on family into five categories: absorptiveness, time, income, world view, and emotional climate.

- Different jobs demand different amounts of time from employees and employees' families. The more time absorbed, the more influence the job has on the family.
- Time can take three forms: the amount of time at work, the scheduling of the time at work, and the amount of time required over some extended period.
- About one-third of women and men agree that work and family interfere with each other but men and women have different concerns about this interference.
- The employment of mothers has a positive effect on their daughters.
- Mothers' employment can have a negative effect on boys but fathers' involvement is a crucial factor.
- The discussion of work and family shows that more time spent on work diminishes feelings of closeness in the family. This relationship holds for girls who work but not for boys, who report greater closeness to their families when they work.
- Chapter 4 illustrated the importance of the decline of income on families.
- Income can also affect the internal dynamics of families. Wives who earn money have more power in families than those who do not.
- Exchange theory assumes that people attempt to increase rewards and hold down costs when they interact with others.
- Exchange theory has been criticized by feminists who argue that gender creates two different kinds of people—men and women—who come to the negotiating table.
- Exchange theory assumes all people are interested most in serving their own self-interest. This may be more true for men in our culture than for women.
- Research comparing gay men, married and unmarried heterosexual couples, and lesbians shows that exchange theory works to explain the behavior of men with other men but loses its validity when explaining women's behavior.
- Occupations can have cultures and those cultures can influence the families of workers who are employed in them.
- One way people are shaped by their work is the creation of gender identities on the job.
- Jobs that are tense, demand interaction, or are boring can affect the behavior of workers in their families.
- Marxist theory is based on the philosophy of dialectical materialism, which emphasizes the mutual effect work and family have on each other: Work affects family. Family affects work.

The Influence of Families on Work

- Families place workers in the labor market, socialize them to be good workers, and provide support for them.
- Families can influence the politics of work.
- Women workers bring their family identities and behaviors to work in a way that can help support the workers' efforts in their struggle over work conditions.

- One response to the dilemma of balancing work and family is to create ways of coping with the problems. Women physicians provide a good example of this.

Can Families Put Pressure for Change on Employers?

- Those people who do not wish to cope with the contradictions of work and family have pressured for changes in work that will allow them to negotiate the two. Flextime is an example of such a reform.
- Flextime could be advantageous for workers but it is also problematic if the flexibility is controlled by the employer or if reduced hours mean lower wages.
- A second reform that has been demanded is family leave. Family leave would allow parents to leave their jobs for ten weeks without pay in order to care for infants, newly adopted or seriously ill children, or older family members.

The Micro-Macro Connection

- The social context is made up of many social institutions. In this chapter we have examined the dialectical relationship between two of them, work and family.
- Work and family have an important effect on individuals and individuals in turn attempt to respond to and sometimes reform the organization of these two social institutions.

Children contribute to their families by doing housework. The amount they do depends on their gender, their age, and whether they live in a one- or two-parent household.

7 HOUSEWORK

Cassandra is a retired school teacher. Her severe arthritis has made it difficult for her to take care of herself and so she lives with her daughter Delena. Delena has been married to Cedric for eight years and they have three school-age children. Cedric works at the auto plant and when their youngest started first grade Delena went back to work as a nurse at the local hospital. Like most people in the U.S. they cannot get by without two paychecks.

Cassandra noticed that at first, Cedric and the children pitched in to do the housework while Delena was at work. But they weren't very skilled. Bologna sandwiches and canned spaghetti seemed to be the most frequent dinner menu and "clean" laundry took on shades of pinks and greys. The arguments and complaints escalated.

Soon Delena found that she spent every weekend and evening freezing meals to be prepared later, cleaning the house, and doing the laundry and most of the other jobs she had done when she wasn't working full-time.

Cedric works hard and feels he needs some time after work to relax so that he can face going into the plant every day. Especially with all of the cutbacks and concessions, he doesn't want to risk losing his job because of poor performance. Cedric and the children feel neglected because Delena is so busy and Delena feels tired, angry, and overwhelmed. Cassandra worries that she is a burden and she feels guilty about not being able to help out more. She also feels sad that the household does not run more smoothly.

- *Is housework work?*
- *Have you ever described a woman as not working when you meant she was a housewife?*
- *Are Delena and Cedric typical in the way in which they divide housework?*
- *Why do people make the decisions they do about how to divide housework?*
- *How is housework tied to gender and gender politics?*
- *How can Cedric and Delena's dilemma be solved? Will sharing help? What about hiring a maid?*

HOUSEWORK, THE INVISIBLE OCCUPATION

Many activities take place within families. Among the most important of these are economic activities, the production and distribution of the necessities of life. Families pool their efforts to provide themselves with food and shelter and all of the other goods and services they need to survive.

The production of many of these necessities in an industrialized society like ours is done outside of families in manufacturing plants, offices, and service industry sites like restaurants. Most of us go to work and earn wages to purchase these goods and services for ourselves and our families.

A significant amount of the activity necessary to keep ourselves fed and sheltered, however, takes place within households and is unpaid. It is called housework.

Chapter 1 outlined several themes in a recent study of families. One of these themes was the linking of families to other social structures. Housework is a good contemporary example of how *economic* activities take place within families, blurring the boundaries between the social institution of the economy and the social institution of family.

A paid worker puts tomato soup into cans in the Campbell's plant. An unpaid worker opens the can and prepares the soup for his or her family to eat. Both are involved in the economic activity of providing members of society with the necessities of life. At these moments, the lines between economics and family cease to exist.

If we allow ourselves to see housework as work, we will then notice that it is the largest single occupation for American adults. Nearly half of the adult female population in the United States are year-round full-time housewives and most other women and many men are part-time houseworkers (Berheide, Berk, & Berk, 1976; Berk & Berk, 1988; Oakley, 1974).

Most people in our society do not consider housework part of the economic system. In spite of the enormous amount of human energy used to do the work and the necessity of the work to our existence, housework is usually not thought of as work. For example, "working women" are women who are not full-time housewives.

Housework is neglected not only in our thinking about the economic system, it is often neglected in our thinking in general. Housework is an invisible activity. In the scenario at the beginning of the chapter, the unpaid domestic work that Delena did at home was invisible until she entered the paid labor force. Then, when she had less time to do the work, everyone in the household started to see the amount of time housework takes and the amount of skill the tasks involve. Housework at Delena and Cedric's home became more visible because it became more problematic.

The invisibility of housework is reflected in the field of sociology where only recently has it been considered a subject for scholarly research (Oakley, 1974). Ann Oakley, a British writer, was a groundbreaker in the movement to acknowledge that housework was work and that scholars in the field of the sociology of work should, therefore, study it as they would other occupations. In the late 1960s when Oakley was studying at the University of London, she tried to register her thesis entitled "Work Attitudes and Work Satisfaction of Housewives" and was met with "frank disbelief or patronizing jocularity" (Oakley, 1981a). She was asked how she could seriously study such a boring subject. But Oakley persisted and since her "discovery" of housework a number of other scholars have investigated it.

This chapter is divided into three major sections. The first is a description of the division of housework in households. It includes information on the question of who does housework and how the division of labor in families has or has not changed in the past few decades. The second section looks at three different theories that attempt to explain what causes housework to be divided as it is. The third section of the chapter addresses the problem of developing ways in which to make the division of housework more equitable.

What Is It Like to Be a Housewife?

Oakley wanted to discover what the job of housewife was like. The women Oakley interviewed described their job as difficult and told her that there were three problem areas in their work. First, regarding the work itself, housewives said that the tasks they performed were monotonous, neverending, and lonely. One woman explained to Oakley:

> I always say housework's harder work, but my husband doesn't say that at all. I think he's wrong because I'm going all the time—when his job is finished, it's finished. . . . Sunday he can lie in bed till twelve, get up, get dressed and go for a drink, but my job never changes. (1974, p. 45)

Second, the housewives complained that the long hours of housework severely limit what women are able to do and what they are able to become. Organizing unions, running for political office, obtaining an education, and being involved in the arts are all activities, for example, that might be hindered by having to do housework.

The third set of problems the housewives mentioned were related to the fact that housework is unpaid. We have a money-based economy, and therefore the value of work is measured by how much it pays. Since housework is unpaid the work is devalued; furthermore, the workers who perform the unpaid housework are devalued and have low social status. Since wages are essential to obtaining the necessities of life, the housewife is entirely dependent on others for her survival because she earns no pay. Her social value is further diminished because it appears that she is unproductive and living off others.

One place where the ideology that those who do not earn money, regardless of the work they perform, are unworthy shows up is in opinions about welfare mothers. Welfare mothers may work full-time as housewives but because that work doesn't count as real work, proposals are constantly being made to "put welfare mothers to work."

Oakley's research on housework and houseworkers is an example of feminist ethnographic research (Reinharz, 1992). Ethnography is a multimethod approach to social research that involves interviewing and observing people, participating in their activities, and examining artifacts in the social context of the people being researched. Ethnography, as I mentioned in a previous chapter, is a methodology that is not exclusive to feminist researchers. Many feminists including Oakley, however, have argued that ethnography has great potential for getting at the reality of women's lives (Reinharz, 1992).

In Oakley's (1974) investigation she uncovered an important facet of the process of ethnographic research. When sociologists conduct research they often do not reflect on or report the relationship between researcher and researched. Oakley, however, noticed in the transcripts of her interviews that the "subjects" asked her questions during the interview and that long-term friendships sometimes developed between the interviewees and herself. She argues that this is at least in part a result of the fact that both she and the "subjects" were women.

Oakley (1981b) concluded that (1) researchers should pay attention to the way in which social research is a social interaction, a two-way street; (2) relationships of camaraderie are especially likely when both interviewers and interviewees are women; and (3) these special relationships produce the most valid kind of data.

Judith Stacey (1988) has questioned this last assertion by Oakley. Stacey says that while it is true that women interviewing women may elicit responses that are more open and self-disclosing, this may be problematic. The methodology provides good data for the researcher but it may also be more exploitative of the "subjects" since the women being interviewed let down their guard in ways they may not really wish to do to further the ends of a sociological study.

Who Does the Housework?

The answer to this question is women. Table 7-1 shows what Shelley Coverman (1989) found when she summarized six famous studies on the division of housework in families in the United States conducted over the period from the late 1960s to the early 1980s. On the left-hand side of the table the studies are listed in chronological order. For each study she has listed the number of hours per

TABLE 7 • 1
Hours per Week Spent in Domestic Labor (Housework and Childcare)

Study	Year Data Collected	Employed Wives	Unemployed Wives	All Wives	Husbands
Walker & Woods	1967–68	33.6	56.7		11.2
Sanik	1977			51.8	11.9
Meissner, et al.	1971	16.1	32.2	33.6	4.2
Geerken & Gove	1974–75	28.6	51.9		11.2
PSID (Davis)	1976	29.3	36.8	32.6	5.0
STU (Hill)	1975–76	26.0	51.0	33.3	11.3
STU (Hill)	1981	22.8		29.3	13.4

Source: Coverman, 1989.

week of housework done by four categories of people: employed wives, nonemployed wives, all wives (includes both employed and nonemployed), and husbands.

These studies show much variation in the number of hours of housework done in different kinds of families by different family members. A general picture, however, emerges from examining these studies that led Coverman to these six conclusions:

1. Wives spend from two to four times as much time in domestic labor as do husbands, even when they are employed.
2. Wives perform about three to four times as many tasks as husbands do.
3. Wives perform about 75 percent of all domestic tasks.
4. Employed wives spend substantially less time in domestic work than do nonemployed wives.
5. Full-time housewives spend over fifty hours per week in housework and childcare.
6. Employed wives average about 26 to 33 hours per week doing housework.

Heidi Hartmann (1981) adds to these observations the concept of "husband care." Hartmann compared the number of hours of housework done by single women with children, to married women with children. The women in these two groups had the same total number of people in their households but the married women did more hours of housework. Hartmann argues that this means that husbands generate more laundry, dishes, and dirt than they clean up. She used the term husband care to refer to the difference between the housework husbands did, and the housework they generated.

Recent Changes in the Division of Housework Has inequality in the division of labor become less intense in the past decade? Some authors have found that there has been some change in recent years toward increased equality in the division of housework between husbands and wives, but change is slow. Joseph Pleck (1983), for example found that men were doing a larger proportion of total housework in 1975 compared to 1965. He points out, however, that the change was mostly due to a decline in the number of hours of housework women were doing. As we observed above, employed women do fewer hours of housework than full-time housewives. As women entered the labor force in the 1960s and 1970s they were doing fewer hours of housework. Although the amount done by husbands remained about the same, the proportion that men did increased. This alteration in the proportion created the appearance that men were contributing more.

Hartmann (1981) cautions that even the small changes toward greater equality that Pleck cites may not be valid. Hartmann argues that Pleck's findings of a trend toward greater equality are based on survey questionnaires in which respondents were asked to estimate the time they spent doing housework. Hartmann maintains that when people are asked to give self-reports they tend to overestimate time spent on housework. When people are actually observed performing tasks, the amount of work done is less than the amounts estimated in the self-reports. Berk and Berk (1979) found, in addition, that wives overestimate the amount of time their husbands spend doing housework.

More recently, Robinson (1988) has presented research that shows that since the mid-1970s men have increased not only their proportion of housework but the actual number of hours doing it. Table 7-2 shows these changes. In this table, inside work refers to jobs we usually call housework: cooking, meal cleanup, housecleaning, and laundry. Outside work refers to other household tasks like repairs, gardening, paying bills, and outdoor work. The table shows that the amount of housework done by women has declined and the amount done by men has increased. In spite of these changes toward greater equality, men still are doing only about 30 percent of the total hours of unpaid domestic work. In addition, the kinds of work done by women and men is different. Women do more "indoor work" and men do more "outdoor work." The different kinds of jobs performed by women and men will be discussed later in the chapter.

Arlie Hochschild (1989) is another researcher who examined these questions in the late 1980s. Her work is different from that reported in these tables because it was qualitative research. She interviewed and observed a relatively small number of couples as they negotiated housework in their households. Hochschild did not have data on change over time. She therefore could not say whether there had been improvement or not. She did find, however, that while there are now some households where women and men share equally, overall women still do the majority of housework. In her study, 20 percent of the men she interviewed shared housework equally, 70 percent did between one-third and one-half of the housework, and 10 percent did less than one-third of the housework. Hochschild concludes that just as women and men continue to have a wage gap in paid work they continue to have a leisure gap in unpaid work. This gap amounts to one extra month of work for women every year (Hochschild, 1989).

The scenario describing the unequal division of labor in Cedric and Delena's household captures the experience for most households. Even though this is the 1990s, the couple are a two-paycheck family, and both the husband and wife see the division as unequal and problematic, Delena is still doing much more housework than Cedric is.

Cedric and Delena are a working-class couple. Does their social class exacerbate the inequality in their household? Research by Ellen Rosen (1987, 1991) suggests that blue-collar families have some of the same kinds of tensions around housework that white-collar families like those Hochschild (1989) interviewed have. But a number of the women factory workers Rosen interviewed

TABLE 7 • 2
Comparing Men and Women and the Hours They Spend Doing Housework Inside and Outside of the House, 1965 and 1985

Hours Spent per Week	1965		1985	
	Women	Men	Women	Men
On indoor work	24.3	2.1	16.1	4.1
On outdoor work	2.7	2.5	3.4	5.9
Total	27.0	4.6	19.5	9.8

Source: Robinson, 1988, p. 26.

said that their husbands took primary responsibility for the housework or at least shared it with their wives. The men were especially likely to take over childcare when the women were at work.

Eric Olin Wright was interested in how social class might affect the division of labor in families and decided to do a comparison. He and his colleagues (Wright, et al., 1992) compared the division of unpaid household labor in couples in the U.S. and Sweden. In their research they divided people into four social classes: small employers, petty bourgeoisie, middle class, and working class. They concluded that social class had little systematic effect on the division of labor in households. They explained,

> There is a common stereotype in the popular media that working-class men do proportionately less housework than middle-class men. The macho worker whose identity is threatened by housecleaning and infant child care is contrasted to the egalitarian and enlightened yuppie who cooks elegant meals and pushes a stroller in the park . . . our data lend no support for such images. (Wright, et al., 1992)

Children's Work in Families Although most children do chores around the house, there has been little research on the subject. In the 1970s, Hedges and Barnett (1972) and Thrall (1978) found that when women worked outside of the home, their children were more likely than their husbands to take up the chores previously done by the women. A more recent study by Lynn White and David Brinkerhoff (1987) shows that children begin doing chores in most households at a very young age and by the time children are ten years old, over 90 percent of them are responsible for at least some housework. At all ages, however, the number of hours most children put into housework is small.

Table 7-3 shows the median number of hours of housework for different age groups and different sexes of children. For all children the median number of hours is 4.0 per week, but there is a range from 1.7 hours for girls under age 4, to 6.1 hours for older girls between the ages of 15 and 17.

White and Brinkerhoff examined the reasons parents gave for why they asked their children to do housework. The most common reason (given by 72 percent of the parents) was that doing chores was an important part of helping children to build character and to learn responsibility. For example, one parent replied that asking the children to do work "gives them a sense of responsibility. Makes them appreciate what they have. I think it helps them grow into responsible adults" (White & Brinkerhoff, 1987, p. 209). Twenty-five percent of the parents said they had their children do housework because it was the child's duty to help out, 23 percent said it was because the parents needed the help,

TABLE 7 • 3
Weekly Median Hours on Chores by Age and Sex

Age Group	Boys	Girls
Total	3.4	4.6
0–4	2.1	1.7
5–9	2.3	4.0
10–14	4.1	4.7
15–17	4.2	6.1

Source: White and Brinkerhoff, 1987.

and 12 percent said that children needed to learn the skills in order to be able to take care of themselves as adults.

In Chapter 2 we examined the historical change of the conceptualization of the value of children from useful to priceless. Zelizer (1985) traced this history from the turn of the century to more recent times. She argues that early in the century children were valued for the contribution they made to the household as workers. Now children are valued for the love and affection they provide and are not thought of as valuable workers. The large number of parents in White and Brinkerhoff's (1987) study who explain that they ask their children to do chores because it helps children develop character and responsibility is an example of contemporary ideologies about children. Housework is not done by children for the benefit of adults as it might have been earlier in the century. Postmodern parents emphasize the value that the activity of doing housework has for the child.

One of the factors that is evident in these data is the difference in the contribution made by boys compared to girls. After the age of five, girls do more than their brothers, on the average, and the difference is especially evident in the comparison between older teens. Older boys continue to do about the same amount of work as younger children, but older girls substantially increase their contribution. This difference is another illustration of gender inequality in which women and girls take more responsibility for housework than men and boys.

This gender difference between teenage boys and girls has an important implication in its possible relationship to the division of housework in future families. If boys are not being encouraged to learn how to perform tasks and are not required to notice and take responsibility for housework, it is less likely that as men they will increase their involvement in housework. This aspect of gender socialization, learning to do housework, may play a critical role in determining whether future families will divide work more equally (Goldscheiter & Waite, 1991). And it appears that in many homes the status quo is being perpetuated.

The data in Table 7-3 indicate differences only by age and gender, but families differ in other ways such as race and whether they have one or two parents. How do these variables affect the amount of housework children do? Goldscheiter and Waite (1991) found that in two-parent families African American children, compared to white children, were responsible for significantly more (21 percent more) housework. African American children were more likely than white children to perform more hours of both less skilled work like washing dishes and more skilled jobs like grocery shopping and baby sitting.

In one-parent households the amount of work performed by children was similar for African Americans and whites. Children in one-parent families, re-

FIGURE 7 • 1
Division of Housework Among Parents and Children

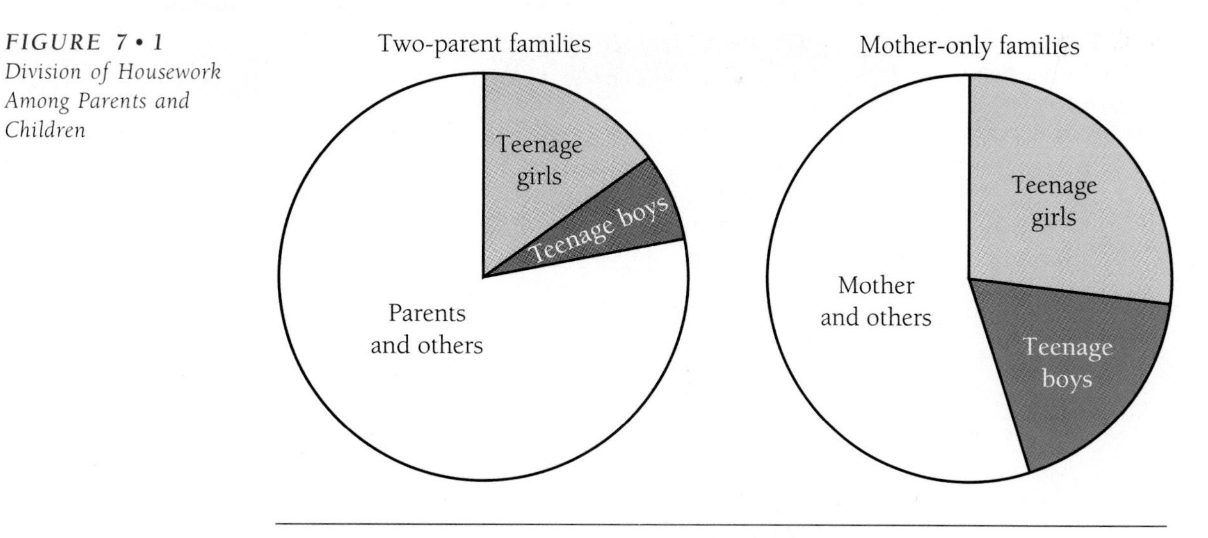

Source: Goldscheiter & Waite, 1991, p. 152.

gardless of race, were likely to perform a much larger share of the housework than children in two-parent households. In Figure 7-1 the contrast between one- and two-parent families is shown. Both boys and girls do a larger share of the housework in mother-only households. Although girls in one-parent families still do more than the boys do, the gap between boys in one-parent families and boys in two-parent families is larger than it is for girls (Goldscheiter & Waite, 1991).

How Are the Tasks Divided?

Table 7-4 shows that within the general category of "housework" men and women do different tasks. There are jobs that are "women's work" and there are jobs that are "men's work." Women are more likely to do cooking, washing dishes, indoor cleanup, laundry, shopping, and childcare. Men are more likely to do repairs and maintenance, gardening, and pet care.

In her interviews, Hochschild (1989) noticed that "men's jobs" and "women's jobs" could be identified by three characteristics. First, women do most of the daily jobs and most of the jobs that are time-bound. For example, women cook dinner, while men change the oil in the car. Dinner must be cooked every day at a particular time. Changing the oil must be done only a few times a year and it may be done any day at any time that is convenient.

Second, women more often do two or more things at once. They make dinner and do the laundry, while men do either the grocery shopping or they mow the lawn.

Third, women are the "time and motion managers," hustling the children through their baths and homework, getting everyone out the door to school and work in the morning. This third characteristic of a woman's housework is especially costly to mothers' relationships with their children. Mothers become "villains" in the process of doing housework and childcare because they are always rushing the children through maintanance tasks like eating, bathing, getting ready for school, or doing homework, when the children may prefer, or even need, to have someone stop and pay attention to them (Hochschild, 1989).

TABLE 7 • 4
Weighted Mean Hours per Week Spent in Domestic Activities by Sex: Study of Time Use, 1975–1976

		Men (n = 410)	Women (n = 561)
Household activities			
meal preparation		1.57	7.25
meal cleanup		.33	2.30
indoor cleanup		.85	5.03
outdoor cleanup		1.59	.56
laundry		.13	2.44
repairs/maintenance		2.14	.68
gardening/pet care		.94	1.00
other household		.92	.72
	Subtotal	8.47	19.98
Childcare			
baby care		.24	.90
childcare		.24	.99
helping/teaching		.07	.15
reading/talking		.07	.30
indoor playing		.13	.18
outdoor play		.06	.12
medical care		.01	.09
babysitting/other		.14	.64
travel—childcare		.23	.50
	Subtotal	1.19	3.87
Obtaining goods and services			
everyday shopping		1.45	2.78
durable/house shopping		.19	.08
personal care services		.06	.35
medical appointments		.15	.37
government/financial services		.15	.19
repair services		.11	.17
other services		.11	.13
errands		.04	.06
travel—goods/services		1.60	2.14
	Subtotal	3.86	6.27
	Total	13.52	30.12

Source: Coverman, 1989, citing Hill, 1985.

Feeding the Family

Much of the discussion thus far on housework has been based on the conceptualization of housework presented by Oakley (1974). Within this framework housework is seen as an economic activity that is comparable in many ways to other occupations. Others (Devault, 1991) maintain that conceptualizing housework as a job is only partially correct and that housework includes other social facets as well (Berheide, Berk, & Berk, 1976).

Marjorie Devault interviewed thirty women and three men about the unpaid work they did to feed their families. She found that (1) the intimacy of the work created requirements that would not have been present if feeding were just a job and not an aspect of familial relationships; (2) the work done by people (most

Feeding the family is more than a mechanism to provide physical sustenance. Determining what to eat and preparing and serving it create a nurturant ritual and a way to create family traditions and to maintain ethnic traditions, as in holiday meals.

often women, as we have observed in the tables above) who purchase, prepare, and serve food to their families is not only work from the point of view of the women, it is an act of love; and (3) feeding families not only nourishes family members, it is a process that helps to socially construct families. Meals are social events that involve patterns of interaction and bring family members together to interact.

Devault argues that although feeding goes on day after day it is not a mechanical task. Especially because of its connection to the personal relationships of family members, it involves coordination and interpersonal work. Before food is prepared it must first be determined what family members wish to eat. Restaurant owners might advertise their food or do a marketing survey to determine customers' preferences, but housewives must interact with family members on an intimate level in order to decide what to have for dinner. Housewives must be aware of individual preferences, often with few direct responses from family members. Devault observed in her interviews, "Many women complained that they could not just ask their husbands what they wanted: 'I'll call the office and ask, 'Is there anything special you'd like for dinner?' and his standard answer is 'Yes, something good.' " (Devault, 1987, p. 183).

Housewives also must coordinate tastes, unlike restaurant owners who can let customers order from a menu. And housewives have a special concern for the health of their families and must consider that in planning meals.

The activity of feeding families is also different for housewives compared to restaurant owners because wives and mothers are concerned with the social interaction that takes place at meals. Many families use mealtime as an opportunity to talk with each other and they also use it as a place to monitor and control children's behavior and model adult behavior. Devault concludes that eating and the work involved in feeding families is "at the center of family life and sociability" (Devault, 1987, p. 190).

The second observation made by Devault is that a link between feeding and love is often noted. One woman explained to her how important the connection between feeding and love is and how feeding involves both labor and love. She said: "In preparing food, you know, there's a lot of work that goes into preparing food. Therefore, for one to commit himself to the work, that's love, that's shared with those people that the food was being prepared for. I think love has a lot to do with it" (Devault, 1987, p. 180).

The third major finding of Devault's work is that in addition to being an act of love, feeding activities also represented social processes that created families. Devault discovered, for example, that working-class women preserved their family and ethnic heritage in the foods they prepared. They also used feeding as a means by which to bring together members of an extended family, for example, at holiday meals.

Devault's work fits into the broad theoretical framework called ethnomethodology. Ethnomethodologists focus their attention on the methods people use to create a shared perception of social order (Chafetz, 1988). "Ethnomethodologists assume that people work constantly and hard at recreating the social order in order to achieve a sense that the world is orderly and predictable. They seek to understand the everyday, take-for-granted implicit rules people use in interacting with one another that allow them to create such a sense" (Chafetz, 1988, p. 20).

The observations Devault (1991) makes of the everyday interactions and activities that take place in feeding a family reveal the way in which women, and sometimes men and children, work to create their vision of a family. Implied in their feeding work are many beliefs about families. For example, the descriptions above reveal beliefs that families are connected by love, that parents should model the behaviors they wish their children to adopt, and that housewives should nurture their families physically and socially.

Ethnomethodologists have emphasized the importance of verbal interaction as a critical process in the social construction of everyday life. Devault's book differs somewhat in that it explores a less well-researched activity, the importance of the work people do in that process. Devault's ethnomethodology examines not only what people say to each other but what they do to and with each other in their everyday interactions.

The Difference Between Full-Time Housewives and Part-Time Housewives

Look at Table 7-1 again and compare the time spent on housework among women who work outside the home and women who are full-time housewives. Why are there such large differences? There are two possible answers to this question. Either full-time housewives do a lot of *unnecessary* work or the quality of life of people in families where women are employed outside the home is diminished because a lot of *necessary* work is not being done.

In Chapter 1, I referred to a book written in the 1960s by Betty Friedan, *The Feminine Mystique* (1963). This book played an important role in revealing the problematic character of the idealized family of the 1950s in which men were supposed to be the primary breadwinners and women were to stay home to care for their husbands and children. Friedan argued that these "ideal" families were more like prisons and she disclosed her discovery of "the problem without a name." The problem she wrote about was the isolation, fatigue, and depression of middle-class housewives whose lives were filled with boring, thankless tasks that were easily undone and never seemed to be finished. Housework filled every moment available. Magazines and television ads told women their dishes should sparkle, their floors should gleam, their laundry should smell like fresh air, and their dinners should be carefully prepared and served to entice every sense. Many women suspected that much of this work was unnecessary but they were trapped with a set of goals and standards that required many hours to achieve.

The difference in the hours spent on housework by women employed outside the home and women who are full-time housewives may be a result of unnecessary housework (added to necessary housework) filling the time available. It is unnecessary, for example, to have floors so shiny "you can see yourself in them." But these kinds of expectations are incessantly promoted by the media. Housewives and others who evaluate their work may come to believe the ads and even the most outlandish expectations can become requirements of the job.

The results of the unstructured, overburdening expectations of the job have been observed to have significant problematic implications for the women who are trying to fulfill the job description of housewife. Gove (1972) and Bernard (1972) argue that the job has created psychological problems for housewives. The housewife role has been cited as a cause of the higher rates of psychological distress among married women compared to married men or single women. One explanation of the gap in housework hours between full-time housewives and women who are employed in the labor market is that at least some of this work is unnecessary but full-time housewives often attempt to accomplish it because of ideals promoted by the media.

The other explanation for the difference in the number of hours of housework done by full-time housewives compared to women in the paid labor force is that the work (at least most of it) done by the full-time housewife is necessary and that the families of women who cannot put in those hours do not have as high a standard of living. Employed women in some households may be able to purchase some products and services, allowing them to replace the labor of a full-time housewife (Berheide, 1984). But in many cases the services may be too expensive or inaccessible and families without a full-time housewife go without. Families without a full-time housewife, for example, may have dirtier houses, less nutritious or diverse foods, less child supervision, or less entertainment, like birthday parties.

Although these kinds of deficits in the quality of life may be felt by all family members, living in a house without a full-time housewife may be especially bad for women in those families. Research shows that working mothers have higher self-esteem and less depression than housewives but they also have more anxiety (Thoits, 1985). For example, Hochschild (1989) found that many of the women she interviewed felt their quality of life was greatly diminished by their

BOX 7 • 1 THE POLITICS OF HOUSEWORK

The idea that relationships between husbands and wives are political and that housework is an arena of political struggle is one that was promoted by the women's movement in the late 1960s. An often-heard slogan of the early feminist movement was "The Personal Is Political."

In 1970, Pat Mainardi wrote a dialogue illustrating the battle of the sexes over housework and the way in which the personal is political. The essay has become a classic in feminist literature. In the following dialogue, Mainardi uses the "He says" line to illustrate the personal kind of comment we might hear from a husband. She then uses the "meaning" lines to explain how those statements have political importance. That is, they express large and pervasive mechanisms for keeping women (and men) in their (politically unequal) places in our society.

1. *He says*: I don't mind sharing the housework, but I don't do it very well. We should each do the things we're best at.

Meaning: Unfortunately, I'm no good at things like washing dishes or cooking. What I do best is a little light carpentry, changing light bulbs, moving furniture. (How often do you move furniture?)

Also meaning: Historically the lower classes (black men and women) have had hundreds of years' of experience doing menial jobs. It would be a waste of manpower to train someone else to do them now.

Also meaning: I don't like the dull stupid boring jobs so you should do them.

2. *He says*: I don't mind sharing the work, but you'll have to show me how to do it.

Meaning: I ask a lot of questions, and you'll have to show me everything everytime I do it because I don't remember so good. Also, don't try to sit down and read while I'm doing my jobs because I'm going to annoy the hell out of you until it's easier to do them yourself.

3. *He says*: We used to be so happy. (Said whenever it was his turn to do something.)

Meaning: I used to be so happy.

Also meaning: Life without housework is bliss. (No quarrel here.)

4. *He says*: We have different standards, and why should I have to work to your standards? That's unfair.

Meaning: If I begin to get bugged by the dirt and crap I will say "This place sure is a sty" or "How can anyone live like this?" and wait for your reaction. I know that all women have a sore called "guilt over a messy house" or "household work is ultimately my responsibility." I know that men have caused that sore—if anyone visits and the place is a sty, they're not going to leave and say, "He's sure a lousy housekeeper." You'll take the rap in any case. I can outwait you.

Also meaning: I can provoke innumerable scenes over the housework issue. Eventually do-

attempts to maintain the house *and* work outside the home. The women spoke of being tired, sick, and emotionally drained:

> Many women I could not tear away from the topic of sleep. They talked about how much they could 'get by on' . . . six and a half, seven, seven and a half, less, more. They talked about who they knew who needed more or less. Some apologized for how much sleep they need—'I'm afraid I need eight hours of sleep'—as if eight was 'too much.' . . . These women talked about sleep the way a hungry person talks about food. (Hochschild, 1989, p. 9)

Hidden behind the numbers that show different amounts of housework done by full-time housewives and employed women are two different possibilities: too much unnecessary housework is done in houses with full-time housewives and too little necessary housework is done in houses with employed women. Behind the statistics, however, in both family types there are similar

BOX 7 • 1 Continued

ing all the housework yourself will be less painful to you than trying to get me to do half. Or I'll suggest we get a maid. She will do my share of the work. You will do yours. It's women's work.

5. *He says*: I've got nothing against sharing the housework, but you can't make me do it on your schedule.

Meaning: Passive resistance. I'll do it when I damn well please, if at all. If my job is doing dishes, it's easier to do them once a week. If taking out laundry, once a month. If washing floors, once a year. If you don't like it, do it yourself oftener, and then I won't do it at all.

6. *He says*: I hate it more than you do. You don't mind it so much.

Meaning: Housework is garbage work. It's the worst crap I've ever done. It's degrading and humiliating for someone of my intelligence to do it. But for someone of your intelligence . . .

7. *He says*: Housework is too trivial to even talk about.

Meaning: It's even more trivial to do. Housework is beneath my status. My purpose in life is to deal with matters of significance. Yours is to deal with matters of insignificance. You should do the housework.

8. *He says*: This problem of housework is not a man–woman problem. In any relationship between two people one is going to have a stronger personality and dominate.

Meaning: That stronger personality had better be me.

9. *He says*: In animal societies, wolves, for example, the top animal is usually a male even where he is not chosen for brute strength but on the basis of cunning and intelligence. Isn't that interesting?

Meaning: I have historical, anthropological, psychological, and biological justification for keeping you down. How can you ask the top wolf to be equal?

10. *He says*: Women's liberation is really a political movement.

Meaning: The Revolution is coming too close to home.

Also meaning: I am only interested in how I am oppressed, not how I oppress others. Therefore, the war, the draft, and the university are political. Women's liberation is not.

11. *He says*: Man's accomplishments have always depended on getting help from other people, mostly women. What great man would have accomplished what he did if he had to do his own housework?

Meaning: Oppression is built into the System and I, as the white American male, receive the benefits of this System. I don't want to give them up.

Source: Pat Mainardi, 1970, 516–518.

problems of women feeling overworked and overwhelmed. Women who are full-time housewives are troubled by the invisibility of their work, its lack of structure, and the enormity of the task. Women who are employed are troubled by their inability to do it all and by the lack of support they obtain from other members of the household.

High Tech and Housework

The most striking feature of the data presented above is the inequality between women and men in the division of housework. But equally impressive is the total number of hours that men and women engage in housework. In our world of fast food, microwaves, and wash-and-wear fabrics, one might think that housework would disappear or at least decrease. Actually, since the the turn of the century, the time spent doing housework has changed little, although the tasks that are done have changed dramatically.

In spite of time saving inventions like microwave ovens and automatic washers and dryers, housework still takes about the same time it did 50 years ago. Part of the reason is that some services that were provided—like the home delivery of milk—are no longer available.

Ruth Cowan (1983) compared housework done in the early part of this century to housework done in the 1970s. She found that housework is really made up of several sets of tasks. Although some tasks have become more efficient and less time consuming, others have become more time consuming. She explains:

> Twentieth-century household technology consists of not one, but of eight interlocking technological systems: the systems that supply us with food, clothing, health care, transportation, water, gas, electricity, and petroleum products. . . . The household transportation system has developed in a pattern that is precisely the opposite of food, clothing and health-care systems: households have moved from the net consumption to the net production of transportation services—and housewives were moved from being the receivers of goods to being the transporters of them. (Cowan, 1989, p. 58)

Cowan (1983) argues that people are more likely now to purchase food, clothing, and health care than to produce these things in their homes. But in order to use these items we have to bring them to us or us to them. Icemen, breadmen, and milkmen no longer deliver; physicians no longer make house calls and gas stations no longer pump gas. Transportation has become the most significant work of the American housewife.

There have also been changes in the systems of water, gas, electricity, and petroleum products. We no longer have to haul water and build fires and we have running water and electric vacuums to help clean. But we also have much higher standards of cleanliness, larger houses, and bigger wardrobes. Cowan explains, "Some of the work was made easier, but its volume increased: sheets and underwear were changed more frequently, so there was more laundry to be done; diets became more varied, so cooking was more complex; houses grew larger, so there were more surfaces to be cleaned" (Cowan, 1989, p. 66).

The end result is that people are doing only slightly fewer hours of total housework than they did fifty years ago. Despite significant changes in technology and the social organization of tasks associated with domestic work the number of hours devoted to housework remains substantial.

A Very Recent Addition to the Tasks of the Housewife Cowan (1983) did her research in the late 1970s and early 1980s. She found that between 1920 and 1980 one activity that diminished within households was the provision of health care. Earlier in the century, families were likely to take care of sick people in their homes. But as the years passed, these activities increasingly were done by paid professionals in hospitals and nursing homes.

Nona Glazer (1990), working in the late 1980s, however, discovered that this trend had been reversed by new federal laws. In 1983, Congress passed the Social Security Amendments ruling that whenever possible, Medicaid and Medicare patients should use outpatient clinics instead of hospitals and that reimbursement for in-home care should be sharply restricted. As a result of these changes, patients are sent home sicker than when they were allowed to stay longer at the hospital to recuperate. Most of these still ill patients are discharged to homes where family members do for free the work once done by paid health service workers. Currently 80 percent of long-term care for dependent elderly people is provided by relatives (Abel, 1986). Furthermore, nearly all of the work that is done for ill or dependent elderly people in households is done by women.

This shift is not an economically small one. "The [insurance] industry estimates a $10 billion savings in wages because of unpaid family work. Of course, the 'family and friends' [doing the work] are mostly women" (Glazer, 1990, p. 488).

Research by British scholars reveals that the ratio of work done by women compared to men in homes that include an elderly dependent person is 19:1 (Abel, 1986). Abel (1986) argues that women take on this responsibility for a number of reasons. First, external sources push them into caring for their relatives. These include a dominant ideology that says women are "natural caregivers" and lower pay for women so that their economic contribution from their paid work is more disposable. In addition, Abel (1986) maintains that women caregivers themselves believe that the sacrifices they make in caring for their parents are necessary and honorable (see also Aronson, 1992).

The tasks themselves are not easy. Glazer (1990) discovered that in many sick people's homes, nonprofessional women use high-tech equipment to deliver treatments for acute and chronic conditions and to treat systemic infection and cancer. They supervise exercise, give mechanical relief to patients with breathing disorders, feed by tubes those unable to take food orally or digest normally, give intramuscular injections and more tricky intravenous injections, and monitor patients after antibiotic and chemotherapy treatments.

One of the themes presented in Chapter 1 was the attention to the interlock between families and other social institutions. Here we see the "boundaries" of families being blurred between the health care system and families of sick people. Identical activities take place in health care facilities and homes. Professional health care providers and housewives participate in the same kinds of work, but in different settings.

The boundaries between family and government are also blurred. Changes in home health care are an example of the government making decisions that directly alter the job definition of homemakers in private homes. Glazer's work in particular shows how Congress can pass a law that "requires" that family members/women provide health care.

Glazer's research, like Cowan's, also shows the way in which the occupation of housework continues to change and that those changes are not necessarily in the direction of a decline in the number of hours necessary to complete it. Glazer's work differs from Cowan's because of the specific time period in which she gathered her data.

WHY IS HOUSEWORK DIVIDED UNEQUALLY?

Thus far this chapter has described the way in which housework is divided in households. We can conclude that while there is historical change and variation from one household to another, housework appears to be a powerful example of gender inequality. In this section we ask the question, "Why is housework divided unequally?"

This section is divided into three subsections. The first describes how a socialization theorist might explain the unequal division of labor in families. The second presents the work of Gary Becker, who would answer this question with the response, "For the most part housework is divided in households on the basis of rational decisions by the members as a working team." The third section presents the work of Sarah Berk and Heidi Hartmann. They would respond to the question by saying that the division of housework is both a reflection and an expression of gender inequality. They would contend that the division of unpaid labor in households is a political question: "He who has the power does not do the dishes."

Socializing Housewives

Socialization is the process by which children learn how to understand people in their society and what is expected of them. Many kinds of socialization take place, and gender socialization is a major type. Boys and girls learn how to be boys and girls as well as what will be expected of them when they become adult men and women. In Chapter 13, we will look in depth at the process of gender socialization when we examine the experience of children in families. In this section we will briefly review some socialization issues that might be related to the question of the division of unpaid work in families.

Spencer Cahill (1983) cites three sources of gender socialization: adult models; cultural artifacts like gender models in books, television, and toys; and peers. According to Cahill, childhood is a time of dress rehearsal in which children act out and refine their notions of gender and the skills necessary for playing the part of girl, boy, woman, or man.

Each of the three sources of gender socialization cited by Cahill (1983) is likely to play a role in preparing girls to be housewives while preparing boys to avoid housework. The first source is adult models. This chapter has shown overwhelming evidence that children are much more likely to see women performing housework. In addition, the section on children and housework indicated that parents differentially assign housework to their sons and daughters.

Second, Cahill suggests that cultural artifacts encourage boys and girls to learn or not learn the skills associated with housework. In a study of the bedrooms of middle-class children Rheingold and Cook (1975) found significant differences in the kinds of toys girls and boys had. Girls' bedrooms had dolls and toy appliances. Boys' bedrooms had athletic equipment, military toys, and building and vehicular equipment. Advertisers and toy stores almost always divide their stock into girl toys and boy toys. In addition, "girl toys" are largely miniature tools of housework: toy stoves, dishes, doll houses and furniture, cooking sets, fake food, ironing boards, vacuum cleaners, and all of the equipment necessary for baby care (Shapiro, 1990; Schwartz & Markham, 1985). Not only do these toys teach girls the mechanical skills associated with housework, they teach the value of nurturance, a major component in the ideology of housework (Miller, 1987). When we reviewed the activity of feeding the family in this chapter we noted the importance that women gave to the way in which buying, preparing, and serving food was a way of showing their love (Devault, 1987).

Finally, Cahill suggests that peer interaction forms an important component of gender socialization. Girls playing house and doctor help each other to learn how to be good housewives. Boys playing football and war do not.

Making a Rational Choice

Gary Becker (1981) is an economist who has studied the way in which households divide all the work done. His analysis of the question of why men and women divide housework in a gendered way is less related to childhood socialization and more concerned with the task faced by adult men and women who are trying to survive/excel in a competitive economic system. Socialization theorists focus attention on what the individuals are like who are dividing the housework and how they came to be the people they are. Becker's work focuses on the decisions those individuals must make. He has examined the ways in which household members divide both housework and market work. Market work refers to employment in the paid labor force.

Becker argues that households work as a team seeking to divide these two types of work in the most rational and efficient manner. Households take into consideration factors like who can earn the highest wages in the paid labor force and who is most skilled at doing housework.

In Chapter l, we looked at the functionalist analysis of family organization. Functionalists assert that in order to maintain an advanced industrial society, the best (most functional) family is one in which men earn the living and women take care of the family.

Becker's analysis is similar to the functionalists' because he too is arguing that housework is distributed to men and women in families in unequal ways because that is the most efficient means of dividing the work. In Becker's assessment there is no consideration of issues like justice or conflict of interest. The fact that women do most of the housework in most families is explained as a result of rational cooperative decisions made by households.

Just as feminists were critical of Talcott Parsons and other functionalist sociologists, feminists have also been critical of Gary Becker's work. Sarah Fenstermaker Berk (1981) has criticized Becker for assuming that *households,* not *individuals,* maximize their interests. She maintains that Becker incorrectly assumes that rational decisions for the group affect all members of the group in the same way.

Take, for example, the case of Cedric and Delena in the scenario at the beginning of the chapter. Becker would argue that since Cedric makes more money and, therefore, his job in the labor market is more valuable to the household, it is a rational decision for the family as a unit to have him do less housework so that he can be rested and able to keep his job. Arranging housework so that Cedric is less responsible for domestic work is further supported within Becker's analytical framework by observing that Delena is more skilled at doing housework. When she does the laundry it does not turn pink and her dinners are more nutritious and appealing. What Becker does not acknowledge about cases like Cedric's and Delena's is that although the division of labor appears to reflect this rational assessment by the household, Cedric and Delena may have very different points of view on the efficacy of the arrangement. The balance sheet that leads to this decision is rational only if the costs to Delena are ignored.

In criticizing Becker, Berk (1985) points out that his assessment of decision making among married couples contrasts with his analysis of decision making by couples before marriage. Berk writes,

> [In Becker's theory of marriage] it is argued that marriage decisions are based on the self-interested assessments of individuals with respect to the best marriage "bargain" that they can achieve. Yet once the bargain is struck and household production begins, a unitary utility function prevails. (Berk, 1981, p. 26)

A second criticism made of Becker by Berk is that he assumes that it is rational and efficient for wives to do more housework because in many households women's time spent in the paid labor force is worth less than men's since they make less money. However, Becker never questions why women make less pay. Gender inequality in the labor market is taken as a given and then used to explain and rationalize gender inequality in households.

Unequal Work and the Politics of Gender

Feminists have developed an alternative analysis to that of writers like Becker. Chapter 1 showed that one of the themes of the recent rethinking of the family is around the question of the undifferentiated experience. Feminists point out that it is not accurate to speak of the family as if it were one single unit because there are different experiences within each family. "Women and men, young and old officially participate in the same families but their experience of them may be quite different" (Rapp, Ross, & Bridenthal, 1979, p. 177).

Furthermore, that differential experience is both separate and unequal. Differential experience in households is organized around relationships between political unequals. According to family scholars who take this point of view, housework is an excellent example of the politics of gender.

Heidi Hartmann has assessed the way in which men and women divide housework and concluded that "the family is a primary arena in which men exercise their patriarchal power over women's labor . . . [and] that time spent on housework . . . can be fruitfully used as a measure of power relations in the home" (Hartmann, 1981, p. 377).

BOX 7 • 2 CAN HOUSEWORK CAUSE GENDER INEQUALITY?

Many people are aware that gender inequality causes imbalanced divisions of labor in families. Berk (1985) has examined the other side of this picture: What effect does an unequal division of housework have on gender stratification?

Berk uses Goffman's (1977, 1979) and West and Zimmerman's (1987) work to try to explain the way in which households are "gender factories." Goffman used the term *gender displays* to describe the sex-specific ways of appearing, acting, and feeling that people use to let others know what gender they are. For example, in our society, as signs to others we wear certain colors and styles of clothes and hair, we move our bodies in "gender-appropriate" ways, and, as Berk points out, we perform certain household tasks.

According to Goffman, these displays serve to establish or reaffirm what gender we are and to "align" ourselves to interact with others in gender-appropriate ways. The gendered division of housework provides a constant way in which people can reaffirm who they are and how they relate to others (West & Zimmerman, 1987).

Men can reaffirm their masculinity by not washing the dishes. Women can reaffirm their femininity by doing them. This division also reaffirms their relationship to each other—women work for men—men dominate, women are subordinate (Berk, 1985).

Every day, several times a day, the division of housework allows us/forces us to practice gender. Then, when we display those behaviors and relationships, it serves to reinforce for ourselves and others the validity of that division of labor and those expressions of gender. Dividing housework unequally between women and men is more than just getting the job done. And it is more than just an illustration of the oppression of women. It is a way in which to strengthen, perpetuate, and create gender in our society.

Natalie Sokoloff (1980) would argue that the unequal division of housework both expresses and promotes the subordination of women. Women do more housework because they are unequal to men. But that is only half of the story. Women are unequal to men because they do more housework. In order to remedy the unequal division of labor, we must devise ways to increase equality in the way housework is divided. We must, however, also address this underlying political relationship—the difference in power between women and men in all corners of our society—in order to make those changes possible.

Heidi Hartmann and Natalie Sokoloff are socialist feminists (Lengermann & Niebrugge-Brantley, 1992). Socialist feminism is a theoretical framework that seeks to synthesize two major theoretical paradigms, Marxism and radical feminism. Socialist feminists draw from radical feminism a concern for gender inequality and the oppression of women. They draw from Marxist theory a concern for the economic factor in understanding the oppression of women. In Chapter 4, I argued that a major component of a Marxist outlook is the recognition of economics as an insufficient but necessary factor in the consideration of all human interaction. Hartmann agrees that understanding of economics is key to unlocking the mystery of gender inequality and the oppression of women. Her work begins with the question of how men and women differ in their contribution to the economic life of the community. Hartmann's work, like that of other socialist feminists, differs from Marxism in its broad view of economic activity, which includes unpaid housework (Lengermann & Niebrugge-Brantley, 1992).

In most Marxist writing, the economy of a capitalist system like that in the contemporary United States is viewed as those activities that directly contribute to the accumulation of capital. When employees go to work and build cars that are worth more than the wages the workers are paid, the surplus gained from the sale of the cars is returned to the owner of the car factory. This return is called profit or the accumulation of capital. Although housework amounts to a lot of work, it does not directly create a profit for anyone because housewives are not employed, the products of their labor are not sold, and no one is able to directly accumulate capital because of their work.

Socialist feminists would argue that housework should nevertheless be considered part of the economic system because it indirectly enhances profitability. For example, a socialist feminist would maintain that when Cedric arrives at work rested and fed and therefore able to work harder, his boss is able to profit from the unpaid labor of Delena, who fed and cared for him. Socialist feminists, furthermore, would draw attention to the way in which both capitalist bosses and men workers benefit from the superexploitation of working-class housewives.

WHAT CAN BE DONE ABOUT THE INEQUALITY IN THE DIVISION OF HOUSEWORK?

The first two major sections in this chapter have looked at the way in which housework is divided and the contrasting analyses of the reasons for this division. In this section the issue we will turn to is: What can be done about the inequality of the division of housework?

The numbers of two-paycheck couples who will be faced with the problem of dividing housework are increasing. Seventy percent of wives and mothers now work outside the home. In a survey of 200,000 first-year college students in 1988, less than 1 percent said they planned to be a full-time homemaker (Hochschild, 1989).

One way to solve the problem is to attempt to resolve it at the micro level between husbands and wives. Hochschild argues that this is a common occurrence and that negotiating housework is an important part of many marriages. The strategies that women and men use to try to resolve the problem of dividing housework create a core dynamic in marriage; the attempts to resolve the problem become a constant source of tension.

Within couples, men and women use different approaches to try to come to some agreement. The strategies women use to resolve the problems include: (1) marrying men who share the work; (2) actively trying to change husbands who do not share the work; (3) passively trying to manipulate their husbands; (4) becoming "supermoms"; and (5) cutting back at work, marriage, self, and children (Hochschild, 1989).

Men use other strategies in the struggle. They share the work equally, or they resist by: (1) disaffiliating themselves; (2) reducing the need for the tasks; (3) substituting offerings; and (4) selectively encouraging their wives (Hochschild, 1989). Hochschild uses the phrase "disaffiliating themselves" to describe the way in which men passively avoid taking responsibility for the work by forgetting things, doing tasks in an unskilled way, and waiting to be asked.

Some men take responsibility for the work but they reduce the need for tasks to be done by redefining the situation. They might argue to their wives, for example, that there is no need for clothes to be ironed or sheets to be changed, and that McDonald's is good enough for dinner.

Other men do not do the work but they try to alleviate the tension by offering substitute support to their wives. For example, this kind of husband might patiently listen to his wife and offer her advice while she struggles to balance her home and paid work. These men do not increase their material support for their wives by doing more of the housework but they increase their emotional support of their wives to make the problem a little less difficult for her. Finally, husbands avoid sharing the work by praising their wives for the efficient, organized way in which the women get the work done (Hochschild, 1989).

Despite great pressure in the households Hochschild examined, as in many households, sharing housework more equitably between husbands and wives does not seem to be occurring. These forms of resistance were more common than equitable distributions of the work.

In those cases where couples did share more equally one variable that Hochschild (1989) found to be important was gender ideologies. Those couples that had more egalitarian ideas about what women and men should be like also tended to share housework more equitably. The work of Wright, et al. (1992) also found small differences of increasing egalitarianism in the division of work in households where people, especially men, had more egalitarian beliefs. This suggests that one way to improve the division of labor in households is to encourage the adoption of more egalitarian gender ideologies, particularly among men.

Hiring a Maid

When sharing the housework does not work, another strategy is to hire someone to do the housework in the home, but this "solution" is also problematic. Some households are wealthier than others and more likely to be able to accomplish this. This solution is, therefore, available only for a limited number of people.

Hiring a maid is also problematic even for those households that can afford it because this solution may create another problem of inequity, this one between employer and maid. Hiring a maid might help to eliminate the inequality between a woman and man in a household but it reveals race and class inequality between the employer and the maid, since the people hired to do domestic work have historically been working-class women, particularly working-class racial ethnic women. Domestic work has been an especially low-paid, devalued, oppressive, and difficult occupation.

Domestic Work and Racial Ethnic Women Until the middle 1960s the largest single occupation of African American women was domestic work. "By 1950, 60% of all black working women (compared to 16% of all white working women) were concentrated in institutional and private household service jobs" (J. Jones, 1985, p. 235).

Until World War II, domestic service was also the most common form of nonagricultural employment for Japanese American women, whether they were American-born or Japanese-born (Glenn, 1990). In 1940 more than 50 percent of Japanese American women in San Francisco were employed in domestic work. In addition, West Indian and Chicana women have been and are likely to work as domestics. "The racial stratification of the occupation is so apparent that Judith Rollins (1985, p. 4) characterizes the present situation as 'the darker domestic serving the lighter mistress' " (Romero, 1988, p. 321).

For women who work as maids, domestic work has meant especially long hours away from their own families. For most of its history, domestic work was a live-in occupation. This meant that domestic workers were able to see their own families only every other weekend. In addition, "work hours were open-ended with the domestic 'on call' most of her waking hours" (Glenn, 1990, p. 450).

One of the most important battles won by domestic workers in the twentieth century was the right to "day work" instead of "live in," which allowed domestic workers to go home to their own families at night (Dill, 1983). This struggle and its importance in the first half of the century were discussed in Chapter 3.

The battle between mistress and maid continues even today. Mary Romero (1988, 1992) has interviewed Chicana women who work as maids. She states that domestic work is characterized by four issues: (1) the work involves the extraction of emotional labor from the maid; (2) mistresses use various methods to maintain superiority and control while creating an informal and apparently companionable relationship; (3) domestic workers use various strategies to minimize the subjugation and unwelcome familiarity they face on the job; and (4) the work involves a constant struggle between employer and employee over the control of the work process.

The extraction of emotional labor refers to the way in which domestic workers are expected to fulfill not only technical roles, but also emotional and caring roles. Chicana domestics told Romero that they often felt called upon to be surrogate mothers for their younger employers.

Because the employer and the domestic worker live in different social worlds the mistress may feel free to tell the maid about her personal problems without fearing the maid will look down on her or retaliate against her. This "sharing" of intimacy, however, is not reciprocal. The employer can choose to obtain consolement but she does not have to return it and the maid is not really free to offer her true feelings. The relationship, thus, only appears to be one of friendship and results in the employee doing additional work as she provides a source of emotional support.

Control over the relationship is maintained by the employer in ways such as allowing the domestic worker to sit only in the kitchen and to eat only after the employee's family has been served; to insist that a maid wear distinguishing clothes (a uniform), at least on special occasions like a dinner party; and generally to demand deferential treatment.

Another way in which the boundaries between maid and mistress are maintained by what appears to be friendly interaction is through the giving of gifts to a maid. Domestic workers frequently receive gifts of castoffs from the employer. Judith Rollins (1985, p. 193) argues that these "gifts" serve to "communicate to the parties involved and to the larger social group who the giver and the receiver are and what their relationship to each other and the community is" (Romero, 1988, p. 333). The message to the maid is that "you are only good enough to wear clothes that are no longer good enough for me." Even the giving of legitimate gifts and bonuses can serve to control employees, if they are given in lieu of benefits and raises.

Domestic workers have not easily given in to this second-class status. They have attempted to retain their control over the work by demanding proper equipment and to be allowed to define the methods of cleaning, standards, and the order and pace of the work. Employees also attempt to have greater control of their work by making agreements with employers based on specific tasks rather than the number of hours worked.

Maids have also tried to achieve more control over their work by eliminating the demands for emotional nurturance. Chicana workers in Romero's (1992) study said that one way to do this is to take positions where they work in homes when the employer is away.

Finally, as we observed in Chapter 3, there is a long history of collective action and attempts to unionize among domestic workers throughout the U.S. The goals of domestic workers who have sought to organize unions "include: raising wages to minimum wage; providing common working benefits such as paid vacations, holidays, sick leave, and workers' unemployment compensation; changing attitudes toward the occupation; and creating awareness among workers about the value of their work" (Romero, 1988, p. 340).

We can see that from the point of view of working-class women, especially working-class racial/ethnic women, purchasing housework by hiring people to do it is not necessarily a good solution to the problem of gender inequality in the division of labor between women and men. As long as the conditions of work remain as they are between maid and mistress this "solution" helps to perpetuate race and class inequality. Hiring a maid may eliminate the problem between a woman and man sharing a household; it can also create or exacerbate other inequalities in those households.

Solving the Problem of Housework Through Policy Change

These two solutions to the problem of dividing housework more equitably—getting husbands to share and hiring a maid—are based on decisions and actions within households. Neither has been satisfactory. Other solutions have been offered that look to larger institutions for change. These include changing laws to require that housewives be paid and socializing housework.

One public policy solution that has been suggested is to require that husbands pay housewives. In a 1985 survey, 42 percent of the women and 26 percent of the men said that men should have to pay for the housework their wives do (Hochschild, 1989). Paying wages is a good way to recognize and reward the extra contribution that women make. The problem of the time constraints for women who also work outside of the home, however, is not addressed by this solution. Requiring that husbands pay their wives for domestic work is also problematic because it reinforces the subordinate relationship between husbands and wives. It also leaves unanswered questions like how the work would be evaluated and by whom. Also, the proposal that husbands pay wives is meaningless for the majority of couples who live paycheck to paycheck and whose combined earnings barely pay the bills for the entire household. In addition, this solution, like the solution of sharing housework, ignores the situation of single-parent households.

A second solution is for the government to pay housewives. Like the proposals above, putting housewives on a government payroll shares many of the weaknesses of the proposal to have husbands pay their wives. First, it does nothing to eliminate the idea that housework is primarily women's work. It also does not solve the problem of time constraints for wives and it ignores single-parent households. Plus, once again, how would the work be evaluated? Who would determine whether the housewife had done her job adequately? And, finally, given the $3 trillion deficit of the federal government, where would the money come from to pay this huge group of workers?

A third option is to socialize housework. Socializing housework means to make housework the responsibility of the entire community rather than the responsibility of each individual household. This effort to spread out the work can be accomplished by radically restructuring the entire society or it can be done in a less dramatic fashion by using the government to facilitate sharing the burden.

Dolores Hayden (1981) has investigated the option of radically redesigning our society, our social relations, and even our architecture and zoning so that workplaces are closer to people's homes, children are raised more communally, and people use community laundries and kitchens. According to Hayden's vision we would share housework, not just among members of our households, but among members of our whole society. Women could work together as a group to ensure that men participate equally. And the actual amount of time necessary to do the work would be diminished for everyone because of the greater efficiency of mass production.

Hayden's solution has been considered by feminists for some time. Around the turn of the century Charlotte Perkins Gilman (1898/1966), a leader in the fight for women's suffrage in the U.S., argued that the housework done privately in separate homes was an intolerable waste of human energy and a basis of gender inequality.

Gilman proposed that houses be built without kitchens and that cooperative dining rooms, laundries, and nurseries be established. Her proposal has remained an unrealized utopia and even today this kind of solution would demand radical political changes. "An egalitarian approach to domestic work requires complex decisions about national standards versus local control, about general adult participation versus efficient specialization, about individual choice versus social responsibility" (Hayden, 1981, p. 28).

A less utopian and perhaps more attainable model for socializing housework is represented by family policy in countries like Sweden (Moen, 1989). In Sweden, socializing housework has meant that everyone pools economic resources in order to allow households, especially those with children, to divide work more equitably. This model includes reforms that provide for parental leave for either a father or mother after the birth of a child; entitle either a father or mother to work six-hour days; and explicitly encourage men and women to share childcare and domestic work (Wright, et al., 1992).

THE MICRO-MACRO CONNECTION

Housework, like many activities that take place within families, is experienced at the micro level as a personal issue. Cassandra, Delena, and Cedric, whose story was told at the beginning of this chapter, feel bad as individuals because of their inability to divide housework in their household in a satisfying manner.

Oakley has explained that housewives feel overworked and devalued. Arlie Hochschild found that women felt tired, overwhelmed, and frequently like a villain in their attempts to keep house, raise children, and maintain outside jobs. The experience of housework for many individuals and families is not a very pleasant one.

In addition, the experience is different for women and men. Compared to women, men do less work, they do different jobs, and they have different strategies for avoiding housework. The experience of housework for men at the micro level, however, is also problematic because in dual-income families housework may go undone and wives are not happy with the way in which the work is distributed.

One possible solution to the problem of housework is to hire people to do the work. But because of the system of class and race inequality in our society this means that domestic work is done mostly by working-class racial ethnic women and the work is difficult, low paid, and oppressive.

These micro-level personal experiences between husband and wife, mistress and maid, are embedded in larger macro-level social structures that include our ideas about work and the way in which paid work is organized. Families are also affected by the macro-level organization of the government, for example, in its attempt to shift costs from taxpayers to housewives through changes in health care funding. Housework is also affected by the system of gender relations, which includes ideologies about what women and men should do, as well as about the differential skills women and men have to do housework and the differential power women and men have to bargain for the division of tasks.

Macro-level social structures of ideologies; race, gender, and class stratification; and the government shape the experience that individual people have with each other at the micro level within their families. People like Cassandra's family are trapped in a web that cannot be altered by themselves alone.

Individual people at the micro level can, however, work together to alter the web. Advocates for government-funded childcare, paying housewives, and socializing housework have offered suggestions about what might be changed at the macro level to make sure that the housework gets done in a more satisfactory and equitable manner.

This chapter also shows another twist on the effect of micro-level activities in its discussion of Berk's notion of the gender factory (see Box 6.2). The gender factory, which is comprised of the micro-level interactions of women and men around housework, helps to create and sustain the macro-level system of gender inequality. Although this book has emphasized the way in which micro-level activities can help to transform the organization of our society, the gender factory reminds us that everyday micro-level activities also serve to reproduce the status quo.

SUMMARY

Housework, the Invisible Occupation

- The story about Cassandra's family shows how burdensome and important housework can be. When their story is duplicated in millions of households it reveals an impressive source of labor and problems. Housework, however, is often invisible and rarely perceived as work.

- Housewives say their work is difficult because it is monotonous, time-consuming, and devalued.
- In Oakley's research on housework she made important discoveries about methods of doing research as well as about the issue of housework.
- Studies over the past three decades consistently show that housework is women's work. Even in households where women and men both work in the paid labor force, women do most of the unpaid housework while men generally do very little.
- There is some evidence that men are doing more housework, but there is much debate on how significant this change is.
- Not much research has been done on children and housework, but what has been done suggests that our reasons for giving children chores is not because it is useful and necessary work but because it is good for their education.
- The division of housework among children is influenced by gender, age, and whether they live in one- or two-parent households.
- In addition to the different amount of time spent by husbands and wives on housework, the work is further differentiated by gender in terms of the kinds of work that is done. Some tasks are considered "women's work" and others are "men's work."
- Devault's work on feeding the family reveals that housework not only accomplishes a necessary survival task, it also helps to create a social world, especially the social world of families.
- Full-time housewives do many more hours of housework than housewives who are also in the paid labor force. This may be because much of the work full-time housewives do is trivial and unnecessary. Or it may indicate a lower quality of life for families that do not have a full-time housewife.
- Both full-time housewives and part-time housewives are stressed by the unpaid domestic work they are required to do.
- Changes in technology have decreased the work required for some household tasks but for others it has increased the work with the end result that housewives work similar hours to women early in the century.
- Housewives' responsibility for their families' health diminished until the 1970s, when laws were passed that sent patients home earlier than previously from health care facilities.

Why Is Housework Divided Unequally?

- There are at least three different answers to this question depending on which theoretical framework is being used: socialization, rational choice, or gender politics.
- Some scholars argue that the division of housework is a result of the way in which children are socialized into gender roles.
- Gary Becker argues that the reason work is divided in an uneven way is because it is more efficient for families as units. Becker sees families as teams working together to maximize their total output.
- Feminists have been critical of Becker's point of view and have argued that within families there are conflicting interests and that the division of labor in families is a political issue.

What Can Be Done About the Unequal Division of Housework?

- Several different strategies have been attempted to try to solve the problem of housework. One of those strategies is to get men to do a larger share.
- A second strategy that has been tried is to hire maids to do the work. This "solution" to gender inequality in households has created greater race and class inequality because women who work as maids are mostly racial ethnic women and the conditions of their work are oppressive. Housework becomes a site of struggle between mistress and maid.
- Other solutions to the problem of housework have also been proposed, including paying housewives and restructuring society so that housework would be done more communally rather than within households.

The Micro-Macro Connection

- Housework is experienced as a problem for many individuals in families. Problems stem from the sheer size of the task as well as the unequal burden it creates for women.
- The way we think about housework and the way we actually divide and conduct the work is affected by the larger social structure—the macro level—in which it is embedded.
- Attempts that have been made to address problems associated with housework fall into two categories: those that are negotiated at the micro level among family members and those that call for changes in the larger social structure such as the government or wagework.

Kissing is a sexual activity that is considered natural by people in the United States. In most cultures, however, kissing on the mouth is considered exotic and unseemly.

8 LOVE, SEX, AND ABORTION

Lori phones her friend Kamika late at night. "Call Stacey on your three-way. I need to tell you two something." Stacey comes on the line and Lori starts to talk. "I started my new job at the phone sex company. You would not believe the guys that call."

Kamika breaks in. "I can't believe you took that job. It's sick. Sex should be for lovers, not strangers on a telephone."

Lori responds, "I can't see the appeal myself, but if it makes these guys happy, what do I care, and besides, it's a job. Sex doesn't have to be only for people in love, you know."

"Well, I'd be afraid one of them might find out who I was or where I live," says Stacey.

Lori replies, "Nobody is going to find out where I live and maybe I'm doing the community a service. Maybe they would be out doing something really terrible in the street if they couldn't have the telephone sex outlet."

"Yeah, maybe you're right. I sure don't want to be telling people what kind of sex they ought to have. Kristy and I get enough of that when we hear people talk bad about us just because we are lesbians. But I don't know, it seems like some of this stuff can get so far out it gets exploitative and doesn't seem like real sex anymore, more like rape or something. Don't you feel kind of used?"

"No, not really. I like sex. It seems kind of fun, you know, taking a risk, trying something new," Lori says.

Kamika shakes her head. "I like sex too. But what's wrong with sex with love and marriage? Personally I do not intend to jump in and out of bed with anyone but my husband and I do not intend to try any of that kinky stuff even with him."

"Well, I gotta go. I need to get up early for the pro-choice rally."

"Yeah, us too, see ya."

- *In approximately what year is this scene taking place? How do you know?*
- *How are the choices these young women face a reflection of the times in which they live? How have the choices changed very recently (in the '90s)? Somewhat recently (since the '70s)? Over a long historical period?*
- *Although each of these women have different opinions about the relationship between sex and love, the fact that this relationship is one to consider is part of their thinking. What do you think is the proper relationship between sex and love? What about the relationship between sex and marriage?*
- *How are gay men's and lesbians' sexual relationships different from those of heterosexuals? How are they similar?*
- *How do each of these women emphasize different aspects of sex, its dangers or its pleasures?*
- *The scenario ends with the young women planning to go to a pro-choice rally. How is the issue of abortion tied to the expression of sexuality?*

THIS CHAPTER EXAMINES THESE ISSUES as they are expressed in theory and in concrete experience. Like all social phenomena, love and sex are dynamic, changing over time. The history of that change is a central theme in the chapter as well. In addition, we will pursue our exploration of the relationships among love and sex and several other facets of our society including politics, economics, technology, and especially families.

The chapter is divided into four sections. The first is a review of the issue of love, including the difficulty of defining the term and its historical development in Western society. The second section reviews the topic of sex and addresses issues like standards for evaluating legitimate sex, laws about sex, and quantitative and qualitative research on the experience of sex in the contemporary United States. In these two sections the emphasis is on the way in which the macro level of social organization affects the micro level of social experience of sex and love. In the third section three theoretical debates are reviewed: theories about categorizing sexuality, theories about women and sex, and theories about the basis of sexuality. The fourth section explores the way in which the micro level has had an effect on the macro level in an examination of the abortion rights movement.

WHAT IS LOVE?

John Lee (1974) created a typology of love that includes six different definitions:

1. Erotic love: romantic, sexual, irrational, and largely based on physical attraction.
2. Manic love: intense, all-consuming, possessive, and fluctuating between joy and despair.
3. Ludic love: egoistic, self-serving, competitive, and based on an unequal relationship between one partner who is highly committed and another who is emotionally uninvolved.
4. Pragmatic love: a rational, practical, fair exchange between two carefully matched partners.
5. Storgic love: the companionate, stable love that emerges from a relationship between friends.
6. Agapic love: the altruistic devotion of one partner for the other (Lee, 1974).

This typology does a nice job of summarizing the variability of love in our culture as well as showing both negative and positive characteristics of different kinds of love. The typology, however, can be criticized for two reasons. First, it is gender blind. That is, it does not recognize the ways in which women and men may, as a group, tend to play one role or another in the love relationships described. For example, are women more likely to be involved in the agapic type of love and men in the ludic type of love relationships? Or are men more concerned with physical attraction as an aspect of erotic love and women with practical factors or pragmatic love in choosing a lover?

Research on love shows that women and men, on the average, do seem to have different perceptions of love. For example, when questioned about what they are seeking in a mate men and women respond differently. Buss (1985) asked 162 young people what characteristics were most important to them in considering a future mate. Table 8-1 reports his findings. The table shows that while there are many similarities, women and men differ in their assessment of the importance of two variables, physical attraction and good earning capacity. Men rate physical attraction higher than women do. Men also define physical attractiveness differently than women (Patzer, 1985). More women are concerned with a man's weight (50 percent); his face is the second most likely feature (10 percent) to be mentioned. More men think a woman's face is the most important physical feature (50 percent) and weight is the second most important feature (10 percent).

Good earning capacity is the second factor for which Buss found a difference between women and men. Women were significantly more concerned with this issue than the men were. Does this mean that for men love is likely to be tied to the erotic category while for women it is tied to the pragmatic category?

TABLE 8 • 1
Characteristics Commonly Sought in a Mate

Rank	Preferred by Men	Preferred by Women
1.	Kindness and understanding	Kindness and understanding
2.	Intelligence	Intelligence
3.	Physical attractiveness*	Exciting personality
4.	Exciting personality	Good health
5.	Good health	Adaptability
6.	Adaptability	Physical attractiveness*
7.	Creativity	Creativity
8.	Desire for children	Good earning capacity*
9.	College graduate	College graduate
10.	Good heredity	Desire for children
11.	Good earning capacity*	Good heredity
12.	Good housekeeper	Good housekeeper
13.	Religious orientation	Religious orientation

*Significance greater than .001.

Source: Buss, 1985, p. 48.

The second problem with Lee's (1974) six-part typology of love is that it makes no judgments as to whether these definitions of love are valid. Although many people may define the doting, martyrlike behavior of agapic love as love, is this valid? And should we perpetuate this definition of love by including it in the typology or should we define it as something else?

Emma Goldman (1910), a feminist and anarchist in the early twentieth century, defined love as:

> . . . the strongest and deepest element in all life, the harbinger of hope, of joy, of ecstasy; love, the defier of all laws, of all conventions; love, the freest, the most powerful moulder of human destiny. (Goldman, 1910, p. 242)

Goldman was appalled that love was assumed to be part of what she saw as the drab, restricted, and economically based relationships between most husbands and wives. Goldman insisted that her definition of love be made part of the goal of political activists—labor organizers, socialists, communists, and feminists—of the early twentieth century. In particular she maintained that love be freed from the fetters of legal restrictions of marriage laws.

Both the question of gender and its relationship to love and the need to redefine love in a way that excludes certain kinds of relationships are ones with which feminists have been concerned. Second Wave feminists in the late 1960s and early 1970s like Shulamith Firestone (1970), Germaine Greer (1970), and Constantina Safilios-Rothschild (1977) focused their attention on the way in which women and men experience love differently and especially on the way in which certain kinds of "love" keep women tied to men in dependent and oppressive relationships.

Firestone wrote that love was "the pivot of women's oppression today" (Firestone, 1970, p. 126). She answered the question "Why are there no great women artists?" by stating that while men were busy thinking, writing, and creating, women were busy loving those men, thereby allowing them to be creative and nurturing their creativity. The dominant form of love among heterosexual couples in the U.S., according to Firestone, was a combination of ludic love on the part of men and agapic love on the part of women. This unequal love relationship was the key to gender inequality and the oppression of women.

Firestone argued that this sexist love relationship has two different sources for women and men. Men, she said, cannot love, perhaps as a result of their male hormones. Women, on the other hand, can only love, and their obsessive, clinging behavior is a result of their social vulnerability, particularly their economic dependence. Firestone maintained that as long as women have no access to expressing themselves they must live vicariously through men. And as long as women live under a system of economic patronage they must endear themselves to men. For women, then, love is a strategy for social and physical survival.

Arlie Hochschild (1983) is another writer who has examined the issue of love. Her perception of the typical love relationship between women and men in contemporary U.S. society is somewhat different from Firestone's. Like Firestone, Hochschild believes that these relationships are frequently imbalanced and that the experience of love is quite different for women compared to men. Hochschild, however, differs from Firestone in two ways. First, the typology into which she places women and men is different from Firestone's. Second, her assessment of the roots of these differences is not the same as Firestone's.

Hochschild's view of women's experience of love is similar to the pragmatic love described by Lee (1974). She argues that a woman first makes a practical decision based on her assessment of the man's social acceptability, especially his ability to provide economically for her. If she finds he is able to pass this first hurdle, she works on developing strong feelings of attraction for him. If a woman finds she has spontaneous feelings for a man who does not pass the test of social acceptability, she works hard at falling out of love. Hochschild refers to the process of talking oneself out of love or creating a love relationship with a man whose primary attraction is his economic status as emotion work. This work includes analyzing and discussing feelings for a lover. In her research, Hochschild found that women, compared to men, were much more likely to engage in this kind of analysis and discussion of their relationships. We will review some of the ideal characteristics of the relationships that women attempt to create in their emotion work in our discussion of Cancian's typology in the next subsection.

Men, on the other hand, were more likely to allow themselves to assess a woman based on their own spontaneous response regardless of the woman's social character. Zick Rubin's (1973) research supports this observation about men and love. He found that men score higher than women on a romanticism scale; they fall in love more quickly and easily than women and they have more difficulty ending a relationship. Kephart's (1967) work indicates the same gender difference. He asked college men and women, "If a man or woman had all of the other qualities you desire, would you marry the person if you were not in love with him or her?" Sixty-five percent of the men said no; only 24 percent of the women did. Using Lee's (1974) typology to analyze these observations, we see that men are more likely to experience the categories of erotic or manic love. In contrast to Firestone's argument that men cannot love, Hochschild asserts that men can love and that in fact men's experience of love may be closer to our romantic ideals than women's experience.

The second way in which Hochschild and Firestone differ is in their analysis of the source of gender differences in love. Firestone places the source of women's experience of love in the social organization of our society that prohibits women from being able to fully express themselves and especially from being able to take care of themselves economically. Firestone places the source of men's experience in their biology. For Hochschild, both women and men experience love in ways that are shaped by the social organization of society. Since women are more likely to need men economically they are careful to make men's economic ability central to their "lovability." Men are less likely to need women economically but more likely to need women emotionally. Men's assessment of women, therefore, is based on their spontaneous emotional responses.

Historical Development of Love

Like many issues in this book, love is a phenomenon that appears to be natural, inevitable, and unchanging but when examined historically is revealed as highly variable from one time period to another. Our ideas about love and our experience of love have changed throughout history. One of the most significant changes concerns the connection between love and marriage. It is only in recent years that this link has been idealized (Hendrick & Hendrick, 1992).

Love existed in the ancient societies of Greece and Rome. In fact, much of the symbolism and language of contemporary love, such as Cupid, Venus, Eros (erotic), and Aphrodite (aphrodisiac) come from the names of gods and goddesses of those civilizations.

In those ancient societies, love and marriage were kept separate. In classical Greece, for example, aristocratic men married women but the lovers about which they wrote were teenage boys with whom they had sexual relationships. In medieval Western society, the aristocracy carefully arranged their marriages as ways to annex land and wealth and to avoid or win wars. Among the peasants, marriages were also economic arrangements, not sentimental ones.

The concept of romantic love emerged during the thirteenth century. Unlike the ancient Greeks, for whom love meant primarily a sexual, not an emotional relationship, romantic love in medieval society was characterized by the frequent lack of physical consummation. The rules of romantic love demanded that the lovers be emotionally devoted to one another but they were not necessarily supposed to maintain a sexual relationship and they would never be married to each other.

During the colonial period in the United States, the economic and political reasons for marrying became less rigid and couples began to have more of a choice in the matter. Both love and sex began to replace business interests as a basis for partnerships and marriage was linked to love and sex.

In the late eighteenth and early nineteenth centuries in the U.S. the agrarian-based economy began to change into an industrial-based one. The family of the industrial period was characterized, at least for white middle-class families, by a split between work and home. Love was idealized as the asexual love between mother and child (Cancian, 1987). The split between work and family was paralleled by a split between women and men. As described in Chapter 2, the Puritan patriarchal household was replaced by the urban middle-class household. Husbands were dominant in both cases but in the modern family of the industrializing nation, women and men led separate lives. Their emotional at-

FIGURE 8 • 1
Blueprints of Love

	Who is responsible for love?	What is love?
Feminized Love		
Family duty 19th Century	woman	fulfill duty to family
Companionship 1920–	woman	intimacy in marriage
Androgynous Love		
Independence 1970–	woman and man	individual self-development and intimacy
Interdependence 1970–	woman and man	mutual self-development, intimacy, and support

Source: Cancian, 1987, p. 31.

tachments were therefore sex segregated. In Puritan society the key relationship was between husbands and wives. In the modern society of the nineteenth century, "the key relationship was an intense emotional tie between mothers and children and raising moral, respectable and healthy children was a woman's major task" (Cancian, 1987, p. 31).

Francesca Cancian (1987) has examined the history of love as it has developed in the U.S. through the nineteenth and twentieth centuries to the present. She has developed a typology of what she calls "blueprints of love" that is shown in Figure 8-1.

The figure shows that the Family Duty blueprint of love was the type that emerged after the transition from the agriculturally based Puritan society to the industrial-based modern society in the nineteenth century. The Family Duty blueprint was eventually challenged by three subsequent blueprints: the Companionship blueprint in the 1920s and two Androgynous forms that emerged in the 1970s, the Independence blueprint and the Interdependence blueprint. Cancian argues that currently these three competing blueprints of love are all part of our society.

The Companionship blueprint differs from the Family Duty blueprint of the Victorian period in its focus on the relationship between the husband and wife as opposed to the relationship between the mother and child. The Companionship blueprint fuses sex, love, and marriage. It remains similar to the Family Duty blueprint in its reliance on the wife as the person primarily responsible for maintaining family relationships and its feminized conceptualization of love.

When Cancian (1987) argues that love during both the Family Duty period and the Companionate period is feminized, she means that the expression of love is identified with activities that are considered feminine in our society. A feminine conception of love exaggerates emotional expression and talking about feelings. A masculine conception of love, in contrast, acknowledges instrumental and physical aspects of love such as providing help, sharing activities, spending time together, and having sex.

Cancian (1987) cites Lillian Rubin's (1976) interviews with blue-collar couples as evidence of masculine and feminine love. A working-class man who is concerned about his wife's complaint about his lack of communication says,

> What does she want? Proof? She's got it, hasn't she? Would I be knocking myself out to get things for her—like to keep up this house—if I didn't love her? Why does a man do things like that if not because he loves his wife and kids. I swear, I can't figure out what she wants. (Rubin, 1976, p. 146)

Another man explains that sex is love for him. He says that after sex,

> I feel so close to her and the kids. We feel like a real family then. I don't talk to her very often, I guess, but somehow I feel we have really communicated after we have made love. (Cancian, 1987, p. 77)

These men are defining love as sexual expression and providing shelter. The women Rubin interviewed describe their view of love in a different way. Love for them involves emotional expression and verbal interaction. For example, one wife says, "Love is sharing, the real sharing of feelings" (Cancian, 1987, p. 69). Another woman says,

> It is not enough that he supports and takes care of us. I appreciate that, but I want him to share things with me. I need for him to tell me his feelings. (Rubin, 1976, p. 146)

Cancian argues that (1) women and men have different ways of defining love, as shown in these quotes, and (2) our society has tended to define love in a feminized way, that is, in a way that is consistent with the definitions given by the women.

This feminized version of love, however, began to change in the 1960s. A more androgynous blueprint emerged as an alternative. Although the Companionate blueprint illustrated by the quotes above remains important, the Androgynous blueprint has become increasingly evident.

The Androgynous blueprint has two variations. Both types are androgynous in two ways: love is the responsibility of the man as well as the woman, and it includes "feminine" emotional support and expression as well as "masculine" material assistance and sex.

The first type of Androgynous blueprint, the Independent, could be called the separate but equal type. The goal of this type of love relationship is to break away from the unequal relationships of Companionate love by having each individual establish his or her separate life, which might be shared with another if it does not interfere with his or her independence. In the 1970s a popular poster appeared stating this philosophy. It read, "You do your thing and I'll do my thing and if by chance we come together it is beautiful."

The second type of Androgynous blueprint is the Interdependent. Here the goal is also to eliminate gender inequality but the Interdependent blueprint also seeks to retain the interdependence of the Companionate blueprint. In this type of love relationship, individuals are in control of their own lives and develop their own interests and goals, but they view their ability to develop themselves as individuals as a product of their relationship. If a poster had been produced to illustrate the Interdependent blueprint it might have read, "You do your thing and I'll do my thing and we'll both do it better because of our commitment to each other."

Cancian (1987) argues that our fear of the changes in love in our society and our perception that the "me generation" will give up on love altogether is rooted in our perception that the only alternative to the obsolete and oppressive Companionate blueprint is the Independent Androgynous blueprint. She maintains that the other type of Androgynous love, the Interdependent blueprint, provides a model that allows for both love and equality.

As you probably noticed, discussing love and sex as two separate issues is somewhat artificial. The tension between love and sex and the way in which the relationship between the two varies historically and from one individual to another make them almost inextricable. In the next section our focus will move to a consideration of sex as a phenomenon that is sometimes tightly bound to love and other social factors as well.

HUMAN SEXUALITY

One characteristic that distinguishes humans from other animals is sexuality. Animal sexuality is "limited, constricted, and pre-defined in a narrow physical sphere" (Padgug, 1979, p. 19.). Human sexuality, in contrast, is marked by its richness, plasticity, and diversity. Human sexuality, like animal sexuality, is involved with biological sexuality, but biology is only a precondition and human sexuality stretches far beyond it in an apparently infinite number of ways (Padgug, 1979; Schneider & Gould, 1987).

Issues like what we view as sex, what we view as sexual, who we perceive to be legitimately involved in sexual activities, and what conditions we believe must be part of valid sexuality vary from one culture to the next, from one historical period to the next, and from one individual to another. For example, among the Ik in East Africa, masturbation is the primary form of sexuality and intercourse is accepted as an occasional extension of masturbation (Turnbull, 1972). Among the Aranda in central Australia, boys are not eligible to marry until they have been initiated into sex through homosexual relationships with older men (Bullough, 1976). These behaviors are quite different from what our society presents as sexual standards. And, on the other hand, activities we might see as normal would be viewed as exotic by others. For example, "Kissing, especially for anything longer than a brief peck, appears to be a feature of only a few, mostly Western cultures. Kissing is viewed unfavorably by many African and Asian cultures" (Reinisch, 1990, p. 103).

Sexual Standards

As in all societies, in contemporary U.S. society certain standards tell us how we are supposed to feel and act sexually (Kelly, 1992). The first five of these standards were proposed by Kelly (1992, p. 230) and the last one comes from the work of Altman (1982, p. 172). These standards can be summarized as follows:

1. *Heterosexual Standard*: We are supposed to be sexually attractive to and attracted by the opposite sex and to desire to be sexually involved with them.
2. *Coital Standard*: We are supposed to view sexual intercourse between a woman and a man, or coitus, as the ultimate sexual act. All other sexual

activities, such as oral-genital contact or manual stimulation, are considered foreplay, implying that they are not really sex but can lead to the real thing. According to this standard, a person may have much sexual experience, heterosexual or homosexual, but consider him or herself a virgin because he or she has not participated in coitus.

3. *Orgasmic Standard*: We are supposed to experience orgasm as the climax of any sexual interaction. This standard has been prevalent for men for a long time in Western society, and more recently has become the rule for women as well.
4. *Two-Person Standard*: We are supposed to view sex as an activity for two people. Masturbation has become more acceptable but it is still seen as a substitute for real sex. Sexual activity involving more than two people is supposed to be perceived as exotic and unseemly.
5. *Romantic Standard*: Sex and love are supposed to be tied together. Love without sex is seen as incomplete and sex without love is perceived to be shallow and exploitative.
6. *Marital Standard*: Marriage must include sex. Sex in marriage is the norm to which other sexual activity is compared. For example, we use the words premarital sex and extramarital sex to describe sex between people who are not married to each other. The use of these terms illustrates how sex in marriage is the "normal" reference point, although other kinds of sexual relationships are recognized and sometimes even accepted as legitimate.

This list of standards is not a formal set of rules by which we must abide but it makes up a dominant sexual ideology in our society and is reflected in the media, in political rhetoric, and in some cases in the law. The standards also describe the actual sexual experience of many people. But many people, probably most people at some time, do not abide by the standards. Which ones are part of your beliefs about sex? Which describe your sexual experience? Do any of these standards hurt people? Is it necessary to have sexual standards? Is it inevitable?

Sex and the Law

Some characteristics of the sexual standards described above have been formally sanctioned in law. For example, currently in over one-half of the states, some forms of nonmarital and extramarital sexual activity are crimes. These include adultery, fornication, sodomy, and oral sex (Stetson, 1991).

The first standard mentioned above, the Heterosexual Standard, is one that has been covered by the law. In 1962 Illinois was the first state to decriminalize homosexual behavior practiced in private between consenting adults. Since then about half of the rest of the states have followed suit (Kelly, 1992). Twenty-two states now entirely exclude all consensual sexual conduct between adults from the criminal realm. Eight other states exclude consensual conduct between husbands and wives.

In twenty states, restrictive laws are still in place for both married and unmarried persons. Sometimes these laws carry maximum penalties of several years in prison. For example, the statute in Rhode Island, a typical one, reads: "Every person who shall be convicted of the abominable and detestable crime against nature, either with mankind or a beast, shall be imprisoned not exceeding 20 years nor less than 7 years (Katchadourian, 1985, p. 514). In these kinds of laws "the abominable crime against nature" is defined as anal intercourse. The laws have been used almost exclusively against homosexual anal intercourse, not heterosexual anal intercourse. The laws may be written to condemn any sexual activities by people not married to one another or to condemn certain activities by anyone, whether heterosexual or homosexual, married or unmarried. Gay men and lesbians, however, are most often the people who are treated as criminals.

Most of the active legal restriction of homosexual behavior is directed toward activities that take place in public, such as movie theaters, parks, or bathrooms. Local police forces use a number of methods including surveillance and decoy operations to arrest gay men and lesbians who are practicing sexual activities in these places.

But private consensual sex between homosexuals and lesbians is not exempt from legal action. The federal government discharges about 1400 men and women (women are much more likely to be discharged) each year from the armed forces for violating a directive to refrain from homosexual activities (Kelly, 1992). College students enrolled in ROTC courses must sign a statement that they have never participated in sex with someone of their own sex and furthermore that they have never even considered engaging in such activity. Civilians are not exempt. In 1986 two gay men were arrested in their home and convicted of sodomy by the state of Georgia. A police officer had come to the house on another matter but charged them with sodomy when he determined that the couple had been having sex. The two men challenged the state court's decision by taking the case to the Supreme Court, which ruled that people do not have the right to engage in homosexual acts if the state chooses to enact laws against it (Kelly, 1992).

Other aspects of the standards listed in the previous section have also been formalized into law. For example, twenty states have laws that provide for misdemeanor charges for adultery, although they are rarely enforced (Katchadourian, 1985). And regarding the last standard listed in the section above, the Marital Standard, the law views sexual intercourse as a required feature of marriage. If a marriage is not sexually consummated, it can be annulled. This means that if sexual intercourse does not take place between a wife and husband, the marriage can be said never to have existed.

Problems in Sex Research

Research on sex has been hampered by several problems. Some of the data are old, some are not representative, and some do not completely reveal the significance and meaning of sexuality (Schneider & Gould, 1987). Scholars still refer to the Kinsey (1948, 1953) reports as the most complete data set on sexuality in the U.S. today, although those data are now more than forty years old.

More recent research may not be representative. In 1974, for example, *Playboy* magazine conducted a widely cited national survey (Hunt, 1974), but only 20 percent of those contacted agreed to participate. Since only 1 in 5 responded the people who did respond may have been unrepresentative in some way. For example, they may have been more interested in the subject than the others, or they may have felt less threatened by the sponsoring company. Blumstein and

Schwartz (1983) conducted another highly regarded study in the 1980s. Their research has the benefit of including a variety of categories of respondents—married, cohabiting, gay, heterosexual. Their sample, however, was drawn from volunteers and data were gathered in a select few cities—New York, Seattle, and San Francisco—and may not be representative of people who would not volunteer for sex research or who live in other regions of the United States.

Finally, sexuality is difficult to quantify because the significance and meaning of events can be critical. Here we will examine some of the statistical information that has been gathered and offer some explanation and critique of those numbers by reviewing qualitative research as well.

Premarital Sex

The issue of premarital sex is one example of an area in which researchers have not maintained long-term, consistent data collection from a large, representative, cross-national population. Instead, different studies have gathered evidence here and there, from one group and then another, and during some years but not in others. Carol Darling, David Kallen, and Joyce VanDusen (1992) tried to overcome this methodological problem by examining a large number of studies on the topic. They gathered as many data sets as they could find to try to piece together a description of the changes that have taken place. They then eliminated studies that had weak methodologies or were of select populations like participants in birth control clinics, patients in psychiatric institutions, or retrospective studies of people who were married at the time of the research. Darling, Kallen, and VanDusen (1992) also were careful to determine when the data were collected rather than when the study was published, since there is often a time lag between these two events that can confuse comparisons from one year to another. In addition they decided to report only the variable of changing status from virgin to nonvirgin, which they defined as participating in coitus. Their final sample included thirty-five studies that they believe represent the nation as a whole and the changes that have taken place in premarital sex from 1900 to 1980.

Their assessment concluded that:

1. Since the turn of the century the numbers of both men and women reporting coital involvement prior to marriage have increased.
2. Gender differences have diminished during this time period.
3. Premarital coital involvement for both women and men now appears to be the norm (Darling, Kallen, & VanDusen, 1992).

In 1903 researchers reported a coital rate of 12 percent for never-married young women (Dickerson & Beam, 1915), and a rate of 36 percent for never-married young men (Exner, 1915). By the 1970s, researchers were reporting rates of 70 percent to 80 percent for never-married men and women (Jessor & Jessor, 1975; Murstein & Holden, 1979). Increasingly larger numbers of both women and men were indicating they had experienced premarital sex and the differences between women and men were becoming nearly nonexistent.

At the same time that changes in behavior were occurring, attitudes toward premarital sex were also becoming more tolerant. Table 8-2 shows how these attitudes changed between 1972 and 1988. These data came from the NORC general survey, a large national survey of a representative sample in the U.S.

TABLE 8 • 2
Attitudes Toward Premarital Sex: Selected Years, 1972–1988

Year	Always Wrong	Almost Always Wrong	Wrong Only Sometimes	Not Wrong At All
1972	37%	12%	24%	27%
1974	33%	13%	24%	31%
1975	31%	12%	24%	33%
1977	31%	10%	23%	37%
1978	29%	12%	20%	39%
1982	29%	9%	22%	41%
1988	26%	11%	22%	41%

Source: Kain, 1990, p. 130.

done by the National Opinion Research Center at the University of Chicago. The subjects were asked, "There's been a lot of discussion about the way morals and attitudes about sex are changing in this country. If a man and woman have sex relations before marriage, do you think it is always wrong, almost always wrong, wrong only sometimes, or not wrong at all?" The majority of people still think premarital sex is wrong at least sometimes. But the percentage of people answering "not wrong at all" gradually increased from 27 percent in 1972 to 41 percent in 1988.

Darling, Kallen, and VanDusen (1992) argue that changes in premarital sex have moved through three stages in the twentieth century. They call the first stage, which lasted from 1900 to the early 1950s, the era of the double standard. During this period, men often experienced sex before marriage but women rarely did.

The second era they label the era of permissiveness with affection. This was the transition period in the 1960s and 1970s during which the proportion of men experiencing premarital sex continued to increase and the proportion of women began to catch up with the men.

The current period they describe as one in which premarital sex is expected of young men and women and in which sex in nonlove relationships is acceptable. This era might be called permissive. The contemporary picture that emerges is one of more sexual activity, fewer emotional restrictions on that activity, and a fair amount of gender equality (Reiss, 1986).

Teen Romance Sharon Thompson's (1984) research on premarital sex among never-married young women depicts a different picture from the one seen by Darling, Kallen, and VanDusen (1992). Thompson (1984) interviewed fifty working-class teenage girls between 1978 and 1983. Her work uncovered three factors: (1) the lower age of first coitus does not necessarily mean that young people are more sexually active; (2) although young women are engaging in coitus at an earlier age, they may not be enjoying their encounters; (3) young women spend an enormous amount of energy infusing romance into their sexual experiences.

The statistics reported by Darling, Kallen, and VanDusen (1992) indicate that young women and men are having coitus at a younger age than in previous decades. But loss of virginity requires only one act of intercourse and research shows that first coitus does not mark the beginning of high levels of sexual activity. In the late 1970s, young never-married women who were not virgins reported having sex an average of 2.5 times per month. For whites the average was 3.0 times per month and for African Americans it was 1.7 times per month (Zelnick, Kanter, & Ford, 1981).

In her interviews Thompson (1984) also found that "having sex all the time is rarely the aftermath of first sexual intercourse for teenage girls" (1984, p. 368). She argues that it is easier to say no after saying yes once because intercourse no longer carries symbolic weight. In another study by Lillian Rubin (1991) two-thirds of the teenage women said they had tried it once and then decided to "put the issue on hold" (Rubin, 1991, p. 62).

Thompson observes that at least for some women, first coitus was perceived more as meeting a challenge or moving through a rite of passage rather than satisfying sexual desire. Lillian Rubin found this to be true among the adults—both men and women—she interviewed. When describing their first experience with coitus, Rubin's subjects used phrases like "crossing the great divide . . . the pressure was off . . . a burden was lifted . . . a hurdle was gotten over" (Rubin, 1991, p. 43).

Thompson (1984) observes, furthermore, that higher rates of coitus at earlier ages does not necessarily mean more sexual activity is taking place. She argues that coitus has taken on such significance in recent decades that other forms of sexual activity may have lessened. In Kinsey's research in the 1930s and 1940s, for example, respondents spoke of extensive noncoital petting and orgasmic sexual play. Her subjects in the 1970s and 1980s, in contrast described their sexual activities as more limited: "fewer partners, a limited repertoire of sexual activities and relatively little pleasure, particularly orgasmic pleasure" (Thompson, 1984, p. 363). Subjects in Rubin's research described the loss of pleasure in their own lives when they moved from "the kid stuff" to the "real thing." One woman who was nineteen when she first had intercourse described it this way:

> I mean it was directed to his coming, no sensuality, nothing soft and pleasurable. The so-called kid's sexual experience I had before, with whatever guy I was with, were much more in the nature of mutual pleasuring. I used to think about those experiences and miss them and wonder why I ever got into the so-called real thing. (Rubin, 1991, p. 57)

The second theme that emerges in Thompson's work is the frequent lack of enjoyment associated with coitus. Some young women spoke of physical discomfort, others just were disappointed it hadn't lived up to their expectations. One teenager said, "I was expecting it to be much nicer than it felt. It didn't hurt or anything. It just didn't feel really good" (Thompson, 1984, p. 365).

The third theme in the interviews Thompson conducted was the elaborate romanticized context the women created around their sex lives. They spoke for long periods of time about the men: details about his appearance, the first time they met, how they broke up, and their feelings for him. They spent hours with their girlfriends rehearsing and embellishing these stories. Unlike the conclusion reached by Darling, Kallen, and VanDusen (1992) that sex in nonlove relationships is increasingly acceptable, Thompson found that romance was the focus of her subjects' discussion of sex.

A final issue is the question of the double standard—condoning sex for boys and condemning it for girls. Darling, Kallen, and VanDusen argue that the double standard has disappeared. Thompson did not address this issue, but Rubin's qualitative work suggests that the issue is more unsettled. She describes the current period as one of tolerance and entitlement where both teenage women and men feel entitled to engage in sex. Important gender differences, however, still emerge around the question of monogamy.

Rubin found that women who "sleep around" are consistently condemned while men who are not monogamous are not often criticized. She observed that even the terms the people she interviewed used for this behavior—"stud" for men and "slut" for women—suggest admiration for the men and disparagement for the women. When Rubin asked college aged–men if they wished women were less sexually available they always said no, but they also frequently complained of women who were "too sexually active . . . too easy . . . or who come on too strong" (Rubin, 1991, p. 118). And 40 percent of the women interviewed said they understated their sexual activity because "they fear people wouldn't understand or would think they were sluts . . . or their boyfriends would be upset" (Rubin, 1991, p. 119).

Quantitative and Qualitative Methodologies in Research on Sex

Sociologists use a variety of methods to examine social phenomena. These methods can be divided roughly into quantitative and qualitative approaches. Quantitative research stresses the compilation of information that can be quantified—turned into numbers. Quantitative researchers frequently use questionnaires and ask questions that have limited possible responses such as yes, no, and sometimes. Qualitative researchers, on the other hand, emphasize the importance of determining the meaning of responses. They are concerned not just with whether a respondent says yes or no but why, from the respondent's point of view, he or she gave that answer.

Darling, Kallen, and VanDusen's (1992) work is based on quantitative data. They ask: How many people participate in certain activities? What age were they the first time they experienced coitus? How many agree or disagree with certain activities? All of these questions can be answered with numbers and proportions.

Thompson (1984) and Rubin (1991) are qualitative researchers. They collected information through interviews that allowed respondents to elaborate on their answers and provide definitions and ways to think about the issues that come from their own experience. The data in their research provide a sense of the meaning of the activities within the framework of the respondents.

Both kinds of research are important. Quantitative data usually provide a stronger sense of the representativeness of the social phenomena because these projects are usually conducted on large representative samples. If we know what proportion of a large population, such as the nation, is affected by the issue we have a better sense of how important it is to the society as a whole. For example, on the question of premarital sex we find that over the past century increasing numbers of people are engaging in coitus at younger ages and that the gap between women and men is closing.

Qualitative research helps us to see the complexity of the issues. Larger numbers can have many different meanings. Thompson's and Rubin's work

TABLE 8 • 3
Marital Coitus: Median Frequency per Week, Male and Female Estimates Combined, 1938/46 and 1972

Age	Kinsey 1938–1946	Hunt 1972
16–25	2.45	3.25
26–35	1.95	2.55
36–45	1.40	2.00
46–55	.85	1.00
56–60	.50	1.00

Source: Hunt, 1983, p. 231.

shows, for example, that the closing gap in premarital coital experience between women and men indicates one feature of sexual experience may be more equal but that many other gender differences remain, and in some cases the gap between women and men may actually be increasing (Schneider & Gould, 1987).

Marital Sex

Tables 8-3 and 8-4 provide data on historical change in frequency of coitus among married couples. Table 8-3 shows information from the Kinsey report collected from the late 1930s until the late 1940s and data from the Hunt research done in 1972. Table 8-4 provides data from the Blumstein and Schwartz research done in the early 1980s. Unfortunately, the two tables are organized differently, making comparisons difficult. Table 8-3 breaks down the information by age of respondent regardless of the number of years they have been married. It also records the median number of times sexual intercourse occurred in a week. Table 8-4 breaks down the data by the number of years of marriage but does not give the age of the respondents. In addition the quantity of coitus is reported in three possible ranges rather than in medians in Table 8-4.

In both tables, couples who are younger or have been married fewer years report more coitus than those who are older or married longer. Table 8-3 indicates that the frequency of coitus increased between the 1930s and the 1970s in every age category. Determining the exact changes between the 1970s (Table 8-3) and the 1980s (Table 8-4) is not possible but it appears that the numbers in the 1980s have remained similar to those in 1972. For example, the youngest age group in Table 8-3 report that 50 percent had coitus 3.25 times a week or more. The group married the least number of years in Table 8-4 (which might loosely relate to age) shows that 45 percent had coitus 3 times a week or more.

Table 8-5 allows us to make another set of comparisons. This table compares the proportion of couples reporting they have sex three or more times a week within four categories: married couples, cohabiting heterosexuals, gay men, and lesbians. The data are further subdivided into groups depending on how long the couples have been together. The table shows that couples who have been together longer are less likely to report having sex three times a week or more. It also shows a smaller proportion of lesbians reporting having sex three times a week or more and a larger proportion of gay men reporting they have sex three times a week or more. The lower rates reported for lesbians compared to the other couples and the higher rates reported for gay men have been interpreted in two ways. Some have argued that men have a biological propensity to desire more sexual activity and therefore in men-only couples more sex occurs because pressure comes from both partners, while for couples

TABLE 8 • 4
Marital Sex: Years Married and Sexual Frequency, 1983

	Years Married:		
	less than 2	2 to 10	10 or more
Sex once a month or less	6%	6%	15%
Sex between once a month and once a week	11%	21%	22%
Sex between one and three times a week	38%	46%	45%
Sex three times a week or more	45%	27%	18%

Source: Blumstein and Schwartz, 1983, p. 196.

with one woman and even more when there are two women, the pressure is less.

Other scholars (Blumstein & Schwartz, 1983) have argued that the difference is not biologically based. They assert that men are more likely to initiate sex when they wish to because that is part of being masculine in our culture. Women, on the other hand, may have learned to feel less comfortable making their sexual needs known. Second, as we observed in the discussion of love above, men have learned to express their feelings through having sex, while women have learned to express their feelings through other forms of intimacy.

The Marriage Bed Like the quantitative data on premarital sex, the statistics on sex in marriage do not tell us what the changes have meant for those who are living through them. Hunt (1974) maintained that these numbers show that Americans were becoming more sexually liberated and that people's sex lives, especially married people's, were improving. He argues that historically in Western society, sexuality was just barely tolerated as legitimate, only a step this side of hell. Hunt (1983) quotes Saint Paul to illustrate the message of conservative Christianity on the question of marital sex:

> I would that all men were even as I myself (celibate) . . . I say therefore, to the unmarried and widows, it is good for them if they abide even as I. But if they cannot contain, let them marry: for it is better to marry than to burn. (quoted in Hunt, 1983, p. 219)

In the 1970s, however, this perception of marital sexuality was altered, according to Hunt: "The principal effect of sexual liberation upon American life has been to increase the freedom of husbands and, even more so, of wives to explore and enjoy a wide range of gratifying sexual experiences within the marital relationship" (Hunt, 1983, p. 222).

Hunt was especially impressed with the number of married couples who incorporated fellatio and cunnilingus into their sexual repertoire. He referred to the increasing percentage of couples reporting they engaged in oral sex as a change of "major and historic proportions" (Hunt, 1974, p. 198).

Hunt's assessment contrasts with Rubin's (1976) observations. When Rubin (1976) interviewed working-class couples she asked them about sex in their marriage. She found that women and men experienced sex differently, especially in

TABLE 8 • 5
Percentage of Couples Answering Yes to the Question: "Do you have sex three times a week or more?"

Years Living Together	Married Couples	Cohabiting Heterosexuals
2 or less	45%	61%
Between 2 and 10	27%	38%
10 or more	18%	very few in this group
Years Living Together	**Male Couples**	**Female Couples**
2 or less	67%	33%
Between 2 and 10	32%	7%
10 or more	11%	1%

Source: Blumstein and Schwartz, 1990b, p. 126.

their expectations about sex. Although married couples may be engaging in more oral sex, husbands and wives frequently disagree about whether this is producing more sexual gratification for both of them. And the expectation that oral sex should be a part of their experience sometimes caused difficulties between the husbands and wives.

On the average, men and women in her study differed in their desire to experiment in sex. As an example, Rubin quotes a husband and wife, who were interviewed separately, discussing oral sex. The husband says,

> I've always been of the opinion that what two people do in the bedroom is fine; whatever they want to do is okay. But Jane, she doesn't agree. I personally like a lot of foreplay, caressing each other and whatever. For her, no. I think oral sex is the ultimate in making love; but she says it's revolting. I wish I could make her understand. (Rubin, 1976, p. 138)

His wife has a different opinion. She says,

> I sure wish I could make him stop pushing me into that (ugh, I even hate to talk about it), into that oral stuff. I let him do it, but I hate it. He says I'm old-fashioned about sex and maybe I am. But I was brought up that there's just one way you're supposed to do it. I still believe that way, even though he keeps trying to convince me of his way. How can I change when I wasn't brought up that way? I wish I could make him understand. (Rubin, 1976, p. 138)

Rubin discovered that part of the reluctance of the women was based on their belief that if they indulged in sexual behavior other than coitus, they might be judged as slutty or cheap by their husbands. Most of the husbands said they would not judge their wife's behavior that way.

The husbands, however, also expressed pride in their wives' sexual naiveté. The husbands frequently made statements like, "She was like an innocent babe, I taught her everything she knows" (Rubin, 1976, p. 142). These comments were made even by husbands whose wives were in their second marriage. Rubin concludes that wives receive mixed messages. They are supposed to be interested in trying new sex techniques but they are also supposed to appear to be innocent and naive.

Women are not the only ones who question the liberalization of sex. Men, too, sometimes express discomfort and apprehension. Their anxiety seems to

TABLE 8 • 6
Respondents Reporting at Least One Nonmonogamous Instance in the Previous Year

	Percentage	n
Husbands	11%	1510
Wives	9%	1510
Male cohabitors	25%	288
Female cohabitors	22%	288
Homosexuals	79%	943
Lesbians	19%	706

Source: Blumstein and Schwartz, 1990b, p. 127.

center around the fear of losing control, since an important feature of the social construction of masculinity in our society is control. Sex can make men feel out of control and at the mercy of a woman. One man told Rubin (1981) in another set of interviews she conducted:

> I'm not always comfortable with my own sexuality because I can feel very vulnerable when I'm making love. It's a bit crazy, I suppose, because in sex is when I'm experiencing the essence of my manhood and also when I can feel the most frightened about it—like I'm not my own man, or I could lose myself or something like that. (Rubin, 1981, p. 107)

In a similar vein another man said,

> Sometimes I can get scared. I don't even know exactly why, but I feel very vulnerable, like I'm too wide open. Then it feels dangerous. (Rubin, 1981, p. 109)

Carol Vance (1984), as I will discuss below, has written that the central tension in sexuality for women is between pleasure and danger. Gad Horowitz and Michael Kaufman (1987) propose that for men the central tension is between pleasure and power. Men seek pleasure in their sexual experiences but they are also supposed to maintain control—control over women, control over themselves, control over the situation. Because sex can involve letting go of one's control of others and even oneself—giving in to the pleasure—some men are frightened of sex in the way these two men describe.

The statistics on sexuality indicate that in terms of quantity the experience of women and men is converging and that the level of sexual activity has increased for both. Rubin's interviews show that similar numbers can hide different meanings and that increasing numbers only tell part of the story of the important social and emotional changes taking place.

Nonmonogamous Activity Table 8-6 shows the percentages of people who say they have had sex at least once in the past year with someone other than their spouse or partner. The table indicates that marriage is a critical factor in determining the extent of nonmonogamous sexual activity. Married men and especially married women have much lower levels of nonmonogamous activity than cohabiting heterosexuals, homosexual, or lesbian couples (Kassoff, 1989).

When asked not just what has occurred in the previous year but what has occurred throughout the marriage the numbers are much larger. In Hunt's survey in the 1970s, married people were asked if they had ever had sex with

TABLE 8 • 7
Attitudes Toward Extramarital Sex: Selected Years, 1973–1988

Year	Always Wrong	Almost Always Wrong	Wrong Only Sometimes	Not Wrong At All
1973	70%	15%	12%	4%
1974	74%	12%	12%	3%
1976	69%	16%	12%	4%
1977	73%	14%	10%	3%
1980	71%	16%	10%	4%
1982	73%	13%	10%	3%
1984	71%	18%	9%	2%
1988	79%	13%	6%	2%

Source: Kain, 1990, p. 132

someone other than their husband or wife while they were married. Among the men who were at least forty-five years old, half said they had. Twenty percent of the women answered yes (Hunt, 1974).

Have these numbers increased? The numbers of people reporting nonmonogamous sexual activity in their marriage in the 1970s is almost identical to the numbers Kinsey reported in the early 1950s: half the men and one-quarter of the women, indicating little change over the period from the early 1950s to the mid-1970s.

During the 1980s the numbers seem to have increased. Smith (1990) reported in 1989 research that 70 percent of the men and 35 percent of the women in his sample had extramarital sex at some time in their marriage. Very recently there is some indication a reversal has taken place. In a 1990 survey (Greeley, Michael, & Smith, 1990) 96 percent of married people said they had had only one sexual partner—their spouse—during the previous year.

Most people believe that married people should have sexual relations only with each other (Reiss, 1986). However, ideology about extramarital sex followed a pattern of greater liberalization during the 1970s and then increasing conservatism in the 1980s. Table 8-7 shows this pattern in the change in opinion about whether extramarital sex is wrong or not from 1973 to 1988.

Some researchers argue that the recent change toward greater conservatism, or at least not increased liberalism, on the question of extramarital sex may be related to concern with STDs (sexually transmitted diseases), especially AIDS. Some people say they are changing their behavior as a result of their fear of AIDS, as shown in Table 8-8.

Given the graveness of the issue, however, the numbers are relatively small for all groups and especially for married people, women, whites, and the middle-aged. It is difficult to interpret the meaning of the numbers, however. The small numbers stating they have changed their behavior could mean that most people, especially people in those categories, are foolishly unconcerned. The data could also be interpreted, however, as evidence that most people, especially those in the low number categories, were not engaged in behaviors associated with AIDS. For example, those who are celibate or who are involved in long-time monogamous relationships in which both partners are sexually exclusive and neither have been exposed to someone else's blood, may feel they do not have to change their behavior. The numbers probably reflect a combination of these factors.

TABLE 8 • 8
People Who Have Changed Their Sexual Behavior as a Result of Their Fear of AIDS

Married	7%	Men	13%
Never married	22%	Women	9%
Whites	10%	Under 25	19%
Blacks	22%	Age 35–50	10%

Source: Greeley, Michael, and Smith, 1990, p. 260

Extramarital sex in this section has been defined broadly as sexual activity of committed partners with people other than their spouse or partner. Research shows that the proportions range from 9 to 79 percent, with the highest numbers among homosexuals and the lowest among wives. There is some evidence that the numbers reporting extramarital activity increased between the 1950s and 1980s and attitudes became more liberal during those years. In the late 1980s and 1990s there seems to be a reversal, which some have linked to the increased fear of AIDS. When people are asked, however, about whether concern about AIDS has changed their behavior, a large majority say no.

SOME THEORETICAL DEBATES ON SEXUALITY

The description of sexual behaviors and ideologies shows them to be dynamic and diverse, changing over time. What exactly is sexuality? Can people be categorized by sexuality? Is sex best identified as dangerous or pleasurable? Is sexuality repressed in our society? In this section we will discuss some of the theoretical debates over the answers to these questions, beginning with the question of sexual categories.

Heterosexuality, Homosexuality, and Lesbianism

In some of the tables discussed in the previous section, respondents were divided into three groups, heterosexual, homosexual, and lesbian. Most people in the U.S. identify themselves as heterosexual, homosexual, lesbian, or bisexual. These distinctions are politically and socially important. A large and important gay and lesbian movement is based upon the recognition of the validity of sexuality that is not heterosexual.

Some theorists argue that distinctions between heterosexuals and homosexuals are biologically or socially determined early in a child's life. This point of view is an essentialist argument: Sexuality is perceived to be an orientation that is given. It is an essential, immutable part of an individual's character.

Other theorists (Blumstein & Schwartz, 1990b) argue, in contrast, that the categories are entirely socially constructed. "Sexuality is situational and changeable, modified by day-to-day circumstances throughout the life course," say Blumstein and Schwartz (1990b, p. 173). They maintain that there are no scientific criteria by which to sort people into these limited number of categories. Human sexuality is much too fluid and diverse to capture in the four slots commonly referred to in our society: heterosexual, homosexual, lesbian, and bisexual. Sexuality includes self-identification, sexual histories, large ranges of sexual activities, physical arousal, opinions, ideologies, fantasy, and physical contact.

For example, many people have had sexual contact with someone of the same sex but identify themselves as heterosexual; others have had sexual contact with someone of the other sex and consider themselves homosexual. Many people's choice of partner varies by sex over time. Some people experience a dissonance between their ideas and their physical reactions and are physically aroused by sexual stimulation they believe to be disgusting. Some people are aroused by viewing the sexual activity of others but do not participate. For example, heterosexual men may enjoy viewing erotic films of lesbians. People fantasize during sex about having sex with someone of a different sex. People engage in group sex, and so on.

The social constructionists believe that attempting to create a sexual typology that could differentiate all the possibilities would result in a large number of categories and would make the whole effort meaningless. Social constructionists point out that the task would be further complicated if one were to look at the issue historically. A review of the history would show that the practice of dividing people into categories based on their sexual activity has not existed for more than a century or so. Karl Kertbeny coined the terms heterosexual and homosexual in 1868 in Germany. The first documented use of the term homosexual in the U.S. was in 1892 (Katz, 1990).

Homosexual and heterosexual behavior existed prior to the invention of these terms. The important change the words indicate, however, is the identification of individuals as a function of their sexual activity. Before the nineteenth century people were known to participate in sexual activities with those of their own sex. This activity, however, was not seen as indicative of a particular type. For example, sometimes men would engage in anal intercourse with other men, sometimes they might perform it on women. In either case they were not seen as therefore being a particular type of person, a homosexual or a heterosexual.

The social constructionist argues that the rational alternative is to accept the diversity and not try to fit people into sexual types. Kinsey wrote:

> The world is not to be divided into sheep and goats. Only the human mind invents categories and tries to force facts into separate pigeonholes. The living world is a continuum. (Kinsey, 1948, p. 689)

This rational alternative, however, ignores the importance of the oppression of people because of their sexuality and the social movements that have developed to try to end that oppression. Altman summarizes this dilemma by stating:

> On one level to love someone of the same sex is remarkably inconsequential—after all, but for some anatomical differences, love for a man or a woman is hardly another order of things—yet society has made of it something portentous, and we must expect homosexuals to accept this importance in stressing their identity. (Altman, 1971, p. 219)

Sexuality remains a salient political arena for gay men and lesbians. And self-identification as homosexual, bisexual, or lesbian is an important aspect of the affirmative character of the gay and lesbian rights movement.

Feminist Debates on Sexuality

For the women's movement also, sexuality is an important political arena and there are a number of theoretical and political splits within the movement over the question of sexuality. A central aspect of the Second Wave women's move-

The feminist debate over pornography reflects important differences between those who emphasize the dangers for women associated with sexuality and those who emphasize its pleasures.

ment was the quest to free women sexually. Women called for a change from their position of sex objects to subjects of their own desires. They explored their own and other women's bodies in workshops on self-examination and they formed consciousness-raising groups to discuss orgasms, cunnilingus, eroticism, and lesbianism (Ehrenreich, Hess, & Jacobs, 1986). As we have seen in the discussions above, women increasingly engaged in premarital and marital coitus. We have also seen, however, that this liberalization of sexuality was not without problems. The sexual relationships among people and particularly between women and men are highly charged with questions of power. And the politics of sexuality are contradictory, offering oppression as well as opportunity, victimization as well as fulfillment and happiness.

In the 1970s some feminists began to question the liberalization of sexuality and its meaning for women. Organizations like Take Back the Night and Women Against Pornography marched under the banner of "pornography is the theory, rape is the practice" and well-known feminist scholars took leadership roles (D'Emilio & Freedman, 1988, p. 351). Andrea Dworkin (1981) and Catherine MacKinnon for example, wrote anti-pornography legislation that was adopted by the City Council in Indianapolis and reviewed in a number of other cities (R. Rich, 1986).

The focus on the dangers of sexuality and on the link between pornography and rape sometimes were coupled with the argument that male sexuality (of which pornography was a part) was usually violent and oppressive, and therefore that heterosexuality for women was dangerous. In addition to being personally dangerous, from this point of view, heterosexuality impeded the ability of the women's movement to end the oppression of women. Rita Mae Brown wrote,

> Straight women are confused by men, and don't put women first. They betray lesbians and in its deepest form, they betray their own selves. You

can't build a strong movement if your sisters are out there fucking with the oppressor. (Brown, 1976, p. 114)

Adrienne Rich (1980) wrote the most well-known piece on the question of the link between heterosexuality and the oppression of women. Rich reviewed the empirical evidence of men's violence against women, especially their sexual crimes against women, rape, coerced prostitution, incest, the objectification of women in sexual encounters, and the ways in which women are battered, restricted, and terrorized in order to keep them at men's sexual disposal (see also Barry, 1981). She then posed the question, "Since men appear to be so problematic in women's lives, why do women continue to seek men as their sex partners?" Rich's answer to this question is that keeping lesbianism invisible, or if necessary, punished by violence or at least ostracism and shame, leaves women, if they are to be sexual, with no option except men. According to Rich, lesbianism therefore is a political act, an attack on men's access to women.

Rich's formulation links sexuality to the oppression of women. In a society like ours where sexuality is defined as heterosexuality and lesbianism is invisible, women's availability is guaranteed to men, and this availability makes them vulnerable to frequent exploitation by men. From this point of view, in order to solve the problem of gender inequality and the oppression of women, sexuality must be reconstructed. And in order to create a social movement capable of ending the oppression of women, women must remove themselves from relationships with men and begin to commit themselves to other women. This commitment can take the form of genital sex, what we usually think of as lesbianism. Rich proposes, however, that we think in terms of a range of activities, a "lesbian continuum." By this she means women sharing a range of emotional, practical, and political support that can include sex but does not necessarily do so.

All feminists, however, do not agree with this emphasis on the dangers of sexuality, especially heterosexuality, for women. An active and sometimes volatile debate on this issue dominated the 1980s (Freedman & Thorne, 1984).

Alternative Feminist Viewpoints Gayle Rubin (1984) takes a position on sexuality that contrasts with that of Rich, Brown, Dworkin, and MacKinnon. She focuses on the pleasure of sex and the need to make many forms of sexuality acceptable, rather than to restrict women's (and men's) options. Although she would agree with Rich's call for making lesbianism visible, she would disagree with discarding heterosexuality as a valid option as well. Her concern is with ending what she refers to as the barbarity of sexual persecution.

Gayle Rubin (1984) has created a visual depiction of the way in which sex is "ranked" in our society with certain behaviors labeled as "good, normal, natural, and blessed" while others are supposedly "bad, abnormal, unnatural, and damned" (see Figure 8-2). She asserts that while sexuality is entwined with gender inequality, the solution to the oppression of women is not to oppress them further by defining valid and invalid forms of sexuality. Our goal should be instead to end both the oppression of women and the oppression of sexual minorities, including those who use pornography.

Carol Vance (1984) tries to tie these two opposing sides together by arguing that the central character of the politics of sexuality for women is this tension

FIGURE 8 • 2
The Sex Hierarchy: The Struggle Over Where to Draw the Line

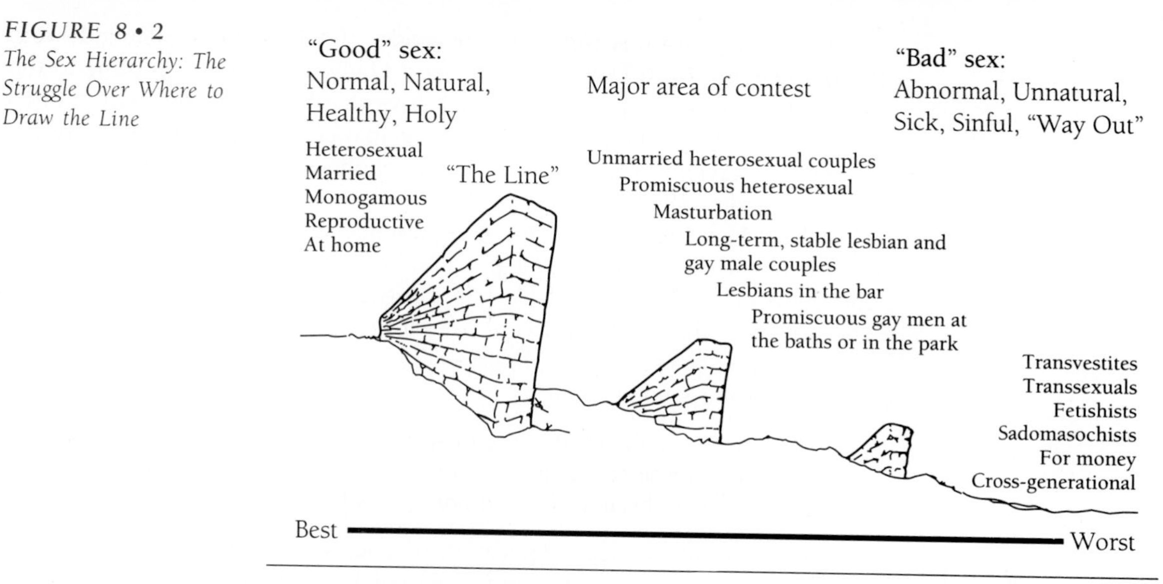

Source: G. Rubin, 1984, p. 282.

between pleasure and danger. The dangers of sexuality include "violence, brutality, and coercion in the form of rape, forcible incest, and exploitation as well as everyday cruelty and humiliation" (Vance, 1984, p. 1). The pleasures of sexuality include "the positive possibilities of sexuality—explorations of the body, curiosity, intimacy, sensuality, adventure, excitement, human connection, basking in the infantile and non-rational" (Vance, 1984, p. 1). Vance maintains that we must consider both sides. Our activism must address the problems of violence and coercion, but it must also fight against "the repression of sexuality, especially women's sexuality, that comes from ignorance, invisibility, and fear" (Vance, 1984, p. 23).

Explaining the Focus on Sex

The discussion of sex in this chapter reveals an assumption in the writings of many authors that sexuality is a powerful natural force controlled by society and that recently those restraints have been lessened. Researchers like Hunt (1974), Blumstein and Schwartz (1990b), and Gayle Rubin (1984) focus their attention on the positive features of unleashing sexuality from the tight grip of society. Others (Rich, 1980; MacKinnon, 1982; Dworkin, 1981) have been concerned with the problems that come with breaking down barriers to the expression of sexuality. All of these scholars, however, share a conceptualization of sex as a human drive that is ever-present and restrained, repressed, or set free by the social systems in which humans live.

In the scenario at the beginning of the chapter one of the comments Lori makes illustrates the popular notion of this idea. She says maybe she is doing the community a favor by helping the telephone callers to release their sexual tensions by talking on the phone rather than in some other less acceptable behaviors. The underlying assumption is that sex is a powerful force that must be expressed in some form and a community should be concerned with controlling and directing its expression.

This point of view is rooted in the works of scholars like Sigmund Freud (1963), Wilhelm Reich (1945), and Herbert Marcuse (1966). According to this framework, sexuality is a strong essential drive that has been repressed in order to establish and maintain civilization. As a matter of necessity, sexuality has become "subservient to labour discipline, to subdue hedonism in the interest of progress" (Brake, 1982a).

Michael Foucault's (1978) work represents an important critique of this point of view. He posits that sex is not a biological urge repressed by society but a socially constructed urge with socially constructed means to satisfy that urge. He would argue that our recent concern with sex is not because it has been set free but because our society emphasizes certain forms of sexuality. Jeffrey Weeks explains Foucault's point of view:

> We give supreme importance to sex in our individual and social lives today because of a history that has assigned a central significance to the sexual. It has not always been so; it need not always be so. (Weeks, 1985, p. 3)

The important political implication of Foucault's and Weeks's position is that we can and must decide what place sexuality should have in our lives and in our society. The debate over sexuality cannot be one of repression versus freedom. Instead we need to do even more fundamental thinking and planning about what sex is and how it should be structured within our society.

According to theorists like Foucault and Weeks who emphasize the importance of social context in understanding sexuality, the key factor in how sex is socially constructed in our society is its commodification and therefore its growth along with the development of consumer capitalism (Altman, 1982; D'Emilio & Freedman, 1988). Sex is a lucrative business. Industries use sex to sell everything from cars to blue jeans. In addition, sex itself is for sale in a vast array of formats including films, television, music, clothing, books, and telephone messages. The commercialization of sex has an enormous impact on the promotion, repression, and all-around shaping of sex in contemporary U.S. society. Foucault and Weeks would see the phone sex business discussed in the opening scenario as an example of this.

This section has reviewed three important theoretical debates on sexuality. In the first subsection we looked at the question of diversity in sexuality. The dominant ideology about sex in our society posits that each individual can be identified as one of four types: heterosexual, homosexual, lesbian, or bisexual. Theorists assert that individual sexuality is much more variable and complex than any system of categorization could capture.

In the second subsection we examined the central tension of the expression of sexuality by women in our society, pleasure and danger. This tension is present not only in individual women's lives but within the women's movement itself. Some feminist theorists argue that popular expressions of sexuality, especially heterosexuality, are so permeated with misogyny and violence against women that women must work hard both to remove themselves from the danger and to reconstruct sexuality in a more woman-identified way. On the other side, feminist theorists maintain that our goal should be to increase the diversity of acceptable forms of sexuality, no matter how repugnant they may be to us individually.

In the third subsection, our attention turned to the question of whether sex is an intrinsic human drive that is always repressed and controlled by society. This view comes especially from Freudian thought. Others, like Foucault and Weeks, maintain that sex is both created and shaped by social interaction at both the micro and macro level.

In the next section we will move to a more concrete issue that has taken center stage in political debates in the United States. The question of whether abortion should be accessible to women is an example of a practical but volatile issue within the debate over how sexuality should be organized and especially how its expression among women should be constructed and controlled.

ABORTION

Thus far this chapter has shown the way in which the macro level of our society has had an important effect on the micro level of our most intimate social relationships, love and sex. An economic system commercializes sex and a political system passes laws to control it. We have also observed the way in which the macro level has been challenged and shaped by the micro forces of individuals interacting with one another to form social movements to recreate sexuality. The abortion rights movement is a particularly important social movement that has organized to allow its view of sexuality, particularly women's sexuality, to exist. The abortion rights movement will be the focus of our attention in this section.

One issue that has become central in the debate over women's sexuality is the legality of abortion. When the Second Wave feminist movement emerged in the late 1960s in the U.S. abortion was not legal. The fight to make it legal was a top priority on the feminist agenda and has continued to be a rallying point (Ellis, 1983). Feminists organized marches, educational programs, and rallies for abortion rights and feminist-identified lawyers argued the *Roe v. Wade* case that made abortion legal in 1973 (Joffe, 1986).

The legal restrictions, prior to *Roe v. Wade,* however, did not stop abortions from being performed. Leo Kanowitz (1969) estimated in the late 1960s that each year over a million illegal abortions were performed and that between 5,000 and 10,000 women died every year from complications from these back-alley procedures.

In 1965 the Supreme Court ruled that it was legal to sell contraceptives to married people. The growing feminist movement of the 1960s joined this battle, claiming that legal access to abortion was intrinsic to the right to control one's own body and was necessary if women were to claim equality (D'Emilio & Freedman, 1988). In 1973 *Roe v. Wade* was hailed as a major victory by the women's movement.

In the *Roe v. Wade* decision, the Supreme Court ruled that states could not pass laws that made abortion illegal during the first trimester of a pregnancy. The decision also declared that states could regulate abortions in the last six months of pregnancy, but those regulations must be based on restrictions related to maternal health. Finally, *Roe v. Wade* stated that only in the last ten weeks could abortion be banned entirely.

Today those who are opposed to legalizing abortion call their movement "Pro-Life." Those who advocate keeping abortion legal call their movement "Pro-Choice." Kristin Luker (1984) interviewed activists on both sides of the issue and found that people in each group shared a cluster of beliefs in addition to their opinion about whether abortion should be legal. The views of the pro-life group included the beliefs that the purpose of sex is to have children; that sex is sacred but is currently being profaned; that making contraception available only encourages young people to be sexually active; and that the way to halt the immoral behavior of unmarried teenagers is to eliminate sex education and the availability of birth control. Pro-life activists also believe that women and men should have equal status but that they cannot be equal because of their biological differences. Pro-choice activists believe that the primary purpose of sex is communication, not necessarily to have children. Their view of parenting is that it must be purposeful and designed to give children every possible advantage. Pro-choice activists believe that women should be equal to men and see abortion as one part of a package that includes birth control and sex education, designed to give women maximum control over their lives. The most important difference between pro-life and pro-choice activists is in the importance of religion in their lives: 75 percent of the pro-choice advocates said religion is either unimportant or completely irrelevant; 69 percent of the pro-life advocates, in contrast, said religion is important to them and 22 percent said it is very important.

Since the passage of *Roe v. Wade*, pro-life activists have organized and lobbied at the state level for restrictions that have subsequently been brought to the U.S. District Court and overturned. For example, in 1984 the state of Rhode Island passed a law (which was eventually overturned) requiring that physicians notify a woman's husband before performing an abortion (Lott, 1987).

At the federal level, pro-life activists successfully lobbied Congress to pass the Hyde Amendment in 1977. The Hyde Amendment allows states to prohibit the use of public funds (Medicaid) to pay for abortions of poor women, unless it can be proven that it is necessary to save the mother's life or that the pregnancy was a result of rape or incest. All but fourteen states (Alaska, California, Colorado, Georgia, Hawaii, Maryland, Michigan, New Jersey, New York, North Carolina, Oregon, Pennsylvania, Washington, West Virginia, and the District of Columbia) passed laws eliminating state funding. This means that if poor women wish to obtain an abortion in the other thirty-six states they must find their own funds to pay for it (Petchesky, 1990).

The *Webster* decision in 1989 dealt another blow to *Roe v. Wade*. Although the Supreme Court did not overturn *Roe v. Wade* in deciding *Webster*, the justices upheld the constitutionality of four aspects of the Missouri law that were under consideration. The Missouri law (1) states that the life of each human being begins at conception; (2) prohibits using public facilities or employees to perform abortions; (3) prohibits public funding for abortion counseling; (4) requires that physicians conduct viability tests on fetuses of twenty weeks or more before aborting those pregnancies (Stetson, 1991).

The issue of parental consent is one that has most recently become an area of contention. A number of states have passed laws requiring that parents be notified if a minor obtains an abortion. In 1990 this restriction was widely publicized after the death of Becky Bell. Becky Bell was an Indiana teenager who died

BOX 8 • 1 WAS ABORTION ILLEGAL IN THE U.S. UNTIL THE *ROE V. WADE* DECISION?

Many people believe that *Roe v. Wade* marked a turning point in American history—making abortion legal for the first time. In fact, prior to the middle of the nineteenth century, abortion was not a criminal issue in the U.S., nor was it condemned by the Catholic or Protestant churches if it occurred in the first few months of pregnancy before "quickening." Quickening was the term used to refer to the moment at which the fetus was believed to gain life and therefore a soul. There were a variety of opinions throughout the centuries as to when this moment occurred and the sex of the fetus affected the moment of quickening. For example, Aristotle estimated that quickening occurred for a male fetus at forty days after conception and for a female fetus at ninety days after conception (Gordon, 1977). Since technology did not exist that could precisely identify the time of conception, in practice quickening was identified as the point at which a pregnant woman felt the fetus move. Removing the contents of the uterus by chemical or surgical means before the woman felt movement was therefore not a crime or a sin. It was thought of as correcting the regularity of a woman's menstrual cycle (Petchesky, 1990).

Abortion services and chemicals that could be used by women themselves to initiate an abortion were widely advertised. During the nineteenth century ads referred to contraceptives as French; for example, condoms were called French letters. The term Portuguese referred to abortifacients. One ad in the *New York Times* in 1871 read, "A great and sure remedy for married ladies—the Portuguese female pills always give immediate relief . . . price $5" (Gordon, 1977, p. 54).

Allopathic physicians are what are now called medical doctors. Although allopaths have become so dominant that we now reserve the title of physician for them, in the nineteenth century they were just one group of health care practitioners among many. During the 1800s, many different kinds of health care providers existed, including homeopaths, osteopaths, allopaths, midwives, and abortionists. During the last half of the nineteenth century allopaths in the U.S. began a campaign to make abortion at any time in the pregnancy illegal. Part of the reason they took on this political crusade was to enhance their position in the competitive field. To wage their battle against abortion, allopaths formed the American Medical Association (AMA), passed a formal resolution against abortion, and then convinced Congress that abortion should be made illegal and that allopaths should be the only legitimate decision makers about whether a therapeutic abortion was justified. The AMA argued that abortion and contraceptives were part of a dangerous "female folk culture" that was destined to destroy The Family and domestic order in general.

Another surprise in the history of abortion in the U.S. is that nineteenth century feminists advocated making abortion *illegal* (Gordon, 1982).

Nineteenth-century feminists supported criminalizing abortion because they felt that legal abortion encouraged promiscuity among men who did not have to take responsibility for the pregnancies they caused (Ginsberg, 1990). The AMA and feminists were joined by the New York Society for the Suppression of Vice, headed by Anthony Comstock. In 1873 Congress passed "the Comstock laws." Among other things, the Comstock laws made it illegal to give information about abortions and contraception or to perform abortions.

from an illegal abortion. She felt forced to have an illegal abortion because in order to obtain a legal abortion in Indiana she would have had to inform her parents first, something she did not wish to do. Table 8-9 shows that young single white women like Becky Bell make up a significant proportion of the people seeking an abortion.

Table 8-9 also indicates that the largest proportion of women obtaining abortions are over twenty, with almost half over twenty-four. Abortion is an

TABLE 8 • 9
Who Has Abortions? 1973, 1988

Age	1973 (n = 743,000)	1988 (n = 1,591,000)
<15	2%	1%
15–19	31%	25%
20–24	32%	33%
25–29	17%	22%
30–34	10%	12%
35–39	6%	6%
>40	2%	1%
Race		
White	74%	65%
Nonwhite	26%	35%
Marital Status		
Married	29%	17%
Unmarried	71%	83%
Number of prior live births		
0	55%	51%
1	15%	24%
2	14%	17%
3	8%	6%
>4	7%	3%
Number of prior induced abortions	**1973**	**1988**
0	NA	57%
1	NA	27%
2+	NA	16%

Source: S. Henshaw & J. Van Vort (eds.), 1992.

issue for a broad range of women, although the issue of parental consent makes it especially problematic for women who are minors.

In addition to the changes in the legal issues surrounding abortion, the 1980s brought a technological change that further intensified the debate (Norsigian, 1990). In 1988, the French government approved the use of a new drug, RU486. RU486 is a steroid that comes in pill form. If taken along with an injection of progesterone by a pregnant woman within forty-nine days of her last menstrual period, it causes her body to expel the embryo. RU486 is an effective replacement for more complicated and sometimes more problematic surgical means of abortion that are currently in use in the U.S. (Lader, 1991). RU486 is still illegal in the U.S. but growing pressure from abortion rights activists is making it an important feature of the battle for choice.

At this point, the conflict over abortion rights exists in a number of arenas: the courts, political campaigns, and the streets. Pro-life advocates have staged civil disobedience actions—such as "Operation Rescue"—and have blocked entrances and fire-bombed abortion clinics. Pro-choice advocates have provided escorts through the blocked entrances, staged the largest protest marches in recent history in Washington, D.C., and launched their own program of nonviolent civil disobedience: "Operation Fight Back."

A Feminist View of Abortion

> Because of its historic link to the legalization of abortion, its contribution of a specific category of abortion personnel, and its ongoing engagement with the defense of abortion in the courts, the feminist movement can sustain a claim to ownership of the abortion issue. (Joffe, 1986, p. 37)

In the nineteenth century, feminists were in favor of restricting abortion, but in the late twentieth century, abortion rights have become a central component of the women's movement. Many pro–abortion rights activists are feminists and most feminists are in favor of keeping abortion legal (Simon & Danziger, 1991).

Ellen Willis (1983) argues that the reason abortion is so important to feminists is because it allows women to break the tie between sex and reproduction. The data presented in this chapter show that women are becoming increasingly engaged in marital and nonmarital coitus. Both unmarried and married women appear to be more sexually active but desire fewer children. Contraceptives are available, but some are dangerous and none are totally effective. Feminists view access to legal abortion as a way to guarantee that women can be heterosexually active and not be forced to continue an unwanted or unsafe pregnancy (Petchesky, 1990). This argument fits into the conceptualization of sex as an act of pleasure. Access to abortion allows women to enjoy the pleasures of sex without the possibility of unwanted reproduction.

As we observed in the previous section, sex involves another component besides pleasure for women and that is danger. Abortion also helps to make the dangers of sex less harmful. Feminists argue that women who are exploited sexually through rape and incest must also have the right to abort a fetus that results from those encounters.

Besides debates between pro-choice activists and pro-life activists, the question of abortion has also elicited debates between Second and Third Wave feminists (Petchesky, 1990). Recall from discussions in previous chapters that Second Wave feminists are those who emerged during the late 1960s. Their emphasis is on the issues they believe are shared by all women in the sisterhood of women. Third Wave feminists are those who have emphasized the differences among women, especially differences of race and class. In the recent history of the debate over abortion, both Second and Third Wave feminists agree that abortion should be kept legal. Third Wave feminists, however, have been critical of Second Wave feminists for stopping there and they have called for an expansion of the agenda (Fried, 1991; Davis, 1991; Walker, 1990).

From the Third Wave feminist point of view, legalizing abortion alone will do little to increase its accessibility to poor and working-class women. Third Wave activists have demanded that abortion rights be promoted as part of a package that includes economic as well as legal access to abortion. Third Wave feminists, furthermore, have insisted that women should have the right to choose not only abortion but also to bring a pregnancy to term within a health care system that provides them with prenatal, maternal, and pediatric care (Rodrique, 1990).

Racial ethnic women have taken a particularly strong stand on this question because of the way in which their right to bear children has been restricted. For example, in 1976 it was revealed that 3000 Native American women had been sterilized after signing inadequate consent forms (Hartmann, 1987). In 1970 a

The battle over whether to keep abortion legal is a heated one. Proponents of both sides—pro-choice or pro-life—have widely disparate beliefs about not only whether abortion should be legal but about what women and men should be like.

federal district court found that "an indefinite number of poor people had been coerced into accepting a sterilization operation under the threat that various federally supported welfare benefits would be withdrawn unless they submitted to irreversible sterilization" (Hartmann, 1987, p. 241). Nearly all the cases heard on sterilization abuse have involved poor women who were African American, Chicana, Puerto Rican, or Native American (Petchesky, 1990).

In the late 1970s, federal reforms were put in place to halt the abuse. They included "more rigorous informed consent procedures, a thirty-day waiting period . . . a prohibition on hysterectomies [as opposed to tubal ligations] for sterilization purposes, and a moratorium on federally funded sterilization of minors, the involuntarily institutionalized, and the legally incompetent" (Hartmann, 1987, p. 241). But evidence remains that abuse continues and recently proposals have been made to require that welfare recipients agree to using contraceptive implants in order to continue receiving their grants (Lewin, 1991).

In 1990 the Food and Drug Administration approved Norplant, a contraceptive that works as long as five years after it is implanted in a woman's upper arm. By 1991 100,000 women had received the implant and it was fully covered by Medicaid—federally funded health benefits for poor people. While its easy use could make it highly beneficial to women it also could be used in a coercive manner. Unlike other forms of contraception the presence of Norplant could be easily monitored by welfare officials or parole officers (Lewin, 1991). In Louisiana, State Representative David Duke, a Nazi Party and Ku Klux Klan member, proposed that welfare recipients be offered $100 a year to use Norplant. The ACLU (American Civil Liberties Union) is concerned that these proposals will escalate (Lewin, 1991).

Third Wave feminists share this concern and have called for a reconceptualization of reproductive rights that is broader in its view. Speaking from the Third Wave feminist point of view Fried (1991, p. ix) asserts,

> The public has been galvanized [by the Webster decision to fight for legal abortion]. But for what—freedom of choice circumscribed by race and class, removed from feminists' demands about women's autonomy, and shrouded in "privacy," or reproductive freedom for all women? . . . We have an opportunity to move ahead with a positive reproductive rights agenda. Doing so requires that we build the kind of movement we have not had in the past—one that is broad based in its membership, its leadership, and its politics; a movement that goes beyond reaffirmation of *Roe* to demand access not just to abortion, but to the full range of reproductive rights; a movement that is based on a class- and race-conscious feminism.

THE MICRO-MACRO CONNECTION

We often use the word choice when we discuss sex. Abortion rights are called pro-choice. Whether a person identifies him or herself as gay or heterosexual is referred to as sexual choice. Our ideas and experience of sex are supposedly personal ones, largely carried in our own minds and only occasionally shared with a few close people whom we choose. This chapter, however, has revealed that although sexuality is experienced by people at the micro level, sexuality is pressured, shaped, and molded by macro-level social forces.

Factors at the macro level that affect sexuality include economic institutions, laws, stratification, and ideologies. Economic institutions that sell sex or use sex to sell other goods affect our ideas and experience of sex. Laws determined by political bodies rule on the legitimacy of various forms of sex. Social stratification, especially by gender, has a critical impact on sex. Ideologies rooted in competing theories about sexuality also influence us at the micro level. When we enter into a sexual relationship with another person, we bring with us to this micro encounter an array of forces emanating from the macro level of social organization.

As I have emphasized throughout the book, the micro level has also had an effect on the macro level on the issue of sex. These ideologies, economic arrangements, and laws were developed through the interaction of individuals with one another at the micro level. Allopathic physicians, feminists, pro-life and pro-choice activists, antipornography activists, and gay rights activists are just a few of the micro-level organizations that have worked and continue to work to influence the organization of sexuality in the contemporary United States.

SUMMARY

What Is Love?

- The opening scenario shows the diversity of opinions about what constitutes legitimate sex and especially about the link between love and sex.
- John Lee divides love into six types; erotic, manic, ludic, pragmatic, storgic, and agapic.

- Ideas about love and the experience of love vary by gender and from one historical period to another.
- Cancian divides the history of love in the U.S. into four types: family duty, companionship, independence, and interdependence.
- The current period is one in which the last three types are in competition. Only the last one allows for both love and equality.

Human Sexuality

- The United States, like every society, creates rules about what constitutes legitimate sex.
- The six standards for sex are: heterosexual, coital, orgasmic, two-person, romantic, and marital.
- Sometimes these standards are written into law.
- Problems exist in much research on sex because the data are old, studies are not comparable to one another, or the information was collected from possibly biased sources.
- Since one of the standards of sex in our society is that it is linked to marriage, sex can be divided into a variety of topics based on its relationship to marriage, such as premarital sex, marital sex, and nonmonogamous sex.
- Premarital sex experience has increased for both women and men and the gap between the two has closed.
- The subjective experience of sex, however, still remains quite different for women and men.
- Qualitative and quantitative researchers have looked at sex and its link to marriage and families in different ways.
- Marital sex appears to have increased in frequency between 1950 and 1970 and has remained about the same since then.
- Comparisons of frequency of sex between committed partners show that it is highest for gay men and lowest for lesbians, with married and cohabiting heterosexuals falling in between.
- Hunt's assessment of recent changes in sexuality between husbands and wives emphasizes positive effects. Rubin cites problematic issues emerging from the changes.
- Nonmonogamous sexual activity appears to have increased through the 1970s and 1980s but very recently has reversed, perhaps because of fear of AIDS.

Some Theoretical Debates on Sexuality

- Besides collecting empirical data on sex, scholars have also constructed theories about sexuality. These include theories about the range of sexual experience and whether it can be categorized into a few variations; the contradictory character of sexuality for women; and the question of whether sex is an essential drive or whether it is socially constructed.

Abortion

- The abortion rights movement is an example of a social movement that has attempted to support the right of women to express their sexuality in a way that breaks the link between sexuality and reproduction.

- The history of abortion shows a number of contending forces working to create laws about abortion. In the nineteenth century the fight to make abortion illegal was led by feminists and physicians.
- The abortion rights movement is now identified with feminists, although these are important debates about how best to conceptualize abortion as an aspect of feminist goals.
- Third Wave feminists have been especially critical of limiting the focus of the reproductive freedom movement solely to legal abortion. They argue that abortion must be part of a total package that guarantees the right to bear children and control reproduction as well.

The Micro-Macro Connection

- Economic factors, technology, and laws are all part of the larger social context that helps shape our intimate relationships of love and sex.
- Gay rights activists, a variety of feminists, pro-choice and pro-life advocates are all examples of people attempting to alter the macro organization of our society in ways they believe will enhance human interaction.

Until the 1960s marriage between people of different racial ethnic groups was illegal in the United States. The numbers have increased since that time, but it is still relatively unusual for people to marry across race lines.

9 MARRIAGE

Sharon Kowalski and Karen Thompson lived together as committed partners in Minnesota. Because the state of Minnesota does not allow people of the same sex to marry, they were not legally married to each other. The two women, however, had fallen in love, vowed their commitment to each other, exchanged rings, and bought a house together (Griscom, 1992).

In 1983, after they had been living together for four years, Sharon was in a head-on car collision in which she was nearly killed. The crash left her almost completely paralyzed and unable to speak. Since the accident Sharon has developed the ability to use a typewriter and even to speak a few words, but she remains nearly immobile.

Immediately following the accident, Karen stayed by Sharon's side for hours at a time read-

ing and talking to her and massaging her muscles so that they would not be damaged by the inactivity. At first Sharon's parents accepted Karen's presence but after a short time Sharon's father told Karen that only family could love his daughter and that Karen should not spend so much time with her. Sharon's father also disagreed with Karen about the care Sharon was receiving. He preferred to place her in a nursing home that offered little rehabilitation service. Karen believed that Sharon was making progress and would improve with sufficient care. But since Karen was not legally married to Sharon she had no legal right to be involved in decisions about her partner's care or even to visit her.

Karen decided to file for guardianship of Sharon. She and Sharon's father went to court to decide who would have control over Sharon's life. After a long series of decisions and counterdecisions the court awarded Sharon's father full guardianship. Karen was not permitted to visit Sharon for three years, from 1985 to 1988. While the court was deliberating its decision, Sharon submitted testimony that she could communicate with a typewriter and that she was gay and wished to live with her lover, Karen Thompson. In her last visit with Karen before her father barred Karen from seeing her, Sharon typed to Karen, "Karen, help me. Get me out of here. Take me home with you" (NC NOW, 1992). However, the court ruled that Sharon was incompetent to testify in her own behalf.

Sharon's father transferred her to a small nursing home with few rehabilitation services. Before the move, Sharon had begun to stand and feed herself. In the new home, Sharon was locked away for three years with a feeding tube and given insufficient stretching and exercise. She was unable to leave the home and prohibited from seeing her friends and her lover. Her physical condition deteriorated but both she and Karen continued their efforts to live together again. In December of 1991, after hearing overwhelming medical testimony that it would be in Sharon's best interest to live with Karen, the court finally decided to allow Sharon to go home.

- *How would this story have been different if Karen and Sharon were a legally married couple?*
- *What is a marriage? Is it only a legal institution or does it consist of other factors like commitment, economics, power, love, or sex?*
- *Is the partnership between Karen and Sharon a marriage?*
- *In answering this question what criteria are you using for defining marriage? What do those criteria imply for both gay and lesbian couples, and heterosexual couples?*
- *How is the institution of marriage problematic for those who are currently legally allowed to marry?*
- *Although many gay and lesbian couples have fought for the right to marry, others have argued that the institution of marriage is so fraught with problems it is not a worthwhile goal. What is your opinion of this debate? Should everyone be allowed to marry? Should no one be allowed to marry? Or is there a middle ground?*

In answering the questions above you have been asked to define marriage: who qualifies as married and what rights and obligations does that entail? Coming up with a definition is difficult because marriage is a many-faceted social institution. It is defined by law but it also includes economic arrangements and sexual, political, and emotional factors like love and commitment. This chapter will investigate the various facets of marriage and their often problematic character.

The chapter is divided into five major sections. The first investigates the different factors that go into defining marriage. The second section presents data on the incidence of marriage and a description of the changes over this century in our experience of marriage. The third section examines the changes in ideas about marriage over the past few decades. In the fourth section a theoretical model that links class inequality, gender inequality, and marriage is reviewed. The fifth section looks at the political debate over whether marriage is a valuable, humane institution or one that should be radically restructured.

Marriage Is a Legal Contract

Marriage is a social institution that has several components. These include legal, economic, emotional, sexual, and political factors. In this section we will explore each of these factors as an aspect of marriage, beginning with the link between marriage and the law.

Marriage is a legal contract in the United States. The macro-level legal organization of marriage has a powerful effect on the micro-level experience of people who enter into marriage with each other. Even though we think of marriage as a private affair, a marriage contract actually gives more control to the government and less to the contracting partners than do other kinds of legal contracts. The duties and rewards are specified by the government, not the parties who enter into the contract, the husband and wife. This means that the particular arrangements a couple makes are legally invalid if they conflict with the duties and privileges specified by public policy (Stetson, 1991).

For example, if you wished to sell a car, you could set up nearly any contract to which you and the buyer agreed. The state would interfere with the contract by requiring that certain taxes be paid and that certain assurances of the mileage and ownership of the car were valid. But issues like who can buy and who can sell, when and where the exchange would take place, and the price and payment schedule would be up to you. The details of a marriage contract, in contrast, are specified and when the contract is ended through divorce, the state makes the final decision over nearly all arrangements.

One important specification of the marriage contract in the contemporary United States is that the couple must be heterosexual. Gay men and lesbians do not have the legal right to marry. This restriction is problematic because it prevents gay couples from making choices about their lives and because it implies that homosexuality and lesbianism are bad or unnatural.

The restriction prohibiting marriage for gay people also creates practical problems. For example, a gay couple may live together for years, pooling their resources. They may not, however, carry each other on health insurance policies or life insurance policies when that benefit is provided by their employer for workers' spouses. As we read in the account of Sharon Kowalski and Karen

Marriage is a legal contract that is tightly controlled by the government. Only certain people can legally marry, and those who do enter into a contract that is largely determined by law rather than by the parties who are marrying.

Thompson their lack of rights to behave as a couple—to choose to live together, to take care of each other, and to make decisions with each other about health care—led to a long and painful court battle.

In a few cities small concessions have been won. In San Francisco gay couples may register formally as domestic partners. In New York a gay person may inherit an apartment lease if a partner dies. But in most of the United States the relationship between gay partners is legally invisible and invalid. And even in those cities where gay people have the right to formally declare themselves a couple, there is no funding to back up their right. For example, it might be illegal for an employer to refuse to allow a gay couple to cover each other on an employee-spouse insurance program. But if the government does not provide funding to educate people about the rule or to provide legal support for those who may wish to contest unfair treatment, the law may go unheeded.

Three Periods of Marriage and Family Law Dorothy Stetson (1991) observes that there have been three major periods of marriage and family law in U.S. history. The first was the doctrine of couverture, which defined marriage as a unity in which husband and wife became one, and that one was the husband. Stetson quotes an early nineteenth-century document to explain what couverture meant. "By marriage, the husband and wife are one person in law: that is, the very being or existence of the woman is suspended during marriage, or at least is incorporated and consolidated into that of the husband" (Blackstone, 1803, p. 442).

Under the doctrine of couverture married women could not own property. They had to turn over their wages to their husbands. If someone wanted to sue a married woman, she or he had to sue the woman's husband.

The second period was marked by the passage of the Married Women's Property Laws, which were first passed in Mississippi in 1839 and eventually were passed in all the states by the end of the nineteenth century. These laws allowed women the right to own property and to control their own earnings. Marriage during this period was perceived as a union between two separate and different but equal individuals (Stetson, 1991). Women and men had different responsibilities and rights in marriage but neither was supposed to overshadow the other. The wife was expected to provide services for her husband. One court case, for example, specified that a wife was "to be his helpmate, to love and care for him in such a role, to afford him her society and her person, to protect and care for him in sickness, and to labor faithfully to advance his interests" (Weitzman, 1981, p. 60). The husband in turn was obligated to provide for the economic needs of his family.

You can see how a separate but equal notion of marriage is closely tied to the ideology of "separate spheres" that dominated during the nineteenth century. In Chapter 2 we observed the problems associated with the ideology and the way in which the system was not one of equality. Although some women gained certain rights during the nineteenth century, they were still very much under the control of men.

In addition, in Chapter 2 we noted the way in which the ideology of separate spheres was class based. The only people who could live up to the ideal were wealthy and middle-class whites. The case that opened the door for married women to own property in Mississippi was a dramatic example of this kind of class and race ethnicity difference. The property over which the woman brought suit and was granted the right to own was a slave.

The separate but equal notion of marriage persisted until the 1960s when the third period of marriage law began to emerge. The third doctrine identified marriage as a shared partnership in which spouses would have equal and overlapping responsibilities for economic, household, and childcare tasks. This development of greater equality in the legal definition of marriage has been welcomed by many. We have seen, however, that this legal equality has not necessarily meant social equality. The way in which housework, market work, and childcare are divided are all influenced by gender.

In addition the third doctrine has created problems in some cases in which equality between women and men is upheld. For example, changes in laws that make women more equal to men in divorce proceedings are based on assumptions that women are equal to men in their responsibility for children and in their ability to earn an income. The laws assume equality between women and men, which is a valuable reform. But since women and men remain unequal in reality, problems have developed for divorcing women, especially when they are awarded the custody of children. We will examine such problems in the discussions of no-fault divorce in Chapter 10.

Marriage Is an Economic Arrangement

In the history of marriage as a legal contract one of the factors that was important, especially in earlier days, was the economic responsibilities that come with entering into a marriage. During the colonial period marriages were more likely to be arranged by a patriarch. A critical issue in his decision was the economic deal that could be struck with the parents of the prospective spouse of his son or daughter. Although the opinion of the potential bride and groom increasingly gained control over the decision as this history unfolded throughout the nineteenth century, economics continued to play a role. The legal contract of marriage is filled with references to issues like whether a married person can be sued and by whom, and to whom the property of a married person belongs.

Today, economics remains an essential feature of marriage. The organization of our economic system at the macro level has great importance for decisions about marrying and the experience of marriage at the micro level. As we observed in Chapter 4, the economic facet of marriage may be especially salient for those in the owning class. Owning-class mothers work hard to ensure that their children interact socially with others of their class, for example, by arranging debutante balls, in order to find a suitable person—a class peer—for a mate. The macro-level organization of class stratification is both expressed and maintained by the micro-level activities of individuals marrying one another.

But even for those not in the owning class, economics plays a major role in marriage around the issue of mate selection. Nearly all people in the United States marry within their own social class and even when people marry outside of their class, they marry someone who is close to their position in the class system (Carter & Glick, 1976). Among those who do marry outside of their class there are interesting gender differences. Women tend to marry up and men tend to marry down. The tendency of women to marry up is called hypergamy. Hypergamy results in decreased marriage opportunities for women at the top of the stratification system and men at the bottom of the stratification system compared to those in the middle. The greater ability of women to marry up through marriage, however, is countered by men's greater ability to move up in the class system through occupation. Furthermore, not all women can use the marriage option. The ability to marry up among white women seems to be tightly tied to appearance. The bigger the step up the more attractive she must be (Elder, 1969). In contrast men's ability to move up through their occupation is related to their I.Q. And black women's tendency to marry up is related to their education (Udry, 1977).

The relationship between economics and marriage was investigated in Chapters 6 and 7 in the discussion of economic questions of the division of paid and unpaid work as a central aspect of intimate relationships. In Chapter 10 we will look at the organization of divorce in our society and see how marriage is most clearly revealed as an economic institution when a couple divorces.

In your assessment of the scenario at the beginning of the chapter, did the fact that Sharon and Karen had bought a house together influence your thinking about the validity of their relationship? Suppose you were in a heterosexual relationship and planning to marry, and your prospective husband or wife told you s/he would not proceed with the marriage plans unless you agreed to sign a contract allowing his or her finances to be kept entirely separate and unknown to you. Would you be surprised or unhappy?

Marriage Is a Sexual Relationship

Randall Collins and Scott Coltrane state that "the core of the marital relationship is a claim to permanent and exclusive sexual possession of one's partner" (1991, p. 638). Marriage law requires that in order to consummate a marriage—to bring the wedding ritual to completion—a couple not only must have sex with one another, they must have a particular form of sex, intercourse. On the other hand, social norms and laws in many areas in the U.S. prohibit sex between people who are not married to one another. Of course the real experience of sexuality is much broader than this.

In Chapter 8 we explored the relationship between sex and marriage. The important point to be considered in this chapter is that sex is tightly bound to marriage, legally, ideologically, and for many people experientially. Marriage is at least partly defined by its link to sex (Reiss, 1986).

In addition, sex is at least partly defined by its link to marriage; sex outside of marriage is invalid. In Sharon and Karen's case at the beginning of the chapter, their sexual relationship was not perceived by Sharon's father and the courts as a valid basis for arguing that their relationship was one that should be supported and that would grant Karen the right to participate in decision making about Sharon's treatment. Their sexual relationship, in fact, was perceived to be a reason to keep them apart. In the court proceedings, Sharon's father argued that he did not want Karen to see Sharon because he believed Karen would sexually molest her. The court initially granted him guardianship even after Karen stated that she was gay and that Sharon was her lover (Griscom, 1992).

Marriage Means Commitment

Jessie Bernard is recognized as a founding mother of family sociology. Much of her work has focused on relationships between women and men in marriage. She writes,

> . . . one fundamental fact underlies the conception of marriage itself. Some kind of commitment must be involved. Without such a commitment a marriage may hardly be said to exist at all, even in the most avante garde patterns. (Bernard, 1982, p. 79)

Part of the meaning of this commitment is often the promise of mutual sexual exclusivity. But the concept of commitment includes other aspects as well. Bernard (1982) reviewed marriage vows to determine the content of commitment. She writes that in the most traditional wedding ceremonies in the contemporary United States, the commitment is to love, honor, and obey. In more modern examples, couples vow to commit themselves to a variable array of values and behaviors including truth, honesty, personal growth, allowing one's mate to grow, and respecting differences.

Bernard (1982) maintains, however, that permanence has historically been the essential component of commitment in marriage in cultures heavily influenced by Christian belief, like the United States. The line in a traditional wedding ceremony, "till death do us part," is a reflection of this definition of commitment.

According to Bernard a definition of commitment centered on permanence has recently given way to another essential component, love. And in some contemporary marriages couples vow to stay together "as long as ye both shall love." In Chapters 2 and 8 we observed the way in which this ideal of the companionate marriage emphasizing love and affection developed relatively recently in the United States. An examination of contemporary Japanese American marriages indicates that the transition to marital commitment based on love is still occurring among some groups of people in the U.S.

First-generation Japanese Americans are called Issei and second generation are called Nisei. Sylvia Yanagisako (1985) interviewed couples from both generations to find out their perceptions of ideal marital relationships. She found that both generations viewed Japanese marriage as based on and maintained by *giri* (duty) and they believed that American marriage was based on and maintained by romantic love. Issei couples sought to model their marriages after the Japanese type. They recognized that husbands and wives may develop strong emotional feelings for each other, but those feelings were secondary in their marriage. The more basic and essential component was duty and ethical commitment. Nisei couples believed that this view of marriage was often emotionally ungratifying. Nisei couples, however, were also wary of what they perceived to be the whimsical and dangerously unstable character of American marriage. They attempted to create an alternative to these two views by combining ideas of duty and romantic love in order to redefine commitment as an expression of both.

Commitment can be added to our list of components defining marriage. Bernard even goes so far as to assert that without it, a marriage does not exist. In trying to capture the meaning of marriage we have looked at the law, economics, sex, and now the emotional feature of commitment. Commitment itself, however, is not easy to define. The term is slippery, variable, difficult to measure, and changing.

How important is commitment to defining marriage? When the story of Sharon and Karen is presented in the press and in this chapter they are described as having pledged their commitment to one another. How important should this act have been to the courts deciding their case?

Marriage Is a Political Arena

The last feature of marriage is the question of power. Claire Renzetti and Daniel Curran write, "In short, marital relations are fundamentally power relations—usually the power of husbands over wives" (1992, p. 136). Power in marriage is a question that has been investigated for some time by family sociologists.

What exactly is power? Recall that in Chapter 1 politics was defined as the "process whereby individuals or groups gain or maintain the capacity to impose their will upon others, to have their way recurrently, despite implicit opposition through invoking or threatening punishment, as well as offering or withholding rewards" (Lipman-Blumen, 1984, p. 6). This definition comes from a Weberian perspective. Max Weber was a nineteenth-century sociologist. He proposed that power was the chance that one actor in a social relationship is able to impose his or her will on another (Fishman, 1978).

Since Weber made his proposal Berger and Luckmann (1966) have suggested another feature of power. They assert that power is more than being able to force someone to do something. It includes the ability to force one's definition of reality on others. In Berger and Luckmann's framework "Power is the ability to impose one's definition of what is possible, what is right, what is rational, what is real" (Fishman, 1978, p. 397).

key →

Both of these kinds of power are a feature of married life. A husband may enforce his will on his wife by insisting that he decide whether they will move from one community to another. He may also successfully convince his wife that his control over the decision is the only rational and legitimate way in which to make family decisions. Keep in mind these two forms of power as we examine some of the research on power relations in marriage.

In 1960 Robert Blood and Donald Wolfe conducted a survey of married couples in Detroit about the division of power in their marriage. This survey has become a classic in family studies and variations of the study have been replicated several times since then (Peplau & Campbell, 1989). Blood and Wolfe used decision making as their measure of marital power. They questioned more than 900 married women about who made various decisions in their families. They asked, for example, who had the final say in deciding to buy a car, to move, which physician a family member should see, and whether a wife should seek employment. The researchers found that husbands usually had more power in their marriage because they had the final say in more decisions.

Blood and Wolfe also found that when wives brought more social resources to the marriage, like a higher education or income, they had greater control over decisions than if they did not (Gillespie, 1971). Letitia Peplau and Susan Campbell conducted a similar study of dating couples and found that 49 percent of the women and 42 percent of the men reported equal power in their current dating relationships. Peplau and Campbell, like Blood and Wolfe, found that when women had more resources they had more power in their relationships with men. Peplau and Campbell identified gender ideologies, career goals, and involvement in the relationship as important "resources" for dating couples. Women who had more liberal gender ideologies, stronger career goals, and less involvement in the relationship had greater power than women who did not. Gender ideology and involvement in the relationship also affected men, although the correlation between liberal ideas about men and women and power in the relationship was inverse for the men. Career goals did not appear to affect men. For example, among women the higher the academic degree to which they aspired—BA, MA, or PhD—the greater their power in the relationship. This correlation between academic aspiration and power in the relationship was not evident for men.

Recall that I mentioned the Blood and Wolfe study in Chapter 6, citing it as an example of an exchange theory perspective and noting some of the theoretical criticisms of exchange theory. Blood and Wolfe's study has also been criticized for four methodological problems.

First, this study and others like it have been criticized because they do not distinguish among different decisions. They give the same weight to choosing a physician and choosing whether a wife should seek employment (Safilios-Rothschild, 1970).

Second, Blumstein and Schwartz (1983) have criticized Blood and Wolfe, arguing that sometimes women make decisions because they have been assigned that task by their husbands. The "power" they have, for example, to purchase furniture comes from the more fundamental decision by the husband that the wife should do so (Curran & Renzetti, 1989).

Third, Blood and Wolfe have been criticized because rather than observing the behavior of the subjects they only asked them what they do. Surveys like theirs may be a better measure of what wives believe their marriages should be like or what they believe they should say their marriages are like rather than an accurate assessment of power and/or decision making in their households (Renzetti & Curran, 1992).

The failure to distinguish between ideals and behavior has been especially problematic in studies of Chicano families (Baca Zinn, 1982). Quantitative research studies that rely on questionnaire data fail to tap the experience of Chicano husbands and wives. Although the image of Chicano families is one in which husbands have absolute authority, the reality may be one in which women have some power and/or at least the question of power is contested terrain (N. Williams, 1990).

Baca Zinn (1982) set out to examine this question using more qualitative methods: focused interviews and participant observation of a small sample of wives. The women she interviewed and observed all supported patriarchy as an ideology. For example, one woman said, "You know I'm slowly becoming equal with Nabor even though he will always be the head of the family. No matter how much money I make, even if I get a master's degree he will always be the head" (Baca Zinn, 1982, p. 73).

In their behavior, however, Baca Zinn (1982) found that the patriarchal model was not always followed. Especially when women were employed outside the home, they were likely to abandon the patriarchal ideal by challenging their husbands, making independent decisions, and taking actions with which their husbands did not agree (see also N. Williams, 1990).

Baca Zinn's (1982) research shows the way in which the two definitions of power, Weber's and Berger and Luckmann's, do not necessarily mesh in real people's lives. According to Weber's definition, power is the ability to impose one's will on another. The husbands in Baca Zinn's sample were not always able to induce their wives to follow their directions. The husbands did not have power over their wives in the Weberian sense. From the Berger and Luckmann perspective, however, husbands did have power. The husbands were able to impose their definition of what was rational and good on their wives' thinking.

Fourth, Blood and Wolfe have been criticized because of the measure they used to indicate power. Decision making may not be the best measurement or even an accurate measure at all of power in marriages. For example, in Chapter 7, Heidi Hartmann was quoted as defining power in intimate relationships as being best indicated by who did or did not do the housework. Others (Gordon, 1988) have suggested that violence is the most critical indicator of power (see Chapters 11 and 13). Patterns of communication is another factor that has been cited as an important indicator of power in relationships between women and men, including marriage (Fishman, 1978).

Power in Communication One of the ways in which researchers have observed the distribution and negotiation of power in marriage is in nonverbal and verbal communication between husbands and wives. In 1964 Berger and Kellner proposed their view of the social process of marriage. They argued that newly married husbands and wives come to a marriage with different biographies and work together to create a mutually shared view of the world (Glenn, 1987). This work is accomplished through "conversation," which includes verbal communication as well as other forms of social interaction. This interpretation of communication in marriage focuses on cooperation and the creation of a

An important component of marriage is relationships of power. These can be expressed in a variety of ways, including communication. Women and men communicate differently. For example, men talk more and are more likely to choose the topics of conversation. These differences can mean that men use verbal interaction to wield power in their relationships with their wives.

consensual view of the relationship and the world. It is an example of interactionism, a theoretical framework closely identified with George Herbert Mead (Glenn, 1987).

Glenn argues that this kind of interactionist theory focuses too narrowly on consensus—what do people share and how do they negotiate a shared vision? This kind of analysis ignores equally salient factors like conflict, inequality, and power. Feminist researchers have been interested in incorporating these issues—conflict, inequality, and power— into their assessments of the process of communication between intimate partners.

Researchers have found a number of differences in the ways in which women and men communicate (Henley, 1977). Men and women talking to each other are not equal. Men talk more, interrupt more, and control the conversation more. Women listen more, ask more questions, and use more qualifiers, like "in my opinion" and "it seems to me" (Basow, 1992).

Fishman (1978) investigated communication between men and women to see whether there were differences or evidence of conflict and power. In her research she asked couples who identified themselves as egalitarian to record spontaneous conversations in their homes. The couples were requested to turn on an audio recorder and allow their conversations to be taped for her to listen to and analyze. The conversations were about mundane everyday topics like what happened at work today or what shall we have for dinner. Fishman looked for differences in the ways in which men and women talked to one another. Like other researchers in this field she found that men talked more and interrupted more, while women asked more questions and used more qualifiers and more passive ways of entering into conversations. For example, men would begin a conversation with a direct statement; women would begin with a tag like "Isn't this interesting," or "let me ask you this."

In her research, Fishman also found that while men controlled the conversation, women did most of the work. By this she means that women offered an array of possible conversation topics. Men chose which topics, if any, they wanted to discuss. Conversations frequently followed the pattern of a woman offering a topic through a question and a man either stopping the conversation by a short response—"yes, no, uh-huh"—or accepting the topic and then expounding on it as the woman interjected supportive words to keep the conversation going. If the man declined the topic by cutting it short, the woman would continue her work, offering other possible topics from which he could select.

According to Fishman, these language differences are signs of power differences. Men do not need to ask to be heard. They are the center of attention. Their communication dominance, in terms of controlling topics and of the time spent speaking, are indications of their social dominance, even in these self-described egalitarian couples.

Another difference between women and men in communication is self-disclosure or the expression of one's feelings, especially one's vulnerabilities. Jack Sattel (1976) investigated this aspect of communication between women and men that he argues is a mechanism by which men maintain power in intimate relationships.

Sattel's assertion is quite different from the ideas we frequently hear in the popular media about the "tragedy of male inexpressiveness." Discussions of women's greater ability to express their feelings is usually presented as an example of a disadvantage men have compared to women (Balswick & Peek, 1971). Balswick (1979) writes, for example, "men who care, often very deeply, for their wives . . . cannot communicate what is really going on in their hearts."

According to Sattel this lack of emotional expression may not be as disadvantageous as researchers like Balswick suggest. Sattel argues that "keeping cool, keeping distance as others challenge or make demands upon you is a strategy for keeping the upper hand" (Sattel, 1983, p. 243). What appears to be a disadvantage for men may actually be an effective method by which men can manipulate and control others and thereby maintain power in their relationships with other men as well as with women.

Both Fishman and Sattel have very different viewpoints from those expressed by Berger and Kellner (1964). All of these scholars however are concerned with how verbal communication provides a way for husbands and wives to interact and to create an interpretation of their social world. These views all fit within the broad theoretical perspective of symbolic interactionism. Recall that in Chapter 4 we examined the way in which families coped with the layoff of a breadwinner from a symbolic interactionist perspective. In this research on communication in marriage we find another example: husbands and wives interacting with words, or symbols. But Fishman and Sattel, unlike Berger and Kellner, emphasize the importance of acknowledging power and the political inequality between women and men entering into these interactions.

Power appears to be an aspect of marriage that takes a variety of forms. Men have the upper hand in many instances, but marriages are an arena of struggle over power and the mechanisms by which power can be maintained or changed. Differences in power between women and men have become less overt at the macro level in marriage law but remain salient and pervasive in every aspect of even the smallest issues of social interaction between husbands and wives. Later in this chapter we will return to the theoretical question of gender politics as a critical aspect of relationships within marriage. But before moving to that issue,

FIGURE 9 • 1
Marital Status, by Race: 1970 and 1991 (Persons 18 years and over)

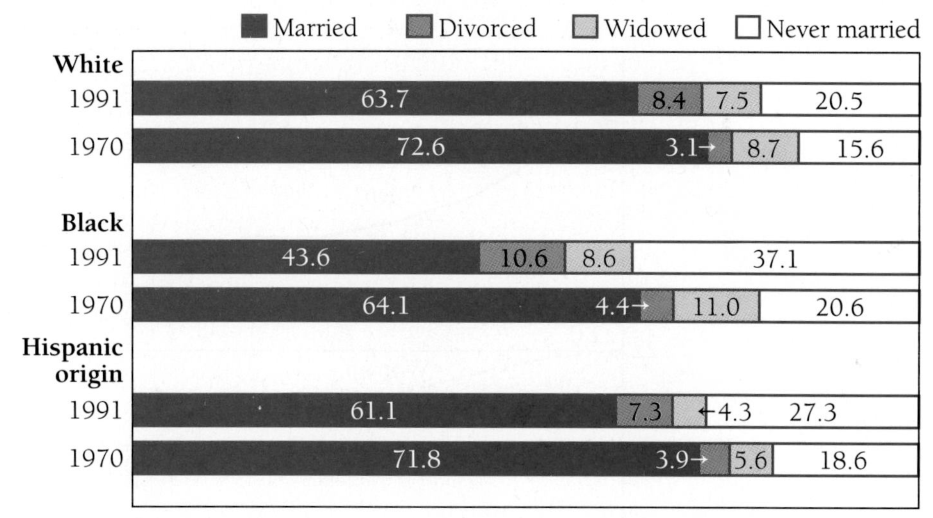

Source: U.S. Bureau of the Census. 1991. *Marital Status and Living Arrangements: March 1991*, p. 2.

let us take a look at data on the incidence of marriage in the contemporary United States.

STATISTICS ON MARRIAGE

In this section we will examine the demographic data on marriage. How prevalent is marriage? How do race and gender affect marriage rates and the age at which people marry?

Nearly all Americans marry at some time in their lives (Cherlin, 1992). Figure 2-2 in Chapter 2 shows the rate of marriage as it has changed in this century. The rate is calculated as the number of first marriages per 1,000 single women between the ages of 14 and 44. The figure shows that the rate of first marriage has fluctuated greatly over the past sixty years, although in the past twenty years the rate has begun to level off.

Figure 9-1 shows the marital status of people in the U.S. in 1970 and 1991. The figure reveals change between those years and it indicates variation by race ethnicity.

Age at First Marriage

Figure 9-2 shows changes in the last century in the median age at first marriage for women and men. (Recall that half the population falls below and half the population falls above the median point.) The figure indicates that the median age at first marriage has risen for both women and men since the 1950s.

Changes since the beginning of the century in age at first marriage have differed for men compared to women. Men's age at first marriage dipped in the middle of the century and recently has risen to the age common in the early part of the century.

FIGURE 9 • 2
Median Age at First Marriage by Sex, 1890 to 1991

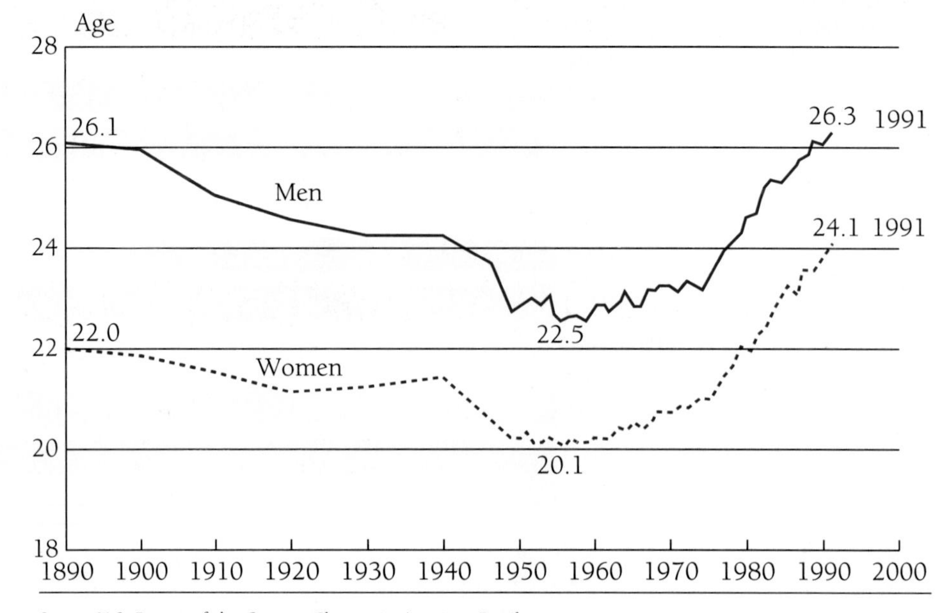

Source: U.S. Bureau of the Census. *Changes in American Family Life*, current Population Reports Series P-23, No. 163, 1989d, p. 5; Census and You, "U.S. Statistics at a Glance" May 1992, p. 10.

Women now are also marrying later than women did earlier in the century. But for women, this later age of first marriage is higher than at any previously recorded time in the United States and is not a return to a previous level as it is for men.

Race Difference in Age at First Marriage Until the 1950s African Americans tended to marry at an earlier age than white Americans. Since the 1950s African Americans have married at increasingly later ages. They now marry later than any other group of Americans. The percentage of African Americans who have not married by the time they are twenty-nine years old is higher than for any other racial ethnic group in the U.S. This proportion, furthermore, has increased in the past two decades.

Table 9-1 divides the population into black, white, and Hispanic and shows the percentage of people who had never married in 1970, 1980, and 1991.

The table indicates that for all three racial ethnic groups the percentage of those who are over 18 years old and have never married has risen over the past two decades. There are differences, however, by race. Whites are least likely to have never married and blacks are most likely to have done so.

Table 9-2 shows these comparisons on the question of whether people are in intact marriage. The category "in intact marriages" excludes both those who have never married as well as those who might have married but are now divorced or widowed. The data on those in intact marriage also show differences by race. Blacks are less likely to be in intact marriages than whites or Hispanics are.

TABLE 9 • 1
Percentage of People 18 and Older Who Have Never Been Married, 1970–1991

	1970	1980	1991
White	15.6%	18.9%	20.5%
Black	20.6%	30.5%	37.1%
Hispanic	18.6%	24.1%	27.3%

Source: U.S. Bureau of the Census. 1991. Current Population Reports. *Marital Status and Living Arrangements: March 1991*, p. 2.

TABLE 9 • 2
Percentage of People 18 and Older Who Are in Intact Marriages, 1970–1991

	1970	1980	1991
White	72.6%	67.2%	63.7%
Black	64.1%	51.4%	43.6%
Hispanic	71.8%	65.6%	61.1%

Source: U.S. Bureau of the Census. 1991. Current Population Reports. *Marital Status and Living Arrangements: March 1991*, p. 2.

Several explanations have been offered for this marriage difference between African Americans and others (Glick & Norton, 1979; Taylor et al., 1992). One explanation is that higher rates of unemployment and underemployment among African Americans causes them to delay marriage longer as they wait to become more financially secure. In support of this theory, research in inner-city Chicago found that employed fathers were twice as likely as unemployed fathers to marry the mother of their first child (Testa et al., 1989). In another study of men and marriage, researchers (Tucker & Taylor, 1989) found that the higher a man's personal income, the more likely he was to marry. For these men, income was unrelated to whether they were involved in a nonmarital romantic relationship. This means that men continue to become involved in romantic relationships with women regardless of their income. Deciding to marry, however, seems to be a step that is inhibited by insufficient income.

A second explanation for why African Americans marry later than whites or not at all is called the "marriage squeeze." In our society, women tend to marry men who are two or three years older than themselves. If the population is increasing, there are more younger people than there are older people. This means that men (who seek younger partners) have a larger pool from which to choose than women (who seek older partners) (Rodgers & Thornton, 1985; Schoen, 1983; Taylor et al., 1992). The birth rates in the U.S. increased from 1933 to 1960 and from 1973 to 1990. Assuming that people will marry sometime in their early twenties, those women who sought a husband between 1954 and 1982 and those who will be seeking a husband beginning about 1997 will have a harder time finding one who is older than themselves.

This problem of pool size exists for all women born during years of larger total births, but it may be especially prevalent for black women because of the higher mortality rates for black men. Not only are there fewer black men born a year or two before the women who will be seeking them as husbands, black men are likely to die young. An African American woman seeking to marry an African American man who is a few years older than herself may find the numbers of eligible people to be insufficient (Guttentag & Secord, 1983).

BOX 9 • 1 DO CONTEMPORARY AMERICAN WOMEN OVER 40 HAVE LESS CHANCE OF MARRYING THAN OF BEING KILLED BY A TERRORIST?

In 1986 a newspaper reporter decided to write an article for Valentine's Day on the state of marriage. She called Neil Bennett at Yale University for information and he agreed to give her some data from a research project on which he was working with Yale colleague David Bloom and Patricia Craig at Harvard (Bennett, Bloom, and Craig, 1989). Based on their research, Bennett said, current marriage behavior showed that white women who were still single at age thirty and who were college graduates had only a 20 percent chance of ever marrying. Women who were still single at thirty-five had only a 5 percent chance of marrying and at age forty their chances were only 1 percent. The reporter wrote the article, which was then picked up by AP wires, and the next day Bennett's data were published in newspapers across the country. The research created a stir. Media attention continued into the summer with a cover story in *Newsweek* in which the author stated that a forty-year-old woman's chances of marrying were less likely than her chances of being killed by a terrorist (Cherlin, 1992b).

American women are marrying later than ever before in recorded history. Does this mean that many will never marry? Bennett's predictions were based on two sets of information. One set of information was the data that showed that many women were reaching their thirties and forties without marrying. The other set of information was the knowledge that in the past women who married did so before reaching the age of thirty or forty. Bennett put these two data sets together and made the prediction that those women who are now in their thirties and forties will never marry. But this prediction is only valid if contemporary women behave in the same way that their predecessors did. Contemporary American women have changed in many ways in recent years. They may decide to begin marrying later in life and demographers will have to readjust their predictive models.

Furthermore, it is relatively unusual for an African American woman to marry a man who is not African American. Until 1967 it was actually illegal in many states for a black person and a white person or an Asian person and a white person to marry. There were 246,000 black–white couples in the U.S. in 1992—four times as many as in 1970. But marriages between blacks and whites continue to be less common than marriages between people from other racial ethnic groups and whites. There were 883,000 marriages between whites and Asians or Native Americans in 1992 (Bovee, 1993). In addition, 42% of Americans still say they disapprove of marriages between black and white people (Gallup & Newport, 1991). A marriage that does occur between a black and a white is more likely to be a case of a black man marrying a white woman than a white man marrying a black woman. In sum, African American women who wish to marry a man who is a few years older than themselves face an insufficient pool of potential mates because of the marriage squeeze if they were born in years of larger numbers of births; a higher mortality rate among black men; and unlikely prospects of crossing racial ethnic lines to marry.

Transition to Marriage

Although the word sounds quaint, sociologists refer to the period between singlehood and marriage as courtship. The preceding discussion of racial ethnic groups and marriage has identified several important factors affecting the transition from singlehood to marriage. The micro-level interaction of courtship is affected by the macro-level issues of economics, population changes, and racial ideologies and laws restricting marriage between members of different racial ethnic groups.

The two most significant characteristics of courtship in the 1990s are the delay of first marriage and the increase in cohabitation (Surra, 1991). These factors have become increasingly important in the past two decades. We will examine them in the next section. But before moving on, let us look at a brief history of courtship in order to get a sense of where our present practices emerged from.

Cate (1992) traces the historical evolution of courtship in the U.S. and note the transition in the early twentieth century from men calling on women to the incorporation of dating into courtship. Calling involved only visiting; dating brought something new to the activity: going out and doing things in public. In the 1970s and 1980s dating changed, becoming more informal, less focused on the goal of marriage, and more likely to be substituted with mixed group activities that did not involve pairing up. Cate argues that two trends in courtship developed throughout the nineteenth and twentieth centuries: (1) the lessening of parental involvement in the process, especially in selecting partners for their children, and (2) a growth in the demand for emotional intimacy as the criterion for continuing a relationship and particularly for deciding to marry (Coontz, 1988).

While courtship continues to change, some factors remain strikingly consistent (Basow, 1992). In 1989 college students were asked to list the content and sequence of events that would occur on a first date. The students were very much in agreement in their responses and the image they conveyed was highly gender stereotyped. The focus of first dates for women was concern about their appearance, making conversation, and limiting sexual activity. The focus of first dates for men was planning activities, paying for the date, and initiating sex (Rose & Frieze, 1989). Other research supports these observations. For example, reviews of personal ads show that men are looking for women with particular physical characteristics while women emphasize a potential male date's status characteristics (Davis, 1990). Women who initiate dates are viewed more negatively than men who initiate dates (Green & Sandos, 1983).

Cohabitation and Domestic Partnerships Table 9-3 shows the proportion of people who report they cohabited before their first marriage and those who have ever cohabited. The numbers vary a little by race. Choosing to delay marriage and to remain single, sometimes indefinitely, has become a sign of our times. But many people who remain single choose to cohabit and an equally important sign of the times is the growing number of heterosexual couples who live together without being legally married. A number of researchers assert that cohabitation is becoming institutionalized as a new phase of mate selection in the U.S. (Gwartney-Gibbs, 1986).

In fact, when the growing numbers of couples who cohabit are considered, the decline in marriage rates is offset to a large extent (Bumpass & Sweet, 1991). The movement away from legal marriage has been accompanied by a movement toward nonmarital unions (Surra, 1991). Cherlin (1992) notes that

TABLE 9 • 3
Cohabitation by Race, 1988

	White	Black	Hispanic
Cohabited Before First Marriage	25.0%	29.3%	25.5%
Ever Cohabited	33.6%	35.0%	32.9%

Source: World Almanac and Book of Facts, 1992, p. 945.

despite the rising age of marriage, young adults are almost as likely now as they were in 1970 to be sharing a household with a partner.

In 1988 2.6 million households included an unmarried couple (U.S. Bureau of the Census, 1989a). Between 1970 and 1979 this type of household grew by 117 percent and between 1980 and 1988 it grew by 63 percent (U.S. Bureau of the Census, 1988).

One of the problems with the data collected by the Census Bureau is that they only report the kinds of people that make up a household. Census data do not report the social relationship the household members have with one another (Surra, 1991). Domestic partners are couples that have a particular kind of social relationship. They are unmarried adults who live together in committed, sexually intimate relationships (Bedard, 1992). The census does not ask respondents to designate whether one is a domestic partner or not. The census data only indicate the sex of the members of a household. If a household has two unrelated adults living together they are counted as cohabitors regardless of their social relationship. This might include, for example, people who are roommates but are not domestic partners.

At the same time the census does not count as cohabitors gay men and lesbians who *are* domestic partners. Researchers estimate that about 10 percent of the population is gay or lesbian (Marmor, 1980), and a number of gay people live together in domestic partnerships. For example, Harry (1983) estimates that about 75 percent of lesbians live in domestic partnerships.

The data collected by the census are also limited because they identify cohabitors only if they share the same address. Steven Nock (1987) has observed that couples frequently live with one another in domestic partnerships while maintaining separate addresses.

Other researchers have tried to get more accurate information than that provided by the census (Surra, 1991). Thornton (1988) asked a sample of people who were 23.5 years old or older if they had ever cohabited and about one-third said they had. In another large survey of 13,000 individuals aged nineteen and over, Sweet, Bumpass, and Call (1988) found that by their early thirties nearly half had cohabited at some time and 4 percent were currently doing so.

The length of time people cohabit with one partner is usually fairly brief. About half of all domestic partnerships dissolve within sixteen months. About 60 percent of the time cohabitation ends with the marriage of the partners; 40 percent break up (Bumpass & Sweet, 1989).

Who are heterosexual domestic partners? Sixty-eight percent are younger than thirty-five. About half had never been married before they cohabited and about one-third are divorced (Surra, 1991). Forty percent of cohabitors live with children in the home, although usually they are children of a previous marriage (Bumpass & Sweet, 1991; Bedard, 1992). Gender also has an influence on cohabitation. The rates of cohabitation are higher for women compared to men (Bumpass & Sweet, 1989).

Education has an inverse relationship to cohabitation. This means that the more education a person has the less likely he or she is to have cohabited (Spanier, 1983). This runs counter to the image of college students as the primary cohabitors. College students did increase their likelihood to cohabit in the 1970s and 1980s but so did most other groups and people with less education were cohabiting at higher rates prior to the 1970s. Cherlin (1992) states that college students were imitators, not innovators of cohabitation.

The intensity of religious belief also has an inverse relationship with cohabitation. People who attend church regularly may regard cohabitation as "living in sin" (Bedard, 1992).

Singles

Most people in the United States eventually marry and a number of those who are not married cohabit within heterosexual and homosexual domestic partnerships, but many people also remain single, at least for some part of their life. In 1990 25 percent of all households were those in which an individual lived alone. As we have already noted, some of those who claim to live alone may in fact be involved in domestic partnerships while they maintain a separate residence (Nock, 1987). On the other hand, other singles may not be revealed in census data because they live with relatives or friends with whom they are not domestic partners.

Many of the 26 million people in the U.S. who are single are single by choice and enjoy single life. Peter Stein (1981) interviewed single people and found that they named freedom, career opportunities, friendship, self-sufficiency, and sexual relationships as being enhanced by their single status. Other researchers have found that single people spend more time having fun (Cargan & Melko, 1982). Compared to married people they go out more often, visit places of entertainment more often, and have more sex partners.

Some single people may remain single although they would prefer to marry. Marriage rates have historically declined when the economy declines. And since, as we have noted in Chapter 4, the current period is one marked by economic difficulty, singlehood may be at least partly a result of inability to marry rather than a matter of choice (Bedard, 1992). Married people tend to be better off financially than singles. Marcia Bedard (1992) argues that it is the ability to meet their economic needs, not the issue of being single, that is the critical key to the happiness of single people.

This section has reviewed the demographic data on who chooses to marry and who does not and how the answers to these questions have changed over time. Changes in the micro level of the experience of women and men entering into intimate relationships have been recorded in the statistical tables compiled by the government and scholars.

While people's experience with marriage has been evolving, their ideas have been changing as well. In Chapters 2 and 3 we examined those changes as they developed from the colonial period to the middle of the twentieth century. In the next section we will look at the way in which ideology about marriage has continued to change through the last few decades.

CHANGING IDEAS ABOUT MARRIAGE: WOMEN'S LIBERATION AND THE MALE REVOLT

In 1970, Bernard (1972) presented her idea that each marriage was actually made of two different marriages, his and hers. Bernard reviewed the literature on the experience of women and men and was struck with how different they were. Furthermore, she noted how much more problematic marriage was for women. Bernard's investigation showed that the psychological costs of marriage were great for women.

> [I]n the data that measure human stress: married women, whatever their claims to fulfillment, and unmarried men, whatever their claims to freedom, rank high on all stress indicators, including heart palpitations, dizziness, headaches, fainting, nightmares, insomnia, and fear of nervous breakdown; unmarried women, whatever their sense of social stigma, and married men rank low on all the stress indicators. (Lengermann & Niebrugge-Brantley, 1992, p. 465)

In Bernard's overview of marriage, she was critical of the institution and concerned about the negative effect it had on women in particular. More radical feminists went even further, declaring that marriage was a form of slavery. Sheila Cronan, for example, wrote,

> The Feminists decided to examine the institution of marriage as it is set up by law in order to find out whether or not it did operate in women's favor. It became increasingly clear to us that the institution of marriage "protects" women in the same way that the institution of slavery was said to "protect" blacks—that is, that the word "protection" in this case is simply a euphemism for oppression. . . . the marriage relationship is so physically and emotionally draining for women that we must extricate ourselves if for no other reason than to have the time and energy to devote ourselves to building a feminist revolution. (Cronan, 1971, p. 329)

It is commonly believed that many of the changes we observed in the data presented in the previous section on the incidence of marriage and the age at which people were marrying were caused by the changing role of women in our society, and that the changing role of women was largely a result of the women's liberation movement led by vocal critics of the institution of marriage. Women were increasingly participating in the labor market. They were demanding greater equality in the home, the workplace, and the law. And these newly radicalized women were frequently delaying marriage, sometimes indefinitely.

The women's liberation movement and changing ideas about what it means to be a woman undoubtedly did have an important effect on marriage rates, but what about men? Were definitions of masculinity changing also, and how were these changes affecting marriage? Barbara Ehrenreich (1983) traced the history of the relationship between changes in ideology about marriage and changes in our definitions of masculinity. She used as her data source popular books and magazines. The central questions of her work were: Why do men marry women? How did the answer to this question change from 1950 to 1980?

> In the 1950's where we begin, there was a firm expectation (or as we would say now, "role") that required men to grow up, marry and support

> wives. To do anything else was less than grown-up, and the man who willfully deviated was judged to be somehow "less than a man." This expectation was supported by an enormous weight of expert opinion, moral sentiment and public bias, both within the popular culture and elite centers of academic wisdom. But by the end of the 1970s and the beginning of the 1980s, adult manhood was no longer burdened with the automatic expectation of marriage and breadwinning. The man who postpones marriage even into middle age, who avoids women who are likely to become financial dependents, who is dedicated to his own pleasures, is likely to be found not suspiciously deviant, but "healthy." And this judgment, like the prior one, is supported by expert opinion and by the moral sentiments and biases of a considerable sector of the American middle class. (Ehrenreich, 1983, pp. 11–12)

Ehrenreich argues that these changes are at least as important as the changes in our ideas about women, yet they are largely ignored. Furthermore, Ehrenreich shows how the ideas about men began to change *before* the emergence of the women's liberation movement. She refers to these changes as the male revolt against the breadwinner ethic—a strong term for a strangely invisible social movement.

In the 1950s, books, magazines, movies, and psychiatric theory promoted the idea that men who were still unmarried in their late twenties were suspect. Adult masculinity was identical to the breadwinner role. According to this point of view, men who were not married breadwinners were, therefore, not adults, not masculine, and probably homosexual.

The goal for men was to be "mature," to settle down and take responsibility for a wife and marriage. Ehrenreich's investigation uncovered cases of those men who did not follow this pattern, who instead spent money foolishly, hated their job, dreamed of owning a sports car, had trouble completing college, hitchhiked, and took midnight swims. They were accused of being stuck in adolescence, a central characteristic, according to the "experts," of homosexuals.

This definition of heterosexual masculinity began to be challenged in the 1950s. The challenge gained momentum when a new magazine, *Playboy,* joined the battle of the sexes in December, 1953. Ehrenreich points out that the first nude centerfold of Marilyn Monroe, which appeared in the magazine and subsequently became legendary, was perhaps not as important as the first feature article, "Miss Gold-Digger, 1953." The article was an attack on alimony, matrimony, and money-hungry women.

Before *Playboy,* the two essential components of masculinity were marriage and a successful career. *Playboy* became the promoter of a social movement that embraced the idea of career success and active heterosexuality, while it simultaneously dismissed the idea of marriage. The magazine described husbands as slaves in the rat race working for masters—their wives. The reader was told that by joining the playboy movement, he could avoid this enslavement and the worry of being labeled a homosexual.

In the late 1950s, playboys and other rebels against the breadwinner model of masculinity were joined by cardiologists, who announced that the breadwinner model was unhealthy for men. During the nineteenth century men outlived women but by 1920 life expectancy was about the same for women and by 1970, women's life expectancy was eight years longer (Ehrenreich, 1983).

Physicians noted that heart disease was an epidemic among male white-collar workers. They argued that the disease was a result of the stress of their work exacerbated by the stress of their responsibility for their nonworking wives. Furthermore, the physicians criticized wives who did not try to alleviate their husbands' coronary problems by feeding them healthier meals and helping them to control their Type A—competitive, intense, workaholic—personalities.

This advice appeared in popular magazines. For example, in an article in 1962 in *The Reader's Digest*, cardiologist Herman Sobol wrote,

> "A wise wife will try to shield her husband from any added stress." He cautioned against "the five or six-o'clock frenzy" that greeted all too many men at the end of the day. Little children should be fed early and older ones dispersed so that Dad can "recover from the day" over his cocktail in a hushed and restful atmosphere. (Ehrenreich, 1983, p. 85)

Psychologists also lent their support to breaking the link between the breadwinner model and masculinity. Abraham Maslow was one of those who proposed changes in personal relationships, including those between women and men. Maslow called for abandoning the old theories of adjustment to replace them with new goals of self-actualization. He argued that psychologists should not demand that people adjust to inhuman roles. Instead individuals should be encouraged to find and develop their own strengths and potential. Nothing should stand in the way of this growth, not jobs, not spouses, not children, not obsolete definitions of masculinity.

By 1980, Ehrenreich says, the new man had emerged. "He is smiling, physically fit, and identified as 27 with zero dependents" (Ehrenreich, 1983, p. 143). According to Ehrenreich, popular ideology declared that marriage was no longer necessary and sometimes even a detriment to leading a properly masculine life.

The demographic data show an increasing number of people delaying marriage, sometimes indefinitely. Relatively short-termed cohabitation as a substitute for marriage has increased. These changes in behavior have been accompanied by changes in our ideas about marriage and about femininity and women. Our ideas about men and masculinity have also changed in equally dramatic ways. In the next section we will investigate the question of why marriage has been arranged as it has and why changes are taking place now.

THEORETICAL DEBATE ON MARRIAGE

This chapter has revealed a number of ways in which the institution of marriage is a problematic one. Some theorists believe that marriage may be the key to the creation and maintenance of gender inequality and the oppression of women, and therefore the institution of marriage must be dismantled in order to restructure intimate relationships, especially those between women and men, more humanely and more equally.

Frederick Engels is an example of an early theorist who believed that marriage was a central factor in the oppression of women. He wrote that the invention of the social institution of monogamous marriage was the moment of the "world historic defeat of the female sex" (Engels, 1884/1970, p. 87).

To those of us living in the twentieth century the idea that marriage was a social invention at first may sound strange, but as we have observed throughout this book, history shows us that relationships among people are shaped by human interaction. Marriage is an institution that exists in some societies but not in others and varies greatly from one society to the next. Therefore, the idea and practice of monogamous marriage must have been created for the first time at some specific "moment" in history. According to Engels, it was created for a particular purpose: to control women and children.

Engels, who was a collaborator of Karl Marx, developed an elaborate theory explaining why women were oppressed and the link of the invention of marriage to this oppression. He argued that early in human history people lived in small groups in which there was no class inequality and in which women were not oppressed. Although women and men had some different responsibilities in these societies, these differences did not include a power differential. In fact, he argued, if anything women may have been more powerful as a group than men because of their importance in reproduction. Women's power in these small societies was indicated by the matrilineal organization of kin relationships in which women and their children formed the core. Men were connected to the system through their mothers and sisters. Men's biological children were considered part of the mother's family and men had stronger social relationships with their sisters' children than with their own offspring.

After describing these ancient matrilineal egalitarian societies, Engels pointed out the striking inequality between women and men in modern capitalist states and asked the question, "How did society change to produce the oppression of women?" He answered by asserting that as the egalitarian hunter/gatherer groups became more efficient at producing their own food and shelter rather than finding it in the wild, they were able to produce a surplus. For example, after animals were domesticated herders could begin to raise more animals than they needed to provide enough food for the clan. When they were not forced to use up the animals every year for food, they could accumulate larger and larger herds. Those people who owned the herds gradually gathered more and more of this surplus and class inequalities began to emerge. Those who owned herds had something to offer or to withhold from others in order to gain control over decisions made by the clan. Furthermore, Engels argued, men, not women, were more likely to be herders and it was men and not women who came to make up this emerging dominant owning class.

The dominance of men in terms of ownership of wealth and the egalitarian matrilineal kinship system did not fit well together. Men who were able to accumulate property through the ownership of their herds and through trading were unable to pass their inheritance to their own children because of the matrilineal kinship system, in which inheritance would pass from a man to his sister's children. In order to create a system of inheritance that allowed the passage of property from a father to his children a mechanism had to be created to ensure that (1) a man's children were in fact his own, and (2) kinship and inheritance would be traced through the father's line. The mechanism that provided for these two needs was patriarchal monogamous marriage. Marriage required that women be sexually exclusive in order to guarantee that any children born of that marriage were the husband's. Marriage allowed property-owning men to consolidate their wealth and to pass it on to their biological and legal heirs. According to Engels, marriage was a critical initial social underpinning of both class inequality and the oppression of women. Marriage aided the consolidation of wealth, locking in class inequality. Marriage guaranteed that men would control women and thereby control the products of their reproduction, children, locking in male dominance of women.

Engels tried to support his theory by citing empirical evidence gathered by nineteenth-century anthropologists. The data he amassed did back up his ideas, but since then other anthropologists have found conflicting evidence as well. Most scholars now argue either that there is insufficient empirical evidence to support or refute his theory, or that the empirical evidence refutes his theory. Anthropologists do not agree about whether Engels was correct in his depiction of the process by which social classes developed or men came to control women. Nevertheless, the framework remains a powerful theoretical argument for analyzing the historical roots of the oppression of women because it suggests a way of understanding the oppression of women in contemporary society (Leacock, 1973; Lerner, 1986; Lengermann & Niebrugge-Brantley, 1992).

The importance of Engels's theory to this chapter is that it points out the centrality of the organization of marriage to the existence of gender inequality and the oppression of women. Throughout this chapter we have noted contemporary authors who argue that the current organization of marriage, despite a history of reform, still aids men in their dominance over women. Engels's theory gives us some clues to understanding why women are oppressed by men and how they are kept under men's control (Rapp, 1982).

Why are women oppressed? Engels proposed, as we have seen, that the reason women are oppressed by men is because in order to maintain the class system, wealthy men need a mechanism for keeping their wealth together from one generation to the next rather than dispersing it. Upper-class men need a family system that allows them to pass their inheritance to their own biological heirs. Therefore they need a means to guarantee their wife's sexual exclusivity and dependence on them. How do upper-class men accomplish this? By maintaining a family system that is patrilineal and that demands that women be monogamous—by maintaining the institution of marriage. Engels maintained that marriage was an institution that served the interest of ruling-class men. He asserted that relationships between working-class women and men were not bound by the same needs and that another kind of marriage, a nonoppressive form, existed or potentially existed between working-class women and men. He argued that as working-class women became economically independent by entering the paid labor force they were freed to enter into egalitarian relationships with working-class men. And this is where Engels has been criticized by Marxist and socialist feminists.

During the emergence of Second Wave feminism, Engels's theories were widely discussed. Socialist feminists agreed that the oppression of women was tied into the maintenance of a class system like the capitalist system in the U.S. and believed Engels's theory provided a convincing argument for how marriage acted as a mechanism to connect these two systems of oppression, capitalism and patriarchy (Eisenstein, 1979). Socialist feminists, however, disagreed with the assertion that working-class men did not benefit from an organization of marriage that placed women in a dependent subordinate position. They also argued that entering the labor force did not eliminate the oppression of women.

The oppression of women served ruling-class men in two ways, by helping to maintain the system of class inequality and by helping to maintain the system of gender inequality. But according to socialist feminists, the oppression of women also served working-class men by granting them a position of dominance in their relationships with women of their class. Male dominance exists in many working-class marriages even when the woman is earning wages outside of the home (Hansen & Philipson, 1990).

The debate over whether class inequality is more salient than gender inequality or vice versa and whether ending class inequality will lead to gender equality has continued for the past few decades. The problematic character of marriage and its importance in the maintenance of class and gender oppression, however, is agreed upon by theorists from Marxist, socialist feminist, and radical feminist frameworks.

This chapter began with a story in which a couple faced a problem that might be solved or alleviated if they had the right to marry. Much of the rest of the chapter has described the problematic character of marriage for those who are allowed to marry in our society. Is marriage the problem or the solution? The debate over this question has been an important one, not only among theorists but among political activists, especially those involved in the gay rights movement. This debate is the topic of the next section of this chapter.

FAMILIES WE CHOOSE

Throughout this book, I have argued that people consistently challenge their society in an attempt to adjust the macro level of social organization. The case of marriage is no different. Individuals interacting with one another have expressed unhappiness with the way in which marriage is organized, especially in the law, and have sought to challenge that organization. Their activity has sometimes taken the form of organized protest, for example, in gay rights rallies. Deciding what the goal of these political activities should be, however, is not a simple task. Should the political battle be to allow marriage for gay people or should gay rights activists lead the challenge to reconstruct intimate relationships?

Andrew Sullivan (1992) argues that gay couples should be allowed to marry. He writes that marriage allows couples to make a deeper commitment to one another and to society and that in return society allows them certain benefits. Those benefits include legal rights like the right to be included on a partner's employer-provided health insurance policy and the right to adopt children. Sullivan also maintains that marriage provides emotional benefits. He writes, "Marriage provides an anchor, if an arbitrary and weak one, in the chaos of sex and relationships to which we are all prone" (Sullivan, 1992, p. 77). Furthermore, he asserts that legalizing and encouraging marriage among gay couples would provide role models for young gay people and it would help bridge the gulf between gay people and their parents.

Paula Ettlebrick (1992) strongly disagrees. She asks the question, "Since when is marriage a path to liberation?" Her concern is that gay couples would be caught in the oppressive trap of the institution of marriage as it is currently organized. In addition, she worries that the more gay people choose to fashion their relationships after conventional heterosexual relationships, the more diversity will be discredited as a valuable feature of our society. Both heterosexuals and gay people who chose not to pattern their relationships on a conservative marriage model would be subjected to social and legal sanction even more than they are now.

Kath Weston (1991) refers to this debate as the dilemma of assimilation or transformation. Sullivan is promoting assimilation and Ettlebrick is criticizing it. According to Weston, the alternative is to identify factors that could and in many cases are transforming intimate relationships among gay people. Weston interviewed gay men and lesbians to talk about their families and what their intimate relations looked like in order to find clues about how a transformed family might look.

She found that gay people often live in social arrangements that include intimate relationships between couples and close familial relationships among other adults and among adults and children. She refers to these arrangements as "families we choose."

Sociologists often refer to two types of families: family of orientation and family of procreation. A family of orientation is the one into which a person is born. A person's ties to a family of orientation is a blood relationship; if the person was adopted, the relationship is a legal one that is modeled after a blood relationship. A family of procreation is a person and his or her spouse and children. The relationships in a family of procreation are based on the legal tie between a husband and wife and the blood tie between parents and their children.

Families we choose may include blood relationship and legal relationships, but they also include equally and often more important relationships that are chosen and consciously built. Weston found that lesbians and gay men were actively building families they chose. Their families of orientation were sometimes part of this effort. The act of coming out frequently put their relationships with their families of orientation to the test. Not all gay people were willing to test their family of orientation, but many were willing to break these ties if necessary to attempt to create stronger, more honest ties. One man told Weston, "If you can't be honest with somebody, then what kind of relationship are you really salvaging? What are you giving up if they react badly and they're gone? What have you really lost?" (Weston, 1991, p. 52).

Gay men and women also created families they chose as alternatives to families of procreation. These families often did include procreation, sometimes by bringing children from previous heterosexual relationships, sometimes by adopting children as a gay couple or single, and sometimes by organizing reproduction through alternative insemination.

When Weston (1991) asked the gay men and lesbians she interviewed whom they considered as their family, they consistently counted lovers as family, usually at the top of the list. They also included children and other relatives of their lovers as family and people who were not their lovers but with whom they shared a household. In addition they often named former lovers as more distant, but still connected parts of the family they chose.

Families we choose expand beyond a couple and their children to include fictive kin who become an extended family. One woman described her family:

> When I go to have a kid, I'm not gonna have my [biological] sisters as godparents. I'm gonna have people that are around me, that are gay. . . . No,

> I call on my inner family—my community, or whatever—to help me with my life. So there's definitely a family. And you're building it; it keeps getting bigger and bigger. (Weston, 1991, p. 108)

One of the problems gay couples and the families they chose faced was that without rituals declaring the validity of their relationship it was sometimes difficult to know how shared the relationship was. For example, one person may consider another as part of his or her family but not know for sure if that person feels likewise. Gay weddings and rituals were one way to try to affirm the connections.

Weston (1991) concludes that gay men and lesbians, in their efforts to create families they chose, present a challenge to an organization of marriage and family fashioned after the conservative model. The first feature of this challenge is the support of diversity. Families we choose are not monolithic; they are individually tailored. Second, families we choose prioritize social relationships over legal or biological relationships. Familial relationships in families we choose are not assumed. Instead they are created, tested, and revised through the social interaction of their members.

THE MICRO-MACRO CONNECTION

This chapter has once again shown us the way in which the seemingly personal character of micro-level family relationships is influenced and sometimes determined by large macro-level social structures and forces. A marriage is a micro-level relationship between two individuals, but that relationship is affected by the macro-level organization of society, including such factors as legal issues, economic opportunities, and ideologies. Who can marry, when they marry, and what a marriage entails are defined by factors beyond the individuals who are marrying.

Both the social structural macro-level factors and the micro-level features of marriage are dynamic and change over time. In recent years marriage laws, the experience of marriage, and ideas about marriage have all changed rather dramatically.

Forces at the micro level in part account for these changes. Feminists have been involved in criticizing the institution of marriage and they have been joined by other activists, sometimes from surprising quarters such as the playboy movement. In particular the emergence of gay people acting in organized political forums or interacting in their own private lives is challenging the way in which marriage is organized.

SUMMARY

Characteristics of Marriage

- The scenario tells the true story of a lesbian couple and reveals the problem of defining marriage and its components.
- Marriage can be characterized by five factors: a legal contract, an economic arrangement, a commitment, a sexual relationship, and a political arena.

- One important indicator of the political relationships in families is verbal interaction.

Statistics on Marriage

- Marriage is a changing institution. The changes appear to be in the direction of larger numbers of people delaying marriage or declining to marry at all. These trends vary by gender and race.
- The marriage squeeze refers to the lack of eligible partners available to women in birth cohorts that are larger than those of men who are a few years older than they are.
- Many people are not choosing to forego intimate relationships but cohabiting without marrying.
- Measuring the number of cohabitors is difficult because census data tell us only who lives with whom, not what the social relationships are within households.
- Singlehood is increasing. Many single people choose to be single, but some may feel unable to marry because of economic circumstances.

Changing Ideas About Marriage

- At the same time that the numbers of people marrying have changed, our ideas about marriage have also changed.
- Some of these changes have been related to changes in ideas about what women should be like and have been connected to the women's liberation movement.
- Men have also changed. Ehrenreich calls the period from the 1950s to the 1980s the male revolt, an equally powerful but less recognized challenge to marriage in the U.S.

Theoretical Debate on Marriage

- Theoreticians have debated why marriage was established and what purpose it currently serves.
- Engels proposed that marriage was originally designed to facilitate both the maintenance of class inequality and the oppression of women.
- His ideas have been criticized by radical, socialist, and Marxist feminists but the central argument he makes about the connection of marriage and the oppression of women is one upon which they agree.

Families We Choose

- Gay rights activists have debated the question of whether gay people should seek to have the legal right to model their relationships after heterosexual marriage or lead the challenge to transform that model.
- Kath Weston advocates following the transformative path rather than the assimilationist one and has described what she sees as the strengths of gay families, which she refers to as families we choose.

The Micro-Macro Connection

- This chapter shows the way in which large social structural factors, especially the law, can have an important effect on the experience of couples interacting at the micro level in intimate relationships.
- The relationship between these macro-level structures and the micro level of social interaction has changed over time and continues to do so.
- People interacting at the micro level have also had an effect on the macro level, attempting to transform the organization of marriage through conscious political work and through the day-to-day acting out of alternative family forms.

Divorce is a legal issue and has been subject to laws and courtroom proceedings in the United States since Colonial days. The 1970s marked an important transition in divorce laws, and changes to what is known as no-fault divorce during that decade have had significant effects on divorcing couples and their children.

10 DIVORCE AND REMARRIAGE

Teresa and Manny have recently been divorced after eight years of marriage. For most of their marriage they were both happy and when their daughter Maria was born they excitedly looked forward to watching her grow up together.

Teresa speaks: "Divorce teaches you a lot about marriage. I had some very romantic ideas about what marriage is and when we couldn't seem to match those dreams I felt so unhappy I couldn't stay married anymore. It has been almost two years since we got our divorce. For me it was a major turning point. Mostly it's the money. We had to sell the house in the divorce settlement and my check is pretty good but not enough to buy a house or even rent in our old neighborhood. Maria has to go without sometimes and it makes me feel bad. Manny is supposed to send child support but sometimes he doesn't. He says his job is not going that well but sometimes I think he doesn't send the check because he likes to keep me on edge. Divorce isn't all bad, though. I've had to take care of Maria and myself and I've done it. That feels pretty good."

Manny speaks: "I liked being married but Teresa and I didn't see eye to eye on what a good marriage is. She was always wanting me to try harder and make more money to get ahead and I wanted to enjoy life a little. At the end I was going out with other women because I needed some sympathy. Divorce has been hard on me. Most of our friends were Teresa's and they stopped seeing me when we split up. Teresa gets mad when I don't send the child support check but work is off in construction and we don't get that much. Plus I feel like I hardly know Maria any more. We get together on weekends and holidays but it's not natural picking her up and making arrangements. Her friends are in her new neighborhood and sometimes it seems like I'm imposing on her new life. I hope I can get back to normal when I remarry next spring. My girlfriend Carol has helped me a lot."

Maria speaks: "I hate it that my parents got a divorce. They don't argue as much but I don't see why they couldn't have worked it out. It's embarassing not to have a father at home. People think he left us or that there was something wrong with my mother or him. And besides, I miss him. When they decided to split up I thought it was my fault because I wasn't doing so well in school. Now I know it was their problems but I keep thinking maybe I could have done something. Divorce is between two adults but it affects us too. Kids have nothing to say about it. When I get married I will be different."

- *How do our ideas about what marriage should be like affect the possibility that we will divorce?*
- *Besides ideas, what other social factors put pressure on marriages?*
- *How is divorce experienced differently by women and men, parents and children?*
- *Divorce is usually a crisis because it creates difficult changes in people's lives. But can those changes be positive as well as negative ones? Can divorce be both a trauma and an opportunity?*
- *Manny looks forward to his second marriage as an opportunity to correct some of the mistakes he made and to start over. While remarriage does provide a second chance, what kinds of problems do couples have who are marrying for a second time?*
- *Since divorce is such a common experience, why do people feel as if they are facing an unexplored and difficult mystery when they decide to divorce?*
- *Divorce is a social problem because it is experienced by so many people. How can our society create ways to ease that experience? Should we make divorce more difficult to obtain? Should we create supports that make it easier to endure?*

DIVORCE IS COMMON in the contemporary United States. Everyone knows someone who has divorced and many of us have lived through a divorce of our own or our parents. Despite its commonness, however, divorce is highly problematic and difficult to understand. In this chapter we will examine the nature of divorce and offer some suggestions for how to use the Sociological Imagination to try to understand why the process is so painful and how it might be made more humane. The chapter is divided into four sections. The first describes the incidence of divorce in the U.S. and reviews the history of change in divorce since the colonial period, with an emphasis on the way in which the macro organization of law has had an important effect on the micro-level experience of divorce. The next section deals with what happens after a divorce and is divided into three subsections: social stigma, divorce as an opportunity, and remarriage. The third section focuses on the importance of gender inequality in understanding the problematic character of divorce. In the fourth section, the effect of the micro level on the macro level is reviewed in a discussion of recent attempts to get fathers to pay for child support.

HOW TO MEASURE DIVORCE

In the introduction to this chapter I asserted that divorce is a common experience and asked that you recall your personal experience as evidence of its pervasiveness. Scholarship, however, demands that additional evidence be provided.

How often does divorce occur? This is a question that demographers have grappled with and the answer is surprisingly difficult to obtain.

Demography is the study of the size, composition, and distribution of human populations. Demographers record and analyze statistics about issues like birth, death, marriage, and divorce. It may seem that the demographers' task of measuring divorce is an easy one. Since divorce is a legal transaction there is a record in the courts of every divorce. Reporting the absolute number of divorces, however, is not very useful unless it is compared to some other number like population size, number of marriages, or number of potentially divorcing people. When demographers wish to report the number of divorces, therefore, they confront difficult choices.

There are five possible ways to report the number of divorces. Each of these methods has its problems (Crosby, 1980).

1. Reporting the number of divorces per year is one method for arriving at the rate of divorce. This measure is unsatisfactory because it does not take into account the changing numbers of people in the population. For example, if we found that the absolute number of divorces was larger in 1970 than it was in 1870 we would not know how much of this increase was a result of the increase in the total population and how much was a true measure of the increase in divorce.
2. Measuring the ratio of current marriages to current divorces is a second way of finding the divorce rate. The problem with this method is that the ratio between marriages and divorces can fluctuate because of changes in the rate of marriage as well as changes in the rate of divorce, and the ratio does not tell us which number is fluctuating. For example, suppose we compared the ratio between the number of marriages and the number of divorces in 1960 to that ratio in 1990 and found that the ratio was higher

in 1960. This can mean that there were more marriages in 1960 than there were in 1990 or that there were fewer divorces in 1960 (or both). But we do not know which of these possibilities are being reflected in the ratio.

3. The crude divorce rate is the number of divorces per 1,000 population. This technique corrects for the size of the population mentioned in method 1. But many of these 1,000 people, those who are single adults and those who are children, are not able to divorce. If the proportion of single adults or children in the total population changes, it will appear that the divorce rate has changed, even though it has remained the same. For example, suppose the total population remains the same but the birth rate rises and children make up a larger proportion of the total population in 1990 compared to 1970. Then we might find that the crude divorce rate falls during that period. But this decline might mean that divorces really had diminished or it might mean that fewer people were married but the incidence of divorce among those who were married was actually more common.
4. The refined divorce rate is the number of divorces per 1,000 married women over age fifteen. This is the measure that is most commonly used and is considered the best available. This measure accounts for changes in the population and eliminates the problem of a fluctuating number of children (too young to marry) in the population and a fluctuating number of singles (not able to divorce). This method of measurement, however, is not completely satisfactory because it does not give any indication of what the possibility is of divorcing at any time in one's life or at any particular age. It does not include information on the relationship of years of marriage to divorce. We do not know from the refined divorce rate how long people have been married. The refined divorce rate tells how many people in a given year are divorced but it does not tell us how many of those who are currently married will divorce at some time in their lives.
5. Data from longitudinal studies would correct this problem by following married couples over their lifetimes to look at patterns of marriage and divorce over time. According to Crosby (1980) a longitudinal study would be the best way to report divorces. This technique, however, is expensive and time-consuming and, therefore, has only been used in small surveys.

Rates of Divorce

Using the refined divorce rate as their measure, demographers Teresa Martin and Larry Bumpass (1989) predict that two-thirds of all first marriages in the U.S. will end in divorce. Their estimate is on the high side of the range of predictions that have been made by sociologists. But researchers agree that divorce is so common that it will be experienced by nearly all of us as spouses, children, or at least friends or relatives of divorcing couples.

The experience of divorce by such large numbers of people is a relatively recent phenomenon. As we saw in Chapter 2, the divorce rate in the U.S. has increased since the turn of the century. Until the mid-1960s the increase was gradual except for a brief but intense increase immediately following World War II. From 1965 to 1975, however, the rates rose dramatically, nearly doubling in

TABLE 10 • 1
Percentage of People Who Were Divorced, 1970–1991

	1970	1980	1991
White	3.1%	6.4%	8.0%
Black	4.4%	8.4%	10.8%
Hispanic	3.9%	5.8%	7.3%

Source: U.S. Bureau of the Census, 1991. *Marital Status and Living Arrangements: March 1991*, p. 2.

those years. After the late 1970s the rate began to level off and decrease modestly.

Table 10-1 shows the percentage of divorced adults in three censuses. This is not the divorce rate but the proportion of divorced people. The table shows that the proportion has increased. It also shows that the proportion of divorced people is highest for blacks and lowest for Hispanics, with whites falling in between.

What Are the Correlates of Divorce?

Divorce is such a common experience in contemporary America that it crosses many social lines such as race, age, class, education, and age. Sociologists, however, have found patterns indicating that particular factors make a couple more or less likely to seek a divorce.

Income has been correlated with divorce (Spanier & Glick, 1981). Correlation means that a relationship exists between two variables, in this case income and divorce. The relationship between income and divorce, however, is not a simple one. People at the lower end of the income range are more likely to divorce. At the upper end of the income scale, couples are less likely to divorce, but only if the husband is the primary source of the household income. In wealthy households if the wife's earnings are a major source of the household income, divorce is more likely than average.

Education is a second factor that has been correlated to divorce. Divorce rates decrease as education increases up to a college education. Women who have more than a bachelor's degree, however, have higher than average rates of divorce. The chances of divorce are especially high for women who begin graduate work after they marry (Houseknecht, Vaughn, & Macke, 1984).

Divorce rates also vary by race ethnicity. Once again, however, the variation is complex. African Americans are more likely than Euro-Americans to divorce (Thornton & Freedman, 1983). The greater likelihood of divorce among African Americans has been linked to other issues such as pressures from discrimination and lower incomes. However, Latinos who share some of these problems with African Americans have divorce rates that are only slightly higher than Euro-Americans. Furthermore, there is great variability in divorce rates as well as other family issues among different groups of Latinos—for example, Cubans, Mexicans, Puerto Ricans, and Central Americans (Baca Zinn & Eitzen, 1990). Although there do seem to be relationships between race ethnicity and divorce, the underlying reasons for these relationships and what the differences mean are not known.

Age at first marriage has the most clear relationship with divorce. Those couples who marry later are less likely to divorce (Spanier & Glick, 1981).

BOX 10 • 1 DO HIGHER DIVORCE RATES MEAN FEWER INTACT FAMILIES?

Sometimes people who see the statistics indicating increasing numbers of divorces jump to the conclusions that contemporary Americans, compared to those of previous historical periods, are much less likely to live in intact families, children are less likely to be raised to adulthood by both of their natural parents, and more families include stepparents and step- and half-brothers and sisters. These conclusions are not entirely valid.

Peter Uhlenberg (1989) argues that while divorce was increasing another demographic statistic, the mortality rate, was decreasing. For example, in 1900, 24 percent of children would lose at least one parent to death by the time they were 15 and one in 62 would lose both parents. In contrast in 1976, only 5 percent would lose one parent and one in 1,800 would lose both. Figure 10-1 shows the contrasting pressure of mortality rates and divorce rates during the twentieth century.

Until the 1970s mortality rates in the twentieth century dropped even more rapidly than divorce rates rose. A husband or wife who married in his or her early twenties was much more likely to be widowed or divorced before reaching a fortieth wedding anniversary in 1900 than in 1976. Even though divorce rates rose during this time period, the net effect of divorce and death actually declined. Writing in 1976 Uhlenberg states, "decline in early widowhood more than offsets the rise in divorce so that the stability of marriage during the childrearing years has actually *increased* over this century" (Uhlenberg, 1989, p. 92).

Since the late 1970s this situation has changed somewhat. Divorces have come to represent a larger proportion of marriages that have ended. Divorce now has caused marital dissolution to rise to higher levels than in earlier years. But the change in the total is still relatively small. The proportion of people today who experience the end of a marriage is not dramatically different from what it was at the end of the last century (Weeks, 1993).

Martin and Bumpass (1989) assert that age at marriage is the strongest predictor of divorce in the first five years of marriage and age continues to have an important effect throughout the marriage. As in the case of other correlates with divorce, researchers are unsure why youthful marriages are more likely to end in divorce (White, 1991).

Historical Changes in Divorce

Although the rates of divorce in the U.S. in the last decade are near record highs, divorce is not new to this country. Glenda Riley (1991) traced the history of divorce in the U.S. from the early 1600s to the late twentieth century and found that the first couple to obtain a legal divorce in North America were granted their decree in 1639. For more than 350 years, divorce has been part of the American tradition.

As we observed in Chapter 2, the Godly Family was established by Puritans fleeing a society in England of which they did not approve. Divorce was part of their new society. Divorce was not available at all in England until the late 1600s and then until 1853 it could only be obtained by a decree from Parliament. In contrast, in the colonies in North America, Plymouth officials in 1620 declared that marriage was a civil rather than an ecclesiastical matter, making divorce possible. This declaration directly "defied English colonial policy that stipulated English laws—in this case anti-divorce laws—be established in parts of the empire that lacked pre-existing legal codes" (Riley, 1991, p. 12).

FIGURE 10 • 1
The Annual Rate of Marital Dissolution by Death and Divorce, 1865–1985

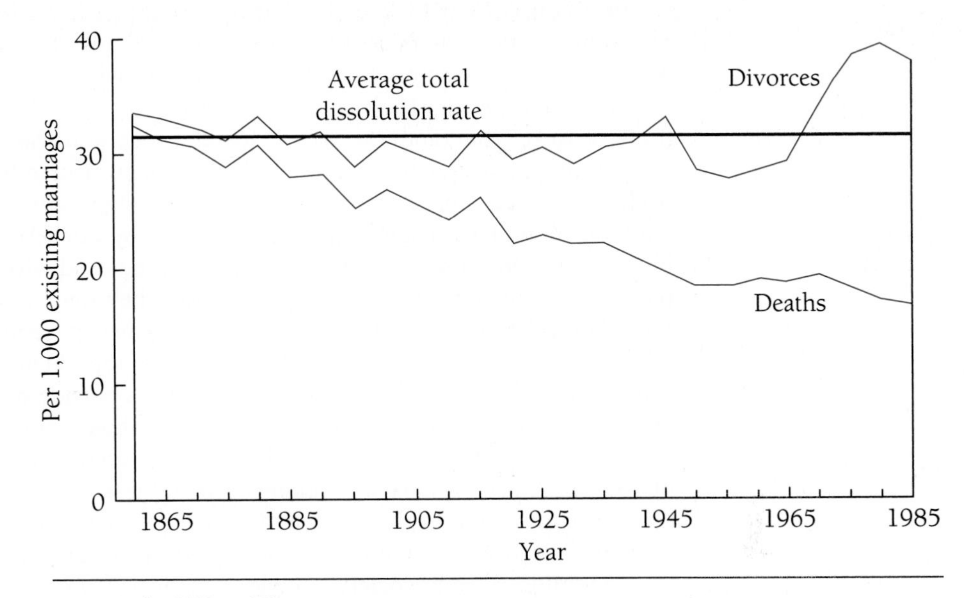

Source: Weeks, 1993, p. 278.

Riley (1991) says, however, that divorce has not been a settled issue in the U.S. Throughout the 350 years of divorce in America, the issue has been debated by two opposing sides, the anti–divorce rights faction and the pro–divorce rights faction. The opponents of divorce rights have usually believed there is only one legitimate reason for allowing divorce, and that is adultery. Sometimes they have also accepted insanity or consanguinity—marriage between blood relatives—as legitimate reasons for granting a divorce.

Pro–divorce rights advocates believe that divorce is a citizen's right in a democratic society. They also believe that divorce is a symptom, not a cause, of underlying problems in marriage and society. Historically, supporters of divorce rights have argued that divorce is sometimes a solution to certain kinds of problems. For example, they hope that making divorce easier to obtain would lead to greater equality in marriage.

Another American History of Divorce Until about the turn of the century, the Inuit in North Alaska, who are commonly called Eskimos, maintained a system of marriage and divorce that was quite different from that of the Euro-Americans who dominated North America. Burch (1970) noted that divorce among the Inuit consisted of the couple ceasing to live with one another and to have sexual relations. A divorce, like a marriage among the Inuit, was accomplished without ceremony. The decision was made by the person who wanted the divorce regardless of whether the spouse agreed with that decision. All that was involved was one or the other spouse leaving or making the other one go. A husband or wife who wished to divorce needed only to walk out the door. A husband or wife who wanted his or her spouse to leave would scatter the spouse's belongings outside the door.

Unlike the legal systems of divorce in Euro-American society, Inuit divorce was not constrained by the problem of giving a reason for the separation or obtaining permission from the community and its institutions. While Euro-Americans argued about who should decide and what the grounds for those decisions should be, the Inuit agreed that individuals should be able to make the decision to separate and any reason for their decision was a legitimate one.

Another characteristic that distinguished Inuit divorce from the legal system of divorce in the United States was the relationship between husband and wife after a divorce. Under the legal system created by Euro-Americans, divorce terminates the marital relationship between husbands and wives, so that, for example, if they decide to get back together again they must remarry. For the Inuit, the relationship between husband and wife is never terminated, but rather is "deactivated." Burch describes this as similar to the phenomenon among Euro-Americans of inactive relationships with relatives outside of the immediate family. For example, we may have second cousins about whom we have heard and we may even know where they live but we have no direct contact with them. In one sense we do not have a relationship with these cousins because we have no contact with them. On the other hand, the potential relationship exists and can often be reactivated if one of the cousins wishes to do so.

In traditional Inuit society, couples deactivated their relationship by ceasing to live with one another. If they decided to get back together they reestablished their relationship by living with each other again.

Solid data are not available on the divorce rate among the Inuit under this system, but Burch (1970) argues that the divorce rate approached 100 percent. Nearly all couples were separated at some time during their married life and many were separated a number of times. Divorce was a pervasive and fluid phenomenon in Inuit society.

Burch also looked at the reasons for divorce among the Inuit. He observed that three reasons seemed most common: infidelity, failure of either a husband or wife to meet economic obligations, and disagreements over childrearing. Burch also observed that the level of interpersonal difficulty caused by these activities was often quite low but resulted in divorce nevertheless. He writes, "Indeed, some of the things we would scarcely regard as justification for an argument were sufficient grounds for an Eskimo divorce" (Burch, 1970, p. 168). The relatively quick transition from problem to divorce in Inuit society was a result of the belief that relationships between husbands and wives were not supposed to contain any strains.

The ease and prevalence of divorce was coupled with a high rate of remarriage among the Inuit. Ex-spouses were obligated to treat each other in a friendly and supportive manner. The jealousy and mutual antagonisms often encountered in our society between new spouses and ex-spouses were not tolerated. New spouses and ex-spouses were required to avoid each other altogether or to get along on good terms (Burch, 1970).

Following a divorce, the children of the divorced couple frequently stayed with their mother, although they could choose to live with the parent they preferred. If there was a very young child, it was frequently adopted by another couple, usually a relative of the natural parents. In these adoptions the child was not seen as having lost its natural parents, but rather as having gained another set of parents with which to live.

Property settlements were also easy. Most property was individually owned with little overlap between the sexes. Men owned the property associated with the economic tasks they fulfilled such as hunting, manufacturing tools, and educating their sons. Women owned the property associated with the womanly work of butchering, making clothing, caring for small children, and educating their daughters. When couples divorced they took with them the tools of their trades.

The difference between Inuit divorce and legal divorce in the U.S. in the nineteenth century and even in the twentieth century is striking. For the Inuit the decision to divorce could be made by an individual and involved minimal action. The decision to "remarry" one's ex-spouse was equally simple. The two difficulties in legal divorce that we face, property settlement and child custody, were also handled in a humane and efficient manner.

Throughout this book I have stressed the importance of observing the link between social context and family organization. The Inuit living in northern Alaska before the turn of the century existed in a social context that is much different from that of most people living in the contemporary United States. The Inuit system for organizing divorce, therefore, could not be directly incorporated into contemporary U.S. divorce law. However, the Inuit do provide us with some insight into the variety of ways in which divorce might be handled more effectively. This discussion also reminds us that the historical pool from which we might draw ideas should include ideas from a small, relatively invisible and disappearing, but equally American culture.

AFTER DIVORCE

What happens after a divorce? This section addresses this question by investigating three important issues: the stigmatization of divorced people; the unfolding of opportunities for personal growth after a divorce; and remarriage.

Social Stigmatization

Social stigma occurs when people are perceived to be different and less worthy than others because of their appearance or behavior. When people are stigmatized, others see them as having something wrong with them. Erving Goffman described stigmatized people as those who are seen and come to see themselves as "of a less desired kind . . . reduced in our mind from a whole and usual person to a tainted, discounted one" (Goffman, 1963, p. 3).

Naomi Gerstel (1987) decided to try to find out if divorced people are stigmatized. Changes in the laws and especially the passage of no-fault laws indicate that people are becoming more accepting of divorce and are less likely to blame divorced people for the divorce (Gerstel, 1987). In addition to changes in the laws, public opinion polls of divorce also show greater tolerance of divorce and divorced people than existed earlier in the century (Veroff, et al., 1981). But is this acceptance reflected in the personal experience of individuals who are divorced?

Gerstel's (1987) interviews with 104 divorced and separated people showed that divorced people continue to be stigmatized in two ways.

First, although disapproval of divorce as a general category has declined, divorced people experience disapproval related to the specific conditions of the divorce. Their divorce is accepted only if it was under certain conditions. Furthermore, these conditions vary by gender. Men who had begun affairs before their divorce experienced disapproval and were stigmatized as "cavalier homewreckers" (Gerstel, 1987, p. 176). Women were more likely to be criticized if they divorced when they had young children. Gerstel (1987) explains that these women were "bad divorcees" when they did not or could not sacrifice for their children.

Second, divorced people still suffer informal relational sanctions. For example, divorced people feel they are punished for divorce by being forced to split friends and being excluded from social interactions. One man explained: "The couples we shared our life with, I'm an outsider now. They stay away. Not being invited to a lot of parties that we was always invited to. It's with males and females. It sucks" (Gerstel, 1987, p. 173). And a woman stated: "Well, I now have no married friends. It's as if I all of a sudden became single and I'm going to chase after their husbands" (Gerstel, 1987, p. 172).

As a result of their exclusion from married couple circles, divorced people, Gerstel found, tended to develop a social life among other divorced people. This is similar to the experience of other stigmatized people. Goffman (1963, p. 21), who studied many different kinds of stigmatized people, found that, in general, stigmatized people often come together to share their stories, offer comfort, and make one another feel at home like "a person who is really like any normal person."

Gerstel concludes that the divorced are no longer thought to be sinful or criminal and many of the friends and families of divorced people do not disapprove of their divorce. She says, "However, a decrease in statistical deviance, a relaxation of institutional controls by church or state, or a decline in categorical disapproval is not the same as the absence of stigmatization" (Gerstel, 1987, p. 183). Divorce continues to set people apart and often to represent a mark against them.

Young Song (1991) has examined the negative results of divorce for Asian American women. She reports that Asian American women experience the stigmatization and self-blame described by Gerstel but in an exaggerated form. Song (1991) argues that the divorced women she interviewed retained stronger ideas than other Americans did about the shamefulness of divorce and its identification with failure, particularly for women. As a result, Asian American women are likely to suffer serious emotional hardships after a divorce. These problems are further exacerbated by the cultural value placed on controlling and hiding negative feelings.

Divorce as an Opportunity

As we have seen in the previous discussion divorce is often accompanied by enormous pain. The difficulties of divorce will be further discussed later in this chapter. Divorce can, however, have a positive effect on people's lives by allowing them to end difficult and sometimes destructive relationships (Lund, 1990). In the scenario at the beginning of the chapter, Teresa says that in spite of the difficulties, divorce has allowed her to test herself and to successfully accomplish tasks she did not know she would be able to do.

Divorce creates such a powerful break it can be used as an opportunity to make dramatic, positive personal changes. Lund explains,

Divorced people often discover that their friendships become stronger as they reach out to others for comfort and advice.

> Divorce can be seen as a traumatic life crisis for any adult. It means the dissolution of dreams, hopes and the security of a socially acceptable and often secure family form. It can also be viewed as an opportunity for tremendous personal growth for women [and men] as they gain a new sense of competence, autonomy and individuation. (Lund, 1990, p. 65)

Catherine Reissman (1991) interviewed women and men who had recently divorced and found that their experience of personal growth after their divorce fell into three categories: management of daily life, social relationships, and identity. Women spoke of their delight in finding themselves competent to do things they had not done before. They talked of learning to fix things, managing their own money, and making choices about entering jobs or school. One woman spoke of her financial abilities:

> I've managed to get almost everything I've wanted. . . . Everything you see here I bought, this is all mine, I have done it in less than two years. I did this, while with him I never got anything. It was either "We can't afford it" or "We need the money for something else." (Reissman, 1991, p. 272)

Another spoke of her newfound skills as a labor organizer:

> [When I first was asked to speak at the strike] I had no voice, I lost my voice, but I was the spokesman . . . they said, "You're the one that's got to go [to the meeting], they're ready to endorse this." And I can't even talk. Well, nobody else could do it, so off we rushed . . . and I spoke to all those people and TV and radio. (Reissman, 1991, p. 273)

Women also talked about positive changes the divorce brought to their social relationships. They felt they could be themselves and that they learned to

FIGURE 10 • 2
For Women, the Probability of Remarriage After First Divorce Decreases with Age

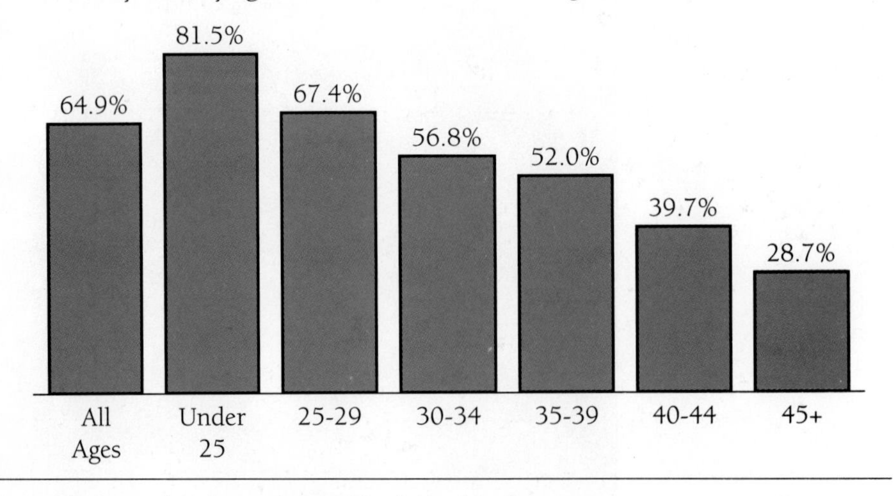

Source: U.S. Bureau of the Census, *Studies in Household and Family Formation*, 1991b.

spend time alone. They also noted how they were forced to develop ties with others in order to survive. Their relationships with friends became closer as they reached out for comfort and advice. One woman explained:

> Marriage assumes that you are, that the couple is self-sufficient. It's harder to reach out. It's funny, I hadn't thought about this but it is true. You're not supposed to be needy when you're married. It's like a betrayal . . . of this myth that marriage provides everything. (Reissman, 1991, p. 269)

Third, women spoke of their new-found identity revealed because of the freedom granted with their divorce. Our society highly values individuality and individual freedom. For many people, especially women, this individuality is submerged in marriage (Bellah, et al., 1985). Divorce, for some, becomes an opportunity to "take back" their lives (Reissman, 1991, p. 270).

Remarriage

Remarriage is a third example of what happens after a divorce: for many, it is followed by remarriage. In 1991, the U.S. Bureau of the Census reported that in more than 40 percent of all the marriages that year the bride or groom or both partners had been married previously. This number has risen in past decades. In 1970, for example, the percentage was 31 percent (U.S. Bureau of the Census, 1989d).

The Census also reports that 70 percent of those who divorce will eventually remarry. There are a number of factors, however, that influence the rate of remarriage. Age is one important variable. Usually young people who divorce quickly remarry; older people are much less likely to remarry (Wallerstein & Blakeslee, 1989). Figure 10-2 shows the proportion of women in different age groups who remarry.

FIGURE 10 • 3
Remarriage Rate for Women, 1921–1984

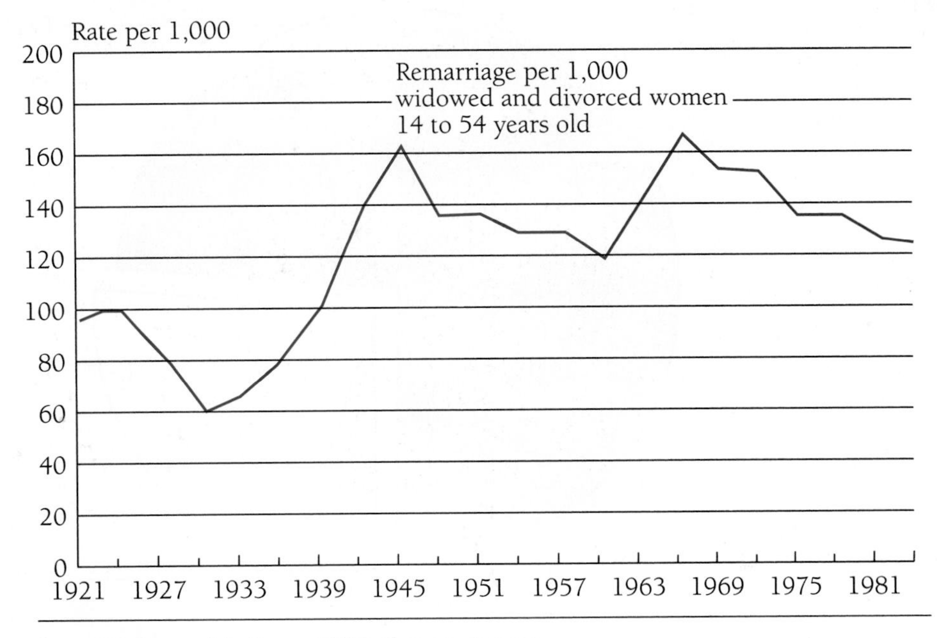

Source: U.S. Bureau of the Census, 1989d, *Changes in American Family Life*, Current Population Reports, p. 7.

Gender and the custody of children are further factors. Eighty-one percent of women without children remarry compared to 73 percent of women with one or two children and 57 percent of women with three or more children (Spanier & Glick, 1981). Men are more likely to remarry than women, and they tend to remarry more quickly (Glick & Lin, 1986). Race ethnicity is also important. In 1988 60 percent of white women who were widowed or divorced remarried; 45 percent of Hispanic women and 34 percent of black women remarried (Census, 1991).

Figure 10-3 shows that remarriage rates have changed over time. This figure indicates that for women, the rates have fluctuated dramatically throughout the century.

The transition from singlehood to first marriage is not entirely unlike the transition from first marriage to remarriage, but there are differences. The couple in a first marriage must work out ways to integrate themselves into their two families of origin—the family into which one is born. Remarried people also have this task. Couples who are remarrying additionally have to integrate their ex-spouses, their children, and perhaps the families of their previous marriage (Ahrons & Rodgers, 1987; Stacey, 1990).

Constance Ahrons and Roy Rodgers (1987) use the term *complex binuclear family* to describe the situation. They described this type of family as it existed for two children in a divorced family they encountered in their research:

> Their family looks like this: They have two biological parents, two stepparents, three stepsisters, a half-brother and a half-sister. Their extended family has expanded as well: They have two sets of step grandparents, two sets of biological grandparents, and a large network of aunts, uncles and cousins. In

FIGURE 10 • 4
Children, by Presence of Parents and Type of Family; 1985
(In percent)

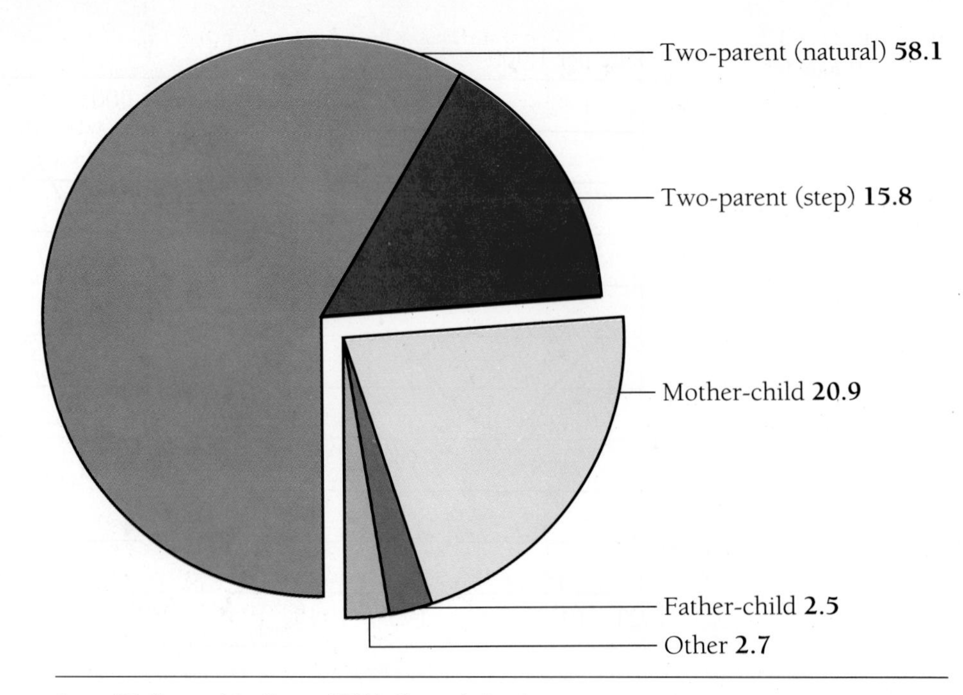

Source: U.S. Bureau of the Census, 1989d, *Changes in American Family Life*, p. 16.

addition to this complex network of kin, they have two households of "family." (Ahrons & Rodgers, 1987, p. 279)

Figure 10-4 shows the proportion of children living in stepfamilies and other families in 1985, the most recent date for which data are available. The figure shows that in the mid-1980s about 16 percent of children lived with a parent and his or her new spouse.

Judith Stacey conducted extensive ethnographic research on two women and their families in Silicon Valley in California. One of her findings was that a new version of extended family had emerged in this working-class community. She describes it as follows: "Your basic extended family today includes your ex-husband or -wife, your ex's new mate, your new mate, possibly your new mate's ex, and any new mate that your new mate's ex has acquired. It consists entirely of people who are not related by blood, many of whom can't stand each other" (Stacey, 1990).

Remarriages are more likely to end in divorce than first marriages. First remarriages, however, have about the same probability of divorce as first marriages. The jump in the probability of divorce occurs after the second remarriage when rates of divorce begin to increase progressively with every remarriage (Cherlin, 1992a).

One remarried stepmother explained to Ahrons and Rodgers:

> When Jim and I decided to get married . . . I was real excited to get married again and give Jamie more of a dad. But it's not working out that way. Jamie is angry a lot about not having time alone with me, which ends

up with Jim and me fighting a lot. Jim feels badly about not spending enough time with his kids and when the kids are together it just seems to be everyone fighting over Jim. And I feel resentful not having enough time alone with Jim. Between every other weekend with his kids and the long hours we both work we never seem to have time alone together. Last Friday we were finally spending an evening all alone and just as I was putting dinner on the table, Nancy [Jim's ex-wife] called. Jim and I spent the next two hours talking about Nancy. It ended up spoiling our evening. (Ahrons & Rodgers, 1987, p. 282)

DIVORCE REFORM IN A GENDER-STRATIFIED SOCIETY

This chapter has presented information on the rate of divorce and the historical development of laws concerning divorce. We have also looked at the experience of people after a divorce. In this section our attention will focus on the relationship between gender inequality and divorce.

In Chapter 5 I introduced the concept of stratification and noted that our society is one that is characterized by gender inequality or stratification of women and men. The issue of divorce, like many in the area of family studies, reveals a striking degree of gender differentiation and the oppression of women. Recent changes in divorce law to what is called no-fault divorce have in particular revealed the enormity of difference in the social experience of women and men in our society.

Before 1970, divorce laws were based on the concept of fault. People who wanted a divorce had to prove that their spouse had committed adultery, cruelty, or desertion (Stetson, 1991). In 1970, California instituted the first no-fault divorce law in the Western world. The new legislation had six major innovations:

1. No grounds are needed to obtain a divorce.
2. Neither spouse has to prove fault to obtain a divorce.
3. One spouse can decide unilaterally to obtain a divorce.
4. Financial awards are no longer linked to fault.
5. New standards for alimony and property awards seek to treat women and men "equally."
6. New procedures aim at undermining the adversarial process and creating a social/psychological climate that fosters amicable divorce (Weitzman, 1985, pp. 15–16).

The focus in the divorce courts shifted from the question of who was to blame to the new question of how the property would be distributed. Reformers hoped that no-fault divorce would eliminate the hostile, undignified, and sometimes fabricated evidence characteristic of divorce proceedings under the fault law. They also thought it would recognize wives as equal partners in marriage and allow for more cooperative postdivorce parenting (Weitzman, 1985).

Lenore Weitzman (1985) studied divorced couples in California to determine if these goals had been achieved. Her findings were shocking. She discovered that no-fault divorce in many cases actually hurt women more than the old laws had. Her investigation showed that in the first year after divorce, women experienced a 73 percent loss in their standard of living while men experienced a 42 percent improvement in their standard of living. Weitzman (1985) also found that this decline was permanent unless the women remarried. Furthermore, because children are usually placed in the mother's custody, divorce also caused impoverishment of children. Other writers have criticized Weitzman's work for exaggerating the economic disparity between women and men in divorce (Faludi, 1991). For example, Hoffman and Duncan (1988) argue that the figures of a 30 percent decrease for women and a 15 percent increase for men in standard of living after divorce are more accurate.

No-fault divorce laws are based on the premise of equality between women and men. Women and men in our society, however, are not equal in marriage or outside of it. Since the premise of equality is invalid, no-fault laws do not work as they were intended and, in fact, actually exacerbate gender inequality (Weitzman, 1985). The change in divorce laws to no-fault altered one important facet of the macro level of social organization but it left other facets intact. In the following discussion we will examine gender inequality in three areas that have interfered with the ability of no-fault divorce laws to work to benefit women: marriage, the labor market, and divorce proceedings.

Gender Inequality in Marriage

In the last chapter we observed a number of ways in which marriage reflects as well as helps to perpetuate gender inequality and the oppression of women. When a couple decides to divorce, inequities in their marriage are especially likely to be revealed. For example, in many households marriage has a different effect on women's and men's earning capacity. One of the most important investments people make is the investment in their careers and in their ability to earn money. Married couples tend to make these decisions in a way that diminishes wives' earning capacity while enhancing husbands' earning capacity (Weitzman, 1985).

For example, a couple may decide that one spouse will work while the other obtains an education, a license, or a membership in a professional organization. They may decide to move in order for one of them to accept a better job or they may decide to stay in an area for one person to build seniority, a professional practice, or a pension. Weitzman (1985) found that these decisions are frequently made to enhance the earning ability of the husband and as long as the couple remain married these decisions enhance the economic standing of the couple. But after a divorce, this "investment" is not seen as part of the property to be divided equally but rather as the "property" of the husband to take with him from the marriage.

Women are more likely to have allowed family commitments to take priority over employment. Although more than 80 percent of the women in Weitzman's (1985) research had been employed at some time during their marriage, only about one-quarter had been employed during the entire marriage. Sporadic or no job experience made it difficult for the other 75 percent to find jobs. Women leave marriage with less ability to support themselves on the average than men do.

Sometimes the "assets" accumulated by the husband are even more invisible, for example, health care coverage. People who drop their health insurance and apply for another policy may be excluded on the grounds of previously existing health problems. A person who divorces a spouse who is the primary health insurance carrier for the family loses that coverage and must apply for another. One of the women Weitzman (1985) interviewed explained how this affected her.

> Since I had always been covered by Bill's policy at Lockheed I never thought about insurance, and no one mentioned it when we drew up the settlement. We agreed to have Bill keep the kids on his policy, but since he was going to remarry I couldn't be covered as his wife. About two years before the divorce, I found a lump in my breast and they removed it and said it was benign. After the divorce, when I applied for individual Blue Cross they wrote in a cancer exclusion because of my history. There was nothing I could do—it was take it or leave it. . . . Then, when I discovered the other lump about eight months later, I went into a total panic. I had to have a radical [mastectomy] and chemotherapy, and there was no way I could possibly afford it. I just wanted to die. . . . I did think of suicide—but I couldn't leave the kids. (Weitzman, 1985, pp. 135–136)

Gender Inequality in the Labor Force

The preceding discussion shows that wives may invest less in their earning capacity and therefore leave a marriage with less ability to support themselves. Some wives, however, do not allow marriage to stand in their way of earning an education and building a career. But even these wives leave marriage at a disadvantage compared to their husbands because education and labor market participation do not necessarily result in equal pay for women.

Table 10-2 shows that women with various amounts of education earn less than men at those same levels. In fact, the chart shows that women with a college degree earn about the same as men who have only a high school diploma. No-fault divorce laws, therefore, make the incorrect assumption that wives are equal to their husbands in the labor market.

Alimony Alimony is money the court requires one spouse to pay to support another after a divorce. Historically alimony was a payment husbands were required to make when separating from or divorcing their wives. Theoretically alimony was a way in which to overcome the inequity between women and men in the labor market. Divorced women were assumed to be unable to earn as much money as their husbands and alimony was a method of closing the gap between their incomes. In reality, however, alimony was always rare and the payments small and since the passage of no-fault it has become even smaller and more short-term.

Many people believe that most women receive alimony when they divorce; even professionals who work with divorcing couples are likely to hold this belief. For example, in 1975 Los Angeles judges estimated that two-thirds of divorced women received alimony awards (Weitzman, 1985). The real number, however, was only 14 percent and that proportion has remained steady since the late 1960s (Stetson, 1991).

14%

The size of alimony payments that are awarded, furthermore, is small and has dropped in the past two decades. In 1978, women who received alimony on the average were granted $4,701 per year in constant 1985 dollars. By 1985 that sum had dropped 13 percent to $3,733 (U.S. Bureau of the Census, 1989b).

TABLE 10 • 2 Median Annual Earnings for Women and Men by Education, 1991 (Year-Round, Full-Time Workers over 25)

Education	Men	Women
Less than 9th grade	$16,880	$11,637
High school	20,944	13,538
High school graduate	26,218	18,042
College	31,034	21,328
Bachelor's degree	39,894	27,654
Master's degree	47,002	33,122
Doctorate	54,626	40,172

Source: U.S. Bureau of the Census, 1992, *Money Income of Households, Families and Persons in the United States: 1991*, Current Population Reports, pp. 129–130.

Alimony has also shifted from permanent awards to transitional ones. In 1968 in California 62 percent of alimony awards were labeled "permanent," "until death," "until remarriage," or "until order of the court." In 1972, two years after the introduction of no-fault, these kinds of designations applied to only 32 percent of the alimony awards. Since the early 1970s the proportion of permanent alimony awards has remained at about one-third of all alimony awards. The median duration of the other two-thirds of the cases in which alimony was granted was twenty-five months (Weitzman, 1985).

Gender Inequality in Divorce Proceedings

The third disadvantage women have when they divorce is in the division of property in the divorce proceedings. In most divorces the largest asset recognized by the court that couples own is a family house. Prior to no-fault divorce, judges were likely to award the house to the parent who was raising the children, usually the mother, in order to allow them a stable continuous residence. Since the passage of no-fault, judges are more likely to divide all property down the middle and to insist that the house be sold and the husband and wife divide the equity between them. "The number of cases in which there was an explicit court order to sell the home rose from about one in ten in 1968 to about one in three in 1977" (Weitzman, 1985, p. 31). Women, of course, are legally allowed to purchase the house from their husbands, which would enable them to remain there with their children. But usually they are financially unable to do this.

In the opening scenario, Teresa says that the house that she and Manny shared was sold as part of the divorce settlement. Not only was she unable to afford to buy her old home, she was unable to rent or buy in the same neighborhood. In addition to leaving the house in which they had lived, she and Maria also were forced to leave the community.

Second, what is defined as property may not include some of the most important assets a couple owns. For example, pensions and education are not treated as marital property (Weitzman, 1985). As I described above, couples are more likely to have made investments in the education and pension plan of the husband in the belief that as long as the man and woman remain married, these things would benefit the entire household. In a divorce, in most cases the state does not recognize these "investments" as property to be divided between the spouses equally and the husband is likely to leave the marriage with more of the assets.

For a brief period in the mid-1980s it appeared as if the problem of dividing family assets more equitably in divorce courts was beginning to be addressed by public officials. In 1983 the Congressional Caucus on Women's Issues pushed a bill through Congress that allowed military pensions to be considered as marital property to be divided in a divorce. Since then, however, the courts have allowed people to convert their military pensions into disability benefits, which can be excluded from property settlements (Stetson, 1991).

A third problem with the division of property in divorce is that it assumes that the property is to be equally divided between the divorcing adults. Since mothers are usually awarded custody of the children, in households where there are children the property may be divided down the middle but it is awarded to two different-sized groups. One-half goes to the husband and one-half goes to the wife *and* children. In the next section we will examine the issue of child custody in divorce.

Child Custody

The history of child custody has changed over the past few centuries. These changes illustrate the way in which alterations in macro-level factors in society like laws have a critical effect on whether the custody of children will be awarded to their mothers or their fathers in a divorce and what the basis of that decision is. Whether a child lives with one parent or the other seems like a choice that should be made at the micro level among those directly concerned with the question. A historical review of the issue, however, shows that what the choices are and whether children are likely to live with a mother or father after a divorce are largely determined by laws that have changed over time.

In the opening scenario, Maria is living with her mother. You may have thought that this was a result of choices she or her parents made. The fact that her mother had custody of her, however, was determined by the social context in which the divorce took place. If Manny and Teresa had divorced in the nineteenth century, the decision probably would have been different.

Until the middle of the 1800s, fathers retained the right to custody of children after a divorce. Unless a father could be proven in court to be a "clear and strong case of unfitness" he was granted custody (Polikoff, 1983, p. 186). In 1848, at the Seneca Falls Convention (mentioned in Chapter 2), one of the demands made by early feminists was that mothers be awarded custody of their children if they divorced.

Throughout the nineteenth century and into the first half of the twentieth century the laws gradually changed, and more and more mothers were awarded custody of their children. These changes were partly based on an increased emphasis on motherhood and a new idea called "the principle of tender years." According to this principle "special, even mysterious, bonds existed between mother and young child which, all other things being equal, made her the preferred parental custodian" (Sheppard, 1982, p. 230). In addition, as we saw in Chapter 2, this historical period was one in which reliance on children for labor was diminishing. Therefore, fathers were perhaps less likely to appeal for custody or to be granted custody because they needed a child to work in the father's business.

More recently, ideas about marriage have changed to emphasize shared partnership and equality in child custody decisions. These changes have resulted

In divorce cases involving children, most children are placed in the custody of their mothers. This may be because mothers are more likely to seek custody. When fathers file for custody they actually have a greater chance than mothers of obtaining their children.

in an erosion of the "tender years principle." Laws in nearly all states now have replaced it with the new principle of "best interest of the child." But 90 percent of children are still awarded to the custody of their mothers (Stetson, 1991).

Organizations like Fathers' Rights of America and the National Congress of Men have taken up the battle against what they see as an unfair advantage of women in child custody cases. Nancy Polikoff (1983) argues that the figure of 90 percent is deceptive. She says that women are more likely to be granted custody because they are more likely to file for custody. When custody is contested, fathers are actually more likely to be granted it.

Weitzman (1985) found that while many fathers say they want custody, few actually file for custody. In her research, 57 percent of the men reported to the researchers that they wanted custody,

> [but] when asked if they ever told their wives that they wanted custody, the 57% drops to 41% of the men who said yes [to the interview question]. And when queried if they ever asked their lawyer if they had a chance of getting custody, the 57% drops to 38% of the men who said yes. Finally, only 13% of the sample of men we interviewed actually requested custody on the divorce petition. (Weitzman, 1985, p. 243)

In 1977 Lenore Weitzman and Ruth Dixon reviewed court documents of California custody cases and found that 63 percent of fathers *who requested custody* were successful. This number had risen from the 1960s and early 1970s when about one-third of requesting fathers were granted custody. Research in New York, Minnesota, and North Carolina has shown that about half of the fathers who sought custody of their children in the late 1970s were granted it (Polikoff, 1983, p. 185).

Although women are currently more likely to be granted custody of their children, control of this decision is in the hands of fathers. Polikoff writes, "The power to decide child custody often lies with the father, not the mother. If he wishes to exercise that power, he is likely to win" (Polikoff, 1983, p. 185).

Fathers' success in the courts is based on two factors: economic advantage and ideologies about parenting. Fathers have an economic advantage in seeking custody because they are more likely to have the money to hire attorneys to seek custody. Fathers are also advantaged because the courts are increasingly choosing to select the parent who they believe can best provide for the material needs of the child.

The courts are also influenced by what is considered "normal" parenting. For example, fathers who hire housekeepers to be primary caregivers to their children are seen as normal, while mothers who work outside the home and can only spend a few hours with their children are perceived as neglectful.

Mothers are also sometimes judged as inappropriate parents because of their sexual behavior. In research conducted by Phyllis Chesler (1986) 48 percent of women who were challenged for custody of their children were challenged solely on the grounds of nonmarital sexual activity. A custody-seeking mother's sexual activity is especially likely to be at issue if she is a lesbian (Rhode, 1989; Lewin, 1984; Erlichman, 1989). Gay fathers are also discriminated against on these grounds.

Child Support

Child support is money to be paid by the noncustodial parent to the parent with whom the child lives. Since most children live with their mothers after a divorce, child support is usually supposed to be paid by men to their ex-wives. Although women are much less likely to be the noncustodial parent they are just as likely as men to be in arrears on child support. Attention has been focused on noncustodial fathers because they are a larger group.

Some researchers have been critical of Weitzman's work, arguing that she has placed too much emphasis on criticizing no-fault divorce laws. They argue that the nonpayment of child support is the most important cause of the impoverishment of women and children following a divorce (Jacob, 1989; Kamerman & Kahn, 1988).

In 1989, the most recent year for which data on child support are available from the Census, almost 10 million divorced single mothers lived with children under twenty years of age. Fifty-eight percent of these women had been awarded child support. This was down slightly from 59 percent in 1981. Of those 58 percent, 51 percent had received the full amount due to them; about 25 percent received no payment and the rest received some but not all of the payment due.

The proportion of women who were awarded child support and actually received the full amount rose only slightly—about 4 percent—since 1981. These numbers are even worse if we look at poor women. Table 10-3 shows the percentages of women who were awarded and received child support for two groups: all women and poor women.

In 1984 the Child Support Enforcement Amendment was passed by Congress. The bill allowed greater power to state agencies attempting to enforce the payment of court-ordered child support. States can now withhold wages and tax refunds from persons who are delinquent in paying child support. In 1988 this bill was further supplemented with policies that will allow states to automati-

TABLE 10 • 3
Child Support Award and Recipiency Status of All Women and Women Below the Poverty Level

	Proportion Awarded Child Support			Proportion Receiving Full or Partial Payment		
	1978	1985	1989	1978	1985	1989
All women	59	61	58	72	74	75
Women below poverty	38	40	43	59	66	68

Source: U.S. Bureau of the Census, 1989b, *Child Support and Alimony*, Current Population Reports, p. 3; 1991c, Current Population Reports, p. 3.

cally deduct child support from parents' wages, beginning in 1994 (Renzetti & Curran, 1992).

Over this same period of time, Table 10-4 shows the average amount of child support awarded to white women is higher than that granted to Hispanics and blacks. A bill passed in 1988 requires that states set formulas for awarding benefits that will address these inequities by requiring fair and consistent awards that do not vary from judge to judge (Renzetti & Curran, 1992).

In addition to not paying court-ordered child support, divorced fathers also do not provide other kinds of assistance to their children living with their ex-wives. Teachman (1991) found that divorced fathers have relatively little contact with their children and are unlikely to contribute to their children in any activities that require direct contact, such as helping with homework or attending school functions. In his sample, 33 percent of the men had not provided any assistance other than court-ordered financial support and 20 percent had provided no support of any kind. About 50 percent had provided either financial support or other kinds of support on a regular basis.

Why Don't Men Pay? One common explanation for why men don't pay is that they do not have sufficient funds to make the payments. Weitzman (1985) and others (Chambers, 1979) argue that noncompliance occurs across several income brackets, including relatively high ones, and that there is only a weak relationship between income and compliance.

Another explanation that has been suggested is that men are retaliating for visitation problems. Weitzman, however, reports that in her research many mothers complained about their husbands' failure to visit children as much as they did about their failure to make child support payments. Another study showed that a large number—40 percent—of children are not visited by their fathers after a divorce (Furstenburg, Morgan, & Allison, 1987). Others (Wallerstein & Blakeslee, 1989) have found little evidence of mothers sabotaging visitation and in fact found that most mothers felt mothers and children benefited when divorced fathers participated in raising their children.

Researchers who have interviewed fathers have found that men report that they do not make child support payments because they feel their relationship with their child is no longer close (Chambers, 1979). In the scenario at the beginning of the chapter, Teresa claims that Manny misses child support payments as a way to get at her. But Manny expresses the feeling suggested by Chambers: He no longer feels a part of Maria's life. Others have found that fathers say they do not pay because they believe that the mother does not spend

TABLE 10 • 4
Mean Child Support Payments by Race

	Mean Payment 1989
White women	$3132/year
Black women	2263
Hispanic women	2965

Source: U.S. Bureau of the Census, 1991c, *Child Support and Alimony*, Current Population Reports, p. 2.

the money on the child (Haskins, 1988) or that the judges are biased and set unreasonably high awards (Greif, 1985).

Weitzman (1985) contends that the lack of compliance lies in the inadequate laws to demand compliance and the failure of judges to use the enforcement procedures that do exist. One example of inadequate laws to force compliance is the ease with which men can avoid payment by moving their residence. If his ex-wife cannot find him, she cannot demand payment from him. Recent changes in the laws have closed this loophole to some extent. All states now have reciprocal agreements to enforce support orders (Stetson, 1991). It still takes time, however, to track down delinquent parents and to file papers contesting their nonpayment.

A second problem, according to Weitzman (1985), is lawyers who contribute to noncompliance by refusing to help their clients get men to pay child support. The lawyers in Weitzman's study argued that once the decree was given the case was out of their hands. Divorced women said that their lawyers told them that enforcement was not their job and they should contact the district attorney's offices. Although there are laws in California that allow the court to attach men's wages, only 1 percent of the attorneys in Weitzman's sample tried to have the man's wages attached.

The third problem Weitzman (1985) found with enforcement of child support awards is the district attorney's office. Here the women faced red tape, constant turnover of personnel, confusing forms, unsympathetic workers, and reluctance of the district attorneys to bother themselves or the judge with nonpayment of child support cases. The delays the women faced were time-consuming and costly.

The fourth problem was the judges, who were uninformed about the variety of enforcement techniques available to them or reluctant to use them. When Weitzman (1985) interviewed judges she presented them with a hypothetical case in which a man refused to pay child support. Eighty-eight percent of the judges said that they would sanction the man, but when asked what they would do, "most judges said they would 'give him a stern lecture and a warning' " (Weitzman, 1985, p. 301).

Weitzman believes that changes in the organization of the legal system and in ideologies among legal personnel would help to solve many of the problems associated with divorce. Others have suggested that changes in custody arrangements for divorcing parents are most critical (Greif, 1979; Bowman & Ahrons, 1985; Rothberg, 1983).

When parents divorce, one or the other may be granted custody of the children or they may both obtain joint custody. Joint custody can be either joint legal custody or joint physical custody. Joint legal custody means that parents have equal rights and responsibilities in regard to their children although the child lives primarily with only one parent. In joint physical custody the child lives intermittently with both parents, trading residence by the day, week, or year (Arditti, 1992). Joint legal custody is the more common form (Emery, 1988).

Joint custody has been shown to have a beneficial effect on fathers. They report a higher frequency of contact with their children and greater satisfaction with custody arrangements (Arditti, 1992). Noncustodial fathers suffer more emotional problems such as guilt, anxiety, depression, and low self-esteem (D'Andrea, 1983; Greif, 1979; Hetherington, et al., 1976).

For mothers the benefits of joint custody include sharing responsibility and physical care for children. Some scholars have suggested that joint custody could improve the record of child support payments—if fathers are granted joint custody they may be more likely to contribute financially. However, this connection is not clear in the research. Custody is only weakly linked to child support compliance (Arditti, 1992).

The disadvantages for both parents are the logistical problems of moving between houses and being tied to a geographical area because of the ex-spouse's residence (Luepnitz, 1982). Another problem that is especially likely to affect mothers (because they usually obtain custody) is that joint custody can be used as a means by which noncustodial fathers can negotiate lower child support payments (Weitzman, 1985).

Surprisingly little research has been done on the effect of joint custody on children (Arditti, 1992). In one qualitative study (Luepnitz, 1982) children in joint custody arrangements said they preferred it because it made their lives "more fun, more interesting or more comfortable" (Luepnitz, 1982, p. 47). Despite problems negotiating two households and two sets of rules, one girl explained why she preferred joint custody: "I love them both and would miss them too much" (Luepnitz, 1982, p. 46). Other children, however, said that joint custody meant they missed one favored parent when they were in the custody of the other.

Weitzman's Research Weitzman's research is an example of feminist multiple methods research (Reinharz, 1992). The project consisted of four separate sets of data, each collected with a different methodology. The first data set came from a systematic random collection of information in court records. From these she determined patterns of property awards and definitions for what was interpreted as individual property and what was considered community property to be divided equally between the divorcing couple.

A second set of data involved interviews with attorneys. In the interviews the attorneys were asked about one hundred questions. In addition they were asked to respond to four hypothetical cases. The use of hypothetical cases allowed Weitzman to determine the way in which attorneys would, at least in theory, deal with different kinds of divorce cases.

In a third set of data, Weitzman interviewed judges with a set of questions and the same four hypothetical cases shown to the attorneys. These data provided evidence for how judges interpreted laws in divorce proceedings. The questions also revealed the ideologies judges brought with them to the courtroom, for example, the unfounded belief described in the previous section that most women are granted alimony.

The fourth set of data consisted of interviews with recently divorced men and women. These data provided evidence on the divorce process, the economic difficulties faced after divorce, and the problems with eliciting support from the courts to implement decisions.

Using a variety of sources for data and methodologies for collecting data is called triangulation. Triangulation is used by researchers who are not feminists, but Reinharz (1992) maintains that there may be a greater proportion of feminist researchers who use the multimethod technique because it fits well with feminist concerns. Reinharz (1992) says that multimethods allow feminists to express their commitment to thoroughness and their desire to be open-ended and to take risks.

Weitzman's book on divorce has been widely cited as an important feminist work because of its methodology and because of its content. Scholars concerned about the growing impoverishment of women in the United States were pleased to be able to cite the research because it supported their belief that gender inequality is an important facet of divorce and that divorce was an important factor in the growing impoverishment of women. Some, however, have been critical of Weitzman's work, arguing that the research methodology is flawed and that recent changes in divorce law, like no-fault, have been progressive steps forward because they allow women to more freely choose to end oppressive marriages (Fauldi, 1991).

Poverty of Women and Children

More women are poor than men and divorce appears to be an important factor in their poverty. Following a divorce women's income is 24 percent of the previous family income, while men's is 87 percent (Weitzman, 1985). While divorcing women anticipate the emotional difficulties of ending their marriage, they are not as likely to anticipate the economic consequences (Arendell, 1986).

In 1983, Terry Arendell interviewed sixty middle-class divorced women. She found that nearly all of them faced serious economic hardship after their divorce: "90% of them found divorce immediately pushed them below the poverty line, or close to it" (Arendell, 1986, p. 231). And their economic decline was not temporary. One woman described her experience:

> I've been living hand to mouth all these years, ever since the divorce. I have no savings account. The notion of having one is as foreign to me as insurance—there's no way I can afford insurance. I have an old pickup that I don't drive very often. In the summertime I don't wear nylons to work because I can cut costs there. Together the kids and I have had to struggle and struggle. . . . There have been times when we've scoured the shag rug to see if we could find a coin to come up with enough to buy milk so we could have cold cereal for dinner. (Arendell, 1986, p. 231)

Because women nearly always have custody of their children, divorce also causes poverty among children. Another woman told Arendell:

> I had $950 a month. . . . My son qualified for free school lunches. . . . We'd been living on over $4,000 a month [before the divorce] and there we were. That's so humiliating. What that does to the self-esteem of a child is

absolutely unbelievable. And it isn't hidden: everybody knows the situation. They knew at school that he was the kid with the free lunch coupons. . . . My son is real tall and growing. I really didn't have any money to buy him clothes. . . . So he was wearing these sweatshirts that were too small for him. Then one day he didn't want to go to school because the kids had been calling him Frankenstein because his arms and legs were hanging out of his clothes—they were too short. (Arendell, 1986, p. 240)

The poverty of women and children is further exaggerated by its contrast to the lives of the ex-husbands. The same woman who was concerned about finding sufficient food and clothes for her son explained how the situation was different in the father's new household:

> His father seldom buys him anything. But his stepmother does. She can give him all these nice things. . . . She's given him nice books, a stereo headset . . . it's a very funny feeling to know that I can't go and buy my son something he would love to have, but this perfect stranger can. . . . But it's kind of ironic—I helped establish that standard of living but I end up with none of it, and she has full access to it. (Arendell, 1986, p. 234)

This section has described the importance of gender inequality in the issue of divorce. Gender inequality is both a cause and an effect in relationship to divorce. Weitzman found that divorce was shaped by gender inequality in marriage, the labor market, and divorce proceedings. No-fault divorce, despite its promise to bring greater equity to men and women in divorce, actually exacerbated some of the problems.

Weitzman (1985) also found that divorce reinforced and helped to create gender inequality. Women and men leaving a marriage found themselves even more unequal in some ways than when they were in the marriage. This gender inequity was especially remarkable in its expression in financial status. Divorce appears to be an important factor in the growing impoverishment of women and their dependent children.

FINDING SOLUTIONS: WHERE WERE THE FEMINISTS?

A number of macro-level factors affect the organization of divorce in our society. In particular, law has played a central role, but economics and gender stratification have also influenced divorce. In turn, the organization of divorce has affected the micro level—the relationships of women, men, and children in families that divorce. Like all macro-level social phenomena, the legal organization of divorce has been shaped by micro-level forces of people working together to try to reform the system. As we have observed in this chapter, changes in the organization of divorce in recent years have been significant with the passage of no-fault divorce laws. Furthermore, these changes have not always benefited women and, according to Weitzman (1985), have operated as a vehicle by which to exacerbate gender inequality and the oppression of women.

How did this come about? Why did recent changes in the laws result in the creation of such problems for women? Herbert Jacob (1989) argues that although divorce reform occurred during a period in which the women's movement was extraordinarily active, "there were few connections between feminist activity and divorce reform'' (Jacob, 1989, p. 482). When legislation was being passed to change divorce laws in the 1970s, feminists were missing in action. Divorce is a critical issue to many women and it has been an aspect of the feminist agenda since the nineteenth century when Elizabeth Cady Stanton spoke out for the liberalization of divorce laws. But during a critical moment in divorce history, the 1970s, feminists failed to mobilize around the issue.

In this section we will examine the way in which the micro level successfully changed the macro level—divorce reformers passed no-fault divorce laws. A central question of our inquiry will be why a feminist perspective was not part of that divorce reform movement.

In 1969 California was the first state to pass no-fault and in 1985 South Dakota was the last state. Jacob (1989) argues that this activity constituted a long-term reform movement in which feminists were only faintly visible. He suggests that there are three possible reasons why feminists were not part of the movement:

1. Women, who Jacob believes may have been more likely than men to hold feminist views, had not effectively penetrated decision-making agencies and, therefore, had no avenue for representing feminist views.
2. Feminist organizations were inadequate to the task, both because they organized too late and because they focused on national rather than state issues.
3. Feminist organizations concentrated on other issues which they believed at the time to be higher priority issues.

The first possibility is that few women were insiders in judicial and legislative arenas in which divorce reform was being proposed and debated. In 1977 only 110 state judges out of 5,940 were women. Although women increasingly gained seats on state legislatures during the 1970s their relative numbers remained small. In addition, Jacob (1989) traced the history of committees that sought divorce reform, especially the National Conference of Commissioners on Uniform State Laws (NCCUSL), and found that not only were women scarce, frequently there were no women present. For example, the original NCCUSL was an entirely male initiative. Jacob found, furthermore, that women who eventually joined the conference did not identify themselves as feminists and showed little concern with the way in which various reforms would differentially affect women and men. In addition, after reviewing the hearings of a number of study panels that eventually led to no-fault divorce reform, Jacob found that there was no testimony by feminist groups. In a few cases—in New York and Wisconsin—Jacob (1989) did find there had been feminist input but he concludes that even when a feminist viewpoint was articulated it did not prevail.

The second factor in the absence of a feminist voice in divorce reform is the possible inadequacy of feminist organization. Jacob maintains that the women's movement in the late 1960s and early 1970s, when divorce reform was taking place, was not organized well enough to demand (as outsiders to the legislative bodies) that their voice be heard. In addition, those women's organizations that were strong enough to make themselves heard focused their attention on federal rather than state laws and divorce reforms occurred at the state level.

The third factor cited by Jacob was the low priority given by feminists to divorce law revision. ''Divorce law revision did not attract the feminist movement because it neither threatened basic values nor promised widespread benefits" (Jacob, 1989). Leaders in the women's liberation movement focused their attention on issues like pay equity, the ERA, and abortion rights, which they believed to be important. The general membership may not have agreed with these priorities. Freeman (1975) observed that the most well-attended workshops at the 1973 NOW meetings were those on family, marriage, and divorce. Divorce reformers may have contributed to the neglect of the issue by feminists because the reformers presented the issues as if they were technical, legalistic, procedural issues of little interest to political organizations (Jacob, 1989). Most feminists did not see divorce reform as a threat to women until after the no-fault legislation was enacted. The subtitle of Weitzman's book—the unexpected social and economic consequences for women and children in America—reflects their belated understanding of its importance.

Since feminists were not central to the movement for divorce reform, who were the divorce reformers? Two groups played a major role in the enactment of no-fault (Jacob, 1989). Lawyers were one group that worked both to initiate the legislation and to testify on its behalf before legislative bodies. Some lawyers saw no-fault as a way to keep divorces in their own states, thereby improving their case load. For example, California lawyers explicitly mentioned that no-fault in California would keep divorcing couples from taking their cases to Nevada.

A second group of divorce reformers were men's rights groups. They were especially active around child custody reforms that increased the likelihood of fathers being granted custody or joint custody. Jacob (1989) argues that they were more likely to have the ear of legislators, who were nearly all men and many of whom were divorced fathers.

THE MICRO-MACRO CONNECTION

Divorce is a critical turning point in family life. The voices of the people we hear in the interview data tell us exactly what the experience of divorce and remarriage feels like at the micro level. Their stories help us to see the human faces behind the statistics and the complex, unforeseen consequences of these events.

An underlying theme in the stories is the mysterious and unknown quality of divorce and remarriage. We hear of people who did not expect to become poor, to be stigmatized, to have their new marriages overshadowed by the old ones, or, on the other hand, to discover unknown strengths and opportunities for personal growth.

All of these micro-level experiences are framed by macro-level structures and events. Ideologies, economic opportunities, mortality rates (that are themselves altered by health care, technology, nutrition, violence, and discrimination), and especially the legal structure have a profound effect on what divorces are like and sometimes whether they can even occur.

History plays a role too, as these macro-level institutions change, sometimes dramatically, over time. Teresa's and Manny's story would be quite different had it been told in 1850 or in 1650.

Two important components of the macro level, the organization of gender and the organization of divorce, have a mutual effect on each other. Gender inequality in marriage, the labor market, and divorce proceedings shapes the experience of divorce. Divorce plays a critical role in maintaining and exacerbating the economic inequality between women and men.

Macro-level institutions have been changed at least in part as a result of micro-level organizing. The two opposing factions on the question of divorce, pro–divorce rights activists and anti–divorce rights activists, have affected our options and our experience of those options. Feminists have played a surprisingly small role in these debates. Increasingly feminists, especially feminist scholars rather than activists, have recognized the importance of divorce reform and the necessity of having a voice in its creation. Their work, however, follows the enactment of sweeping changes and the passage of no-fault in all fifty states, which means that feminist activists must work to further alter the laws so that they more effectively benefit all parties, including women.

SUMMARY

How to Measure Divorce

- The opening scenario shows the social, economic, and emotional difficulties faced by husbands, wives, and children in divorced families. The story is one that is retold in millions of homes because of the large numbers of couples that divorce in our society.
- Divorce is surprisingly difficult to measure adequately. But demographers have developed methods to measure changes in the rate of divorce over time and conclude that it rose steadily over the century except for a brief downturn following World War II.
- Rising divorce rates, however, do not necessarily mean fewer intact families, since mortality has fallen at the same time that divorce has increased.
- In addition to changing rates of divorce, laws have also changed as political advocates struggled with each other over reform.
- Divorce in the U.S. has also been marked by diversity, as illustrated by the organization of Inuit society in the late nineteenth century.

After Divorce

- One of the consequences of divorce is the stigmatization of divorced people, despite their large numbers.
- Some scholars have argued that divorce can also have beneficial effects. Divorced people speak of personal growth.
- Remarriage commonly follows divorce and the numbers of people who remarry has risen in the past decades.

Divorce Reform in a Gender-Stratified Society

- Gender inequality is both a causal factor and a consequence of divorce in our society. The passage of no-fault divorce laws in the 1970s present an especially revealing case of the connections between divorce and the oppression of women.
- Gender inequality in marriage, the labor market, and divorce proceedings influences the experience of divorce.
- Prior to the late 19th century fathers were nearly always granted custody of their children after a divorce. Women are now much more likely to ob-

tain custody, but this may be mostly because fathers rarely file for custody.
- Child support is granted in an unequal manner and frequently goes unpaid.
- Weitzman's research on divorce has been especially important and is a good example of triangulation of multiple methods.
- The end result for many divorced women and children is poverty.

Finding Solutions: Where Were the Feminists?

- Recent reforms in divorce law have been cited as an important factor in the increasing impoverishment of women. Why was this legislation enacted without feminist input?
- There are three possible answers to this question: women were not insiders in the decision-making bodies that passed the divorce reforms; feminist organizations were not well organized enough to make their voices as outsiders heard in those bodies; and feminists did not understand the importance of the reforms and therefore paid them little concern.

The Micro-Macro Connection

- The organization of divorce is affected by macro-level forces such as ideology, mortality rates, and laws.
- Individuals are profoundly affected by divorce and despite the high rates of divorce are surprised by the emotional, economic, and social changes.
- Micro-level organizing by lawyers, men's organizations, and others have had an effect on marriage and divorce policy, but the feminist voice in these debates has been largely absent.

The police have been a focal point in the attempts to address the problem of woman battering. Before the recognition of battering as a social problem, police were likely to try to calm the couple down, and they did not treat domestic violence like other assault cases. Laws are changing, however, and police departments are increasingly required to arrest batterers.

11 BATTERING AND MARITAL RAPE

Linda married Ron when she was twenty-three. She was a teacher and he was an engineer. Before they married, Linda noticed that Ron became jealous if she spoke to any other men. But he was also attentive and charming when they were together, and she wasn't bothered too much by his bursts of anger; they were not directed toward her and they seemed to be part of his aggressive, self-confident style, which she admired. And, anyway, she figured he would mellow after they were married and he was assured that she was all his.

After they were married for two years, Linda became pregnant and Ron's outbursts became physically abusive. At first it was a shove or a slap, but by the time the baby arrived, Linda found herself making excuses to her friends and her obstetrician about the bruises.

Linda did not know what to do. She hoped she could change Ron and she blamed herself—maybe if she had been a better wife, a better lover, a better mother . . . Linda also thought she was unique. This kind of thing doesn't happen in good families like hers.

- *How unusual is Linda's marriage?*
- *Is it reasonable for Linda to blame herself for the violence inflicted on her?*
- *Is it useful to try to solve the problem by changing Ron only?*
- *Linda says she is not sure what to do yet. What might happen to convince her to take action?*
- *What social structural factors might be related to Linda's plight?*
- *Why do we admire men who are physically aggressive, self-confident, and domineering?*
- *What social structural barriers might Linda find if she tries to leave Ron?*
- *What help might Linda find from a battered women's shelter? Were these shelters available for her mother's generation? Where did they come from?*
- *If violence against women in families has been present in society for hundreds or thousands of years, why is it only recently being recognized as a social problem?*

AN EVENT AS PERSONAL as a husband beating his wife is difficult to conceptualize as a *social* problem—a problem that is rooted in the organization of our society, not in the individuals who experience the problem. When we hear about battering or experience it in our own family, we tend to think about it as an individual problem and we ask, Why doesn't Linda leave? And what is wrong with Ron?

As we have observed in this text, in order to answer these questions we must look at the interplay between social organization and individual people's lives, between history and biography. We must ask questions that examine social structure and its role in creating or allowing violence in families. In order to understand the abuse of Linda we need to understand the way in which the macro level of social organization creates opportunities to abuse women and barriers to stopping the violence. We also need to discover the way in which the micro level, individuals interacting with one another, can alter the macro level to allow us to address the problem of battering and to end it.

The chapter is organized into three main sections. The first presents information on the prevalence of battering and marital rape. The second section discusses two theoretical issues that emerge in the analysis of violence in families. One is the debate among three competing theories on why violence occurs in families. These three theories differ in their assessment of the way the macro level of social organization—economic decline, the isolation of families, ideologies, and gender stratification—affects the micro level of individual heterosexual couples interacting in abusive ways. The other theoretical issue explored in this section is the question of why battering has recently become identified as a social problem.

The third section investigates the impact of the micro level of social organization on the macro level. How have people worked to create solutions to violence in families by reshaping social institutions like the police and shelters?

PREVALENCE OF VIOLENCE IN FAMILIES

We live in a violent society. We only have to turn on the television, read the newspaper, or look at our history to see that violence is very much part of the American experience. Families, as part of that society, are affected by the social system in which they are embedded.

Figure 11-1 illustrates violence as it is experienced by women. The figure gives the total reported acts of violence between 1979 and 1987. The figure shows that more than 10 million acts were by strangers and almost 12 million were acts by acquaintances or intimates.

Violence permeates the macro level of our society and is played out in the microcosm of intimate relationships like those in families. But we are surprised when we hear about violence in families because we think about families as a place where people love each other and treat each other well. We watch the AT&T commercials and "The Cosby Show" on television, and we celebrate Mother's Day and Father's Day. Our media and our culture are filled with images of loving, gentle families.

If we stop to think about it, however, if we have ever been hurt by another person, that person was likely to have been someone in our family. Probably all of us have been hit by a parent or sibling. Probably few of us have been hit by a stranger, a boss, a teacher, or a co-worker.

Because we think of families as private places and because, in fact, families are frequently isolated and secluded, it is difficult to find real numbers when it comes to child abuse, incest, woman battering, husband abuse, and marital rape. Researchers have used a variety of methods to try to find out what the numbers are. They have gone to police records, hospital emergency rooms, family court records, divorce records, and random surveys (Anderson, 1988). In this chapter we will explore the problem of violence in families among adults. In Chapter 13 we will investigate family violence between adults and children as it is expressed in child abuse and incest.

Woman Battering

Michele Bograd (1988, p. 12) defines woman battering or wife abuse as "the use of physical force by a man against his intimate cohabiting partner. This force can range from pushes and slaps to coerced sex to assaults with deadly weapons."

Murray Straus, Richard Gelles, and Suzanne Steinmetz were the first sociologists to systematically study physical violence in families and they have published more than any other social scientists on the physical abuse of wives by their husbands (Kurz, 1989). Their research shows that more than two million Americans have been abused by a spouse. If that number is divided by the number of seconds in a year, it reveals that a woman is beaten every eighteen seconds in the U.S.—a statistic commonly used in public education about abuse.

FIGURE 11 • 1
Violence Against Women, 1979–1987

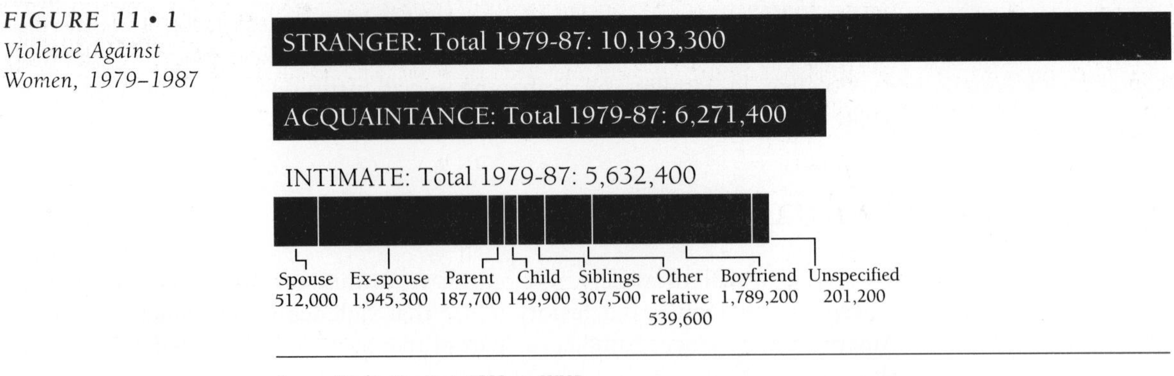

Source: Washington Post, 1992, p. WH5.

In one of the large surveys Straus, Gelles, and Steinmetz (1980) conducted, they interviewed 2,143 married people and found wife beating in every class and at every income level. Sixteen percent of their sample, or one in every six couples, had experienced some form of violence in their marriage during the previous year. And 28 percent of the couples reported at least one incident of violence at some time during their marriage. Table 11-1 shows the specific acts asked about in the survey (Straus, Gelles, & Steinmetz, 1980).

Straus, Gelles, and Steinmetz's (1980) surveys were done on a random sample. A random sample is one in which researchers choose subjects without regard to any of their individual characteristics. Researchers then argue that it is statistically probable that the sample is not unique and that it can be said to represent the larger population from which it was drawn. Based on the assumption that the sample is representative, sociologists who conduct large surveys on random samples argue that the percentages of people who were involved in various kinds of activities in the sample can be extrapolated to the entire population. For example, if 7 percent of their sample of 2,143 people had experienced one spouse throwing something at the other, then we are safe to assume that 7 percent of the total 250 million or so people in the U.S. also experienced this form of violence.

Therefore, although the percentages appear relatively small, if they accurately represent the entire population in the United States, the actual numbers are very large. For example, the percentages indicate that "over 1.7 million Americans had at some time faced a husband or wife wielding a knife or gun, and well over two million had been beaten up by a spouse" (Straus, Gelles, & Steinmetz, 1988).

Straus, Gelles, and Steinmetz (1980) also suggest that these percentages are probably underestimates. They maintain that "it seems likely that the true rate is closer to 50 or 60 percent of all couples than it is to the 28 percent who were willing to describe violent acts to our interviewers" (Straus, Gelles, & Steinmetz, 1980, p. 36).

Straus, Gelles, and Steinmetz describe three important sources of underestimation in their research. First, for many people, these kinds of violent events are so much a part of their normal family life, they might not even be remembered. Second, other people in the sample who do acknowledge the problematic character of such behavior might feel too guilty or ashamed to admit they had

TABLE 11 • 1
Rate at Which Violent Acts Occurred in the Previous Year and Ever in the Marriage

Violent Act	% in Previous Year	% in Marriage
Threw something at spouse	7%	16%
Pushed, grabbed, or shoved spouse	13%	23%
Slapped spouse	7%	18%
Kicked, bit, or hit with fist	6%	10%
Hit or tried to hit with something	4%	10%
Beat up spouse	2%	6%
Threatened with a knife or gun	1%	4%
Used a knife or gun	1%	4%

Source: Straus, Gelles, and Steinmetz, 1988, p. 303.

been either perpetrators or victims. Finally, in the particular study done by Straus, Gelles, and Steinmetz that is reported in Table 11-1, only couples who were currently living together were interviewed. Couples who had been divorced—perhaps because of violence—were excluded. The sample, therefore, eliminated many couples that were potentially high violence cases. In another study Gelles (1976) found that among applicants for divorce 23 percent of middle-class people and 40 percent of working-class people listed "physical abuse" as a primary complaint.

In 1985 Straus and Gelles (1986) repeated the survey they had conducted in 1975. They found that violence by husbands against wives had decreased 6.6 percent during that decade. Violence by wives against husbands had increased 4.3 percent. These changes are not statistically significant. Sociologists use the term statistically significant to mean that the differences were probably not a result of the fact that a relatively small group of people (2,143 in 1975 and 3,520 in 1985) were chosen to represent a very large group of people, the entire U.S. population. Stating that the differences found between 1975 and 1985 were not statistically significant means that they very well could have been a result of the small sample of a large population and did not, therefore, measure real changes in the incidence of violence over time.

Marital Rape

Marital rape is another form of family violence. The difficulty lawmakers have in defining marital rape shows the depth of the problem. Definitions vary from state to state and in many states forced sex—what most people would call rape—is not considered a crime in marriage. In those states, husbands cannot rape their wives because a wife is legally obligated to have sex with her husband whenever he desires it. What is defined as a felony when it occurs between a man and a woman who are not married to each other—rape—is defined as just another sexual technique when the man and woman are married.

Table 11-2 lists the states and notes whether they have exemptions to their marital rape laws or not. In the ten states that have no exemptions husbands can be charged with the rape of their wives in all or most cases. In nine states husbands are fully exempt and can never be charged with the rape of their wives. In all other states various exceptions are made as to who can bring charges. For example, states sometimes exempt wives who have not filed for a divorce and are currently living with their husbands. Other states exempt

TABLE 11 • 2 *Marital Rape Laws: States Allowing No Exemptions, Some Exemptions, and Total Exemption of Husbands*

No Exemptions Allowed	Husbands Fully Exempt	Some Exemptions Allowed
Alabama	Illinois	All other states
Alaska	Mississippi	
Florida		
Georgia		
Hawaii		
Maine		
Massachusetts		
Nebraska		
New Jersey		
New York		
Oregon		
Vermont		
Wisconsin		

Source: National Center on Women and Family Law, 1987.

women who are are not legally married but are cohabiting. In other states the wife can bring charges only if the rape is accompanied by severe physical injuries. No one yet has examined the historical development of these laws in order to determine why certain states differ so dramatically on this question from other states. One important question raised by the state-to-state variation is: Why is there no federal law protecting women from marital rape?

In spite of the difficulties in determining the extent of marital rape, researchers estimate that about 10 percent of all married women have been raped by their husbands. In a random survey conducted by Diana Russell (1986) of 930 women over the age of eighteen, 14 percent were victims of at least one attempted or completed rape by their husbands or ex-husbands.

Finkelhor and Yllo (1983) identified four types of coercion used by husbands to force their wives to have sex. One is social coercion, which refers to situations in which wives participate in sex although they do not wish to because they believe it is their wifely duty. A second type is interpersonal coercion: a husband threatens to leave, to cut off money, or to publicly embarrass his wife if she will not have sex. Third is threat of physical violence. Statements about what a husband says he will do if she does not comply or the memory of a previous beating may coerce a woman. Fourth is physical violence.

Diana Russell's (1986) work is an example of feminist research methods (Reinharz, 1992). She argues that large surveys conducted of carefully selected random samples are an important tool for feminists interested in social change. She wanted to be able to present the strongest case possible to the public and to lawmakers on the prevalence of rape in marriage. She hypothesized that the incidence of marital rape was much higher than commonly believed. Based on this assumption, her goal was to confirm the hypothesis in a way that convincingly established the magnitude of the problem.

In addition to using a random sample, Russell also took great care in the process of data collection. She matched interviewers and subjects by class, gender, race/ethnicity, and age because she felt that then women would be most likely to disclose sensitive information about their experience with rape. Inter-

BOX 11 • 1 HOW DIFFERENT ARE BATTERED WOMEN WHO KILL THEIR HUSBANDS FROM BATTERED WOMEN WHO DO NOT?

The proportion of murders that occur among family members is much higher in the U.S. than in other nations. About one-quarter of murders are by a relative; about one-eighth are by a spouse. Wives are the victims in two-thirds of those cases. Men are more likely to murder their wives than vice versa. The number of wives who kill their husbands is significant, however, because women are so much less likely to commit murder in general. Only about 15 percent of all homicides are committed by women. Women are seven times more likely than men to have committed murder in self-defense. In addition, women charged in deaths of their husbands have the least extensive criminal records of any female offenders (Browne, 1987).

Angela Browne (1987) interviewed forty-two battered women who killed or attempted to kill their husbands, to examine the circumstances of their cases and to compare their situations to those of 205 battered women who had not used lethal means to end the violence.

Their stories were typical of abusive relationships. The battering usually started after marriage. What first appeared to be concern and care by the men escalated to extreme possessiveness and jealousy. The women sought help from ministers and were told to be better wives. They sought advice from lawyers and were told to see marriage counselors (Browne, 1987).

Browne (1987) found few differences in the two groups of women—those who murdered their husbands and those who did not. The biggest differences were between husbands who were murdered and those who were not.

The men who had been killed (or nearly killed) were significantly more likely to have threatened to kill their wives, battered their wives more than once a week, caused more severe injuries (multiple and to the head and body rather than to extremities), abused children also, used drugs and/or alcohol daily, and raped their wives.

Earlier research shows that murder weapons differed by gender—women used weapons to kill their husbands, men beat their wives to death. In Browne's (1987) sample 81 percent of the women used guns, 7 percent used knives, 7 percent used a car, and 5 percent used something else.

Previous reviews of the cases of wives who murder their husbands show they are more likely to be charged with first or second degree murder—rather than lesser charges—and receive longer sentences (Browne, 1987). In Browne's sample of the thirty-six whose husbands died, twenty-seven were charged with first-degree murder and nine were charged with second-degree murder. Twenty received jail terms, twelve received a suspended sentence or probation, nine were acquitted, and one case was ruled self-defense and charges were dropped.

viewers were extensively trained for sixty-five hours in both interviewing techniques and sensitivity in talking about sexual assault. Russell also verified 22 percent of the interviews. In most surveys about 10 percent of the interviews are recalled a second time by supervisors in order to check the accuracy of the information.

Russell argues, however, that in spite of the care taken in the research these numbers are probably underestimates for two reasons. First, as Finkelhor and Yllo (1983) suggest, sex can be coerced through social factors. A number of women told Russell it was their duty to submit to sex even when they were repulsed by their husbands. These women may have been forced to have sex although they did not define it as rape because they felt obligated to submit. Second, the sample did not include women in mental hospitals, prisons, nursing homes, and shelters, who may represent a category of women especially vulnerable to all types of violence, including marital rape.

Battered women sometimes end the beatings by killing their husbands. Wives who murder their husbands are likely to receive longer sentences than other murderers.

Among battered women, the numbers of women raped by their husbands are even higher. "Studies of battered women regularly show that anywhere from a third to a half of them are victims of marital sexual assault" (Finkelhor & Yllo, 1989, p. 387).

Marital rape appears to be a chronic constant threat, not a single event. In Russell's research (which included both battered and nonbattered women), one-third of the wife rapes were isolated cases, one-third of the women had been raped between two and twenty times, and one-third had been raped more than twenty times by their husbands. David Finkelhor and Kersti Yllo (1989) found that 50 percent of the battered women they interviewed had been raped twenty times or more by their husbands.

When the rape was associated with battering it took one of two forms. Sometimes the sex was part of the battering episode, occurring simultaneously with punches. In other cases, the rape would occur when the man thought the battering was finished and now wanted to "make up" by having sex. His exhausted beaten wife would refuse and he would force himself on her.

In their research on marital rape, Finkelhor and Yllo (1985) discovered that the effect of marital rape was even more devastating than rape by a stranger for four reasons. First, women felt they could not talk with others to get outside support. Second, they were forced to continue to live with the rapist. Third, the rape was often repeated, sometimes for years. Fourth, they came to doubt their own judgment about intimate relationships and for a long time after leaving their husbands the women felt they could not trust themselves to make decisions about entering into new relationships.

Judith Herman (1992) has also looked at how violence affects victims. She argues that one of the first problems victims of trauma face is convincing others of the validity of their complaints. When people who have been abused present their case to others, they challenge the listeners' beliefs about the specific abusers and about goodness and justice in the world in general. Hearing what has been done to one human by another and accepting its implication is difficult.

According to Herman (1992) the terror, powerlessness, and isolation experienced by battered women cause psychological responses similar to those experienced by victims of other kinds of trauma such as war or natural disasters. Victims of domestic violence report feeling as if they were separate from their own bodies. They also experience depression, flashbacks, nightmares, and amnesia.

Victims of prolonged trauma of battering and marital rape can recover by moving through three stages, which must take place in the context of supportive social relationships (Herman, 1992):

> First survivors must establish a feeling of safety. Second they must remember and mourn the traumas they have undergone; through mourning, grief slowly diminishes. Finally they must make new connections with ordinary life. (Lewin, 1992, p. 16)

This section has described the incidence and impact of battering and rape. In addition to providing descriptive information about violence in families, scholars have also developed theories to explain the phenomenon. In the next section we will explore some of these explanations for the data we have reviewed.

THEORIES ABOUT VIOLENCE IN FAMILIES

People who are concerned about the problems of battering and rape wish to see changes that will stop the violence. To aid them in this work they have developed ways of understanding the problem—analyses of the problem—that allow them to determine ways in which to best address it.

A number of scholars have attempted to develop theories about how to understand violence in families in our society. One set of theoretical issues centers around the question of how a social issue comes to be identified as a social problem. Woman battering and marital rape appear in ancient documents and seem to have been part of many human societies for a long time. Why has it been only recently that they have been identified as social problems? This is the question that some theorists have addressed. We will investigate their ideas in the second half of this section on theory.

In the first half of this section on theory we will examine another set of theorists who have examined the question of why violence occurs. Three important frameworks have been developed to address this question. The first argues that the characteristics of the battered woman and the battering man are the crucial factors. The second focuses on the importance of social structure. The third also regards social structure as crucial but emphasizes that gender inequality is a determining factor. We will begin with an examination of these three competing theoretical frameworks.

Why Does Violence Occur?

Explaining Woman Battering as Learned Helplessness What causes violence in families? In trying to answer this question one set of theorists has developed an analysis that focuses on the individuals who are battered or who are batterers. This kind of individualistic understanding is sometimes called a therapeutic model (Dobash & Dobash, 1992). Within this framework battering is defined as a phenomenon rooted in the genetic makeup or learned characteristics of individuals who experience the problem. Sometimes the man is blamed, sometimes the woman is blamed, or both may be seen to be at fault. Within this framework battering is caused by men who are either inherently violent or have been socialized to be violent, cruel people. Women within this framework are perceived to be either inherently weak or socialized to be weak, ineffective, passive victims.

Lenore Walker (1979) is one of the most well-known advocates of this point of view. She argues that men and women have learned to play the roles of batterer and battered. Based on her interviews with battered women and their husbands, she claims that as a result of their social experience many women have become unresourceful downtrodden victims. Walker named the concept "learned helplessness" or "the battered woman syndrome."

Walker maintains that some girls are socialized to behave in a passive helpless manner. These behaviors make them likely to become victims of violence. They do not know how to tell their husbands what they want and how they believe he should behave and they are unable to make decisions and take action that would remove them from violent relationships.

While girls are learning to be passive and helpless, some boys have learned to take advantage of them. Boys who become batterers have been socialized to expect a woman to make their lives less stressful, to expect violence in homes, to expect women to behave in gender-sterotyped ways, and to intermittently use charm and anger to manipulate others' behavior (Walker, 1983).

The socialization of men into the role of batterer takes place during their childhood. The socialization of women into the role of battered woman takes place during two time periods in their lives. First, young girls are socialized by their parents. Second, adult women who may not have learned these behaviors as children may be socialized by their husbands. Wives who do not necessarily enter a marriage helpless and passive may become so as a result of the pressure placed on them by their abusive husbands.

Walker argues that both the battering men's and the battered women's behavior is rooted in the internal psychology of the individuals in three ways. First, it is rooted in their feelings about themselves—both battering men and battered women tend to have low self-esteem. Second, it is rooted in their repertoire of behaviors. Although women and men have learned different behaviors, they are similar in the limits of the range of their behaviors. The women only know how to be dependent and ultimately responsible for everyone else's happiness. The men only know how to be jealous, possessive, and extravagant. Third, their behavior is rooted in their beliefs about gender. Both believe strongly in gender stereotypes that women should be pretty, ladylike, and pampered and that men should be virile and reign supreme.

Walker's theory has gained enormous attention among professional and community organizations in the United States. The prevalence of a therapeutic model perhaps should not be surprising in a country that employs one-half of the world's clinical psychologists (Dobash & Dobash, 1992). Her theory, however, is controversial. She has been criticized for emphasizing internal psychology too much. Although psychological characteristics may play a role, the prevalence of wife abuse suggests that its primary source is not psychological

(Bograd, 1988). The large numbers of women who are battered by their husbands and the way in which battering permeates every sector of our society indicate that the problem is a systemic one—a problem that is embedded in our society. According to Walker's critics, the focus must be shifted from an individualistic framework to one that acknowledges the social context in which violence in families occurs. Social context factors such as unemployment, for example, may play a more important role than the psychological state of individuals.

Some of Walker's critics have used her own findings to point out what they see as the weakness of the theory in its overemphasis on internal psychological states to the omission of external social structural causes. For example, Dobash and Dobash (1988), two of Walker's critics, write, "the analysis of Walker's own data demonstrates that the explanation of why women remain in a violent relationship lies much more in social and economic circumstances (employment, housing, number of dependent children, and so on) than in individuals' supposed helplessness" (Dobash & Dobash, 1988, p. 64).

Other critics have been especially concerned with the way Walker's theory easily falls into blaming the victim. Dobash and Dobash (1988), for example, have argued that the theory of "learned helplessness" has become part of the ideology that maintains that women are inferior and the source of their own problem. If the problem of battering is essentially a problem of "learned helplessness," then the solution must be for women to change, which once again blames women for their victimization.

Violence-Prone Family Structure in Our Society A second group of theorists has shifted their examination of battering to focus on the social structural factors that may create social contexts in which even strong women may find themselves being battered. Theorists who take a structural point of view blame the organization of our society for the violence in families and therefore call for changes at a societal level as a way to address the problem. Straus, Gelles, and Steinmetz (1988) are examples of those who take a structural view of violence in families. They describe three structural factors as key causes of violence in contemporary American families: stress, privacy/isolation, and the cultural acceptance of violence (Kurz, 1989). This framework is sometimes called the family violence model.

Stress includes unemployment, poverty, and health problems. Chapter 4 on families and the economy showed that unemployment and poverty are common experiences of contemporary American families. Furthermore, in our society, families are the social institution that are frequently called on to provide a lifeboat in an economic crisis. For example, when young adults are unable to find work, we expect their parents to provide shelter for them rather than requiring the government to do so. The simultaneous demand placed on families to provide for economic survival and the inability of large numbers of families to do so adequately creates a stressful situation that may cause violence to erupt.

Stress may also result from critical events more internal to families such as the birth of a child, the death of a family member, illness, divorce, and remarriage. These events are considered most appropriately handled by individual families, rather than other institutions in the community. And when people run into any difficulty in handling them, it is expected that families will be the major source of support. According to Straus, Gelles, and Steinmetz (1988), if a family cannot cope with the crisis, a situation is created that can push families toward violence.

For example, in the scenario at the beginning of the chapter, Ron becomes more abusive of Linda when she is pregnant. Having a baby puts a lot of pressure on parents. There may be excitement and anticipation but parents-to-be may also worry about how successful they will be as parents, whether they will be able to afford the child, and whether their spouse will still have time for them. In Chapter 12 you will learn that one of the characteristics of parenting in contemporary American culture is the lack of training for the job. Without an opportunity for learning what exactly is expected of them, parents-to-be may feel stressed.

Isolation, the second causal factor, is both an ideological and a structural characteristic of U.S. families. That is, many of us hold strong beliefs that families should be private and separate from the rest of the world. And we frequently experience isolation of our families.

Isolation may be especially characteristic of families where violence occurs. "All forms of abuse have been associated with families that are isolated and that have few community ties, friendships, or organizational affiliations" (Finkelhor & Yllo, 1989, p. 407). Isolation causes violence by diminishing the possibility of outside interference that could stop conflicts from developing into physical violence.

Dobash and Dobash (1979) interviewed 109 Scottish women who were in relationships that involved battering. One of the common patterns they found among the women was that as the couple became more committed to each other and the relationship more serious, the women spent less time with their own friends.

Sometimes the ideology that families should remain private can inhibit victims of abuse from seeking help. Some battered women may go so far as to try to keep the abuse hidden. For example, in the scenario at the beginning of the chapter, Linda expresses the belief that her abuse is a personal problem that she should be able to resolve. Although she is unable to develop a solution on her own, she has not yet sought help from people outside of her family.

In addition, even if the victim does wish to seek help, if she is kept secluded and isolated from others she may be unable to make contact with someone who could help her. Linda says she does not know what to do about Ron's abuse. This implies that perhaps if she had more social contacts or more knowledge about the community she might discover a range of options about which she is currently unaware.

The isolation of a family also means it is unlikely that someone who could offer help unsolicited will even know about the abuse. The people who are in contact with Linda, her physician and her friends, have seen only the bruises and so far Linda has been able to convince them the marks are the result of something other than purposeful abuse. The violent interactions between Ron and Linda are apparently taking place in private and others may not know about them.

Finally, even if the violence is not private and others know about it, if those others believe that it is essential to maintain family privacy, outside interference is unlikely to occur. In some cases real physical isolation of families may not exist but if people believe that "outside interference" is inappropriate they may not act to intervene. Perhaps Linda's friends do suspect she is abused but they believe, like her, that husbands and wives should settle their own problems and that friends should not pry into their family affairs.

Isolation, therefore, can inhibit the stemming of abuse by being part of family organization. Or it can inhibit it by just being part of our ideas about what families ought to be like.

The third social structural factor frequently cited as a cause of family violence is the acceptance of violence as a means of solving conflicts. Straus, Gelles, and Steinmetz (1980) found in their research that 25 percent of the wives and 33 percent of the husbands thought that "a couple slapping one another was at least somewhat necessary, normal and good." Slapping children was even more acceptable. Eighty-two percent of parents thought that slapping a twelve-year-old child is necessary, normal, or good.

We learn that violence is legitimate from movies, cartoon shows and other television programs, music, and even nursery rhymes. We also learn it from our parents who practice violence. Many researchers have discussed the way in which the exposure of children to family violence may play an important role in causing them to behave violently as adults (Bernard & Bernard, 1983; Fagan, Stewart, & Valentine, 1983; Spinetta & Rigler, 1972; Walker, 1979).

A Feminist Social Structural View on Battering and Marital Rape Chapter 1 introduced five themes. One of these themes was that feminists pay attention to the way in which different family members experience differently events that occur within their families. The most dramatic example of differentiation in families is violence in families. The experience of violence in families may be radically different for different family members. One person may be a batterer, another a victim, and another unaware of the abuse. The differences in terms of who is a batterer and who is a victim are not random but tend to lie along lines of gender and generation. Batterers are usually men; victims are usually women. Abusers are usually parents; the abused are usually children. The description above of three structural sources of violence in families shows us that men, women, and children who call themselves a family can treat each other in cruel ways without being pathological individuals or even unusual or deviant people. The source of the violence is found both in the way we organize families—for example, as isolated units—that allow and sometimes even encourage people to treat each other badly, as well as in the social context of families such as during an economic crisis. Demie Kurz has named the theory that takes a social structural approach, which does not emphasize gender, as the family violence approach. The family violence approach does not entirely ignore the importance of women's subordinate position in family violence situations; they believe it is only one of several contributing factors (Kurz, 1989).

Some scholars have argued that the family violence approach does not pay sufficient attention to differentiation within families. Kurz, for example, asserts the family violence approach in some ways is correct but insufficient. She maintains that gender inequality and the oppression of women is the central feature of violence in families and must always be considered in our analysis.

Kurz (1989) names Murray Straus, Richard Gelles, and Suzanne Steinmetz as the scholars who are best identified with the family violence approach and compares their point of view with other social structuralists who have made gender more central in their assessment of violence in families. Kurz (1989) cites Russell (1986), Dobash and Dobash (1979), Bowker (1986), Yllo (1988), Pagelow (1981), and Stark, Flitcraft, and Frazier (1979) as examples of scholars who take a feminist approach.

Kurz notes that the family violence approach uses language to describe the problem that does not emphasize the gendered quality of the violence. For example, people writing from the family violence perspective use terms like family or domestic violence instead of battering or wife rape and they use the term spouse abuse instead of wife abuse. These words "indicate that it is the family, not the relationship between women and men, which is the central unit of their analysis" (Kurz, 1989, p. 492).

Kurz (1989) argues that a feminist approach is different from and better than that proposed by the family violence theorists. This approach makes the politically unequal relationship of men and women the center of the analysis and views gender inequality as the most critical factor in violence (Kurz, 1989). According to the feminist approach, in attempts to understand violence in families and to develop solutions to the problem, special attention must be paid to the unequal political relationship between women and men both in families and outside of families.

The family violence researchers note the importance of cultural norms that condone violence in general. Feminists point out that additionally, there are norms (and historically even laws) that condone violence against women. For example, in the nineteenth century many states had laws specifically approving wife beating (see Dobash & Dobash, 1979, for history of battering).

Feminist researchers, furthermore, argue that battering is not just a reflection of the inequality between women and men, but is a conscious strategy used by men to control women and maintain the system of gender inequality (Kurz, 1989). For example, Walby (1986) asserts that the oppression of women by men is so tightly linked to woman battering and marital rape that as we attempt to solve the problem of inequality and as women are successful in removing themselves from abusive relationships and establishing their independence from abusive men, we may see retaliation against women in the form of an increase in violence against them.

Theorists who take a feminist approach also argue that gender inequality enters into the problem of violence in families by creating barriers to stopping the violence. Economic inequality between women and men, for example, keeps women tied to men who are batterers. Mildred Pagelow (1981) looked at the characteristics of couples in which women were able to leave violent husbands. She found that wives with the most resources were the first to leave physically abusive husbands.

Family violence theorists and feminist theorists share some of their ideas. They both maintain that social structure, not individual psychology, best explains the problem of woman battering and marital rape. But they are quite different in another respect. "[F]or family violence researchers, violence is the primary problem to be explained, while for feminists an equally important question is why women are overwhelmingly the targets of the violence" (Kurz, 1989, p. 498).

TABLE 11 • 3
Comparing Husbands' and Wives' Use of Violence

	Husband	Wife
Threw something at	2.8%	5.2%
Pushed, grabbed, shoved	10.7%	8.3%
Slapped	5.1%	4.6%
Kicked, bit, hit with fist	2.4%	3.1%
Hit or tried to hit	2.2%	3.0%
Beat up	1.1%	.6%
Threatened with gun or knife	.4%	.6%
Used gun or knife	.3%	.2%

Source: Straus, et al., 1980, p. 37.

The Controversy Over the Battered Husband Syndrome In 1980, the debate between these two points of view—the family violence model and the feminist model—made the news. Straus, Gelles, and Steinmetz, spokespersons for the family violence perspective, reported that husband abuse is equally as common as wife abuse and their findings were picked up by the popular press. Table 11-1 shows the incidence of violent acts in marriages. Table 11-3 shows the same information broken down by gender.

Straus and Gelles (1986, p. 472) argue that "violence by wives has not been an object of public concern. There has been no publicity, and no funds have been invested in ameliorating this problem because it has not been defined as a problem."

Feminists strongly criticized Straus, Gelles, and Steinmetz's conclusion that similar reports by husbands and wives meant that husband abuse is a problem equal or nearly equal to wife abuse (Dobash & Dobash, 1988; Pleck, et al., 1977–1978). The first criticism of the family violence researchers' assertions about husband battering centered on how the researchers categorized the answers to their survey questions (Dobash & Dobash, 1988). According to the feminist critics, the family violence researchers' data were collected with a scale called the Conflict Tactics Scale in which respondents were asked whether they participated in certain behaviors. The researchers then categorized the behaviors in terms of how dangerous they were. But the critics argued that their categorization may have been faulty:

> Included in the "high risk" category is "trying to hit with something" and excluded from it is "slapping." Yet, our own research, which does examine injuries sustained from particular attacks, demonstrates that a slap can result in anything from a temporary red mark to a broken nose, tooth, or jaw and that trying to hit with something never results in anything unless the blow is landed. (Dobash & Dobash, 1988, p. 59)

Self-defense was another issue of concern of the feminist critics. They claimed that issues like who initiated the attack and who acted in self-defense were ones that were not taken into consideration in the data that were used to make the argument that husbands are victims of spouse abuse as often as wives are. Dobash and Dobash wrote,

Indeed our research and virtually everyone else's who has actually studied violent events and/or their patterning in a concrete and detailed fashion reveal that when women do use violence against their spouses or cohabitants, it is primarily in self-defense or retaliation, often during an attack by their husbands. On occasions, women may initiate an incident after years of being attacked but it is extraordinarily rare for women to persistently initiate severe attacks. (1988, p. 60)

Finally, feminist critics point out that Straus, Gelles, and Steinmetz's work is based on self-reporting. Data obtained from sources that report the effects of violence, like police and hospital records, show that nearly all violence between spouses is directed at women by men (Kurz, 1989, p. 495). For example, every investigation of police calls shows women as overwhelmingly the victims in domestic disturbances: 96 percent in Minnesota, 95 percent in New York, 93 percent in Detroit, 92 percent in Ohio and numerous other areas (Dobash & Dobash, 1992). Schwartz (1987) found that thirteen times more women than men sought medical care for spousal assault.

Berk and his colleagues (Berk et al., 1983) investigated 262 cases of "domestic disturbances" to determine whether these incidents were primarily mutual combat or woman battering. They coded the severity of injuries on an eight-point scale from no injury to high injuries (combinations of concussions, internal injuries, broken bones, and damaged sense organs). Their sample showed that 43 percent of the women were injured and 7 percent of the men were; when only injury-producing incidents were examined, women comprised 94 percent of the victims. This percentage is similar to the 95 percent figure reported in the National Crime Survey. Furthermore, in 39 percent of the cases only the woman was injured, while in 3 percent of the cases only the man was injured. Berk et al. (1983, pp. 204–207) conclude "This implies that if there is anything to the battered husband syndrome, it either has nothing to do with injury or is relevant to very few couples. . . . When injuries are used as the outcome of interest, a marriage license is a hitting license, *but for men only*."

Family violence scholars respond to this kind of information by arguing that the reports of battered husbands are exceedingly low because men hesitate to come forward, fearing stigma (Dobash & Dobash, 1992). Feminists reply that women are also unlikely to report being battered. Dobash and Dobash (1992) estimate that of 35,000 cases of woman battering they examined, about 2 percent were reported. Others (Rouse, Breen, & Howell, 1988) have compared women's and men's response to being assaulted by a spouse and found that men were *more* likely than women to call the police. Kincaid's (1982, p. 91) review of 3,125 cases found men were also more likely to press charges: "While there were 17 times as many female as male victims, only 22% of women laid charges, while almost 40% of the men did so." In addition, men were four to five times more likely to pursue a complaint and less likely to drop charges (Kincaid, 1982).

Three Points of View, Three Strategies for Policy The review presented above of the literature on violence in families shows three distinct points of view: an individualistic framework, a family violence framework, and a feminist framework. The individualistic framework is illustrated by Lenore Walker's work on the battered woman's syndrome. She focuses her attention on the individuals involved in violent relationships. For her, the cause of violence in families is the men and women who have been socialized to behave in gender-differentiated ways. Men have been socialized to be violent, aggressive people seeking to control their wives. Women have been socialized to be passive, incompetent people who are unable to express their desires, to control others' behavior, or to remove themselves from violent relationships.

The second point of view is the family violence framework represented by the work of Suzanne Steinmetz, Richard Gelles, and Murray Straus. Their work looks to the larger social structures of society rather than to the individual in their assessment of the problem. These larger social structures include ideologies, economic distress, the difficult tasks assigned to families like taking care of birth and death and the isolation of families.

The feminist framework is the third point of view, represented by a number of people cited in the section above. Like the family violence model, the feminist approach emphasizes the importance of social structural issues as opposed to individual failings. Both the family violence model and the feminist approach regard the macro level as the critical factor in the production of violence at the micro level between women and men. Feminists also share the family violence framework's point of view on the kinds of social structural, macro-level issues that are critical to causing violence in families. They differ from the family violence approach in their prioritizing of gender inequality and the oppression of women as being the most important social structural factors contributing to violence in families.

These three different points of view have important implications in terms of policy. The individualistic framework implies social policy that seeks to change the individuals who live in violent families. For example, in the scenario at the beginning of the chapter, the individualistic framework would call for resocializing Linda to be more assertive and to gain a stronger sense of self-worth. At the same time, Ron would be advised to learn to hold less dominating beliefs about relationships and more appropriate ways in which to express his anger.

The family violence approach seeks policies that call for changes in the social structural system. For example, family violence approach theorists would call for reforms that alleviate problems like unemployment. They might also work to eliminate ideologies that promote violence, for example by protesting media images that glorify and romanticize assaults and other violent behavior.

The feminist approach calls for similar changes in social structure but insists that each of these changes be made in light of the gendered character of violence. For example, they would call not just for eliminating violence in the media but for eliminating the glorification of woman battering and rape in particular. For example, the American Academy of Pediatricians recently reported that 56 percent of the videos shown on MTV, which young people watch an average of two hours a day, portrayed acts of violence, often against women (Kellman, 1989).

Feminists also observe the need for an elimination of the oppression of women in all areas of society as essential to ending violence in families. For example, from the feminist point of view, policies that could help to reduce battering and rape would include equal pay for women to allow them to afford to leave violent relationships. Feminists would point out that even though Linda is a professional, as we have observed in Chapter 10, if she chooses to leave Ron she will probably see a drop in her standard of living. Abuse would be more likely to end if Linda were better able to live independently from Ron and she would be able to live independently from him more easily if she were financially self-sufficient.

The History of the "Discovery" of Violence in Families

The second set of theoretical issues that has emerged around the study of violence in families revolves around the question of why now? Woman battering and marital rape have been around for a long time. Why have they just recently been discovered as important social problems? Before examining the theoretical issues related to this question, let us first take a look at the recent history of the emergence of battering as a social problem.

Woman battering and marital rape are issues that have gained much attention in the past two decades. Until the women's movement brought these problems and their underlying causes to the public eye, they were not part of our public consciousness. For example, even the terms battered women and marital rape did not exist twenty years ago (Kelly, 1988). Although many of us had experienced assault in our families or knew people who had, like Linda in the opening scenario, we usually thought of these events as aberrations, something out of the ordinary.

If woman battering was noted publicly, it was trivialized, as in popular television shows in the 1950s like "I Love Lucy" and "The Honeymooners," which included battering as part of their stock of humorous themes. Ralph would threaten to send Alice to the moon and the audience would laugh hysterically.

The struggle to bring woman battering to the public's attention and to stop the violence began with five hundred women and children and a cow in 1971. The women and children and their cow marched through the streets of an English town in protest of a proposed reduction in free milk for schoolchildren. The milk program was eliminated but the women's organization that had formed stayed together to organize the Chiswick Women's Aid center where the first shelter for battered women was established (Dobash & Dobash, 1979).

The center grew out of a commitment to end gender inequality and the oppression of women. R. Emerson and Russell Dobash explain:

> A small group of women working to put the principles of the women's movement into practice . . . set up a community meeting and advice center for women in a small derelict house in Chiswick, a London borough. The problem of assaults on women soon became apparent as women began talking about brutal and habitual attacks by their husbands or cohabitants. Although the house was meant to be used during office hours, the women obviously needed a 24-hour refuge where they and their children might escape from violence and the center quickly became a refuge for battered women. (Dobash & Dobash, 1988, p. 52).

Erin Pizzey was one of the people who worked at the Chiswick Center in Britain as it was first being established. She listened to the women who came there and based on what she heard she wrote *Scream Quietly or the Neighbors Will Hear You*, the first contemporary book about wife abuse, in 1974.

In the United States, the first shelter also grew out of the women's liberation movement. One of the tactics of the women's movement in the 1960s was the organization of consciousness raising (CR) groups. CR groups consisted of small groups of women meeting to discuss feminist literature, to talk about their lives, and to plan ways to fight for equality. In 1971, one CR group in St. Paul, Minnesota organized a shelter for battered women and thereby started the battered women's movement, which spread quickly across the U.S., aided by numerous rape counseling centers and other feminist networks (Schechter, 1982; Tierney, 1982; Johnson, 1981).

Since the emergence of the Second Wave of the women's movement in the 1960s and 1970s, woman battering and marital rape have been recognized as serious social problems. It is no longer as acceptable to treat them lightly. More and more of us acknowledge that they are social problems—not unusual individual experiences. We see that they are now and have been for some time an important, intrinsic part of our society (Gordon, 1988).

The physical abuse of women by their husbands has been a part of many societies, including the United States, for centuries. Yet we see in this brief history that the concepts of wife abuse, battered women, and shelters for battered women are a recent development. How did this apparently old problem become visible? This is a question in which sociologists with a social constructionist perspective have been interested (Loseke, 1992).

Social constructionism is a theoretical framework that posits that what we perceive to be real is a product of social interaction among many people. In the case of the issues of wife abuse and battered women, the social constructionist would argue that these issues were not only invisible before the 1970s, they actually did not exist. "Most certainly some husbands always have victimized their wives—the historical record is clear on that—but assault could not be an instance of 'wife abuse' until the label was available" (Loseke, 1992, p. 38).

The label that allows us to begin to recognize and understand the phenomenon of wife abuse is created by many people and groups of people interacting with one another. The ongoing activity of the social construction of wife abuse and battered women has been an important development because it allowed us to begin to think of ways to address the problem now that it is recognized. But much work had to be done to create the images and make the claims that battering did exist and that it was a serious social problem.

In the review of the history we have seen that feminists have played an important role in constructing wife abuse as a social problem. But many others have been involved as well, including battered women, social service agencies, police departments, and scholars. Because wife abuse and battering are socially constructed by many different people, the meanings of the concepts and the issues behind those concepts are contradictory and dynamic. The character of wife abuse and its causes and solutions had to be hammered out by these various interests and many debates remain about what wife abuse or a battered woman are.

For example, is it wife abuse to "just" slap a woman? If a woman gets drunk, slaps her husband, and then is beaten up by him, is she a battered woman? What about a woman who has not yet been physically assaulted by her husband but has been threatened and believes the problem could easily escalate to physical violence?

These questions are ones that demand that we continue, as a society, to revise and reconstruct our ideas about battering. Loseke (1992) observed the operation of a shelter in California and found that this effort of socially constructing battered women is a continuing one by those in the front lines, the workers in the shelters. Every day the employees and volunteers at shelters for battered women must make decisions based on their understanding of what wife abuse is and who a "real" battered woman is. They make these decisions in social contexts interacting with other workers and residents in the shelters, funding agencies, and the public. Their task of socially constructing wife abuse, battered women, and shelters for battered women is difficult. It requires a lot of thought and creates frustration, anger, and sometimes shame if shelter workers are unable to come to an acceptable understanding. The decisions they make, based on their construction of the problem, can have profound consequences for the women seeking shelter.

The issues of battering and marital rape are so unsettling and so pervasive it is sometimes difficult to figure out how to understand them, not to mention how to stop them. Social constructionist theory helps us to think more clearly about battering and to see the most important factors. Social constructionist theory suggests that we need to be sure to consider the dynamic and complex process by which battering and rape and their causes and consequences are socially constructed. Who is involved in socially constructing the problem? What is their relationship to the problem? What is their relationship to each other?

CATALYSTS FOR CHANGE

Thus far this chapter has reviewed the incidence of violence in families and has discussed a number of theories about violence. The emphasis has been on the way in which the macro level of society shapes and determines the micro-level relationships between women and men in marriage and especially how those relationships are often marked by violence. People have also resisted violence in their micro-level relationships. Sometimes their resistance has taken the form of micro-level changes such as removing themselves from the relationship or altering the relationship. In addition, people have organized themselves at the micro level to put pressure on the macro level of society. They have worked for alterations in social institutions like the police force and have invented a new social institution—the battered women's shelter—to help stop the violence. In this section we will move our attention to these efforts, beginning with an examination of efforts made by individuals to alter their social context at the micro level and then moving to two examples of how the micro level affects the macro level.

As we have seen in the discussion of the prevalence of battering and marital rape, determining the numbers of women, men, and children who are victims of violence in families is a difficult task. Even more difficult is determining the numbers who solve the problem for themselves. Our image of the battered woman is of a person who is under attack, in crisis, and undecided about what to do about it. Linda in the opening scenario projected such an image. At the particular moment we meet her she is confused about what has happened to her and ambivalent about what to do. Many women, however, have determined what has happened to them and have decided what to do. They have taken steps to resist further abuse by leaving the abuser and establishing an independent and unbattered life. What are the catalysts for women making this move?

Kathleen Ferraro and John Johnson (1990) interviewed 120 women who passed through a shelter and asked them how they had decided to leave. They found six factors that served as catalysts for making a change. The first was a change in the level of violence. The kind of change that was especially important was a sudden one in the direction of increased abuse and a change that the women perceived as a direct threat, not just to their physical well-being but to their lives. For example, one woman explained that the deciding factor in her decision to leave her abusive husband was when he held a gun to her head.

The second catalyst was a change in resources. Women began to reinterpret the violence as unacceptable when other forms of support became available to them, such as jobs and shelters for battered women.

A third catalyst was a change in the relationship. Several researchers describe the pattern of abuse as one that includes intermittent periods of violence and then remorse on the part of the batterer (Walker, 1979). The women who spoke to Ferraro and Johnson said that their decision to leave was triggered when the periods of remorse shortened or disappeared.

The fourth catalyst Ferraro and Johnson discovered was despair. For some women, the hope that things would get better prevented them from leaving. When this last hope seemed to disappear their despair forced them to make a change.

The fifth catalyst was a change in the visibility of the violence. "Battering in private was degrading, but battering in public was humiliating" (Ferraro & Johnson, 1990, p. 113). One woman explained:

> He never hit me in public before—it was always at home. But the Saturday I got back [returning to her husband from the shelter], we went Christmas shopping and he slapped me in the store because of some stupid joke I made. People saw it, I know, I felt so stupid, like, they must think what a jerk I am, what a sick couple, and I thought, "God, I must be crazy to let him do this." (Ferraro & Johnson, 1990, p. 113)

The sixth catalyst was external definitions of the relationship. When women were offered responses from others that encouraged them to see the abusive man as intolerable, they were more likely to leave. Ferraro and Johnson (1990) argue that this is an indication of the importance of shelters because they provide just such an outside opinion.

The women Ferraro and Johnson interviewed took it upon themselves to resist the violence against them by leaving their abusive husbands. These women have taken action at the micro level to make changes in their own lives.

Violence in families has also caused activity at the micro level with the intention of changing macro-level social institutions. For example, resistance against violence in families has taken a place at the community level in the form of social movements that have demanded two important changes in the availability of community resources for battered women: the get tough campaigns in police departments and the provision of shelters for battered women.

The Get Tough with Abusers Campaign

Since the recognition in the early 1970s that violence in families is a crucial social issue, significant changes have taken place in the laws, making it easier and sometimes mandatory that the police arrest men who batter women in family fights.

In 1984, Lawrence Sherman and Richard Berk studied the effect of police arrest in Minneapolis as a deterrent in domestic violence. They concluded:

> [A]rrest was the most effective of three standard methods police use to reduce domestic violence. The other police methods—attempting to counsel both the parties or sending assailants away—were found to be considerably less effective in deterring future violence in the cases examined. (Sherman & Berk, 1984, p. 262)

Donald Dutton and his colleagues (1992) have examined the question of why arrest reduces the incidence of repeated wife assault. They propose two reasons. First, the arrest acts as a deterrent—something the men do not wish to have happen again. Second, an arrest alters the power dynamic between the men and their wives. The batterers who were interviewed perceived a power shift emanating both from the wife's willingness to call the police and from the criminal justice system's willingness to defend her.

Furthermore, an arrest prevents batterers from "minimizing and rationalizing" the abuse because it makes the incidents public and subject to others' perceptions and evaluation, which might be quite different from the abuser's. Some batterers, for example, minimize and deny the effects. James Ptacek (1988) interviewed batterers and found a number of them claimed they were not violent—their wives just bruised easily. Others rationalized the violence as, for example, a result of some unique personal flaw such as an inability to use self-control. One batterer explained, "When I got violent, it was not because I really wanted to get violent. It was just because it was like an outburst of rage" (Ptacek, 1988, p. 143).

Others explained the behavior as unusual and out of character, caused by peculiar circumstances such as being drunk. One man said, "I've been involved with A.A., and that's why I'm much better. And a lot of my problems—not all of them, but most of my problems at the time were due to that. And it's just amazing to know there was a reason for the way I acted" (Ptacek, 1988, p. 142).

Some men rationalized the violence by claiming it had been provoked. One man blamed his wife this way:

> She was trying to tell me, you know, I'm no fucking good and this and that . . . and she just kept at me, you know. And I couldn't believe it. And finally I got real pissed and I said wow, you know . . . you're going to treat me like this? Whack. Take that. (Ptacek, 1988, p. 144)

Police departments were made aware of Sherman and Berk's (1984) findings in a publicizing effort unprecedented for a scholarly work in criminal justice (Binder & Meeker, 1992). Few departments, however, implemented mandatory arrest in domestic violence cases. A court case brought against the police in 1983, however, turned the tide (Halsted, 1992).

In 1982 and 1983 Tracy Thurman called the police or came to police headquarters at least seven times begging protection from her husband, Buck. Although she signed several sworn statements, the police hesitated to act. On the last call, the officer took twenty minutes to arrive because he stopped to go to the bathroom. When he arrived he witnessed Buck stabbing and kicking Tracy. He convinced Buck to give him the knife but did not handcuff him. Buck then ran into the house to get their son and began kicking Tracy in the head, in front of the officer. Six more officers arrived but still did not arrest Buck. They restrained him only after he again assaulted Tracy as she lay on a stretcher being carried to the ambulance.

Tracy Thurman sued the police department and won $2.3 million in compensatory damages. The judge ruled in 1985 that Tracy Thurman had been denied equal protection under the law simply because she was married to the assailant (Halsted, 1992).

Figure 11-2 shows that by 1990 nine states had mandatory arrest statutes (Connecticut, Iowa, Louisiana, Maine, Nevada, New Jersey, Oregon, Washington, Wisconsin). All but two states (West Virginia, Alabama) allowed warrantless arrests. Mandatory arrest means police officers are required to arrest batterers. Warrantless arrest statutes allow police to arrest batterers if there is due cause. This means if, for example, an officer sees an injured woman who has called for help, the officer can arrest the batterer without having actually witnessed the assault.

Although these changes in the law were welcomed and have undoubtedly saved women's lives, there are two problems with them. First, requiring that the police arrest the batterer regardless of the woman's wishes removes any choice she may have. Furthermore, it ignores the validity of her fears that the abuser may retaliate even more harshly against her if he is arrested. Second, even if the laws have been altered, the actual policy being implemented by the police in these cases may have changed little (Ferraro, 1989).

Kathleen Ferraro (1989) has investigated this second issue. She was interested in whether change in policy did create a change in police behavior in her study of the Phoenix police. She and her assistants rode in police cars and witnessed police handling of domestic disputes.

In Phoenix in 1984, both the law and the stated policy of the police chief was to arrest batterers. The researchers found, however, that the laws and departmental policy are only one set of factors influencing police decisions to arrest in family fights. The police were also influenced by legal, ideological, practical, and political considerations.

An example of a legal consideration was the question of property damage. In several incidents, the women had not been physically injured but their doors had been kicked in, their tires slashed, and their windshields smashed. The officers did not arrest the men, explaining to the women that Arizona is a community property state. In a community property state all property acquired during a marriage belongs to both the husband and wife. When married men destroy property that they or their wives have acquired, the men are destroying their own property and the act is therefore not illegal (Stetson, 1991).

Racist, classist, and homophobic ideology was also a critical issue. Officers believed that arrests were a waste of time if the people lived in housing projects, or were Mexicans, Indians, or gay. The police referred to them as "low life" and

FIGURE 11 • 2
Domestic Violence Arrest Laws, 1990

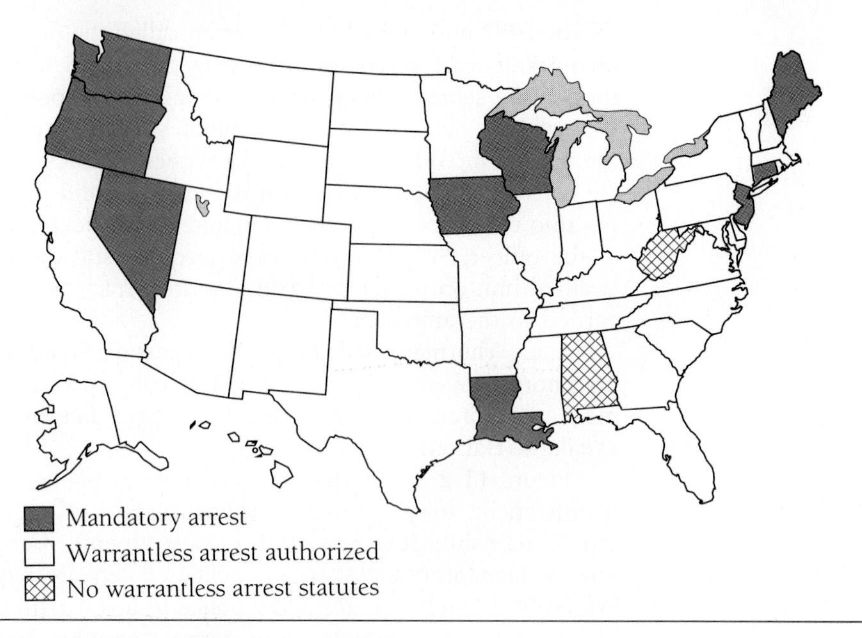

Source: Adapted from Halsted, 1992, pp. 156–157.

"scum" and insisted that "arrest was meaningless because violence is a way of life for them" (Ferraro, 1989, p. 67).

Practical considerations were the third factor interfering in making arrests. If an arrest involved finding shelter or information (for example, about legal rights) for the women and children, the police would often choose to avoid the arrest. Transporting women and children and processing them through the bureaucracies of shelters, foster care, and legal and social services is time-consuming and often requires long-term involvement. The police preferred to take the most expedient course that required the least additional involvement. Hirschel and Hutchinson (1989) add that another practical consideration is the presence of the abuser. In half the cases they examined, the batterer left before the police arrived and in some cities policies require that arrests be made on the scene. In all cities finding him would create much additional police work.

Finally, the police were influenced by their perception of political matters. Since there were no explicit rewards for compliance with the new policies nor punishment for noncompliance they interpreted the policies as just talk and relatively unimportant. They decided, therefore, that making arrests was not politically important to them and unnecessary to maintaining their good standing in the department.

Ferraro (1989) concluded that changing laws is not enough. Her review of all the factors that entered into the decision to arrest indicates that creating a mandatory policy might be a good idea but its implementation is likely to be disrupted by these four factors.

Many other reports show that Ferraro's findings are not unique (Dobash & Dobash, 1992). In a 1985 survey of 173 cities, nearly half relied on a policy of leaving the response to a domestic violence call up to the police officer. "Seventeen percent preferred mediation, 6 percent were still sending one of the parties out of the house and only one third indicated arrest as the preferred policy" (Dobash & Dobash, 1992, p. 208).

The Battered Women's Movement

Police departments are not the only social institutions that have been called on to address the problem of battering. Another institution—the battered women's shelter—was invented for that purpose in the early 1970s. The development of the battered women's shelter is a second example of the way in which micro-level activity of people working together has had an impact on the macro level of social organization.

Shelters are residential refuges that offer a free place to stay to battered women and their children who are fleeing an abusive home. Most shelters in the United States limit the stay to four to six weeks. Shelters usually provide services for residents like personal counseling, job counseling, legal assistance, assistance in obtaining AFDC and food stamps, and information on education.

A little over twenty years ago shelters for battered women did not exist. Susan Schechter (1982, 1988) has examined the history of these past two decades and concludes that shelters for battered women originally were firmly rooted in the women's movement. This tie was a natural one since feminists saw violence against women, like battering, as "an integral part of women's oppression; women's liberation its solution" (Schechter, 1988, p. 302).

In the movement to build shelters, feminists provided theory helping to expose, describe, and explain battering. Based on their ideology of sisterhood, feminists also supplied shelters with physical and emotional support, buildings, funds, materials, and thousands of hours of labor, often for no money.

The feminist vision, however, demands that people organize, not only to support shelters, but to change society in order to stop the violence. One feminist expressed this vision when she told Schechter: "our shelter only holds thirty people at once; it is not enough. The only way to change the system is to build a strong women's movement that will organize and demand change" (Schechter, 1988, p. 308).

By 1980, the battered women's shelter movement was not entirely in the hands of feminist political activists. In a national survey in 1981 of 127 shelters, "fewer than one half of the existing shelters were either founded by or directly related to feminist groups or ideology. By contrast 25% were started by church groups, and perhaps another 25–30% by YWCA and other local civic organizations" (Johnson, 1981, p. 831).

The provision of shelters by nonfeminist organizations has created a change in ideology. For example, social service providers take a more therapeutic view of the problem of battering and tend to focus on the woman and her immediate surroundings while ignoring structural and political issues (Johnson, 1981). Schechter quotes a professional who does not see herself as a feminist and who does not see sheltering as a part of the feminist movement:

> Politics are messy and unnecessary. In fact they might hurt us as we try to reach our goals. We might unite to share skills, find more resources and reform institutions and laws to enhance the quality of battered women's lives, but we are not part of a women's liberation movement. (Schechter, 1988, p. 307)

BOX 11 • 2 DO RESTRAINING ORDERS HELP?

One legal procedure now often used by battered women is a restraining order. These are orders given by the court to make abusive men change their behavior. For example:

> 1. An offender is ordered to move out of a domicile or not to use certain property like a car, even if the title to the property is in the name of the offender.
> 2. An offender is ordered to refrain from further physical or psychological abuse.
> 3. An offender is ordered to refrain from any contact with the victim.
> 4. An offender is ordered to enter counselling to finish the program.
> 5. An offender is directed to pay support, restitution or attorney fees.
> 6. Provisional custody of minors is granted to the victim and drafted onto the protective order. (Halsted, 1992, p. 153)

These orders can have important positive affects. Chaudhuri and Daly (1992) found that they increased police responsiveness, increased arrest rates of repeaters, reduced the chance that the man battered again, and empowered women to end abusive relationships. The researchers caution, however, that one in ten men retaliated by beating or threatening their wives for obtaining a restraining order and one-third of the men ignored the order.

This point of view is met with criticism by feminists who view the trend toward social service–oriented programs as a sign of cooptation of the issue and of the movement (Tierney, 1982; Johnson, 1981; Wharton, 1987). The critical issue seems to be the problem of finding funding for shelters while maintaining a perspective of the problem of woman battering that acknowledges the need for systemic change as well. Two activists write:

> Perhaps the greatest issue facing our movement is cooptation. We recognize the need to rely on traditional institutions to keep our shelters operating. However, these institutions have political structures and goals that are antithetical to ours. . . . These agencies do not understand the social or political context of battering. . . . [Their] purpose is to maintain themselves not to provide any real services or social change. . . . They will try to turn us into more traditional models structurally and politically and in terms of the goals of our services. (Andler & Sullivan, 1980, pp. 14–15)

Another debate within the movement is over the role of battered women (Wharton, 1987). In many shelters, battered women are treated as "other," as clients who are in need of help and as women who are different, incapable victims. At the same time a group of experts, who are predominantly not battered women, has emerged to define the problem and the solutions: "While the feminist ideals underlying the movement for battered women suggested that victims of wife assault could be the only 'experts' regarding their problems . . . ironically, but not surprisingly, the movement was accompanied by the emergence of experts on battered women" (Loseke & Cahill, 1984, p. 296).

In other shelters battered women have played an active role in advising shelter providers and in providing shelter and other services. Increasingly battered women are uniting in task forces to formulate and make demands on shelters and other community institutions, insisting that their voices be heard

Shelters for battered women are a new invention that developed with the recognition of battering as a social problem. Shelters like this provide a temporary residence for women and their children who are escaping an abusive man.

(Schechter, 1988). The shelters themselves, however, have become more professionalized (Wharton, 1987).

THE MICRO-MACRO CONNECTION

We began this chapter with a story about Linda and Ron. Linda thought that the troubles she was having were hers alone and that the violence she was experiencing existed apart from the rest of society. If we use the sociological imagination we see many connections between Linda's personal troubles and social structures.

The social context shapes the behavior of Linda and Ron and limits their ability to eliminate the battering in their lives. Social structural issues like poverty and unemployment cause stress. The demands placed on families in our society to resolve or cope with these stresses may make families violence prone.

Ideologies, another factor in the social context, that condone violence against family members, and especially ideologies that condone violence against women and children, "justify" battering and child abuse. Furthermore, isolation of families and the lack of economically feasible alternatives to violent marriages for battered women limit the possibility of individual women and children escaping the violence. Macro-level social organization both causes the problem and limits its solution.

Gender inequality, a feature of several social institutions at the macro level, such as the labor market and the legal system, plays a central role in the creation and perpetuation of battering and marital rape. Furthermore, violence against women at the micro level of social interaction is an important way in which gender inequality is maintained at both the micro and macro levels.

Individuals also affect their social environment. Over the last two decades we have seen changes in thinking, policy, and social organization in attempts by people to stop the violence. Individual women at the micro level have sought to change their immediate surroundings by leaving the abusive situation. People organized at the micro level have also created changes in the macro level of the law and in the way in which the police are supposed to handle battering.

Another change has been the invention of shelters for battered women. In the history of the development of shelters we see micro-level forces vying with each other over a definition of the problem and its solution.

The Sociological Imagination allows us to see both the personal suffering from violence and the social forces causing and perpetuating that violence. Most importantly, it also shows us a way out. It will not be easy to make the necessary changes to stop the violence but understanding problems is a crucial first step to solving them.

SUMMARY

Prevalence of Violence in Families

- The story in the opening scenario of Linda, a battered wife, is a common one. Researchers have found that a large number of marriages report violence.
- Marital rape, another form of family violence, is often invisible because it is not recognized as a crime in a number of states.
- Battering and rape may have profound effects on victims, who suffer psychological trauma similar to trauma suffered by victims of war or natural disaster.

Theories About Violence in Families

- One theoretical issue is the question of why battering and rape occur. Three frameworks have been developed to answer this question: an individualistic framework, a family violence framework, and a feminist framework.
- The individualistic framework blames violence in families on the socialization of men to be batterers and women to be victims.
- The family violence framework looks at social structural factors like economic distress, isolation of families, and ideologies that condone violence as the source of the problem.
- The feminist framework also looks at these social structural factors but with an emphasis on the importance of gender inequality.
- A second question around which theory has developed is why violence in families has recently been recognized as a social problem.
- Social constructionist theorists explain how many people interacting together "create" a social problem by recognizing it as such and making that case to the public.

Catalysts for Change

- In the search for solutions, three areas were examined: individual change, change in police handling of battering, and the establishment of shelters for battered women.

- One way individual women resist battering is by leaving their husbands. Six factors influence their decision to leave: change in level of violence, change in resources, change in relationship, despair, change in visibility of violence, and external definitions of the relationship.
- Resistance has also taken the form of advocating reforms in police handling of domestic violence calls. Police are increasingly required to arrest batterers but a gap remains between policy and actual procedure.
- A third form of resistance is the invention of shelters for battered women. The social movement associated with shelters has included a number of conflicting voices seeking to define the problem and its solution.

The Micro-Macro Connection

- Family violence illustrates how macro issues affect the micro experience. Violence is caused by social structural factors such as unemployment and ideologies that condone violence. The attempts to end the violence are inhibited by social structural factors such as poor pay for women and inadequate shelter.
- Micro-level activity has also altered macro-level organization in the struggle to eliminate violence through changes in police policy and the provision of shelters for battered women.

The relationship between mothers and children is idealized in American society and is especially strong in Japanese American families.

12 PARENTS

Kimberly and Michael (from Chapter 1) have been married for six years and they now have three-year-old twins, Lauren and Amanda. While the girls play in the makeshift play area, Kimberly busily stuffs envelopes at the campaign headquarters of Lynn Wilson for Congress. The centerpiece of the Wilson campaign is her support of a childcare bill that would create more high-quality daycare centers, raise the pay for workers at the centers, and federally subsidize children from middle- and low-income families.

Another volunteer asks Kimberly how she got involved. Kimberly says, "When I first got married I planned on having children and staying home with them until they went to kindergarten, but I've been forced to change my thinking since then." She explains that when the twins were born she was totally unprepared for the amount of work required to care for them. She also did not anticipate how children change a household—the sleep interruptions and eating schedules, the mess and lack of privacy, and the shifting of activities from those for adults to those that include children. After the initial shock wore off and she and Michael established a schedule, she began to feel lonely and isolated and she realized that the bills were piling up. She decided to go back to work. But finding someone to take care of the twins was more of a challenge than she had counted on.

"I never gave childcare much thought. I noticed the daycare centers around town and figured it was no problem finding a place. But when I went out to look around for myself I discovered some were dreary establishments with no learning program, and since the girls weren't toilet trained, they weren't eligible for many centers. Plus the expense for two children meant that nearly all my take-home pay would be eaten up in childcare."

The other volunteer asks her, "What about Michael? Can't he help?" Kimberly explains that Michael is in sales and is out of town a lot. He spends time with the kids on the weekends to give her a break and he is getting better with them but he really had no experience taking care of babies.

"That's why I'm here. Lynn Wilson is a mother herself and that will help her bring a fresh outlook to Washington. I know she will work to make the government pay attention to parents and children," Kimberly concludes as she excuses herself to track down Lauren and Amanda.

- *When we last heard from Kimberly in Chapter 1 she had different ideas. How is the real experience of parenting different from the anticipated one?*
- *What is it about parenting that makes it different from other activities and how does the experience differ for women and men?*
- *How does the experience of parenting influence our thinking about other activities?*
- *Kimberly is married and middle class. How would her experience be altered if she were single, a teenager, a lesbian, or poor?*
- *Some people argue that the division of childcare between Kimberly and Michael is inevitable because it is rooted in biological differences. What are the weaknesses of that argument?*

PARENTS AND PARENTING

In this chapter and the next we will examine parents and children. Parents and children create a relationship within families. Separating the two, therefore, is somewhat artificial. On the other hand, parents and children are separate categories and they do have different reponsibilities and interests in their relationship with each other. Nevertheless, when you read about parents and parenting, keep in mind how they affect children and childhood, as well as how they are shaped by children.

This chapter is divided into three sections: parenting, the question of biological destiny, and the political debate around childcare. The discussion begins with a summary of the characteristics of parenting and then moves to examining the way in which gender affects parenting—motherhood and fatherhood. Since parenting can be so readily divided into two categories by gender, the question that emerges is whether this distinction is biologically determined. Are mothering and fathering different because of physical differences between women and men? The theoretical debate over the answer to this question is reviewed. The final section of the chapter concerns an important political debate on the question of childcare.

Transition to Parenthood

Being a parent is unlike any other activity in which we participate. But what exactly is unique about the role of parents compared to other roles in our society? Alice Rossi (1992) suggests that there are four salient features of parenting: (1) cultural pressure to assume the role; (2) inception of the role; (3) irrevocability of the role; and (4) preparation for parenthood.

Much pressure is placed on people to assume the role of parents. This is especially true for women. Rossi (1992) states that in order for men to secure their status as adults they must be employed. For women, adulthood is marked by maternity, whether or not they are employed. But both women and men in the United States are usually expected to become parents.

By inception of the role, Rossi means the way in which parenthood begins. Unlike many activities in which we participate that have an equal impact on our lives, becoming a parent almost always includes an element of surprise. Most pregnancies are unplanned, few are timed exactly, and even when children are adopted there is a broad range of time in which the child arrives in the new parents' home. Parenthood, therefore, may be intentional or unintentional but the moment at which one becomes a parent is unpredictable. Furthermore, the transition to pregnancy and from pregnancy to parenthood is an abrupt one.

The third feature of parenting noted by Rossi is its irrevocability. Once a child is born or adopted there is little opportunity for terminating one's status as a parent. Parents can and do place their children for adoption. But this is relatively unusual and does not necessarily end the relationship for the parent or the child.

The recent case of Gregory K, a young boy who initiated a legal suit to allow him to "divorce" his mother and be legally adopted by his foster parents, showed the legal and social difficulties associated with revoking the relationship between children and birth parents. Gregory K won, the first instance of such a case succeeding, perhaps as a sign of greater flexibility on this question in the future. Rossi's assertion, however, that the symbolic and emotional relationship of parenthood is irrevocable probably still holds true, even for Gregory K and his biological mother.

The last factor Rossi cites as intrinsic to parenthood is the lack of preparation parents are given for their new role. There are few informal and even fewer formal opportunities for learning how to be a parent. And even if parent education does exist there are no guidelines for evaluating the validity of the content of that education.

Scholars who have studied Asian American families would add a fifth feature to Rossi's (1992) list—primacy of the parent-child relationship. Sylvia Yanagisako (1985) says that a difference between traditional Japanese families and contemporary white middle-class American families is the importance given to parent-child relationships in contrast to husband-wife relationships. When she interviewed Nisei (second-generation Japanese Americans) she found that they felt they were in the middle of these two cultural groups. Nisei believed their parents' generation overemphasized the parent-child relationship and ignored the relationship between husbands and wives. They described their mothers and fathers as having "lived for their children and in turn of having expected too much of them" (Yanagisako, 1985, p. 110).

Nisei, in contrast, believe that the conjugal relationship between husband and wife should be the closest. One woman explained, "You have to have more than kids, because the kids are going to leave; you have to think not only of now, but later. You need marriage not only for now but later for the companionship" (Yanagisako, 1985, p. 110).

Nisei, however, were also critical of "American" marriage because it goes too far in the other direction. Nisei believed that "American-style parents are too eager to leave their children at home while they seek entertainment and pleasure for themselves. Their egoistic selfish concern for their own interests and recreational pursuits may lead to the neglect of children—a neglect they pay for in problem children, juvenile delinquency, rebellion and ultimately rejection" (Yanagisako, 1985, p. 111).

These five factors give us some good ideas about the general structure of parenting in our society, but they lack one other ingredient—gender. Men and women are similar in some ways in their experience of parenting but whether one is a mother or a father creates significant differences. In the following sections we will look at parenting as a gendered experience, examining the experience of mothering and fathering and the differences and tensions between them.

The Motherhood Mystique

Raising children can be a delightful and gratifying experience. Because of these attractive features of parenting, most women in our society would like to have children—the favorite number is two. Genevie and Margolies (1987) asked 1,100 women in a nationally representative sample about their experience being mothers. About 25 percent of the women said the experience had been mostly positive and 20 percent said the experience had been mostly negative. The majority of women (55 percent), however, said they felt ambivalent. Although there were wonderful aspects to being a mother, they also had experienced disillusionment when real mothering did not match the ideal. They were especially disappointed when mothering was boring and burdensome and their husbands were not as involved as they wished.

Although there are real, positive aspects to mothering, in our culture motherhood is idealized. Our ideas about motherhood ignore the problems and sometimes overemphasize the benefits and pleasures. As a result, when women become mothers they may be unprepared for the real experience and they may feel guilty and inadequate when their own experience does not match the ideal (Lazarre, 1986). Michele Hoffnung (1989) calls the idealized version of motherhood in our culture the motherhood mystique. She says there are four characteristics of this ideal:

1. Women achieve their ultimate fulfillment by becoming a mother.
2. The body of work assigned to mothers—caring for child, home, and husband—fits together in a noncontradictory manner.
3. In order to be a good mother, a woman must like being a mother and all the work that goes with it.
4. A woman's intense, exclusive devotion to mothering is good for her children.

Hoffnung (1989) shows how each of these factors is inconsistent with reality. First, the argument that motherhood is the ultimate fulfillment for a woman denies the multifaceted character of women's lives. The demand that mothers become focused on one aspect of their lives is different from expectations about fathers. "[Fathers] are expected to be interested in and delighted by their children but [their] other interests and employment are expected to continue" (Hoffnung, 1989, p. 162).

Not only is it unlikely that women can truly abandon all of their other interests and activities when they become mothers, it would not be healthy for them to do so if they could. Helena Lopata (1971) argues that for most middle-class American women, the birth of a first child creates a break in their lives that is greater than any other event. She recommends that women preserve some continuity in their lives by continuing other activities, like paid work, after the birth of their child (Mercer, 1986).

Rich (1976) notes that another feature of the demand that mothers focus their full attention on mothering is that they must do this in relative physical and social isolation. The ideal mother is supposed to devote herself to her children to the exclusion of other kinds of activity. Furthermore, she is supposed to isolate from others her relationship and activities with her children. Rich (1976) argues that this isolation of mothers and their children leads to feelings of anger, frustration, and bitterness.

The second aspect of the motherhood mystique is that the ideal mother must fit together the tasks of being a wife and mother in a noncontradictory manner. Real mothers, however, discover that the birth of a child, especially a first child, interferes with their relationship with their husbands. For example, one study found that after the birth of a first baby, husbands and wives talked to each other half as much as before the baby was born (Pohlman, 1969). And in general parents have less time to spend with each other than married couples who do not have children (MacDermid, et al., 1990).

The third aspect of the motherhood mystique is that in order to be a good mother, women must enjoy all aspects of their work. Since the activities of mothers are expected to cover such a large range, it is unlikely that any single person would be skilled at all of these activities and would enjoy them all.

Finally, the motherhood mystique asserts that the full-time devotion of mothers to mothering and the nearly exclusive mothering of children by mothers is supposed to be the most natural and best way of raising children. In human history, exclusive mothering by biological mothers is actually unusual. In precapitalist societies mothering is likely to be shared within the community in a variety of ways (Rosaldo & Lamphere, 1974; Reiter, 1975). In the next section,

BOX 12 • 1 ARE EMPLOYED WOMEN WORSE MOTHERS THAN NONEMPLOYED WOMEN?

Many people believe that employed women are worse mothers than women who are not in the paid labor force. Table 12-1 reports responses to a poll taken by the *New York Times*. The respondents were asked whether employed women are better than, worse than, or equal to nonemployed women as mothers. A substantial number of men and nonemployed women felt employed women were not as good mothers as nonemployed women. Of the employed women, one-quarter thought employed women were worse mothers. And apparently some employers think mothers are worse employees. "Between 1985 and 1990, new mothers were 10 times more likely to lose their jobs than other employees on leave" (Harris, 1992, p. 50).

TABLE 12 • 1
Opinions on Mothers Who Hold Outside Employment

"Do employed women make better or worse mothers than nonemployed women?"	Better	Equal	Worse
Men	22%	22%	40%
Nonemployed women	15%	35%	39%
Employed women	42%	26%	24%

Source: Basow, 1992, p. 242.

when we discuss Nancy Chodorow's and Dorothy Dinnerstein's work, we will see how the isolated, exclusive mothering of children by women may be a source of tremendous psychological and social problems for individual men and women and for the society in which they live.

The Effect of Mothering on Children

In the next chapter we will examine Freudian theory and its approach to gender socialization and incest. In both cases we will see that feminist scholars have been highly critical of Freud. All feminist scholars, however, have not entirely abandoned Freudian analysis. Nancy Chodorow (1976, 1978b) is one theorist who has taken some of Freud's ideas and transformed them to create a new way of looking at parenting, sexuality, and gender.

Freud maintained that humans are born neither gendered nor heterosexual. They become gendered men and women and sexually identified (as either heterosexual or homosexual) through a process, which he referred to as the oedipal stage, that involves children's discovery of male and female genitals. Within a Freudian framework boys notice that they have a penis but their mother does not. They jump to the conclusion that their mothers have been castrated. Boys love their mothers but are afraid their fathers will castrate them. This fear causes them to turn away from their mother and look for another woman for whom they will not have to compete with their father.

Girls proceed through this process in a slightly different way. They notice that they do not have a penis and they blame their mother. Girls then seek to vicariously obtain a penis by finding a man with whom to have sex and by bearing a son.

Chodorow agrees with Freud that this formula provides a powerful framework for explaining how children become gendered and heterosexual. She differs from Freud on her assessment of whether this is inevitable or not. Freud assumes that this process is biologically determined and rooted in the biological difference in genital structures between males and females. Chodorow asserts that the key issue is the fact that it is a woman who is the mother and that the primary responsibility that women take in parenting is socially and not biologically determined. Raising children could take any number of forms: shared parenting by fathers and mothers; shared parenting among many members of the community; or even parenting by fathers alone.

In our society and most others, Chodorow asserts, mothering is done by women and this has important consequences for boys and girls. Furthermore, Chodorow maintains that the consequences are different for boys and girls.

The formation of the masculine personality in boys is characterized by three factors:

1. denial of attachment, especially dependence on another
2. defining masculinity as that which is not female
3. repression and devaluation of femininity on both psychological and cultural levels (Chodorow, 1976; Sydie, 1987).

In the oedipal process, boys completely repress their attachments to their mother because they are strongly motivated by the fear of castration. Chodorow argues that this repression becomes a foundation of masculine personality. As a result of their boyhood experience, men are more likely than women to repress their relational needs and sense of connection. Girls, in contrast, do not repress so absolutely their attachment to either their mothers or their fathers because girls do not have the strong motivation of fear of castration.

> Mother-monopolized childrearing produces women who are able to and will want to mother in their turn. This situation contrasts with males who have a separate sense of self and who lack the capacity or desire to nurture others. (Sydie, 1987, p. 151)

A second difference between girls and boys has to do with sexuality. Because women mother and are therefore the first love of their children, their sons' first experience of sexual attraction is heterosexual. Daughters' first sexual attraction, however, is not heterosexual because they are the same sex as their mother. Not only must daughters transfer their love from their parent to another person, as boys do, daughters must transfer their love from a woman to a man. Because this transfer is more complex, it is not as complete for women. "While she is likely to become and remain erotically heterosexual, she is encouraged by this situation [having a woman as her first love] to look elsewhere [to women] for her emotional needs" (Chodorow, 1976, p. 464).

The feelings that daughters retain of a connection with their mothers is reciprocated by their mothers. Mothers, because they too have been mothered by women, feel a sense of oneness with their daughters. The strength of the bond for both mothers and daughters is strong.

Based on these observations, Chodorow argues that both women's and men's primary emotional attachments are with women. Women do not have as strong an emotional need for men as men do for women. Chodorow cites the work of Zick Rubin (1973), who studied men's and women's response to breaking up a love relationship. Rubin found that the men were much more distraught than the women. As we observed in Chapter 8, these findings are also apparent in other studies that indicate that contrary to popular notions, men are actually more romantic and emotional about their relationships than women. According to Chodorow this is because the women do not have as great an emotional investment in these relationships as the men. And the women have other resources—women friends—to support them after their loss.

Just because men need women for nurturance does not mean that they like women. Another feature of the oedipal process is that boys must become engendered. They must learn to behave in appropriately masculine ways. In societies that isolate mothers with their young children, most of the contact boys have with adults is with a woman and their father is only a distant figure. Boys learn to be men by learning not to be women. Learning not to be women can slip into learning not to like things that are womanly and even not to like women themselves.

What Problems Result from the Exclusively Female Mothering of Children? This oedipal process creates two different kinds of people. It produces men who are emotionally repressed and unable to nurture children or to provide emotional support for their wives, and even unable to like their wives, but who are highly dependent on women for their emotional needs. The process produces women who are not as emotionally repressed and are able to nurture children, and who are also dependent on women for their emotional needs.

A problem emerges when these two different kinds of people—women and men—form marriages. Because of their different needs and different ways of connecting with others they inevitably have difficulties in these heterosexual relationships (Chodorow, 1976).

A second problem is that the process restricts the fully human being from developing. Instead two "subhuman" beings are created (Dinnerstein, 1977). The subhumanity it creates in men is especially ominous because it leaves them unable to connect with other humans in an effective manner (Rothman, 1989a; Dinnerstein, 1977). To overcome men's subhumanity, Rothman suggests that we "involve men fully in childcare, enabling boy children to experience the continuity, connectedness, womanliness in themselves that would make them whole" (Rothman, 1989a, p. 213).

The exclusive nurturance of young children by women creates difficulties for individuals whose personalities are deformed. In addition it creates difficulties for couples who attempt to form heterosexual relationships. Dinnerstein (1977) argues that it also causes problems for our society as a whole. She maintains that the creation of two types of people—men and women—has been accompanied by the dominance of one type in world affairs. That one type—men—furthermore, have repressed their feelings of human connection and instead have learned to seek themselves in the accumulation of things and power. War, violence, and environmental degradation are the results of a society dominated by the masculine psyche.

Comments on Chodorow's Theory Chodorow's theory has been criticized because it does not allow for any way out of the present system. Chodorow concludes her research by stating that the central political goal of feminists must be to restructure parenting to remove the monopoly of women over childcare. Since mother-monopolized parenting prepares girls but not boys to be mothers, it is unclear how to break into the pattern because according to this theory, men are not psychologically prepared to be mothers (Chafetz, 1988).

Another problem of bringing men into childrearing is getting men to participate (Polatnick, 1983). Given the assumptions of Chodorow's framework, men see themselves as different from and better than women and it is not clear how men could be convinced to agree to participate in "womanly"—and therefore degrading—activities like childraising (Sydie, 1987).

A second criticism of Chodorow has come from Third Wave feminists like Collins (1990), who question the validity of her theory because it ignores diversity in both the organization of childraising and the value placed on childraising. As we observed in Chapter 3 on the history of African American families and in Chapter 5 on the organization of contemporary racial ethnic families, "mothering that occurs within the confines of a private nuclear family household where the mother has almost total reponsibility for child-rearing is less applicable to the black family" (Collins, 1990, p. 43). African Americans and Latinos are more likely to share childcare among biological mothers, grandmothers, sisters, aunts, cousins, and friends (Stack, 1974). Native Americans also rely on a variety of kin and nonkin for childcare (Red Horse, 1980). Furthermore, Carothers (1990) asserts that theories like Chodorow's ignore not only such family organizations and the kinds of relationships that develop among mothers and children in the black community, they ignore the benefits imparted. Carothers says shared mothering facilitates a more empowered model of women and motherhood and instills values of social justice and race responsibility.

Lesbian mothers have also questioned Chodorow's insistence that men be involved in raising children. They ask, Is the dual parenting by two lesbians not a valid way in which to raise children? (Bart, 1983).

In addition to neglecting the variety of ways in which childraising is organized, Chodorow's theory has also been criticized for assuming that mothering is always a devalued activity. In Chapter 3 we saw that the social position of African American women was enhanced by the work they did as mothers in the slave community.

A third set of criticisms have come from those who believe that Chodorow's psychoanalytic framework places too much emphasis on micro-level psychological processes and ignores macro-level structural sources of gender inequality (Lorber, Coser, Rossi, & Chodorow, 1981). For example, Judith Lorber (1981) notes that the reason women mother and men do not may be less a result of their early relationship with their mother and more related to ideologies in our society that expect these gender differences. Furthermore, Lorber (1981) states that when inadequate childcare facilities force parents to decide to have one person stay home to take care of the baby, married couples may make rational choices to divide labor this way because husbands, on the average, can earn more in paid jobs. For Lorber (1981) it is these macro-level factors, gender ideologies and inequalities in wages, that channel women into mothering.

Alison Jagger (1983) is also critical of Chodorow for laying too much blame for gender inequality and the oppression of women on the fact that women mother. She writes, "A male-dominated society is always going to be misogynistic, no matter who rears the children. It may be that women rear children because they have low status, rather than they have low status because they rear children" (Jagger, 1983, p. 321). Both Lorber and Jagger contend that structural sources of gender inequality outside of families are equally if not more to blame for the fact that only women mother.

Childbirth

Throughout most of the twentieth century, childbirth for many women in the United States was a medical event in which they played what seemed to be a minor part. And the part they did play was primarily to stay out of the way of the real actors, the physicians (Rothman, 1982; 1989b). In the early 1970s Nancy Stoller Shaw (1974) observed birthing in Boston hospitals. Based on her observations she described a typical American birth:

> The woman was placed on a delivery table similar to an operating table. The majority were numbed from the waist down, while they lay in the lithotomy position (legs spread apart and up in stirrups), with their hands at their sides, often strapped there, to prevent their contaminating the "sterile field." The women were unable to move their bodies below the chest. Mothers were clearly not the active participants. That role was reserved for the doctor. (Rothman, 1989b, p. 156)

Rothman (1989b) defines ideology as the way in which a group looks at the world and organizes its thinking about the world. The ideology that underlies the scene above of the woman in labor is a medicalized one. The birth is a modern medical event. The woman is a patient under the supervision of the physicians. She is in a position—on her back with her feet lifted and spread apart—that is to the advantage of the physician. She is in a hospital where if the birth does not go in the manner described in the medical texts, further medical equipment is available. She is in a sterile environment, hooked up to an IV and monitors to assess her condition.

This medical view of birthing has been widely criticized in the past two decades and other birth formats have been allowed to creep back in. These alternative birthing experiences are informed by alternative ideologies. Alternative birth settings are homes and birthing rooms. The women are not lying flat; they may not even be lying down. Birth may occur with the woman in a squatting position using gravity to help her push. She is surrounded by people she knows, not hospital workers. She becomes the centerpiece of the event rather than just a birth canal.

Rothman (1989b) claims that the changes in birthing that have occurred as a result of the demand for midwives, home births, and birthing rooms have had a radicalizing effect on the health care workers who attend births and have called into question so-called "facts" about giving birth. Rothman explains four examples of these kinds of challenges.

First many physicians had believed that vomiting and nausea were a natural and common part of labor. Midwives who do not work in a medical setting and who do not forbid women in labor to eat and drink, as the hospitals do, have found nausea is uncommon. Rothman asks, "Is nausea caused by labor or lack of food?" (Rothman, 1989b, p. 180).

A second example is the likelihood of infection if much time passes between the rupture of the membranes and the birth. In hospitals, where physicians and nurses, as a matter of regular procedure, perform frequent vaginal examinations after the membranes rupture, infection of the mother is likely. Therefore when any delay occurs the physician is likely to advise inducing labor. Midwives working outside of hospitals have sometimes avoided infections in the mother by not performing vaginal exams after the membranes rupture.

Third, Rothman (1989b) discusses lactation. In a hospital setting it is considered normal for a woman's breasts not to produce milk until three days after the birth. In the nonhospital setting where the mother and baby have immediate and unrestricted access to each other, milk commonly comes in between one and two days after birth.

Fourth, in the medical setting once full dilation of the cervix occurs, if the mother does not immediately begin pushing the baby out, the situation is called second-stage arrest and is considered pathological. The physician uses drugs or mechanical techniques to end the interruption and continue the birth. In the nonhospital setting, midwives have observed women who were exhausted after reaching full dilation take a nap before resuming labor, and deliver healthy babies (Rothman, 1989b).

All of these examples, and there are many more, show that the medicalization of birth has created medical problems, which have then been reformulated to be the normal, natural experience of birth. Demedicalizing the experience and working with the mother as a unique individual and the central figure will improve not only the emotional and social experience of birthing but the physiological one as well.

Demedicalizing birth, however, continues to be a political question. Midwives provide an alternative to medicalized birth. Midwives treat pregnant and birthing women as people who must be seen as part of the whole social context, not as draped bodies that arrive at the delivery table. "Midwives do not 'deliver' babies. They teach women how to give birth" (Rothman, 1989b). In the U.S. midwives commonly assisted mothers until the nineteenth century, when physicians led a battle against midwives. They were successful in eliminating midwifery in all but the most remote rural areas by 1950 (Weitz & Sullivan, 1986). Since that time there has been a small but growing movement to reject obstetrical interventions in birth like shaving the pubic area, inducing labor, using drugs, breaking the amniotic membrane, and performing episiotomies. Some mothers are opting for home births and midwives are being reintroduced.

The medical community has responded by implementing reforms such as birthing rooms in hospitals, childbirth classes for prospective parents, and allowing family members to attend births. Diane Eyer (1992) argues that the latest ploy by the medical community to maintain their control over birthing is the invention of the concept of bonding. She asserts that the idea that mothers and infants must bond is not supported by empirical evidence, but it remains a popular concept and is actively promoted by health care professionals and hospitals. Furthermore, hospitals advertise their services as skilled in regulating the bonding procedure.

Heavily medicalized births in the traditional sense continue as well. For example, between 1970 and 1989 the rate of cesarean sections increased from 5.5 percent to 23.8 percent (U.S. Bureau of the Census, 1992).

Physicians continue to denounce home births and midwifery, saying they are too risky. It is difficult to test this contention by comparing hospital births to home births because hospital births include many of the high-risk groups, such as breech births, twins, and medical complications that midwives would not handle. Mehl, et al. (1980) designed research that matched mothers in order to more carefully compare their deliveries. They found that in comparing home births to hospital births there were no differences in birth weight, perinatal mortality, and other major complications. The home births, in fact, were actually better when factors like Apgar scores, meconium staining, postpartum hemorrhage, birth injuries, and need for infant resuscitation were considered.

Since there is evidence that demedicalizing birth would be healthier, why is there so much resistance by the medical profession? Rothman (1982) argues that it is because of a faulty conceptualization of women's bodies. She maintains that in our society because of women's lesser social value the activities of their bodies are compared to the male norm. For example, menstruation is seen as a complication in the female system contrasted to the reputed biological stability in the supposedly noncycling male. Pregnancy and childbirth are also seen as potentially problematic and risky processes that need constant watchfulness and can easily become pathological. According to Rothman (1982) normality must be redefined within the context of a female reproductive system. That will take a redefinition of the social value of males and females in all arenas.

The Effect of Mothering on Mothers

Several examples in this book have shown how ideas influence our behavior. Behavior can also affect our ideas. Sara Ruddick (1982) argues that when mothers participate in certain activities as part of their role in mothering they develop a point of view that she calls "maternal thinking."

> A mother engages in a discipline. That is, she asks certain questions rather than others; she establishes criteria for the truth, adequacy and relevance of the proposed answers; and she cares about the findings she makes and can act on. (Ruddick, 1982, p. 77)

Three activities compose the practice of mothering. The first is preservation, which means ensuring the physical survival of the children. The second is fostering their physical, emotional, and social growth. The third activity is making sure that the children's development occurs in a way that will make them acceptable in their society. The child must fit into and meet the needs of its society.

Ruddick's (1982) ideas on mothering are similar to those of Ann Oakley (1974) on housework. Oakley, as I described in Chapter 7, worked to reconceptualize housework as an occupation rather than an identity. Ruddick (1982) makes the same arguments about mothering. Women may come to identify themselves as mothers but mothering also includes specific responsibilities and skills. This is an important point because it reminds us that even though it is mostly women who mother, mothering is not part of our genetic makeup. Mothers are not a natural, instinctively directed group. Ruddick shows us how mothering is an occupation, a set of activities, in which people may participate. And while the occupation is dominated by biological mothers it can include men and women who are not biological mothers.

TABLE 12 • 2
Percentage of Births to Unmarried Women as a Percentage of all Births, 1970–1989

	1970	1980	1989
Black	38%	55%	64%
White	6	11	19
All	11	18	27

Source: Statistical Abstracts of the United States, 1992, p. 69.

Ruddick's (1982) assertions about the way in which the activities of mothering create a way of thinking are similar to the argument in Chapter 6 on work and families about how various occupations have different cultures. For example, I argued that one cultural value of those employed in car sales is respect for competition and individual achievement. Spending time selling cars can cause people to think about themselves and their relationships with others in ways that are consistent with their activities on their job. According to Ruddick (1982), involvement in mothering activities also creates a way of thinking.

Ruddick argues, furthermore, that because the activities of mothering are nurturant and demand a concern for the future and safety of children, the ideologies that emerge from mothering are consistent with peacemaking. Her argument implies that if mothers (not just biological mothers, but men and women who care for children) were in positions of political power, their decisions would be guided by different—less militaristic—values than the values of those who are currently in power.

This argument is similar to that of Dinnerstein (1977). Dinnerstein asserts that our world is dominated by people who have learned to repress their feelings of human connection and therefore make decisions that are inhumane. Ruddick argues that political decisions are made by people who are not mothers and are therefore less capable of creating peaceful solutions. The difference in the two points of view is that Dinnerstein (1977) roots the behavior of leaders in the psychoanalytic process and Ruddick (1982) takes a more sociological point of view, rooting it in their activities as parents.

Choosing Single Motherhood Single mother–headed households are a rapidly growing category of family (see Figure 12-1). In 1989, 58 percent of black families with children and 19 percent of white families with children were headed by single women.

One reason for the sharp increase in births outside of marriage is a decreasing tendency for single women to marry if they become pregnant. In the early 1960s 52.2 percent of unmarried women between 15 and 34 who became pregnant married before the baby was born. In the late 1980s this proportion was only 26.6 percent (U.S. Bureau of the Census, 1991).

In 1970 11 percent of all births were to unmarried mothers and in 1980 18 percent of all births were to unmarried mothers. In 1989 27 percent of all births were to unmarried mothers (U.S. Bureau of the Census, 1992). Table 12-2 shows that these numbers vary by race.

Race also differentiates the marital status of single mothers. Most black single mothers have never been married, while most white single mothers are separated or divorced. In 1991 54 percent of households headed by single black women were never-married. For whites the proportion of never-married mothers was 19 percent.

FIGURE 12 • 1
The Changes in One-Parent Families

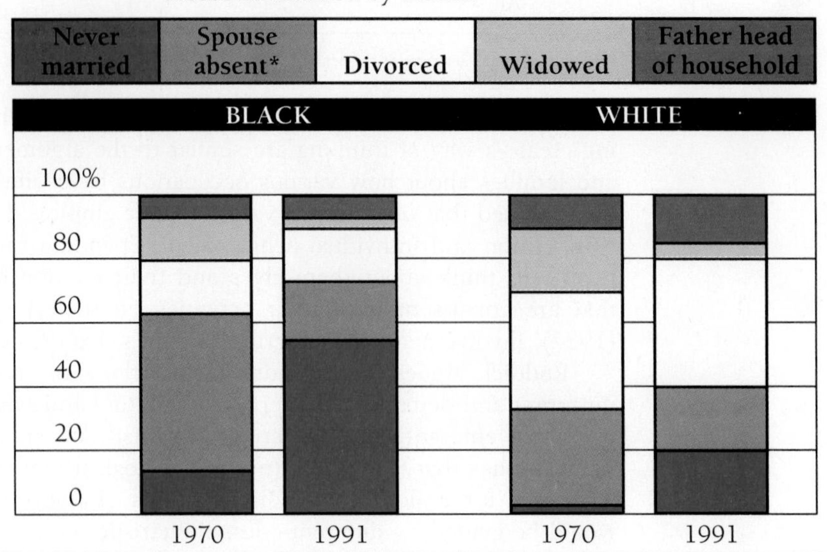

Source: The New York Times, 1992, p. A12. *Includes separated.

Poverty characterizes single mothers. In 1990 nearly half of all female-headed households lived below the poverty level (Furstenburg, 1990). Recall from Table 5-3 in Chapter 5 that single parents with young children were much more likely to be poor than married couples with young children.

Single mothers also face greater stress than other parents. McLanahan (1983) examined the Michigan Panel Study of Income Dynamics and found three types of stress in families: the presence of chronic life strains, the occurrence of major life events, and the absence of social and psychological supports. She compared two-parent families to single-parent families and found a higher incidence of major life events among the single-parent families. These included voluntary and involuntary job changes, decrease in income, moves, and illness.

One important source of stress for single mothers is much lower levels of psychological support. Others (Belle, 1982) have suggested, however, that sometimes support can involve strain as well. For example, if single mothers seek support among others who themselves are in need of material and social support, they must be willing to reciprocate with the few resources they do have.

As a result of these stresses, poor and single mothers suffer psychological effects (Mednick, 1992). These include depression, lower self-esteem, lower feelings of efficacy, and greater pessimism.

Why do women choose single motherhood? Some women do not make a real choice to be a single mother. They "choose" from limited options when they find themselves with an unplanned pregnancy and a reluctant groom or an abusive husband they wish to divorce. Others find themselves single parents through no choice of their own when their spouse dies.

Some women, however, actively choose to be single parents. Jean Renvoize (1985) interviewed women who had chosen to be single mothers. Many of the women were active in an organization called Single Mothers by Choice that worked to support single mothers, to educate the public about the validity of single-parent families, and to help women thinking about becoming single mothers to make their choice. One woman explained to her why she had made the choice: "I felt that I could go through life without being married, I could be fulfilled without having a man in my life, but I knew I couldn't be fulfilled without at least having experienced a pregnancy and raising a child" (Renvoize, 1985, p. 91).

Regardless of the way in which people became single parents, those who move from being parents in two-parent households to being single heads of households face a difficult transition. Polly Fassinger (1989) investigated the experience of women and men who became single parents when their marriage ended by divorce. She was interested in finding out how the type of marriage individuals had affected their experience as single parents.

Fassinger divided marriages into four types: segregated, modified segregated, integrated, and primarily wife-shaped. Segregated marriages were ones in which husbands and wives controlled different sets of tasks—men were breadwinners and decision makers, women were homemakers and mothers. When these couples were divorced and women found themselves single parents they responded to their new and drastically different situation with feelings of doubt and being overwhelmed but also with some satisfaction as they became more experienced making decisions in the areas that had been dominated by their husbands.

One woman in this category explained,

> I was a dependent wife. . . . My first thought was how am I going to survive? Not financially, but who is gonna make my decisions? . . . But when he was gone, I wondered, "Now, how is this gonna work? Where is my decision maker?" I wasn't prepared for it. (Fassinger, 1989, p. 170)

Men from segregated marriages who became single parents after a divorce felt confident and claimed they saw little difference in their role.

Modified segregated households are similar to segregated ones except the wife played the role of a junior partner. The experience of the single parents from these households was similar to that of the parents from segregated marriages. The wives had feelings of self-doubt and uncertainty but much less than wives from segregated marriages. Husbands in this group said they did not think there had been much change, although they were slightly more likely to acknowledge the loss of aid from their partners.

Integrated marriages were those in which husbands and wives shared relatively equally in tasks and decision making. Fassinger's sample did not include any wives who had come from this kind of marriage. The single men said that the transition from an integrated marriage to single head of household was significant. They were not sure at first if they would be able to handle it and regretted not having someone with whom to check their decisions. They were the most hesitant and doubtful group of single fathers.

One man in this group said,

> I think the worst part of that is being wholly responsible for the decisions and nobody really to try out those decisions on. To test them out and talk about them. . . . That's really one of the tougher things. That really wears you down after a while. It's always on you. (Fassinger, 1989, p. 176)

The fourth category is the primarily wife-shaped marriage. Although husbands in these marriages retained significant influence and veto power, the wives were active decision makers and handled finances. There were no men in Fassinger's sample that came from this type of marriage. The women who had come from primarily wife-shaped marriages felt little disruption in their transition to single parenthood and expressed satisfaction and enjoyment in their new role.

A woman in this category described her experience as a single parent:

> I think I felt real stifled living with this person. So many times when I would suggest something it was laughed at or, "We can't do that." . . . Now all that's not there, I don't have to worry about it. And if I want to do something, I do it. (Fassinger. 1989, p. 172)

Fassinger concludes that gender combines with marriage type to create patterns of reactions to single parenthood. Gender is a key factor in the organization of marriage and together with the variation of marriage types has a critical effect on the experience of the divorced women and men who move from those marriages into single-headed households. Women from segregated marriages and men from integrated marriages have remarkably similar reactions to that transition. And women and men from segregated marriages have dramatically different reactions.

Lesbian Mothers Many lesbian mothers are also single mothers but lesbian mothers sometimes avoid the stresses of single parenthood when they share parenting with another woman. Lesbian mothers, regardless of whether they are single parents or not, face problems because of homophobia.

Homophobia is the hatred and fear of homosexuals and lesbians. Homophobia is often expressed in the courts. In divorce and custody decisions lesbian mothers must face assertions that they are not adequate mothers even though research indicates they are similar to other mothers and they appear, in fact, to be more child-centered. Lesbian mothers are likely to lose custody in divorce cases (Falk, 1989; Robson, 1992).

Lesbian mothers and their children also face overt hostile remarks and ostracism by homophobes in other settings such as schools and the community. Rohrbaugh (1979) points out, however, that such adversity is not necessarily all bad because it allows opportunities for the children of lesbian mothers to learn about standing up for one's convictions.

Teenage Mothers Between 1955 and 1988 the adolescent fertility rate dropped by 41 percent (U.S. Department of Health and Human Services, 1991). That's right, the proportion of teenagers who bore children has declined in the

TABLE 12 • 3
Birth Rates for Teenage Mothers, 1955–1988, by Race

	1955	1960	1965	1970	1975	1980	1985	1988	% Decline
White	79.1	79.4	60.6	57.4	46.5	44.7	42.8	43.7	45%
Black	167.2	156.1	144.6	140.7	111.8	100.0	97.4	105.9	37%

Source: U.S. Department of Health and Human Services, 1990, *Vital Statistics 1988*, pp. 7–8.

past four decades. Table 12-3 shows the decline over the years 1955–1988. The birth rate is the number of live births per 1,000 women. Table 12-3 shows the birth rate for all women 19 years old and younger by race. The rate is higher for black women. The rate has declined for both white and black women.

At the same time that the birthrate has been declining, concern over adolescent pregnancy and childbearing has grown.

> Between 1975 and 1985 Congress created a new federal office on adolescent pregnancy and parenting; 23 states had set up task forces; the media had published over 200 articles, including cover stories in *Time* and *Newsweek*; American philanthropy had moved teen pregnancy into a high priority funding item; and a 1985 Harris poll showed 80% of Americans thought teen pregnancy was a serious problem. (Luker, 1992)

Why is this so? Why is concern growing while the incidence of teen pregnancy is declining? There are several different points of view on why teen mothers are seen as so problematic. Some scholars maintain that health problems are more common in pregnancy and childbirth for young women. "Compared to mothers who have babies later in life, teen mothers are in poorer health, have medically more treacherous pregnancies, more stillbirths and newborn deaths, and more low-birthweight and medically compromised babies" (Luker, 1992, p. 164).

But this view has been questioned by some. First, they argue that solid evidence of a link between health problems and youth is lacking because few studies have been done (Lamb & Elster, 1986), and those that have tend to overstate the negative aspects of early motherhood (Furstenberg, Brooks-Gunn, & Morgan, 1987).

Second, they maintain that the health problems that have been associated with young women having babies are more likely a result of socioeconomic factors than age. According to this perspective, poverty, not youth, should be blamed for most health problems of teen mothers and their babies. Carlson, et al. (1986) explain that research indicates that when young mothers are provided with good nutrition, prenatal care, and pediatric care the outcomes for them and their children are good. And in other research (Furstenberg, Brooks-Gunn, & Morgan, 1987) that compares younger and older mothers and finds better outcomes among those who are older, when socioeconomic variables are considered the differences are greatly reduced or disappear (Phoenix, 1991).

A third critique of the argument that teen childbirth is medically problematic is that positive outcomes for younger mothers have been ignored. Arline Geronimus, for example, reported to the American Association for the Advancement of Science that at least some health problems may actually be greater for women beyond their teens. Her research shows that if poor women delay childbirth they risk greater health problems and possible sterility. Furthermore, Geronimus notes that social support from the pregnant teens' mothers on whom they rely for help in raising their babies may not be as accessible if those grandmothers are older (Charlotte Observer, 1990).

Fourth, even if health problems are a result of youth, we are still left with the question of why teen pregnancies are perceived to be so much more prevalent and problematic in the 1980s and 1990s than they were in previous decades. Elise Jones (1986) has examined that question.

She argues there are five reasons why concern emerged and has grown. First, because of the baby boom, in the 1960s and 1970s teenagers made up a large proportion of the total population and anything they did became a matter of controversy and concern.

Second, while the proportion of teens who bear children has declined, it has not declined as rapidly as the fertility rate for nonteens. This means that of all the babies born in the U.S. the proportion of babies that are born to teenage mothers has increased. Furthermore, compared to other industrialized Western countries, the rate of teenage pregnancy in the U.S. is much higher.

Third, pregnant teenagers are more visible because until 1972 they could be legally expelled from school. Since 1972, however, it has been illegal to expel pregnant students and therefore they remain more active and visible in the community.

Fourth, unmarried teenage mothers are now more likely to keep their babies rather than to have them adopted. This also makes their fertility rate more visible.

Fifth, while the birthrate among teens has declined in the past forty years, the rate of pregnancy has not. For example, the rate of pregnancy among women age fifteen to nineteen rose from 82.1 per 1,000 women in 1973 to 96 per 1,000 women in 1981. The growing gap between pregnancy rate and birthrate is explained by an increasing number of abortions for young women. About one-quarter of all abortions performed on women in the U.S. are for those under the age of nineteen (Basow, 1992).

Finally, fewer teen mothers are getting married. In the 1950s many teenagers became pregnant but were quickly married, which "hid" them. Unmarried teen mothers are no longer hidden. In 1970 one-third of teen mothers who gave birth were single. In 1986 almost two-thirds were single (Luker, 1992).

Ann Phoenix (1991) argues that it is this factor that is most responsible for making teen childbirth a problem. She asserts that the difference between teen pregnancy in the 1950s and in the 1980s is that it makes more visible the sexual activity of young *unmarried* women. This creates a moral dilemma, especially for growing conservative forces that define legitimate sex in a narrow way.

In addition, because young single women are less likely to have access to jobs that will allow them to earn enough to financially support their children, they make visible another politically difficult question, "Who should be responsible for poor children?" In Chapter 4 we observed that poverty is a growing problem in the U.S. and that young people are especially likely to be poor because they are attempting to enter the labor market during a time when jobs are diminishing.

Young mothers are trapped in the position of being poor *and* responsible for children, a combination that is condemned. The debate continues between

The incidence of births among teens has actually decreased in the past 40 years, but concern for this issue has increased. Part of the reason is that teen mothers were less visible because women who became pregnant were not allowed to attend public school until the 1970s. Now they are allowed to continue their education, and teen mothers are often seen in high schools.

those who advocate eliminating their poverty and those who advocate eliminating their childbearing. The new federal office on adolescent pregnancy, the numerous state task forces, and media articles are examples of the former. They maintain that young women's childbearing must stop. They suggest policies that punish young women who bear children by placing their children in foster homes and preventing the mothers from obtaining welfare (Luker, 1992).

On the other side of the debate are those such as welfare rights organizations and the young women themselves (Phoenix, 1991), who maintain that the elimination of poverty should be the target. They argue that sex education and contraceptives should be provided to those who prefer not to have babies. But young single women should be allowed to have children and the health care and sustenance for those babies and mothers should be guaranteed. Furthermore, the structural factors such as low-wage jobs and unemployment that make young women unable to support themselves should be addressed (Williams & Kornblum, 1991).

Delaying Childbirth At the other end of the spectrum from the teen mothers are the increasing numbers of women who are delaying childbearing. In 1976 19 percent of all births were to mothers over thirty. In 1988 33 percent of all births were to mothers over thirty (U.S. Bureau of the Census, 1989a).

Figure 12-2 shows the proportion of women who have given birth by the time they were 30. The dates at the bottom of the figure indicate the year of the woman's birth. The figure shows that women who were 30 by 1990 have one of the lowest birth rates in recorded history in the U.S., with 69 percent of them having borne a child.

Those women who delay childbirth, when compared to other mothers, have more education, more prestigious jobs, higher incomes, are more likely to have

FIGURE 12 • 2
Percentage of Women Having Their First Birth by Age 30, by Year of Woman's Birth, 1890–1959.

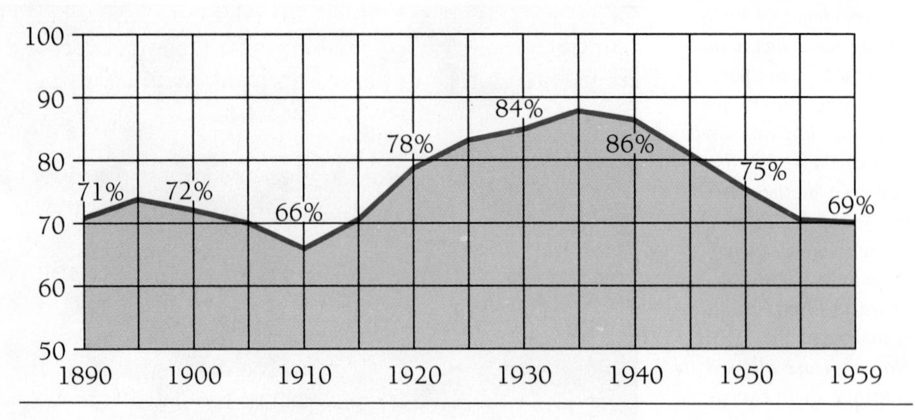

Source: U.S. Bureau of the Census, 1992. *Census and You*, Vol. 27, 2/92, p. 7.

planned the pregnancy, have fewer children, and spend more money on their children (Baldwin & Nord, 1984).

When the larger proportion of "older" women bearing children and young women bearing children are taken together, a trend becomes evident. Mothering has changed in its link to a particular age. Women as a group appear to be extending their childbearing over a greater period of time. This change in itself is unsettling to those whose ideas about motherhood limit it to a specific age range—early twenties to early thirties. When these mothers are poor or unmarried it is viewed as even more problematic.

High Tech and Mothering A central theme of this book is the recognition of the importance of social context on the organization of families. The impact of advanced technology on mothering is a dramatic example of this relationship. For many centuries people have used contraceptives and performed abortions, which are examples of the application of technology to mothering. In recent years, the use of these applications, which are now called new reproductive technologies (NRTs), has been stepped up.

NRTs can be divided into three categories: (1) those that inhibit the development of new life; (2) those that monitor it; and (3) those that create it (Achilles, 1990). The inhibition of new life includes contraception, abortion, and sterilization. The general character of technology used for this purpose has not changed dramatically although recent applications are often more effective and efficient. Some are also more dangerous, like the pill and some IUDs.

NRTs that monitor new life include ultrasound, amniocentesis (sampling fetal cells by extracting some of the amniotic fluid), chrionic villi sampling (sampling fetal cells from the pregnant woman's cervix to determine abnormalities and the sex of the fetus), fetal monitoring (monitoring fetal heartbeat, especially during labor), and fetal surgery (Achilles, 1990).

For whose benefit these techniques will be used is a thorny question. Surveys of physicians in the U.S. and Canada show support for court-ordered obstetrical procedures and they are being ordered in increasing numbers (Kolder,

Gallagher, & Parsons, 1987). In addition, even when they are not ordered by someone other than the mother, these techniques serve to further medicalize the event and to transfer control from mothers to medical technicians and physicians. And as was discussed earlier in this chapter in the section on childbirth, this can be problematic.

The third type of NRT, concerned with the creation of life, is the newest and has produced the most difficult ethical and political questions. These techniques include alternative insemination (sperm is inserted into a woman's vaginal canal by some means other than sexual intercourse), sperm banks, surrogate motherhood, in vitro fertilization (an egg is fertilized outside of a woman's body and then implanted in her uterus), frozen embryos, and surrogate embryo transfer (Achilles, 1990). In the discussion of Baby M a little later in this chapter, we will review one case illustrating some of the difficulties associated with one example of this type of NRT, surrogate motherhood.

Advances in NRTs' ability to overcome infertility can also create problems for the "infertile" individual or couple. Infertility affects up to 15 percent of couples in the U.S. and Canada and it appears to be a growing phenomenon. In one-third of the cases the infertility is attributed to the man, in one-third to the woman, and in one-third to either unknown causes or those shared by the couple (Achilles, 1990).

The procedures sound straightforward, technical, and unproblematic. The social meanings of the activities and their product, however, are much messier (Stanworth, 1990). For example, the Catholic Church considers alternative insemination adulterous. Others may not take this position but are troubled by the way in which sperm donors are chosen. Currently, sperm donors are actually sperm vendors who may sell their sperm for about $25 per ejaculate. One vendor in Achilles' (1990) research had donated 240 times. In the distribution of the sperm, screening for genetic diseases was inadequate, there was little concern for incestuous matings, and recordkeeping was haphazard (Curie-Cohen, Luttrell, & Shapiro, 1979).

A second set of issues arises from the way in which these NRTs redefine infertility. A woman who in previous years would have been told she could not have children is now told she might be able to if she is willing to go through sometimes enormous physical procedures. Some women may welcome any opportunity to conceive. Others may feel pressured to try to overcome infertility. Some of the procedures used include hysterosalpingograms (dye is injected into the fallopian tubes and uterus), endometrial biopsies (surgical removal of samples of tissue from inside the uterus), and drugs to regulate ovulation or sustain pregnancy (Achilles, 1990). In vitro fertilization involves taking drugs to stimulate egg production, daily blood tests, pelvic examinations, ultrasound, and when the technique is successful, an increased number of multiple births.

Third, NRTs create problems for children. The product of the procedures is a child. Since the relationships among surrogates, gamete donors, and recipient donors is a financial one, the child is a commodity. If the commodity is "imperfect," "damaged," or "defective" the consumers may find it unacceptable and the child may be abandoned by all parties. Such a case was documented when a handicapped child was born to a surrogate mother (Rosenblatt, 1983).

What About Fathers?

"Like mothering, fathering should be thought of as a social relationship" (Rothman, 1989a, p. 536). In the discussion on Baby M later in the chapter, Rothman (1989a) explains her assertion that the link between children and parents should not be regarded as primarily a genetic one. She argues that it is our social relationship with children that creates parenthood. For biological mothers this relationship begins with the physical contribution women make to their children's lives in pregnancy and birth. For fathers it begins with the care a father provides for his child through his care for the pregnant mother and for the child after it is born. Fathering, according to Rothman (1989a), cannot be defined as impregnating a woman who will then raise the child. Fathering must involve "doing the work of attentive love" (Rothman, 1989a, p. 537).

Bringing Fathers into Parenting Many people share the opinion expressed by Rothman (1992) above that men should be more involved in childraising (Rotundo, 1985; Pleck, 1987). Pleck (1987) notes that ideas about fathers have changed over time and recent images of fathers include a stronger role for fathers in the care for their children. According to Pleck, in the late nineteenth and early twentieth centuries the dominant image of the father was as distant breadwinner. Then from 1945 until the mid-1960s fathers played the role of sex role model. Pleck (1987) maintains that since 1966 a new model of father has emerged: father as nurturer.

Ralph LaRossa (1992) reminds us that this change has been more a change in the culture of fatherhood than the conduct of fatherhood. By this he means that on the average ideas and rhetoric about fathering have outstripped real behavior changes. In Chapter 7, for example, we discovered that women are still responsible for the majority of unpaid domestic work, including childcare.

Lamb's (1987) work measures the changes in the last two decades in the amount of time mothers and fathers spent with their children. Lamb defines engagement as "time spent in one-on-one interaction with a child (whether feeding, helping with homework, or playing catch in the back yard)" (LaRossa, 1992, p. 524). According to Lamb (1987), between 1976 and 1981 the amount of engagement time increased among fathers by 26 percent while it increased for mothers by only 7 percent. It is necessary to look at the actual number of hours for mothers and fathers in order to get a complete picture, however. The number of hours of engagement time for fathers went from 2.29 to 2.88 hours per week. For mothers, engagement time went from 7.96 to 8.54 hours per week. For both mothers and fathers this amounts to an increase of about 5 minutes per day and mothers still spend much more engagement time with their children than fathers do.

LaRossa and LaRossa (1981) had similar findings in their interviews with fathers. Their work showed that fathers' level of engagement with children was smaller than mothers'. They also found that fathers spend a larger proportion of their engagement time playing with their children, while mothers spend more time in caregiving activities. The kinds of play that fathers engaged in often were those that could be carried out in a semi-involved manner, such as watching television together. Finally, they found that fathers sometimes viewed their involvement with their children as a job, an activity in which they had to participate, and they prided themselves in discovering ways to make childcare as noninterfering as possible with their other activities. For example, one father explained that the hours for which he was responsible for his infant son were hours that the baby slept and he could get two hours of his own work done while watching the baby sleep.

Our expectations about fathers are increasingly calling for greater involvement of men in all kinds of care for their children. The behavior of fathers, however, sometimes falls short of those expectations. What happens when ideas about fatherhood and the experience of fatherhood are disparate? LaRossa and LaRossa (1981) refer to this as asynchronous social change. They argue that the result is mixed messages for children who are spending more time with their fathers but who may become aware of the burdensome joblike character with which their father views that time. Marital conflict and guilt-ridden fathers are also problematic results.

New Fathers in Word and Deed Nevertheless, in some households men have attempted to bring their conduct as fathers in line with their beliefs and have begun to play an important part in the raising of children. Diane Ehrensaft (1987) was interested in investigating their experience.

Ehrensaft (1987) defines mothering as an activity having two components in which men and women can participate. First, it involves the activity of the day-to-day primary care of a child. Second, it includes a consciousness of being directly in charge of the child's upbringing. She chose a group of heterosexual couples with children who shared mothering and asked them about their experience.

She found that shared mothering is difficult to implement. Both men and women find there are barriers to shared mothering within themselves and from the outside.

Men are rewarded for mothering by their increased access to children. The tradeoffs are difficult, however. Men who mother must give up the privilege of being the distant father and they must do work like changing diapers that is regarded as debasing in our society. Fathers who mother are also likely to be responded to negatively by others who feel they are behaving in an unmanly manner or shirking their real duties when they are caring for their children.

Women benefit from relinquishing some of their reponsibilities for mothering because it allows them more time to participate in other activities. Women may also find, however, that sharing mothering is difficult for them as well. First, although they are able to share the physical tasks of mothering they may still feel guilty for not doing them. Second, they may discover that mothering is an important part of their identity and if they are unable to establish themselves as successfully as they would like in "extramother" activities like paid work, they may feel caught between identities, without an arena of success.

Ehrensaft claims that despite these difficulties shared mothering is a worthy goal because it can have important positive effects on parents, children, and our society. She lists seven specific results of having both women and men mother:

1. Liberates women from full-time mothering.
2. Affords opportunities for more equal relationships between women and men.
3. Allows men more access to children.

4. Allows children to be parented by two nurturing figures and frees them from the confines of an "overinvolved" adult who has no other outside identity.
5. Provides new socialization experiences and possibly increases the likelihood of less gender-stereotyped behaviors and ideologies in children.
6. Challenges the myth that women are better equipped biologically for parenting.
7. Puts pressure on political, economic, and social structures for changes such as paternity and maternity leaves, job sharing, and freely available childcare facilities (Ehrensaft, 1987, p. 45).

Gay Fathers It has been estimated that 10 percent of the U.S. population is gay (Barret & Robinson, 1992). Bozett (1993) estimates that between 1 and 3 million gay men in the U.S. and Canada are fathers. In addition to those gay men who are married natural fathers, single gay men and gay couples have adopted children or have used alternative insemination and surrogate mothers to have children.

There are few studies of gay fathers. Bozett (1988) is one of the few researchers to have interviewed gay fathers about their experience. His work indicates that gay fathers try hard to maintain good relationships with their noncustodial children. Like lesbian mothers, gay fathers face the problem of homophobia. In addition, gay fathers are less likely than lesbians to have support from the gay community (Basow, 1992).

Bozett (1984) interviewed children of gay fathers to examine their experience. He found that while the children accepted their father's homosexuality they worried others would think they were gay because their father was. The children used a number of strategies to address this dilemma. Some children attempted to control their father's behavior, for example, by not allowing him to appear with a lover in their presence. Others would not allow certain friends to meet their father. Bozett also found that gay fathers frequently tried to protect their children by avoiding disclosure of their homosexuality or advising their children to refer to the father's lover as uncle or housemate. Some fathers who had custody of their children placed them in schools outside of the neighborhood in order to give the children more privacy. The fathers Bozett interviewed, however, were also concerned that their children understand that although the wider society disapproves of homosexuality it is not a negative attribute and that the gay fathers are as moral and virtuous as other men.

Single Fathers Much of the literature on fathers compares fathers to mothers in two-parent families. One in every six single-parent household is headed by a father. Although the percentage is small—2.6 percent of all households—the numbers are substantial—1.4 million households. In addition, single father–headed households is a rapidly growing family type. The number has grown faster than either single-mother families or two-parent families since 1959. Over that time period, single-father families grew 300 percent, mostly since 1973 (Meyers & Garasky, 1991).

What happens when fathers are left to parent on their own in single-parent households? Barbara Risman (1987) investigated this question using a questionnaire survey examining parenting in four family types: single mothers, single

Gay couples are increasingly raising children. Parents who are gay face special problems because of the homophobia in our society.

fathers, two-parent households where the father was in the paid labor force and the mother was not, and two-paycheck households.

Risman (1987) measured variation in parenting by looking at three factors: time spent in housework, parent-child intimacy, and overt affection. She found that single fathers, single mothers, and housewives spent more time than married fathers or married employed mothers on housework. Single fathers did not hire others to do the work for them and were forced to take responsibility for these household tasks.

The second factor, parent-child intimacy, was measured by how often children shared their emotions—sadness, loneliness, anger, happiness, pride—with their parent. Here she found that femininity was the most important predictor of parent-child intimacy. Parents who considered themselves more feminine on a range of issues were more likely to report higher rates of parent-child intimacy. This was true regardless of the sex of the parent. Fathers who were more feminine (according to personality scales) were more likely to report parent-child intimacy than mothers or fathers who were less feminine. Single fathers reported levels of femininity similar to those of employed mothers, suggesting that the activity of parenting may create different expressions of personality, which in turn help fathers to build parent-child intimacy.

The third factor was overt affection, which was measured by the amount of physical contact—hugging, cuddling, wrestling—children had with their parents and the kinds of interactions that took place when the parent and child were alone with each other. This factor was affected by the sex of the parent and the parental role. Mothers, regardless of household type, reported displaying more overt affection. But parents in two-paycheck families were more likely to report overt affection than single mothers or fathers and married fathers in one-paycheck families. In this case both single fathers and single mothers report less overt affection than married mothers in two-paycheck families.

Risman concludes that the activities of parenting have an important effect on the behavior and personalities of people who do that parenting. Fathers who become primary parents are capable of caring for their children and they begin to behave similar to women parents. LaRossa and LaRossa (1981) argue that even if men (and women) have the best intentions of sharing parenting, their intentions remain largely unexpressed when women are available as primary caregivers. Risman suggests that the change cannot begin with ideas but must be socially structured in ways that insist/allow men to become primary caregivers as well.

Childless by Choice

"Virtually all societies are, of necessity, pronatalistic to some extent, in that parenthood is normative in all societies, and is generally defined as a moral imperative for married persons" (Veevers, 1980, p. 5). The importance of parenthood in defining one as a well-adjusted adult is most salient for women.

J. Veevers (1980) argues that the negative meanings given to remaining childless can be divided into three categories: the diagnostic model, the deprivation model, and the labeling model. He describes the diagnostic model as one in which a person who chooses to remain childless is by virtue of that fact alone perceived to be maladjusted.

The deprivation model maintains that people without children are not maladjusted because they have chosen childlessness. They become maladjusted, however, because they are not allowed the experience of having children. Having children is argued to be a critical factor in developing full emotional and sexual maturity.

The third model, labeling, claims that people who choose to remain childless are not necessarily maladjusted and that the experience of having children is not necessary to fully maturing. Remaining childless, however, does put a person at a serious social disadvantage because of the way in which others perceive childless people.

In our society we are surrounded by messages that say we must bear children in order to be normal adults. We are also surrounded by people having children. In 1990 84 percent of all women in the U.S. between the ages of 40 and 44 had borne at least one child (U.S. Bureau of the Census, 1991). A substantial number of people, however, still remain childless by choice. What is their experience?

Veevers (1980) interviewed 156 people, married women and men who had deliberately avoided having children, to see what their experience had been. He looked at the factors involved in their decisions, the pressures they felt, the way in which they coped with pronatalism, and the character of their childfree marriages.

The couples Veevers studied followed two paths in their decision not to have children. In one-third of the couples at least one person had decided before marriage that he or she did not want to have children and an agreement was made between the man and woman as part of the marriage "contract." In two-thirds of the couples, the decision was made after marriage in a series of postponements.

Those who postponed having children moved through stages. First they postponed pregnancy for a specific period. Then they postponed it indefinitely. Next they deliberated the pros and cons of parenthood, and finally they accepted permanent childlessness. Four factors tended to accelerate moving from one stage to the next. One factor was pregnancy scares. An unplanned pregnancy scare made people assess their real feelings about having a child. When the wife determined that she was not pregnant after all and was relieved to discover this, the couple was encouraged to decide more permanently to remain childless.

Adoption was a second factor. As the wife became older and the couple feared they would need to make a decision, the possibility of adoption emerged as a way of further delaying childbearing. Veevers (1980) found that most couples did not seem to have much knowledge about adoption as a real option. Symbolically, however, the possibility of adoption allowed them to decide not to have children when they could claim they had not really made that decision yet.

The third factor was achievement of goals originally set as criteria for having children. Some couples set goals, such as achieving some status in their career or becoming financially secure, as the point at which they would have children. When they reached that point but still preferred not to have children, they moved to the stage of indefinite postponement.

The fourth factor was concern over the stability of their marriage. Some of Veevers's subjects told him they were motivated to increase their commitment not to have children because they were not sure their marriages would last. This issue was less evident than the other three, although it was expressed by a few couples.

Couples faced many outside forces attempting to compel them to have children. Pressure to have children was especially intense in the first six or seven years of their marriage. Their sexual ability was questioned, especially the husband's. They were required to explain themselves to friends, physicians, acquaintances, and would-be grandparents. Childless couples found some ideological support in maintaining their lifestyle from organizations that were concerned with population problems, such as Zero Population Growth.

What are relationships like between husbands and wives in childless couples? Veevers (1980) found that the most salient feature of these marriages was the intensity of the relationship. And although Veevers had no way of determining whether these marriages were happier than those with children, the people perceived them to be happier. Veevers also noted that the childless couples he interviewed had more egalitarian relationships.

IS BIOLOGY DESTINY?

Thus far we have examined a number of issues related to parenting. Gender appears to be an important factor in determining the character of that experience. The question of why gender is so salient is an important theoretical issue. Are differences in mothering and fathering biologically rooted? In this section we will review the theoretical debate around this question.

This book has presented much evidence that families vary by race, class and historical period. The idea, however, that some features of families are natural, unchanging, invariable, and essential is a powerful one in our culture. Thinking about parents and children is one issue that seems especially vulnerable to assumptions that some aspects of families are not socially determined. Perhaps lingering in your mind is the belief that there is something intrinsic to the relationship between parents and children that cannot be altered by time or place or the society in which that relationship exists. This belief is called essentialism and is tied to a belief in biological determinism.

Recall from Chapter 8 that essentialism is a philosophical viewpoint that posits that there are certain phenomena that are essential; they cannot be omitted or altered but must be accepted as a given. In the case of parenting, essentialism is the belief that biological parents have a biologically determined (or at least partly biologically determined) social relationship to their children.

Essentialists argue it is in our genes for biological parents to care for their young and if they do not our species is threatened. Because males and females are biologically different, within this framework they have different biologically determined roles they must play in raising the next generation. In sum, this point of view asserts that biological mothers must raise children because they are physically best equipped to do so and that if women do not take this responsibility the survival of our species is at risk.

In sociology, this argument is put forward by sociobiologists (Wilson, 1975). Alice Rossi has been an important figure in the sociobiological view of parenting. Rossi (1964) was one of the early writers in the Second Wave feminist movement in the late 1960s. In 1977 Rossi published an article that was met with strong criticism from other feminist scholars (Breines, Cerullo, & Stacey, 1978).

Rossi (1977) argued that the responsibility of women for childcare and cooking are "the products of an innate physiological disposition supported by natural selection and a long evolutionary heritage" (Breines, Cerullo, & Stacey, 1978, p. 46). In other words, women have been genetically programmed to be effective parents and cooks.

A second factor in Rossi's argument is that men have not been biologically programmed to be as proficient in these activities. Regardless of social factors, therefore, men cannot be as good at parenting as women can. She states, "Unisex socialization has proven insufficient to remove whatever sex differences currently exist" (Rossi, 1977, pp. 4–5).

Third, Rossi asserts that because human babies have such a prolonged dependency on their parents, they require the formation of strong bonds between themselves and their parents in order to survive and flourish psychologically and physically. Therefore, it is necessary for mothers, because they can best provide the relationship infants need, to take responsibility for this essential activity. Childraising cannot be left to the inferior parents, fathers.

Rossi's work, for the most part, ignores men (Breines, Cerullo, & Stacey, 1978). Other than being inferior parents they are not a problem, in fact they are quite invisible. Other sociobiologists (Wilson, 1975; Barash, 1977) have proposed a theory about men as well as women.

Sociobiologists argue that because they are biologically different, men and women have a different way of contributing to the perpetuation of the species. Women are genetically programmed to be careful when selecting a mate and they need only have sex every couple of years in order to best preserve their genes.

Men, on the other hand, are genetically programmed to invest little in the raising of children because they are not as skilled. Instead, men must go for quantity. According to sociobiologists the best mating strategy for men is to impregnate as many women as possible.

> For males a different strategy applies. The maximum advantage goes to individuals with fewer inhibitions. A genetically influenced tendency to "play fast and loose"—"love 'em and leave 'em"—may well reflect more biological reality than most of us care to admit (Barash, 1977, p. 48).

What Is Wrong with the Sociobiological Model?

Ruth Bleier (1984) is a biologist who has carefully analyzed the sociobiological model and discovered three problems with its premises. First, sociobiology presumes certain human behaviors and certain differences between males and females are universal, when in fact they do not exist in all times in all societies or even among all people in a particular time and place. The behaviors that sociobiologists define as masculine, like competitiveness, aggression, and selfishness may be an important part of men's character in white upper-class American and British society but they are not important in many other cultures (Bleier, 1984; Sydie, 1987).

A critical issue for the argument Rossi makes about mothering is the variation of the separation of childbearing from childcare. Although it is always women who bear children, there is enormous variation in the role that fathers, mothers, and others play in childrearing.

Second, sociobiologists often call on research on other animals as evidence for their arguments. For example, Rossi (1977) cites evidence from primate studies of maternal deprivation. This practice is questionable for several reasons. It is impossible to show how translatable one set of behavior is from one species to another. How significant is it to explaining human behavior if we show that gorilla females are more likely to take care of their young than gorilla males are? Even though primates are our closest contemporary relatives, the first hominids split off from apes 5 to 15 million years ago and both branches continued to evolve over that period. The species from which *Homo sapiens* evolved are extinct. Making comparisons even with our nearest relative involves a huge evolutionary leap. Sociobiologists, however, are often quite unconcerned with even larger leaps, like comparing humans to birds. For example, among mallards, a female is chased by several males who then appear to roughly jump on her and "force" her to allow them to copulate. This behavior has been cited by some sociobiologists as evidence of rape among animals and therefore evidence of the naturalness of rape by male animals, including humans (Judd, 1978).

Another problem with extrapolating from the behavior of one species to another is the arbitrariness with which examples are chosen (Bleier, 1984). For example, among the silver-backed gorilla a male may "dominate" the movement of the troop. In other species, however, like the Japanese macaques, females form the core of the hierarchy. And in many primate species dominance hierarchies do not exist at all. Sociobiologists are especially prone to omitting species that do not fit their model.

Using data from other species is also problematic because it assumes that animals do not have cultures and do not socialize their members. Perhaps part of a gorilla's behavior is a result of what it has learned (Bleier, 1984).

Another problem with the practice of comparing humans to other species is that the behaviors that are identified in some members of another species may not be universal within that species (Bleier, 1984). For example, animals may have different ways of handling childcare or territory disputes in different communities. This is particularly important because a number of studies cited by sociobiologists are based on research on animals held in zoos, which subsequently have been determined to behave quite differently in the wild (Lewontin, Rose, & Kamin, 1984).

The third area Bleier (1984) identifies is the problem of separating biology from other factors. She says it is not possible to tease genetic and other factors from learning and environment in human behavior. From conception, humans are affected by their environment. For example, pregnant mothers can respond to social factors by different hormonal output, which can then affect the fetus. Or mothers may have access to food or health care depending on their social situation, which can affect their children even before birth. And after birth, Bleier (1984) declares, biology and social and physical environment are not polar opposites to be unwound. They are fused in the lives of humans.

If Biology Is Not Destiny, Are Men and Women Identical as Parents?

Liberal feminism is a theoretical position that takes the opposite position of sociobiologists. Liberal feminists posit that inequality between women and men "results from the organization of society, not from any significant biological or personality differences between women and men" (Lengermann & Niebrugge-Brantley, 1992, p. 462). The goal of liberal feminism is to make the world see how similar women and men are and to organize ways in which to treat them equally. For the liberal feminist,

> A woman lawyer is exactly the same as a man lawyer. A woman cop is just the same as a man cop. A pregnant woman is just the same as . . . (Rothman, 1989a, p. 248)

Rothman's quote illustrates a problem in liberal feminism. In nearly every human activity, women and men are the same, or the differences between one woman and another woman are as great as or greater than they are between a woman and a man. But pregnancy and childbirth make some women different. And pregnancy and childbirth have an especially important effect on parenting.

Both women and men contribute equal amounts of genetic material to a fetus, but it is only the female's body that contributes nine months of gestation and the labor of giving birth. Rothman (1989a) argues that this difference is important and should play a role in legal battles, like that in the case of Baby M.

Rothman tells the story of Baby M to explain her ideas. Bill Stern wanted a child that came from his sperm but his wife Betsy had multiple sclerosis and had been told that bearing a child might worsen her condition. The Sterns decided to hire Mary Beth Whitehead as a surrogate mother to carry Bill's child. For a fee Whitehead allowed herself to be alternatively inseminated with Stern's sperm and agreed to give the baby to him. Whitehead, however, changed her mind after the child was born and hid herself and the baby for four months while the courts deliberated the case.

The court said that the contract between the Sterns and Whitehead was not binding. However, because the child was genetically related to Bill Stern he had the right to sue for custody. According to the court he and Whitehead were equally parents. The case was then decided on the basis of who would be a better parent: Stern, a wealthy professional, or Whitehead, a working-class wife and mother. In the proceedings, evidence was presented that Whitehead had been a go-go dancer, had dyed her hair, and had trouble with her husband (Rothman, 1989a, p. 24). The court decided in Stern's favor.

According to Rothman (1989) the mother's link to the child was incorrectly reduced to that of a father's, thereby making them equal parents. Bill Stern and Mary Beth Whitehead had an equal genetic link to Baby M. The baby was a product of equal parts of Stern's and Whitehead's genes. But Rothman argues that the fact that Whitehead had the additional connection to the child of being a pregnant and birthing mother should not have been dismissed as irrelevant. The court's decision was based on the liberal feminist assumption that men and women are equal. Even though men and women are not equal in their contribution to reproduction, the factors that make them unequal were ignored and the one way in which they are equal—their genetic contribution—was the only criterion acknowledged.

Rothman (1989a) demands that a new framework be developed that can capture the need for women to be treated equally to men while not dismissing the unique character of the relationship between biological mothers and their children. She proposes a policy in which the contribution biological mothers make in pregnancy and childbirth and the contribution fathers, mothers, and others make in caring for children after they are born is factored into equations leading to decisions about questions like custody. Specifically, Rothman suggests that policy should be based on the following rules:

> [1] Infants belong to their mothers at birth because of the unique nurturant relationship that has existed between them up to that moment . . . genetic ties will not give parental rights. . . .
>
> [2] Adoption can only occur after birth and the birth mother is given 6 weeks in which to change her mind. . . .
>
> [3] All custody cases after 6 months will be determined by the amount of care provided by the adult with joint custody if this role is fully shared. . . .
>
> [4] There is no such thing as surrogacy under this system. Every woman who bears a child is the mother of the child she bears, with full parental rights regardless of the source of the egg or the sperm. (Rothman, 1989a, pp. 254–260)

CHILDCARE

In this chapter our focus thus far has been on the micro level of parenting and how the micro level is affected by the macro level. Our attention has been on the diverse experience of parenting and how social factors, especially the factor of gender, shape that experience. The micro level has also affected the macro level. People have tried to influence the organization of parenting and the social context that affects it. One of the most direct examples of this is the political battle for increasing public responsibility for childcare.

In 1988 56 percent of mothers with children under the age of six—25 million mothers—were in the paid labor force. If the current trends continue this proportion will rise to 75 percent by 1995 (Hofferth & Phillips, 1987).

Table 12-4 shows the types of childcare parents use. The table shows that 36.8 percent of the children of employed mothers were cared for in someone else's home by relatives (12.6 percent) or nonrelatives (24.2 percent); 28.9 percent were cared for in their own homes by a relative (24.4 percent) or a nonrelative (7.5 percent); 24.4 percent were in a daycare center; and 5.6 percent were cared for by their mothers at her workplace.

Only about one-third of employed women with children under the age of fifteen report making cash payments for childcare (U.S. Bureau of the Census, 1989c). Many parents who cannot afford to place their child in a licensed daycare facility rely on unlicensed babysitters, who are paid less than a minimum wage and who do not claim these earnings on their federal income tax form. Since several aspects of these arrangements are illegal they probably are not reported and therefore the number of people reporting cash payments is less than the number who actually make cash payments (U.S. Commission on Civil Rights, 1981). Of those that say they do make cash payments, the median payment per week was $38, which equals $1,976 per year (Marx & Seligson, 1990).

The number of slots available for parents seeking daycare is insufficient in some areas, especially for infants and very young children. But availability of childcare is a less significant problems than the quality of care available and the cost (Hofferth & Phillips, 1987).

One of the problems with the quality of childcare is the rapid turnover of personnel in daycare centers. Daycare workers have a turnover rate about twice as high as the national average. This is undoubtedly related to the demanding work coupled with the low pay. On the average, daycare workers have nearly two more years of education than other workers in the U.S. but they earn about half of what the rest of the workforce earns (Hartmann & Pearce, 1989). And since the mid-1970s daycare workers have seen their wages decline from about $14,000 to $11,000 in the early 1990s. In addition, the workload for daycare workers has increased during this period with a rise of 25 percent in child-to-staff ratio (Taylor, 1991). Other issues related to the quality of care are the problem in finding childcare for sick children and the lack of an educational component at childcare centers that provide full-day care.

On the average, families spend 11 percent of their income on childcare. For poor and moderate-income families, however, this proportion can rise much higher, especially if there is more than one child in the family. For example, a single parent with two children, earning the median income for a single parent, could pay as much as 36 percent of his or her gross income for childcare (Marx, 1985).

Currently the federal government supports childcare in two ways. First, the government provides direct subsidies through Social Service block grants to states. Second, the government allows tax credit to parents who spend money on childcare and to employers of parents. Few employers take advantage of

TABLE 12 • 4
Child Care Arrangements in the U.S., 1988

Characteristic	Number of children (in thousands)	Care in child's home				Care in another home			Group care		
		Father	Grand-parent	relative	Nonrelative	Grand-parent	relative	Nonrelative	Nursery or pre-school	Day care center[1]	Mother, while working
		Percent distribution[2]									
All children[3]	13,259	12.9	6.0	2.6	7.6	8.7	2.6	21.3	23.4	7.8	4.8
Age and school status											
Under 2 years	3,772	15.4	8.8	2.5	10.2	9.7	3.7	28.1	—	11.8	6.6
2-3 years	4,609	12.5	4.7	2.3	7.5	9.3	2.8	21.1	28.8	3.8	4.5
4-5 years, not in school	3,421	9.6	3.1	2.6	4.5	7.0	1.3	14.7	49.7	2.7	3.6
4-5 years, in school . . .	1,323	16.4	9.5	3.8	8.7	7.6	1.5	19.1	—	25.7	3.3
Race											
White	10,854	13.2	4.7	2.1	8.2	7.6	2.3	22.9	23.6	7.6	5.4
Black	1,830	9.2	11.0	5.8	3.3	16.2	4.3	15.2	21.4	10.5	1.4
Hispanic origin											
Hispanic	1,352	10.1	8.0	8.1	6.7	10.2	4.9	22.7	21.1	5.1	1.9
Non-Hispanic	11,331	12.8	5.6	2.0	7.7	8.6	2.3	21.2	23.8	8.1	5.2
Family income											
Less than $10,000	1,119	12.6	6.7	5.5	5.5	13.0	3.4	13.1	25.1	6.5	5.5
$10,000-$24,999	3,635	17.9	5.7	3.6	5.5	10.2	2.9	21.9	18.2	6.0	6.1
$25,000-$39,999	3,635	13.4	4.7	1.5	6.4	9.3	2.6	23.7	22.7	9.0	4.6
$40,000 or more	3,613	8.5	4.8	1.8	10.7	6.2	1.6	22.1	28.7	8.5	3.5
Geographic region											
Northeast	2,242	17.0	8.3	1.8	10.5	9.8	2.0	17.5	20.0	6.9	4.5
Midwest	3,492	14.7	5.1	2.3	7.8	7.2	2.0	26.5	20.1	6.1	5.5
South	4,596	10.0	5.8	2.7	6.2	10.8	3.3	19.1	25.8	10.2	3.8
West	2,913	12.1	5.5	3.4	7.1	6.1	2.5	21.3	26.2	6.9	5.6
Place of residence											
MSA:											
Central city	4,035	10.9	8.9	4.2	7.5	9.9	2.3	18.9	24.1	7.3	3.4
Not central city	6,182	13.5	4.9	1.5	7.6	6.9	2.6	21.4	25.3	9.0	5.0
Not MSA	3,042	14.3	4.1	2.7	7.6	10.6	2.8	24.1	18.7	6.3	6.1
Mother's education											
Less than 12 years	1,488	16.5	6.8	9.4	6.6	10.3	3.6	15.3	19.7	3.7	4.2
12 years	5,308	13.3	7.7	2.5	6.5	9.8	3.5	21.5	21.4	7.2	4.1
More than 12 years . . .	6,446	11.7	4.4	1.1	8.7	7.4	1.6	22.4	25.9	9.3	5.4
Mother's employment status											
Employed	10,060	15.6	6.2	2.6	7.5	9.8	2.8	24.2	16.0	8.4	5.6
Not employed[4]	2,033	0.9	3.2	0.7	6.4	4.0	0.9	8.8	62.7	5.2	0.1

Source: World Almanac and Book of Facts, 1992. 1991, p. 945.

these tax incentives. Of the 6 million employers in the U.S. only about 4,000 provide any form of childcare assistance, usually in the form of referral services, information seminars, scholarships, and discounts for use of community-based childcare. Only 600 employers provide on-site daycare (Friedman, 1988).

Parents do take advantage of tax credits. The program, however, is regressive. This means that higher income families benefit more from the tax credits than low- and moderate-income families.

Thus far neither the private sector nor the federal government have come forth with adequate support for childcare but the issue is increasingly being raised as a critical one and a matter of public concern. In the 100th Congress alone childcare was the focus of 100 bills and the subject of congressional hearings, government reports, and planks in both the Democratic and Republican party platforms (Marx & Seligson, 1990).

THE MICRO-MACRO CONNECTION

Despite the popular notion of the essential character of parenting, we can see that the social context and the interplay between the micro and macro levels of the social context have a defining influence on parents. An exploration of parenting shows that the micro-level experience of parenting is affected by macro-level social structural factors of the legal system that define parents as those who are genetically linked to children. The micro level is also affected by the macro features of gender stratification, technology, essentialist ideology, and the organization of health care.

In the opening scenario in Chapter 1, Kimberly and Michael look forward to marriage and parenting with little knowledge about the kinds of limitations that will be placed on them as they begin these activities. The technology of reproduction and birth, the ideologies that promote the idea that women should be the primary parents, and limited support from the community around needs like childcare all influence their experience of parenting, the problems they face, and the possible solutions available to them.

Micro-level forces have also worked to have some effect on the organization of parenting by contesting the medicalization of birthing and developing alternative birthing methods. People have also worked to resist the negative effects of technology by proposing different ways of defining relationships between children and adults. The political battle of people at the micro level attempting to alter the macro level of government is most clear in the effort to promote legislation to bring more public support to childcare.

SUMMARY

Parents and Parenting

- The opening scenario shows Kimberly explaining her experience with the surprisingly complex, sometimes difficult, and always gendered experience of parenting.
- Becoming a parent is different from other transitions for four reasons: cultural pressure to assume the role, inception of the role, irrevocability of the role, and lack of preparation for the role.
- Scholars who study Japanese Americans add that the centrality of parent-child relationships is another important factor.
- The motherhood mystique refers to the contrast between the idealized image of mothers and the real experience.
- According to Chodorow the fact that women mother affects boys and girls differently and may be a source of some of the problematic identities of

adult men and women and their interpersonal relationships. Chodorow's view has been criticized by a number of scholars.
- Giving birth is a socially structured event in mothers' lives. In the contemporary U.S. birth is medicalized and often oppressive for women.
- The activity of mothering may have important consequences for the people who mother.
- Single mothers are a growing population. They have met with criticism if they are young and poor and challenges if they move from sharing parenting to single parenting.
- Delaying childbirth is another growing trend with larger numbers of women waiting to have their first child.
- New technologies such as alternative fertilization are another recent change in mothering.
- People increasingly believe fathers should be more involved in parenting but ideas seem to be changing more quickly than behavior.
- Some fathers, especially single fathers, are becoming more involved in childcare and these changes could result in significant changes for parents and children.
- Not all couples have children. This growing population faces strong outside pressure to have children but has developed a process to make that choice.

Is Biology Destiny?

- The gendered character of parenting creates an important and volatile theoretical question.
- Sociobiologists argue that a division of labor that makes women nearly exclusively responsible for childcare is our biological destiny.
- Critics of sociobiology have questioned the assumptions, logic, and data sources of the sociobiologists.
- If biology is not destiny, are women and men identical in their relationship to children? This question is one that is debated among liberal and radical feminists.

Childcare

- The political struggle over provision of childcare is one that affects many parents and is increasingly seen on the agendas of politicians at the federal level.

The Micro-Macro Connection

- Numerous macro-level forces such as technology, the health care system, the legal system, and gender stratification have an effect on the micro-level experience of parenting.
- Micro forces have resisted some of those effects by organizing alternative birthing and parenting and by calling on the government and business to help parents provide childcare.

Socialization is the process by which people learn what to expect from their society and how they in turn are expected to behave. Gender socialization teaches children how to be masculine or feminine.

13 CHILDREN

Mrs. Gonzalez teaches sixth grade at an elementary school in Chicago. The school year is just beginning and as a way to help the students get to know one another, she asks them to talk about their summer vacation. The school is located in a neighborhood that includes a broad range of classes and cultures.

Rodriguez quickly raises his hand. "We had a lot of excitement at our house this year. Our house burned up. Lucky for us we weren't in it. My brother and mother and me had to move in with my grandma. My mother says we're sure poor now."

Mrs. Gonzalez calls on Amy next. She says, "My brother and I both got to go to camp this summer. My father said that just because he and my mother were at work every day we weren't going to spend the summer in front of the TV.

My brother chose basketball camp and I chose to go to ballet camp. We had to practice every day and at the end there was a beautiful recital."

Neu is next. "My parents got separated this summer. I had to help. Every day when my mother went to work, I stayed home and took care of the two babies. I also had to make the dinners and do the dishes. Every few days I took the clothes to wash them at the laundromat. That was pretty fun."

"My aunt made me and my cousin go to summer school," says Jamal. "Last year I did real well in math and they made me take enrichment classes at the Afro-American Center this summer so that I could get into the advanced classes."

Mrs. Gonzalez notices Jennifer sitting quietly at her desk. Two years ago Jennifer's older sister Melissa was in her class. Melissa had confided that she was being sexually abused at home and Mrs. Gonzalez had helped their mother work with protective services to have the father removed from the home. She wonders if Jennifer has been molested as well. Mrs. Gonzalez notes to herself that she will need to make a special effort to get to know the quiet girl.

- *Which of these children have better lives, in your opinion?*
- *How do you define "better" and from where do your ideas come about what children should be like and how they spend their time? Would those ideas be different if you lived in 1900? In 1400?*
- *What different relationships do children have in families?*
- *What are all of the roles children may play in families? Are they students to be taught? Workers? Burdens? Victims?*
- *Why are families an important arena for gender socialization? What about other kinds of socialization? Are families the only place children are socialized?*
- *How do class, race, and family organization affect children's lives?*

Children's lives are diverse, frequently difficult, and always shaped by the social context in which they live. Most children grow up in a family. But the kind of family in which they live can vary by race and class and by household arrangements—who they live with and what their relationships are to those people. Some children face great difficulty, for example, if they are abused or poor. Children's lives are influenced by factors at the micro level, the relationships and interactions in which they are involved with family members and others they see day to day. Children's lives are also affected by large macro-level social structures like the media, the organization of education, the government, the economy, and the social system of stratification.

This chapter will examine the diverse and often problematic experience of children in the United States. The chapter is organized into three main sections. The first is a review of the three dominant images of children in our society: social learners, threats to adults, and victims. The second section examines two theoretical issues: the way in which children and childhood are socially constructed and the insufficient character of our current images of children. The third section introduces political activities that attempt to improve the quality of life for children.

MODERN AMERICAN CHILDREN

Barrie Thorne (1987) has examined the place of children in our society, especially their relationship to adults. She asserts that contemporary American images of the relationship between children and adults fall into three categories: (1) children as learners of adult culture; (2) children as threats to adult society; and (3) children as victims of adults.

Thorne argues that these three images fall short of capturing reality because they ignore many other aspects of children's lives. In the second section of this chapter we will examine the question of whether we need to broaden our conceptualization of children and what theoretical tools might be most useful in that effort. In this section we will focus our attention on the three dominant images of children, beginning with the image of children as learners.

Children as Learners: The Case of Gender Socialization

The first image of children is that of learners of adult culture. Families are an important site of this learning activity, which is called socialization. Socialization is the process by which people learn to behave in an acceptable manner in their society—they learn what behaviors are considered appropriate or inappropriate and what they might expect in response to those activities. Through socialization we learn how to express our feelings, how to respond to events and people, and how to understand other people's behavior.

One type of socialization to which social scientists have paid special attention is gender socialization. Gender socialization is the process by which people learn what it is to be masculine and feminine. Chapter 5 described how gender is socially constructed. One of the ways in which people work to construct gender is by teaching each new generation how to behave in "masculine" and "feminine" ways. Learning how to be masculine or feminine requires a lot of work because there are so many rules and nearly every facet of our lives is gendered. Gender socialization includes a wide variety of lessons like learning how to walk, talk, play, and express emotions.

Masculinity and femininity also differ from one culture to the next and across classes. Some rules that exist in one society may be exactly the opposite in another. For example, in the United States we learn to display gender through our clothing. One rule is that skirts and dresses are feminine and for women only. But in Scotland male soldiers wear kilts and in Saudi Arabia men are expected to wear long gowns. Since the rules about how to be masculine and feminine are complex and arbitrary, intense socialization is required for us to learn the lessons.

When we learn about gender through socialization we learn not only what to do but also how to do it. We are not outside observers. The lessons are absorbed and become part of us. *Sex typing* is the term that psychologists use to refer to the acquisition of gender-appropriate preferences, skills, personality attributes, behaviors, and self-concepts.

Although gender socialization takes place in families, many other people and institutions socialize children. For example, in the scenario at the beginning of the chapter Amy tells about how she and her brother attended separate camps where he learned about a masculine activity—football—and she learned about a feminine one—ballet.

In contemporary American society the media are recognized as another institution that plays an important role in socializing children. Children spend many hours watching television and learning about our culture.

The average American child watches about thirty hours of TV every week (Lyle & Hoffman, 1972). Staples and Jones (1985, p. 17) estimate that "by the time the average American child reaches eighteen years of age he [or she] has watched 22,000 hours of TV . . . and has seen 350,000 commercials."

Elliot Medrich and his colleagues (1982) surveyed a large sample (n = 764) of sixth graders to find out how they spent their time. "In 61% of the households the children reported that the TV was on most of the afternoon and in 84% of the households most of the evening" (Medrich, et al., 1982, p. 200). In 59% of the households the set was also on during dinner.

Boys and girls watch TV about the same amount of time and households of different social classes do not vary much in terms of television watching. Race ethnicity does seem to have an influence on how many hours are watched. In their study Medrich et al. (1982) found that about half of the African American children and nearly one-half of the other non-Asian minority children watched three or more hours a day. Only about one-quarter of the white and Asian children were in this heavy viewing population (Medrich, et al., 1982).

One important set of ideas that children learn by watching TV is gender ideologies. The Women's Institute for Freedom of the Press has done extensive research on the messages about gender promoted by the media. In one study of 20,000 programs they found that gender difference continues to be a salient issue on TV (Women's Institute for Freedom of the Press, 1986).

For example, men are more visible on TV. During prime time women continue to be outnumbered by men, with little change since the 1950s. Female characters are younger, less mature, and less authoritative than males. Female characters are also less diverse in appearance than male characters. Although both men and women on TV tend to be young, thin, and attractive, men are more likely to vary from this model than women are.

Men and women also behave differently on TV. TV women are more emotionally expressive and seven times more likely than men to use sex or romantic charm to get what they want.

The settings for women's activities are also different from those of men. Women characters are more often shown at home. Even when women characters are employed outside of the home, they usually are not shown on the job and the plots revolve around family matters and interpersonal relationships.

The activities in which men and women participate are different on TV. Men are much more likely than women to appear in action/drama programs and to be problem solvers. Men are also more likely than women to be violent or criminal, especially if they are nonwhite, with Latinos most frequently cast as criminals (Condry, 1989).

Race stereotypes are combined with gender stereotypes, resulting in the portrayal of African American women in a negative manner. African American women

> are depicted as unskilled, unpolished and lacking decorum, rarely in control of situations; financially dependent on a parent or even a white family; never giving information unless it is about child care or housekeeping; and, if single, desperately preoccupied with finding a husband. (Renzetti & Curran, 1992, p. 116)

Age stereotypes are also combined with gender. Women, especially older women, portray victims of crimes. Condry (1989) found that older women are victims of violent attacks on television thirty times more than they are in real life.

When we think of the socialization of children we tend to emphasize the importance of families. We can see from the enormous amounts of time children are exposed to TV that forces outside of families also play a critical role in socializing children. The huge amount of time spent watching TV and the consistency of the images on TV are important socializers of children.

Television is a feature of the macro level of social organization. The huge television industry is itself shaped by governmental regulation and business pressures. This review of TV in children's lives shows the way in which the macro level of society has an important effect on children, if only in terms of the amount of their time it consumes. But in addition, the images undoubtedly have an effect on the thinking and behavior of children as well. A child's individual family is an aspect of the micro level of social organization. The micro-level interaction that takes place in families is a critical factor in socialization. But the macro level, in this case TV, plays an important role as well.

Gender Socialization in Families

Sandra Bem, a feminist psychologist, has been interested in sex typing, the acquisition of gender through socialization, in families. Her goal is not only to understand the process better but also to develop ways for parents to short-circuit sex typing in order to eliminate gender restrictions in their children's lives.

Bem (1983, p. 598) explains this short-circuiting as "raising gender-aschematic children in a gender-schematic society." To accomplish this it is necessary to understand how gender-schematic children are currently being raised, or how gender socialization takes place.

Bem (1982) reviews three dominant theories of how gender socialization takes place: psychoanalytic theory, social learning theory, and cognitive-developmental theory. She then presents her own theory, gender schema theory, which describes a way for parents to eliminate or at least resist gender socialization of their children (Bem, 1983).

Because she is a psychologist, Bem's work emphasizes the internal psychological processes and the social interaction between individuals at the micro level that create gendered people. Our discussion will begin with her framework and the four models she has reviewed. Then we will look at two more sociological theories about gender socialization.

Psychoanalytic Theory of Gender Socialization Psychoanalytic theory was initially developed by Sigmund Freud in the 1920s and 1930s. Recall the discussion in Chapter 12 that explored Freud's ideas about mothers, children, and gender socialization and the importance he placed on the stages of children's psychological development, especially their transition through the phallic stage and the oedipal crisis.

Large pieces of Freud's work have been rejected. Later in the chapter, for example, we will see critics of Freudian theory in the discussion of incest. Freud, however, was a prolific and creative scholar. And as we saw in Chapter 12, some of his theories have been accepted and revised by scholars like Chodorow (1978b). Freud's insights have had a tremendous effect on twentieth-century psychology and continue to play a role in the field.

Freudian theory has been criticized on several counts. First, some critics argue that little empirical evidence exists to support the argument that the psychological development that Freud proposed actually takes place. Second, the model does not account for the large numbers of children who grow up in single-parent families and who become "properly" gendered (Maccoby & Jacklin, 1974).

Third, Freud asserted that females are in a subordinate position in our society because of inferior genital structure rather than because of sexism. He maintained that girls are disappointed when they discover they do not have a penis and develop penis envy and a disdain for women. Researchers question whether this is a valid conclusion.

Empirical research shows that, in fact, girls more often prefer to engage in "masculine" activity than boys prefer to engage in "feminine" activity (Connor & Serbin, 1978). But is this because girls feel their genital structure is inferior? A strong case could be made for an alternative explanation. Girls' preference for boys' activities could be a result of the greater social value placed on boys, men, and masculinity rather than the superiority of male genitals. "By age six, children consistently attribute more social power to the father; they consider him smarter than the mother and the boss of the family" (Kohlberg, 1966).

Basow (1992) has suggested that another explanation for why girls might prefer "masculine" toys and games is because the "masculine" activities are intrinsically more entertaining. She writes, "Engaging in sports or playing with an Erector Set simply may be more fun than playing with dolls" (Basow, 1992, p. 119). Research of children's toys supports this assertion. Boys' toys are more varied, encourage activity outside of the house, and are more challenging (Vaughter, 1976).

Social Learning Theory of Gender Socialization Social learning theory is a second model that has been proposed to explain gender socialization. This model for explaining how we are gender socialized grew out of stimulus-response theory or behaviorism (Stockard & Johnson, 1992). Social learning theorists argue that children learn gender through reward and punishment. They display those behaviors for which they are rewarded and avoid displaying those behaviors for which they are punished.

Social learning theorists believe that sometimes children are directly taught that some behaviors are good and others are bad. For example, girls may be told by their parents that their behavior is not ladylike or boys may be told not to play with dolls or wear lipstick because those things are for girls. If the girl persists in her "unfeminine" behavior she is punished. The boy who insists on wearing lipstick or playing with dolls may be ridiculed or he might have the "girl toys" taken away. Fathers seem to have stronger expectations of gender-stereotyped behavior (Anderson, 1988).

Sometimes children are taught these rules more indirectly through modeling. Modeling is the process of learning by seeing others behave in appropriate ways and being rewarded or seeing them punished for aberrant behavior. For example, boys would frequently see their mothers wear lipstick but not their fathers. And they might hear others comment on how pretty their mother looks with lipstick or how lipstick is taboo for men. TV is an important source of models and as we have observed television programs include gender and race stereotypes. Books are also important in this regard and they too are dominated by male figures and present more positive images of male characters.

Critics of social learning theory note that people stick with gender-appropriate behavior even if it is not rewarding. For example, even though men are paid more than women and are more likely to hold political office, few women are willing or able to behave "like men" to obtain these rewards. Social learning theory, however, implies that any behavior, including those associated with gender, is readily learnable and unlearnable.

Social learning theory has also been criticized because it assumes that children are entirely passive in the process. Psychologists (Damon, 1977) and sociologists (Cahill, 1983) have found that children actively participate in learning gender.

Cognitive-Developmental Theory of Gender Socialization The third theory of gender socialization is cognitive-developmental theory. Cognitive-developmental theory improves on social learning theory by recognizing the child as an active participant in the process. This theory also differs from social learning theory because it proposes that age is a critical factor; exposure to "rules" is experienced differently by different age groups. According to cognitive-developmental theorists children must be at the appropriate age to be able to receive the message being sent to them.

Cognitive-developmental theorists argue that children move through stages of psychological maturity. Children who have reached the appropriate stage identify themselves unalterably as either a boy or a girl. This judgment of their own sex then determines whether they will accept or reject punishments and rewards. For example, once a boy determines that he is a boy, those activities and behaviors that he perceives to be masculine become intrinsically positive regardless of their real effect on him. He may feel uncomfortable wrestling on the playground and prefer to play with dolls, but he may perceive the wrestling as consistent with his sex and therefore good, and the playing with dolls as inconsistent and therefore bad.

Rewards or punishments are not as important as consistency between sex and gender. According to cognitive-developmental theorists "because of the child's need for cognitive consistency self-categorization as female or male motivates her or him to value that which is seen as similar to the self in terms of gender" (Bem, 1983, p. 601).

Bem criticizes cognitive developmental theory. Her key criticism is that the theory ignores the importance of society in the "choice" children make to use sex as a central organizer of their lives. Children could, for example, choose size, color, or age as the way to organize their world and the basis for selecting certain behaviors as consistent or not. Cognitive-developmental theory does not ask why children decide sex is going to be the central organizer. Bem, however, does ask this question and answers it by pointing out that our society's emphasis on sex causes our children to use sex as the most important way to determine their place in the world and their decisions about how they should think, feel, and act.

Gender Schema Theory of Gender Socialization Bem (1981, 1982, 1983) proposes gender schema theory as the fourth and best theory to explain gender socialization. Like cognitive-developmental theory, gender schema theory argues that children are active participants in their socialization. Children are exposed to rewards, punishments, and models of various behaviors, and they interpret and select based on their belief that gender is an important organizer of people's lives.

Gender schema theory is different from cognitive-developmental theory, however, because it assumes that the "choice" to make gender the central factor is itself determined by the culture in which the child is living. It is therefore not inevitable. One could imagine a society in which gender is not a critical factor and that is what Bem emphasizes in her conclusions.

In a culture like ours, however, gender is enormously important, so children use gender as the main criterion for determining if a behavior is good or bad for them. Children look around and see that virtually everyone can be distinguished as either masculine or feminine, and that many factors in their environment fit one category or the other—names, colors, games, jobs, clothes, language, interests, and so on.

This implies that if we wish to eliminate gender inequality, we must introduce children to the idea that gender could be a much less important factor. Children could learn that it is good for both males and females to behave in a variety of ways. More importantly, children could learn that there are only a few instances in which sex should play a role at all in assessments of how they should think, feel, act, or relate to one another.

For example, if a man wished to find a person with whom to conceive a child, it would be important to pay attention to the sex of the other person. She would need to be a woman. (She would also have to possess a number of other characteristics: fertility, interest in having a child, and interest in having intercourse with him or being alternatively inseminated by his sperm.) Very few other activities would necessitate acknowledging sex.

Gender schema theory goes beyond suggesting that we teach children that it is all right for boys to like nail polish and girls to like football. Gender schema theory suggests that we need to teach children that gender is an irrelevant issue in determining what is good, acceptable, or fun.

Bem explains,

> Gender schema theory thus implies that children would be far less likely to become gender schematic and hence sex typed if the society were to limit

> the associative network linked to sex and to temper its insistence on the functional importance of the gender dichotomy. Ironically, even though our society has become sensitive to negative sex stereotypes and has begun to expunge them from the media and from children's literature, it remains blind to its gratuitous emphasis on the gender dichotomy itself. (Bem, 1983, p. 609)

Bem emphasizes the role parents play in gender socialization and concludes her review by posing the question, "What can parents do to try to eliminate gender restrictions in their children's lives?" Bem suggests that we need to take an active part in presenting alternatives to our children. If we just ignore gender, children will learn the dominant gender ideology in our society, which insists that gender is a central way to organize people and that the two "types," girls and boys, must learn many "rules" to differentiate themselves.

Bem has two specific recommendations for counteracting this message and raising gender-aschematic children. First, parents should emphasize the biological distinction between males and females. If children are aware of real physical distinctions between males and females and the limited significance those differences have in nearly all situations, they will be better able to distinguish real differences from socially determined ones.

Second, parents must propose an alternative set of criteria for assessing how people should think, feel, and act. Bem suggests that parents create new materials like stories, games, and toys that do not teach gender stereotypes and instead stress commonalities among people and individual nongender-related variation. Parents must also censor cultural messages in television programs and books.

In creating these alternative "rules" parents need to accentuate the importance of individual differences among people, as opposed to categorical differences between males and females. The new "rules" can also be supported by showing differences across cultures among different groups of people.

Finally, Bem argues that teaching alternatives is not enough. Children must be shown how to actively resist gender socialization and gender inequality. She asserts that parents must teach their children to be "morally outraged by and opposed to whatever sex discrimination she [or he] meets in daily life" (Bem, 1983, p. 615).

One of the central themes of this book is the importance of the "resistance" movements of people working together at the micro level to change their immediate social circumstances and especially to change the macro organization of their society. Bem's suggestions for how to raise gender-aschematic children provide a specific plan for challenging gender inequality, especially at the micro level. Parents and children in micro-level, face-to-face relationships are encouraged to interact differently in a way that diminishes the role of gender in their relationships.

Socialization or Something Else?

How far will Bem's recommendations take us in eliminating gender inequality? If we assume that socialization is the basis of gender in our society and that families are a key site of this process, then the solution to the problem of gender inequality is to change the socialization of children within families. This theory is attractive. Early in the Second Wave of feminism this was the goal that was promoted: Eliminate gender inequality and sexism by socializing our children to be gender-free.

Socialization has since been recognized as just one of many sources, rather than the only source, of gender inequality. Social interaction and social structure have been added to socialization to expand the conceptualization of the sources of gender inequality.

Social interaction theory, like socialization theory, emphasizes the way in which micro-level social interaction creates and sustains gender. Spencer Cahill (1983) is an example of a social interactionist. He shares Bem's assessment of the way that gender-based behaviors and self-concepts are transmitted from one generation to the next. His own work includes the two factors that Bem finds important. First, he emphasizes the active interaction between adults and children during the socialization process. Second, he acknowledges the system of inequality that makes up the social context of that interaction.

Cahill differs from Bem in the importance he gives to peer interaction among children. He argues that while adults guide the gender development of children, it is from their peers that children actually learn to perform gender-specific behavior (Chafetz, 1988). Cahill's view implies that change in gender socialization must come from change in institutions of socialization both inside and outside families.

Another difference between a socialization framework and a social interactionist framework is the acknowledgement of the dynamic character of gender and the breadth of its influence. West and Zimmerman (1987) refer to this as "doing gender." They argue that gender socialization theory conveys the idea that gender is achieved at an early age by people doing things to us and we in turn responding. A social interactionist approach, in contrast, notes the way gender is an active, ongoing, everpresent process. We are almost constantly practicing gender and we practice it not just as children and not just in relationships that entail socialization. In Chapter 7, for example, we discussed Sarah Fenstermaker Berk's (1985) work examining housework from a social interactionist perspective. She argued that in doing housework, women and men were doing gender. They were practicing and displaying the supposedly appropriate activities of masculine and feminine people.

Socialization theory implies that gender is an aspect of who we are—I am a woman, therefore I behave in certain ways. Social interactionist theory conceptualizes gender as what a person does and does recurrently in interaction with others. Socialization theory suggests that we can alter patterns of gender by changing socialization practices. Social interactionist theory demands that we alter all of our social interactions that include doing gender, that these make up a large proportion of our encounters, and that they are as common to adults as they are to children.

Another critique of socialization theory comes from those who focus on social structure as the source of gender. Rosabeth Moss Kanter (1977a) is an important figure in this work. In her research on women and men in corporations, Kanter noted that the women workers often decorated their offices in "feminine" ways with pictures and homey decorations. She also saw how women organized domestic activities at the office with birthday parties and pot-lucks. One explanation for why women behaved in this way would be socialization.

One might argue that women have been socialized to create homelike environments. When they take jobs outside of the home they bring their upbringing with them and try to recreate these same surroundings at work. Kanter insisted that socialization was not a sufficient explanation.

She proposed that the reason women decorated their offices and tried to make them homelike was because their jobs were more boring and less satisfying than the jobs the men did. Women attempted to make up for the lack of control they had over their work and the drudgery and repetitiveness of clerical activities by sprucing them up with decorations and social events.

When she looked at women and men in similar positions of power she found that the differences between them diminished. Women who were executives were not more likely than men executives to participate in these "feminine" behaviors. Kanter concluded that the social structure of work creates certain behaviors among certain employees, regardless of their gender or their upbringing.

Kanter's theory is called segmented labor market theory. Its main assertion is that jobs create worker characteristics (Chafetz, 1988). The specific mechanism that creates differences in the behavior of people in "women's" jobs and people in "men's" jobs is called blocked mobility. Blocked mobility is the result of the hierarchical structure of the workplace that prevents women workers from moving into all jobs at all levels. One of the responses to blocked mobility is to increase one's commitment to peers and sociability. These behaviors expressing sociability are identified with femininity in our society. The segmented labor market theory, however, emphasizes that the roots of the behaviors at work are in the organization of the workplace.

Bem, Kanter, and the social interactionists represent three different theoretical models for answering the question "How is gender inequality created and maintained?" Probably all three answers are partially true.

The difference among the three is in the factors each emphasizes. Kanter emphasizes the macro level of social structure. Bem recognizes the macro level of social structure but emphasizes the micro-level interactions of parents and children. Social interactionists also recognize the macro level of social structure and the micro-level interactions of parents and children, but insist that we pay attention to the broad array of social interactions in which people "do gender" throughout their lives.

Racial Socialization Gender inequality and the oppression of women is an important social problem. Bem tells us that socialization can operate as a mechanism of social control or as one that can generate new ideas, new behavior, and challenges to the organization of gender in our society.

As we have seen throughout this book, inequality by race ethnicity is also an important social problem. The following poem entitled "Incident" by Countee Cullen (1947) captures the pain that racism can inflict, especially on children. It also shows that both the victims and the sources of racism can be children. Socialization in families can help to counter the effects of racism. Through racial socialization, children can learn what racism is and how best to confront it. And like gender socialization, racial socialization can often take place in families.

Once riding in old Baltimore,
Heart-filled, head-filled with glee,
I saw a Baltimorean
Keep looking straight at me.

Now I was eight and very small,
And he was no whit bigger,
And so I smiled, but he poked out
His tongue, and called me, "Nigger."

I saw the whole of Baltimore
From May until December;
Of all the things that happened there
That's all that I remember.

Countee Cullen wrote his poem more than forty years ago. Has racism disappeared from African American children's lives since then? Marie Peters wrote in 1988 that racism is still a salient factor and one about which black parents are very much aware: "An inescapable aspect of the socialization of Black children is that it prepares them for survival in an environment that is hostile, racist and discriminatory against Blacks" (Peters, 1988, p. 237).

In response to the hostile environment, African American families have devised ways to buffer some of the demeaning messages African American children receive outside of the community (Ogbu, 1978; Scanzoni, 1971; Richardson, 1981; Willie, 1988; Peters, 1985, 1988). Marie Peters (1985) refers to the special attention African American families give to preparing their children for being an African American in the U.S. as "racial socialization" (Peters, 1985).

Peters (1985) followed the childrearing practices of the parents of thirty African American children for two years to examine racial socialization. All of the mothers in the study were conscious of racism and they offered examples of their experience of discrimination in the course of their daily activities. They discribed incidences of being ignored or treated badly or unfairly by employers, store clerks, waitresses, and bank officials (Peters, 1985; see also Sigelman & Welch, 1991).

Because of the parents' own experiences they felt responsible to teach their children to understand that they might be discriminated against and how to survive in a hostile environment. One mother explained: "It's most important for me to teach my sons how to deal with a society as it is—to let them know they're protected as long as they're at home, but when they get out there in the world, they're not protected anymore" (Peters, 1985, p. 164).

Parents also attempt to counteract racism by presenting an alternative. The parents actively tried to instill pride, self-respect, and assurance of love as protection against negative images the children might encounter. Another mother explained to Peters: "I'd like them to have enough pride, because if you have enough pride or self-confidence in yourself, you'll let a lot of things roll off your back" (Peters, 1985, p. 165).

The racial socialization strategies of the African American parents in Peters' (1985) study in some ways sound like the proposals made by Bem (1983) in her discussion of gender schema theory. The African American parents teach their children to be aware of race discrimination and to develop alternative ways

African American parents celebrate holidays and rites with African roots like Kwanzaa as a way of socializing their children into the African American culture. Peters refers to this as racial socialization.

of seeing themselves and of confronting inequality. Bem calls for teaching children to become morally outraged at sex discrimination and to oppose it in their lives.

Some African American parents have gone beyond relying solely on informal socialization within families. They are creating formal rituals to socialize children into their community. The Passage (Hare & Hare, 1985) is one example.

The Passage is an initiation rite created by Nathan and Julia Hare (1985) for African American boys (one for girls is in the process of being prepared). The rite consists of a year of activities beginning with the eleventh birthday and culminating with a celebration on the twelfth birthday. The rite is primarily guided by family members but the wider community is also involved. The purpose of the Passage is to socialize African American boys into the black community.

During the year, the boy is expected to accomplish a number of tasks, including:

1. Keeping a log.
2. Reading about African American culture and history.
3. Developing a list of his relatives and their residences.
4. Providing service to the community.
5. Adopting a senior citizen.
6. Preparing for future education.
7. Practicing courtesy and making contact with public officials to begin taking leadership in his community.
8. Preparing for the Passage Celebration, which includes fasting, presenting a speech, and responding to questions from a panel of adults.

Children as Threats

The review of the image of children as learners shows that there are important differences among children by gender and race in the experience of socialization. All children and their families, however, are influenced by the image of children as threats and victims and by the problems children and parents face in their relationships with one another.

Research shows that children may have a detrimental effect on parents' mental health. In an extensive review of the literature, Sara McLanahan and Julia Adams (1987) conclude that parents with children at home are worse off than nonparents on a number of indicators of psychological well-being. Compared to nonparents, parents are less happy and satisfied and they are more worried, depressed, and anxious.

Mothers who are primarily or solely responsible for children are further negatively affected by them because of the additional work they create. Mirowsky and Ross (1989) examined the depression levels of husbands and wives. They found that unemployed wives without children and employed mothers whose husbands shared responsibility for childcare had lower than average levels of depression. But wives with small children, paid jobs, difficulty finding childcare, and sole responsibility for childcare had much higher than average levels of depression. When we looked at the division of childcare and housework in families in Chapter 7, we saw that this last type of situation is the most typical one.

Children may also have a negative effect on their parents' marriages. "Satisfaction with marriage decreases with the birth of the first child, and does not return to prechildren levels until all the children have left home. . . . As the number of children, especially young children, increases, marital satisfaction decreases" (Mirowsky & Ross, 1989, p. 103).

Figure 13-1 shows how marital satisfaction is affected by children of different ages. The line traces the marital happiness in families from early years in the marriage before children are around, through the years of childrearing, to the stage in family lives when children have left their parents' home. The figure shows that marital satisfaction is higher before and after children are in the home, and that satisfaction hits its low point when the children are teenagers.

The negative effect children have on parents' psychological well-being may be related to the decreased amount of time husbands and wives have to spend together. In addition, even when parents do get time alone without the children, they must spend much of it considering issues that relate to the children (Mirowsky & Ross, 1989; White, et al., 1986).

Children may also cause distress for their parents by increasing the economic strains on families (Mirowsky & Ross, 1989). With every addition to a household, each dollar must be stretched further to cover food, clothes, and all the other necessities. More people in a household also increases the pressure for a larger—more expensive—place to live.

Table 13-1 shows the cost of raising a child for different kinds of families. The total ranges from $91,920 to $176,860. The biggest expense is for housing.

Given the inadequacies of childcare in our society, which is discussed in Chapter 12, having children frequently makes it more difficult for all of the adults in a household to be in the paid labor force, which further increases the financial strain. Hofferth and Phillips (1987) found that the birth of a first child had a greater negative effect on family finances than the birth of subsequent

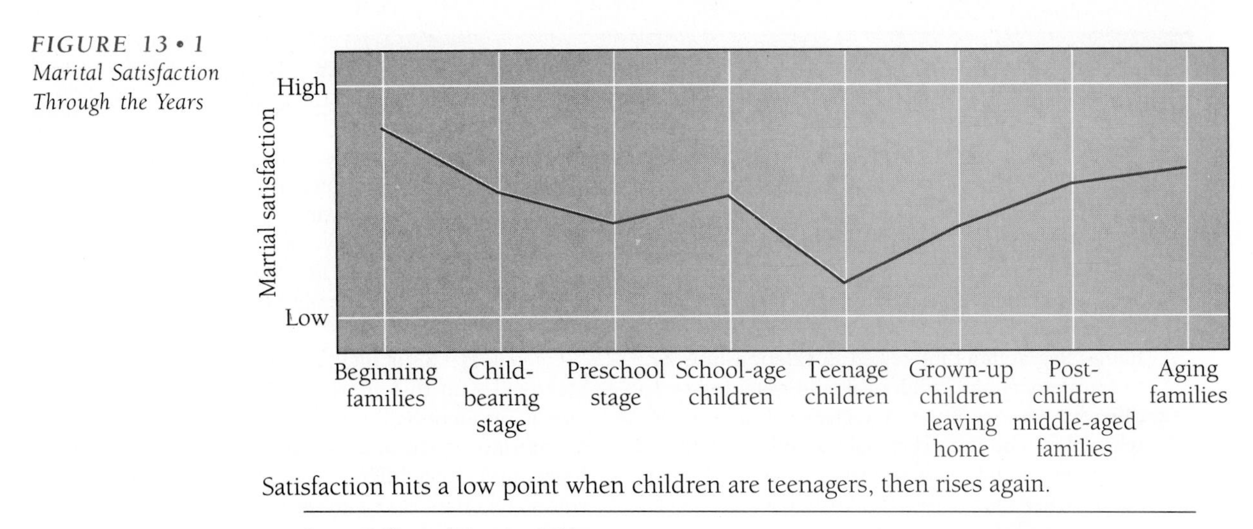

FIGURE 13 • 1
Marital Satisfaction Through the Years

Source: Rollins and Cannon, 1974.

TABLE 13 • 1
What's It Going to Cost to Raise a Child from Birth to Age 17?

	Total	Housing	Food	Transport.	Clothing	Health	Education/ Child care
Couples with income less than $30,700	$91,920	$30,870	$18,480	$15,090	$8,790	$5,070	$13,620
Couples with income between $30,700 and $49,700	$126,900	$40,680	$22,920	$23,160	$11,070	$6,450	$22,620
Couples with income more than $49,700	$176,860	$60,840	$27,480	$30,600	$13,290	$7,980	$36,670
Single-parent family with income less than $31,200	$94,140	$31,110	$18,120	$22,890	$6,990	$3,030	$12,000
Single-parent family with income of $31,200 and up	$174,480	$61,710	$27,120	$37,440	$9,420	$6,720	$32,070

Source: U.S. Department of Agriculture, 1991, *The Cost of Raising a Child*, 1990, pp. 11, 14, 16.

children and that reductions in income with each birth are related to reductions in the labor force participation of their mothers. Or if all of the adults are in the paid labor force, the cost of childcare is added to the monthly bills.

The economic strain of children in families may be currently increasing as young adults delay moving out of their parents' homes. In 1990 the U.S. Census reported that 77 percent of eighteen- to twenty-four-year-olds have never married and that two-thirds of those 19.6 million never-married persons were residing with their parents (U.S. Bureau of the Census, 1990).

When Thorne (1987) proposed that children are frequently thought of as a threat in our culture, she noted examples such as images of gangs and youthful

BOX 13 • 1 A LIFE CYCLE APPROACH TO FAMILY SOCIOLOGY

Figure 13-1 shows the way in which families change over their life cycle. The figure indicates variation from one stage of a family's life to another. It shows that marital satisfaction in families that have children decreases as the family ages until children are adolescents and then increases after the children grow up and leave home. Researchers interested in relationships between parents and children would obtain very different findings if they interviewed parents when their children were babies compared to when their children were teens. A series of snapshots rather than a single snapshot would better capture the relationship.

Research that pays attention to changes over time in families is called a life course perspective. In this book we have observed the importance of historical change at the macro level. For example, in Chapters 2 and 3 we noted the way in which Euro-American and African American families changed over several centuries from the 1600s to the 1900s. The life course perspective pays attention to the change over time at the micro level within families.

Glick (1977) is one scholar who has been identified with this approach (Mattessich & Hill, 1987). He identified critical stages in family life as marriage, birth of children, children leaving home, and "postchildren" couples. Other life course theorists have proposed anywhere from two to twenty-four stages. These stages are frequently identified with some phase of childrearing or childhood. The approach, therefore, is especially pertinent to issues related to children.

criminals that are common in the media. Her emphasis was on the way in which we perceive children to be a threat rather than the way in which children might pose a real threat to adults.

Research by other scholars on the mental health, marriages, and financial situation of parents, however, shows that the image of children as a threat is not entirely without an empirical basis. Research indicates that when children are in families psychological distress, marital unhappiness, and financial strain increase. People raising children do seem to have problems in these areas that are not as intense for nonparents.

The question, however, is whether these problems are caused by children or whether they are a result of the way in which social institutions like the government and business fail to support parents or children. For example, is the financial decline experienced by families with children caused by the children or is it caused by an economic system inadequate to the needs of all of its members? Is the marital strain in families with children caused by the children or caused by a government and business community that provides insufficient childcare options for employed parents?

The Positive Effects of Parents on Children This section has focused attention on the problematic way in which children can affect their parents. Children can also have a positive effect on their parents as well. Elise Boulding (1980) has investigated empathy among children and the way they nurture their parents. Boulding defines empathy as the ability to understand how another person is feeling without actually experiencing what the other person is experiencing. She argues that children develop the capacity to empathize at a remarkably early age.

Boulding asked young people from adolescents to college students to give an example in writing of when they had been supportive to their father or mother before they were fifteen and one example from after they were fifteen. Nearly all the respondents were able to provide examples. They described incidents from when they were as young as five years old. Incidents included everyday situations like making a Father's Day card to providing support during family crises like illness, job problems, death, and divorce. The respondents remembered performing chores, providing companionship, listening, and counseling. One woman remembered at age nine taking care of two infant brothers so that her mother could have some time to herself. Another remembered running with her dad while he was training for the Boston Marathon. A man remembered dropping out of school for a year to help his father rehabilitate after surgery. Boulding concludes that her findings reinforce the need for sociologists to pay more attention to the positive character of children's contribution to their parents' lives and, perhaps even more importantly, to reconceptualize children as active, autonomous social beings.

Beverly Purrington is another scholar who investigated the positive effects of children on their parents. She interviewed seventy parents, asking them how they perceived their children had affected their lives. The responses to this question fell into two broad categories. Parents said that having children necessitated commitment and decision making that made them feel good about themselves, like responsible adults. In addition, parents said that having children allowed them to be more childlike by legitimizing more open reactions to experiences and more heightened emotional responses to their surroundings.

One man explained how being a parent made demands on him that he felt good about being able to meet. He said, "It's . . . fun to just have the ability to be a protecting force in somebody's environment—it gives you a sense of accomplishment and importance . . . it's easy to get [that] with a kid. It's much easier to get that with a kid than an adult" (Purrington, 1980, p. 124).

Several parents told stories of how having a child allowed them to be more open to participation in activities they otherwise might have felt were unseemly for adults, such as watching cartoons and playing on swings. One father said,

> suddenly 'cause you have a kid you have a license to do all of these things again. And that's fun, that's really terrific . . . one of the good things, the greatest joys of having kids is that you can be a kid again, you're suddenly allowed to do things that when you're grown up you're no longer supposed to do, you know . . . suddenly Christmas is Christmas again. You know, you remember Christmas as a kid was great joy, and then it got sort of dull, 'cause the miracle had gone out of it and then suddenly you have kids and oh boy, it's just a miracle to them and then you can enjoy it that way. (Purrington, 1980, p. 64)

In addition to giving access to certain activities, children also provided access to memories. Parents spoke of remembering their own childhoods only after seeing their children involved in activities and situations that had been part of their childhoods.

Parents also believed that having children allowed them to experience an intensity of emotion they would not otherwise have felt. A mother described the way being a parent had helped her to express love. She said, "there's real warmth there that I never felt with any other person . . . but I guess that's been a positive influence in learning how to love" (Purrington, 1980, p. 107). A father described the intensity of emotion he experienced at the birth of his children. He said,

> I was there in the delivery room right there when they were both born and that is just the most fantastic emotion you have ever experienced. . . . It's just such an amazing feeling of relief and joy and happiness—it's like nothing else you've ever experienced . . . you both just start crying. (Purrington, 1980, p. 100)

Boulding's and Purrington's work is unusual because of its focus on the autonomy of children and their beneficial effect on their parents. The perception of children as problematic and threatening to their parents' well-being is much more prevalent in the literature.

Why are children perceived to be threats in our society? And why are they blamed for what could just as easily be blamed on inadequate social institutions? Letty Pogrebin (1983) argues that the perception of children as threats is rooted in an ideology that prevails in our society. She argues that "America is a nation fundamentally ambivalent about its children, often afraid of its children, and frequently punitive toward its children" (Pogrebin, 1983, p. 42).

Pogrebin has named this ideology pedophobia, from *ped* meaning child and *phobia* meaning fear, dread, aversion, or hatred. Pogrebin argues that pedophobia is expressed not only in an ideology that perceives children as a threat, but also in the victimization of children, the third image that Thorne observed was dominant in our culture.

According to Pogrebin the victimization of children can be categorized into two types: individualized pedophobia and institutionalized pedophobia. Individualized pedophobia is the cruelty of parents who physically and sexually abuse their own children. Institutionalized pedophobia is the public policies, practices, and institutions that keep children impoverished, malnourished, unhealthy, and frequently unable to survive.

Children as Victims: Child Abuse

Children are frequently victims of a number of types of abuse in families. Table 13-2 shows the different kinds of maltreatment of children reported for different age groups. The Ns indicate the number of incidents reported for all children in 1982. Since then the numbers have increased dramatically. The National Committee for Prevention of Child Abuse (NCPCA) estimates that the reported cases of child abuse reached 2.5 million in 1990. This is about 39 of every 1,000 children in the United States (NCPCA, 1991).

The Ns for different kinds of abuse show that deprivation of necessities is most common and major physical maltreatment is least common. The percentages within each maltreatment category show variation by age. For example, very young children are more at risk for major physical maltreatment than teens are. Teens are more at risk for sexual maltreatment than other children.

In the Child Abuse Prevention and Treatment Act of 1974, Congress defined child abuse as overt acts of aggression such as excessive verbal derogation, beating or inflicting physical injury, and sexual abuse. Child neglect means failing to provide adequate physical and emotional care. Emotional abuse or neglect

TABLE 13 • 2
Type of Maltreatment and Age of Involved Child (N = 381,168)

	Age 0–5	6–11	12–17	Total
Major or Major with Minor Physical Injury (N = 8,800)	64.1%	19.8%	15.9%	100%
Minor or Unspecified Physical Injury (N = 71,884)	37.3%	32.6%	30.0%	100%
Sexual Maltreatment (N = 27,714)	24.8%	34.3%	40.6%	100%
Deprivation of Necessities (N = 192,223)	48.6%	33.2%	18.2%	100%
Emotional Maltreatment (N = 24,808)	34.0%	34.2%	31.9%	100%
Other Maltreatment (N = 15,917)	47.0%	26.8%	26.2%	100%
Multiple Maltreatment (N = 39,822)	41.5%	31.7%	27.0%	100%
Percent of All Involved Children (N = 717,315)	43.3%	32.5%	24.1%	100%
Percent of All U.S. Children (N = 62,580,000)	34.5%	30.7%	34.7%	100%

Source: American Association for Protecting Children, *Highlights of Official Child Abuse and Neglect Reporting* (Denver, CO: American Humane Association, 1985), p. 16.

means being overly harsh and critical, failing to provide guidance, or being uninterested in the child's needs (Lamanna & Riedmann, 1991). Although all of these cause harm to children, the public's attention has been almost exclusively focused on physical abuse and there is little discussion of or political activity around the other "lesser" forms of abuse. In this section we will examine the problem of physical abuse of children and then move to a discussion of the sexual abuse of children in families.

Although every state now has compulsory child abuse reporting laws, it is difficult to obtain accurate statistics because much abuse still goes unreported. In addition, government agencies that used to fund nationwide compilation of these data no longer do so. Between 1974 and 1986 an annual survey was funded by the Federal Children's Bureau. The study was reduced in 1987 and has not been funded at all since then (Popple & Leighninger, 1993). Even when research is done, some forms of maltreatment are difficult to measure, such as emotional abuse and neglect, and much physical abuse goes unreported. One way of determining more valid rates of abuse is to choose a random sample of people, determine the rates of the activity in the sample, and then extrapolate that proportion to the entire population.

For example, in their research, Dibble and Straus (1980) found that 3.8 percent of children between the ages of three and seventeen in their sample had been physically abused in the past year. Since they had obtained this percentage by asking a large random sample they argued that the proportion was representative of the entire United States. Thus, 3.2 percent of all children in the U.S. between the ages of three and seventeen equals 2 million cases of child abuse a year (Dibble & Straus, 1980).

In research done by Gelles and Straus (1987), 63 percent of the parents admitted to having used some form of physical violence on their children. Table

BOX 13 • 2 HAS THE U.S. GOVERNMENT ALWAYS SOUGHT TO PROTECT CHILDREN FROM ABUSE?

Although child abuse has occurred throughout U.S. history, until about twenty-five years ago most physicians, even pediatricians, denied that abuse occurred. The legal system also failed to protect children by either providing no laws restricting child abuse or insufficiently supporting laws that did exist (Johnson, 1991).

Early American common law stated: "If one beats a child until it bleeds, then it will remember the words of its master. But if one beats it to death, then the law applies" (Radhill, 1968, p. 4). While murder was illegal, anything short of murder in the treatment of a child seemed to be acceptable.

John Johnson (1991) describes a famous case that illustrated this assumption and created the impetus for establishing the Society for the Prevention of Cruelty to Children:

> In 1875 the American Society for the Prevention of Cruelty to *Animals* (ASPCA) in New York City was asked to intervene for the purpose of protecting Mary Ellen, a nine-year-old girl who had been neglected, beaten and even slashed with scissors by her foster parents. Earlier efforts to intervene had failed, because the parental right to discipline had been heretofore considered absolute by law. So the ASPCA was asked to intervene to protect Mary Ellen on the argument that she was a member of the animal kingdom and thus the legitimate recipient of laws already on the books to protect animals. (Johnson, 1991, p. 673)

As recently as the 1960s child abuse and neglect were still issues that were invisible. Among a few professionals their existence had been suggested but dismissed as invalid accusations. In 1962, the first study of child abuse was done by the American Humane Association and in 1963 the government finally began to take the initiative to both study and address the problem. By 1967, 49 states had new laws mandating reporting child abuse (Johnson, 1991).

13-3 shows the kinds of violence that were used just in one year prior to the interviews.

The table does not indicate what the result of the violence was but homicide is the fifth highest cause of death among children (Gelles & Cornell, 1985). And in their research in the late 1970s Straus, Gelles, & Steinmetz (1980) found that more than 2,000 children a year were killed by their parents.

Parents are not the only ones involved in violence against children in families. Sibling violence is also common, although it is rarely reported. Gelles (1977) found that 80 percent of children who had sisters and brothers reported they had tried to hurt a sibling during the previous year. Fifty percent said they had kicked, punched, or bitten a sibling. Forty percent had hit a sister or brother with an object and 15 percent had beaten one up. Straus, Gelles, and Steinmetz (1980) found that parents reported their children used on average twenty-one acts of violence a year on their siblings. Steinmetz, Clavan, and Stein (1990) argue that if these percentages are extrapolated to the entire population, 36 million attacks occurred between siblings that legally could have been considered assault. And finally, 1.5 percent of the homocides committed in 1983 were people who had been killed by a sister or brother (U.S. Department of Justice, 1985).

Child Abuse and Women About half the reported cases of physical violence against children involve women and about half involve men (Gordon, 1988).

TABLE 13 • 3
Kinds of Family Violence Against Children, 1985

Type of Violence	Percent of Families Who Reported Violence in the Past Year
Slapping or spanking	55%
Pushing, grabbing, or shoving	31%
Hitting child with something	10%
Throwing something at child	3%
Kicking, biting, or hitting child with fist	1.3%
Beating up child	.6%
Burning or scalding child	.5%
Threatening to use or using knife or gun	.2%
Overall violence	62%
Severe violence	11%

Source: Gelles and Straus, 1987, p. 84.

The large proportion of child abuse for which women are responsible can be explained partly by the greater amount of time women spend with children. If children spend most of their time with their mothers, and fathers are still responsible for half of the child abuse, women must be significantly less likely to abuse children than are men. Judith Martin's (1983) work supports this contention. She found that mothers are as likely as fathers to abuse young children, but three out of four abused adolescents had been abused by their fathers. Adolescents are also likely to spend more time than infants with their fathers.

The 50 percent proportion may also be a result of the way in which research has been conducted (Martin, 1983). For example, nearly all of the research on single-parent families and the incidence of abuse is on mother-only families and in general researchers who study child abuse have tended to focus on mothers.

Nevertheless women must take responsibility for a significant proportion of the physical abuse of children. Very little, however, has been written on this issue from a feminist perspective (Washburne, 1983). Linda Gordon (1988) is an exception. Her work is discussed in the last section of this chapter.

Incest Incest is another form of child abuse. Incest refers to sexual contact between close kin. The word incest can refer to sexual relationships that are not abusive. For example, a sister and brother might choose as adults to have sexual contact in a mutually consenting relationship. In most of the literature in contemporary sociology, however, the word incest is used to mean sexual abuse by a relative.

The prohibition against incest is so widespread in human societies that the incest taboo is called a cultural universal (Murdock, 1949). In the United States (and probably most other societies) the taboo, however, is much stronger in ideology than it is in actual behavior. Relatively large proportions of people report incestuous experiences. For example, Diana Russell (1986) found that 16 percent of a random sample of 930 adult women had been sexually victimized as children by a relative. In another random survey in 1979 of 796 college students regarding childhood sexual activity, 10 percent of the students (19 percent of the women and 9 percent of the men) reported cross-generational incestuous experience (Finkelhor, 1979). Furthermore, the actual incidence of incest is probably substantially higher than the reported rate.

Gender plays an important role in incest. For example, in Finkelhor's survey 93 percent of the people who said they had experienced incest said it had involved an older man and a girl (Finkelhor, 1979). Russell (1986) found that uncles were the largest category of perpetrators, accounting for 25 percent of the incidents. Uncles were followed closely by fathers/stepfathers, who accounted for 24 percent of the incidents.

The incidence of reported child abuse including sexual abuse as well as physical injury, emotional abuse, and neglect increased from about 700,000 in 1976 to almost 2 million in 1985. During this same time period, incest comprised an increasingly larger proportion of this growing total. In 1976 about 3.2 percent of the reported cases of child abuse and neglect were for sexual abuse. By 1985 sexual abuse accounted for 11.7 percent of the total (American Association for Protecting Children, 1985).

Although incest is illegal in every state, the legal system is organized in a way that hinders the ability of victims to protect themselves. During a trial, a child may be cross examined by an attorney or even the father or other accused molester. In some states victims must provide other witnesses, which is of course difficult to do.

> In a study of 256 known cases of sexual abuse of children involving 250 offenders conducted by the Brooklyn Society for the Prevention of Cruelty to Children, parents and family found the police process so trying and frightening to the children that 76 cases were dropped, leaving 174 cases eligible for prosecution. Once charges were officially made, the number of interrogations and court appearances resulted in such trauma to the children and their families that another 77 cases were so discouraged that they too dropped the charges. This left 97 cases. Of this number, 39 offenders either absconded, were acquitted, or were left to take a lesser charge such as assault rather than rape or incest; and 5 were committed to mental institutions. Of the 53 found guilty (excluding the 5 sent to mental institutions), 30 escaped jail sentence by suspension or fine, 18 were sentenced to jail from 6 months to a year, and 5 received indeterminate sentences. (Begus & Armstrong, 1982, p. 243)

Incest is an important social problem. It affects a large number of people and it involves families, legal institutions, and social service agencies. Finding a solution or even a way of approaching the problem is a difficult task. Understanding the issue is a critical step to ending incest. In the next section we will investigate two markedly different ways to conceptualize incest.

Two Competing Views of Incest Incest is an issue about which there are competing points of view. Two that present a particularly important contrast are a conservative psychiatric model and a radical feminist one. The conservative analysis is much older and has been identified with Sigmund Freud. Freud's ideas are still held by many clinicians today and the debate between contemporary Freudians and radical feminists continues (Butler, 1979).

The first model for understanding incest is the psychoanalytic framework first offered by Freud and subsequently adopted by Freudian psychoanalysts. Freud is a highly controversial figure, as we described earlier in this chapter's discussions of mothering gender socialization. Another area of contention has been around his work on incest. In his practice, several young women told him they had been sexually abused by older men. Freud interpreted this to be a fantasy on their part. He wrote that the women wanted to have sex with their fathers and that they had turned these childhood fantasies into memories. In 1937 Freud expressed this opinion in a letter to his colleague Fliess (Herman & Hirschmann, 1977): "There was the astonishing thing that in every case blame was laid on perverse acts by the father, and realization of the unexpected frequency of hysteria, in every case of which the same thing applied, though it was hardly credible that perverted acts against children were so general" (Freud, 1954, p. 215).

Those who continue today to subscribe to a conservative model for understanding incest, like Freud's, assume that many children make up stories about sexual experiences with adults. Critics of the Freudian model argue that the statistics on incest and our knowledge of the psychological difficulties caused for girls who are sexually exploited by their fathers indicate that there is a good possibility that the patients who came to Freud's office were not lying or fantasizing but had been, in fact, molested.

In some cases the abuse is known to so many people that it cannot be discounted. A second characteristic of the conservative model is that when incest is proven to have occurred, the child is at least partially blamed for being the one who wanted the contact in the first place (Russell, 1986).

Third, the conservative model disclaims the damage done to the victim. Wardell Pomeroy, who was one of the researchers in the Kinsey studies, expressed this opinion in 1976. He said:

> Incest between adults and younger children can also prove to be a satisfying and enriching experience, although difficulties can certainly arise. (In any case I want to emphasize that in no case would I condone incest—be it between adults or between children and adults—when force, violence, or coercion are involved.) (quoted in Russell, 1986, p. 8)

And even when it is acknowledged that harm may have been done, the way in which "damage" is assessed is not in terms of the victim's interests, but rather in terms of her ability to adjust and function "normally":

> If an adult woman who was sexually victimized as a child is "frigid," "promiscuous" or lesbian—all seen as sexual maladaptations—the incestuous assault is seen to have had a deleterious effect on her natural development as a woman. When these young girls grow into women who marry and as wives remain silent, such behavior is seen as an appropriate model for women's mental health and no "damage" is believed to have resulted. (Butler, 1979, p. 57)

Fourth, the conservative model blames nonmolesting mothers for the abuse of their daughters (Jacobs, 1990). Mothers, who have not participated in the abuse and who may not even have known about it, are still held responsible: "These interpretations are grounded in the premise that mothers are responsible for maintaining the family unit in a state of balance and equilibrium . . . if she withdraws from that role by choice or necessity, then all that happens within the family unit is her fault" (Butler, 1979, p. 58).

Within this framework, mothers are further blamed as the cause of the assault if they do not provide a warm sexual response to their husbands. Whereas mothers are blamed from this point of view, fathers are given sympathy. Fathers are portrayed in the conservative literature as victims themselves, insecure, stressed, immature, and poorly mothered.

The second and very different model for understanding incest is that offered by radical feminists. To counter the Freudian assertion of the lying child, radical feminists point out that if children lie about incest, it is likely to be in the direction of denying the assault, since the cases of reported incest are so low compared to the cases revealed in surveys of adults who were molested as children.

The second factor in the conservative model is the assertion that seductive children are the instigators of the incestuous contact. The feminist model refutes this assertation by showing evidence that children are frequently threatened, bribed, or at the very least tricked into keeping the secret because the one who is doing the assaulting is so powerful in their lives. Furthermore, they have shown that often girls who have been molested have resisted the assaults (Gordon, 1988).

Linda Gordon (1988) investigated 542 cases of incest, child abuse, and child neglect recorded in social service agency documents between 1880 and 1960. In her research she determined the numbers of girls and boys who had attempted various forms of resistance to abuse. She found that incest victims were more likely than victims of other kinds of abuse or neglect to use every form of resistance (including fighting back and telling the police or others) except one. "Attempting to flee" was the only form of resistance incest victims were less likely to have tried. Fifty-four percent of the victims of child abuse had attempted to flee, while 40 percent of the victims of incest had. Like Jennifer's sister in the vignette at the beginning of the chapter victims of incest have a long history of fighting back.

The third component of the conservative model is the argument that incest does no harm. Feminist researchers have made careful studies of adult women who describe the great pain and feelings of powerlessness, degradation, and wrong-doing they carry into adulthood (Russell, 1986).

For example, Russell (1986, p. 139) interviewed a sample of 151 women who had been incest victims as children. Twenty-five percent reported great long-term effects from the abuse, 26 percent reported some effects, 27 percent reported few effects, and 22 percent reported no long-term effects. The most frequently mentioned long-term effects were negative feelings about men in general and about the perpetrator; negative feelings about themselves, such as shame and guilt; and negative feelings in general such as fear, anxiety, depression, and mistrust.

The mother's culpability is the fourth factor in the conservative model. Some mothers sexually abuse their children but in the conservative model mothers are blamed for "allowing" their husbands to abuse children. Feminists dispute the validity of blaming mothers for abuse of children by their husbands. Radical feminists argue that mothers may themselves be victims of abuse in families where fathers are molesting their daughters. Women may also be constrained from removing themselves and their children by economic factors, as we discussed in the chapter on divorce and the impoverishment of women.

Feminists also question the validity of blaming incest by fathers on inadequate or sexually inattentive wives. Even if the mothers are weak or inadequate wives and mothers, it is not they who are the sexual abusers.

Feminists argue that abusive men—fathers, stepfathers, grandfathers, and uncles—may have problems and may be victims in their own lives, but they are still responsible for the assault on their children. The critical point of the feminist analysis is not that complex relational issues are nonexistent or unimportant but that they are secondary to the sexual assault of children by male relatives.

Radical Feminist Theory on Incest In the discussion above the term radical feminist has been used to describe the work of scholars who have reviewed and critiqued the conservative model of understanding incest. "Central to radical feminism is an intense positive valuation of women and as part of this, deep grief and rage over their oppression" (Lengermann & Niebrugge-Brantley, 1992, p. 474). Radical feminism sees society as permeated by oppression of many types—race, class, age, and sexuality. But gender oppression is the most salient and the focus of their attention. From the radical feminist point of view, the most basic power resource that men hold in their oppression of women is physical force, with violence as the last line of defense in the maintenance of the system (Lengermann & Wallace, 1985).

Radical feminists use the term patriarchy to describe the system of oppression of women that they believe underlies all other systems of oppression (Firestone, 1970). In Chapter 2 I discussed the dual nature of the term patriarchy. One use of the term refers to a historically specific form of male dominance. For example, patriarchal families in Puritan communities were ones in which fathers dominated entire households, including younger men, women, and children who were both kin and nonkin. In a second, more general sense patriarchy means the dominance and control of women by men in nuclear families and extended households, as well as in nonfamily social institutions such as government and work.

A number of radical feminists have done research on the organization of this broad form of patriarchy and especially its link to violence against women (Barry, 1981; Dworkin, 1987; Frye, 1983). Herman and Hirschmann (1977) have used the concept of patriarchy to explain the particular problem of incest.

Herman and Hirschmann (1977) have examined the relationship between incest and the structure of patriarchy and conclude that (1) the social sanctions against father-daughter incest are relatively weak in a patriarchal family system like that of the contemporary United States; and (2) within individual families, incest occurs most frequently in families characterized by extremely dominant fathers or stepfathers. Incest, they conclude, is a result of patriarchal families within a patriarchal society (Herman & Hirschmann, 1977, p. 741).

In further research Herman (1981) began to specify the characteristics of patriarchal incestuous families. She compared families where fathers and daughters had seductive but not sexual relationships to families where incest had occurred. She found that incest was more likely to occur in families that (1) rigidly conformed to traditional sexual roles, (2) had fathers who dominated through the use of force and expressed no contrition for their behavior, and (3) had mothers who were physically or psychologically disabled (Gordon & O'Keefe, 1984).

Our images of children tend to fall into three categories: children as learners, children as threats, and children as victims. Children actually play a much broader range of roles, including that of political activists when society's problems affect their lives.

Poor Families, Poor Children A second type of victimization of children is through poverty. Table 5-3 in Chapter 5 shows the proportion of families with children under age 6 that live in poverty. The figure shows that the proportion is large and has been growing in the past two decades. It also shows that race and family type have an especially important effect on whether children are poor or not.

Table 5-2 in Chapter 5 shows that in 1990, 15.1 percent of white children were poor, while 44.2 percent of black children and 39.7 percent of Hispanic children were poor.

The Children's Defense Fund (CDF), an organization in Washington, D.C. under the direction of Marian Wright Edelman, conducts research and publishes reports on issues related to the well-being of children. The CDF also monitors federal and state policies and provides assistance for educating and organizing advocates for children.

As part of its work, the CDF published a summary of each state's successes and failures in addressing the problems children face. This "report card" concludes that while the U.S. earns an "A" in its capacity to care for our children, it earns an "F" for performance (Children's Defense Fund, 1990).

One in five children live in poverty and that number is growing. Between 1979 and 1988, the proportion of children living in poverty grew by 23 percent. Every night an estimated 100,000 children go to sleep homeless (Children's Defense Fund, 1990).

Table 13-4 shows that the chances of children being poor are greatly increased if they are black or Hispanic, if they are living in a female-headed family, if they or the head of their family is young.

The U.S. does not rank well in comparison to other nations. In one study of eight industrialized countries—Australia, Canada, Sweden, Switzerland, United Kingdom, United States, and West Germany—the U.S. had the highest rate of

TABLE 13 • 4
U.S. Children's Chances of Being Poor

If white	1 in 7
If black	4 in 9
If Hispanic	3 in 8
If in a female-headed family	1 in 2
If younger than 3	1 in 4
If 3 to 5	2 in 9
If 6 to 17	2 in 11
If the child is also a parent	7 in 10
If the family head is younger than 25	1 in 2
If the family head is younger than 30	1 in 3

Source: Children's Defense Fund, 1990, p. 28.

childhood poverty. The U.S. child poverty rate was two to three times as high as in the other nations (CDF, 1990).

Every day in America twenty-seven children die from poverty-induced causes. Although this number is shocking, it is not surprising when we look at the U.S. infant mortality rate, especially as it compares to other nations. When the infant mortality rate of the world's nations is ranked from the least to the most, the U.S. ranks eighteenth.

The infant mortality rate is the number of children that die before their first birthday among every 1,000 babies born. Infant mortality is related to factors like nutrition, housing, and access to health care. All of these factors are related to poverty. Because the U.S. is such a wealthy nation it is surprising to find that the infant mortality rate is so high.

Despite the wealth of the nation as a whole, 10 babies died for every 1,000 live births in the U.S. in 1988 (Children's Defense Fund, 1990). This rate is higher than in nations that are much poorer than the United States, such as Singapore, Hong Kong, Spain, and Ireland.

Black American children are at especially high risk. "A black child born in the inner-city of Boston has less chance of surviving the first year of life than a child born in Panama, North or South Korea or Uruguay" (Children's Defense Fund, 1990, p. 6). The poverty, inadequate shelter, and access to health care that result from racism take a heavy toll on black children.

But children do not have to die to be devastated by poverty. In 1991 the Stanford Center for the Study of Children, Family and Youth interviewed homeless children to find out how homelessness affects them. The research team asked the children what they would wish for if they had three wishes. One child said, "I wish we could be less poor. Not rich. But have enough to eat" (Stanford, 1991, p. 22). Another wished for a car, a bed, and a kitchen for his family. One boy listed: my father having enough to buy food, money to rent an apartment, and if there was money left over, a bike. The children were also asked what they liked most about themselves. One ten-year-old girl answered, "That I haven't gone crazy yet" (Stanford, 1991, p. 25). The comments of these children show that poverty has deprived them of the most fundamental human needs and that their hopes and even their ideas about themselves have been dashed by homelessness.

How does the United States compare to the rest of the world in its policies on children and poverty? The CDF compiled a list of contrasts on several issues related to the economic well-being of children in the U.S. and other countries.

- Seventy nations provide medical care and financial assistance to all pregnant women. The U.S. does not.
- Sixty-one nations insure or provide basic medical care to all workers and their dependents. The U.S. does not.
- Sixty-three nations provide a family allowance to workers and their children. The U.S. does not.
- Seventeen industrialized nations have paid maternity/parenting leave programs. The U.S. does not.

The Children's Defense Fund argues that the well-being of children cannot be left to their parents alone, but that the public and its representatives must act to raise children out of poverty. To encourage officials to take action to invest in our children the CDF asserts that we must write or call the president and our state and national legislators. The CDF suggests that we educate our representatives by bringing them to visit children in daycare centers, public housing, schools, and homeless shelters to see the problem for themselves. If our representatives do not respond to this type of education, the CDF suggests that we might run for office ourselves and that we must at least monitor public officials and use the ballot box to make our views known.

This point of view implies a focus on sources outside of families to take responsibility for children. In the scenario at the beginning of the chapter, Rodriguez told the class how his mother had found a way to survive the fire that destroyed their family's house and belongings. The solution she found was an individual one and it was only partially successful. Rodriguez says they are sure poor now. Should his mother be left to solve these economic problems on her own? Who is responsible for children? Should their fate be left in the hands of parents alone or should we, as the CDF suggests, call on the community and the government to provide for their well-being? These questions will be further examined in the next chapter.

THEORETICAL PROBLEMS IN THE STUDY OF CHILDREN

Thus far this chapter has described three dominant images of children in our society. In this section we will turn our attention to two theoretical issues that emerge in our attempts to understand who children are and where they fit in families and in their community. The first issue addresses the question of how we conceptualize children and childhood and how that has changed historically. The second issue addresses the problem of the limiting character of the three dominant images and how feminist theory might help us to uncover a fuller and more realistic picture of children and childhood in our society.

How Do We Think of Children and Childhood?

Most of us think of childhood as a natural process that is similar in all societies and historical periods. Human beings go through a fairly long period of being generally smaller and weaker, frequently less knowledgeable about their society and the world, and certainly less powerful than adults. In our society, we take it for granted that the people who fit into this category should be set off from the rest of us and given a special name: children. We also assume that these child-humans should be treated differently: in dress, responsibility, activities, and authority.

Historical evidence suggests that it is not valid to assume there is a natural category of children. Furthermore, it is not valid to assume there is an unchanging way to experience being a child—childhood—in all societies. Childhood varies from one society to another and from one historical period to the next.

Sociologists speak of childhood as being socially constructed. This means that, although it is a physical fact that human children are younger, smaller, weaker, and generally less experienced, they do not necessarily constitute a separate category that must be treated differently. They must wear clothes, for example, that are smaller, but they do not need to wear clothes that are different from adults' in any other way. They cannot do all of the tasks that adults can do because they are not as strong, but they need not necessarily be restricted from working in factories, mines, fields, and offices.

When we treat children differently from adults (as we do in our society), we have constructed a social category called children. Along with the category, we have also constructed a set of characteristics appropriate for the people in the category and a set of relationships others should have with those people.

One of the strongest pieces of evidence to support the argument that the concepts of children and childhood are socially constructed and not inevitable is the way in which children and childhood vary historically. Phillip Aries (1962), a French historian, was one of the first modern scholars to investigate the concept of childhood. In his research on Europe he discovered that both the perception and the experience of childhood changed over time.

Nancy Mandell (1988) argues that children's status has moved through three distinct stages. The first stage characterized European society before 1500 and might be called the period of indifference toward childhood. The second stage ran from the 1600s until the mid-1800s and could be called the emergence of childhood. The third period is the ascendency of childhood, which dates from the mid-1800s and continues to develop.

John Demos (1986) examined childhood in U.S. history during the period of the emergence of childhood. He discovered that childhood was quite different during the colonial period from what it is today. Children were more similar to adults and the period that we would recognize as "childhood" was much shorter:

> If childhood is defined as a protected state, a carefree period of freedom from adulthood responsibilities, then a Puritan childhood was quite brief. Childhood abruptly came to an end around the age of seven. (Mintz & Kellogg, 1988)

Puritan children worked alongside adults doing the same kinds of work that their elders did—caring for domesticated animals, gardening, spinning, making candles, and preparing food. Sons were miniature versions of their farmer fathers and daughters were models of their mothers (Demos, 1986).

In addition to doing the same work as adults, children in this period also looked more like their parents. Before age six both boys and girls wore frocks or petticoats. At around age six they began to wear the same clothes as their parents. Boys wore breeches, shirt, and doublet; girls wore caps, chemises, bodices, petticoats, and skirts (Mintz & Kellogg, 1988).

Children were not seen as equal to their parents, but as morally and physically inferior. The difference between children and adults, however, was not in what they did but how well and how much they performed (Demos, 1970). On the other hand, children were not seen as a special group with their own needs and interests, as they are in our society (Demos, 1970).

Throughout the last three centuries children have increasingly been differentiated from adults. Labor laws prohibit them from taking most jobs. Huge industries in children's clothes, toys, and media, like children's books and cartoons, have been created. Organizations like the Boy Scouts, Girl Scouts, and Little League have been developed to allow children to participate in children's activities.

Public education separates the lives of children from adults and further segregates children by age in class levels and in schools—elementary, middle, junior high, and high school. Childhood in our time is perceived as a period that is very different from adulthood, and childhood itself can be differentiated into many stages.

In the scenario at the beginning of the chapter, one girl, Neu, tells how her parents' divorce has meant that she must spend a lot of time working in her family, caring for her brothers and sisters and doing the housework. Because we think of childhood as a period in which children should participate in "children's" activities and not those of adults, we might respond to her story with pity for her and criticism for her parents. How are our responses shaped by the dominant ideas about childhood in our society? Are other conceptualizations of childhood, like that of Puritan society, wrong or inhumane?

The review of this literature suggests that ideas about childhood and children are shaped and even determined by the social era in which they exist. In this chapter we have observed that contemporary ideas about children in the United States are dominated by three images. A theoretical point of view that acknowledges the way concepts of childhood and children are affected by historical context suggests that these images are a product of our times. The images are an aspect of the macro level of society that influence our thinking and our experience but the images in turn are created by other macro-level factors that predominate in this particular historical period.

How Accurate Is Our Image of Children?

The second theoretical issue that has emerged in our examination of children is the question of the accuracy of the images of children in our society. Thorne (1987) argued that the three images (of children as learners, threats, or victims) do not capture the range of experience and influence that children have.

Medrich and his colleagues (1982), for example, found that children's lives are more varied than the three dominant images suggest. They found that in addition to the time they spend in school, children spend time in five other kinds of activities: activities with their parents; working both at home and outside the home; participating in organized activities like sports and cultural programs; watching TV; and spending time on their own in unstructured activities. On an average weekday children spent about seven hours a day in school; three to four hours watching television; two to three hours on their own (this category included activities with friends or alone and excluded watching TV) playing, reading, doing homework, and relaxing; one to one and a half hours with their

TABLE 13 • 5
Percent of Total Population Under 19, 1970–1991

	1970	1980	1991
Under 5	8.4%	7.2%	7.5%
5–9	9.8%	7.4%	7.3%
10–14	10.2%	8.1%	6.9%
15–19	9.4%	9.3%	7.2%
Total Children	37.8%	32.0%	28.9%

Source: Statistical Abstract of the U.S., 1992, 1991, p. 19.

parents; less than an hour on chores; and a little less than an hour on organized activities like lessons and athletic teams. Medrich and colleagues' (1982) findings, as well as those of Purrington and Boulding, suggest that children are involved in many activities about which we know little and which do not fit easily into the categories of socialization, threats, and victimization.

Medrich and colleagues' (1982) work, however, is unusual. Part of the limited nature of the three dominant images may be a result of the lack of research directed toward understanding children and childhood. Family scholars have focused their attention much more on the experience of adults than on that of children. For example, in 1985 Viviana Zelizer noted that the latest edition of the *International Encyclopedia of the Social Sciences* had only two listings under *child*: child development and child psychiatry.

Table 13-5 shows that the proportion of the population that is under 20 is declining but children remain a significant percentage of our population. And before the 1990s people under the age of twenty never constituted less than 30 percent of the American population. The study of children would seem to be an obviously important area. But the study of children and childhood has been much less attended to than other issues by many scholars, even those who study families.

Thorne (1987) argues that feminist family scholars, in particular, need to pay more careful attention to the experience of children because the fates and definitions of women have been so closely tied to those of children. Children are intertwined in many women's lives and in the images, expectations, and controlling mechanisms that exist for women in our society. Scholars who are interested in women's lives must look at the relationships between children and adults, especially women, in order to unravel the question of gender inequality.

Children must also be studied in their own right, and the insights that feminists have brought to the sociological study of women can be useful in helping us examine the experience of children. Thorne (1987) says that feminists have re-visioned women, acknowledging that (1) women are not just passive victims but are active agents in society; (2) women are diverse; and (3) gender exists in all corners of our society, in all social institutions. These principles need to be applied to our study of children, recognizing their agency, diversity, and activity throughout society.

In this chapter we have focused our attention on the three images of children in the U.S., as learners, threats, and victims. Most of the research on children fits into these three categories. If we take Thorne's advice, however, we will need to broaden our vision of children to examine more fully their position in society as sexual beings, workers, athletes, teachers, artists, leaders, and in any number of other roles, including as political activists.

RESISTANCE AGAINST CHILD ABUSE AND INCEST

This chapter has reviewed much research on children in the United States and has offered some suggestions about the kinds of theoretical questions that have been uncovered and need to be studied further. In this section we will look at a few examples of the political activities that have developed in the effort to improve children's lives. The discussion of child abuse and incest reveals some of the most devastating information about the condition of children in our society. Not surprisingly, this is where political activists have focused their efforts.

Resistance against child abuse and incest has a long history. In Gordon's (1988) investigation of records of incest, she found that girls who were victims of incest were likely to resist by running away, fighting back, or informing the police or others. In addition to these individual acts of rebellion, organizations have also been active in seeking recognition of the problem and in proposing specific policy changes to increase the power of children in our society so that they might be better able to protect themselves from assault and so that they might be better able to see that their assailants are punished.

As we noted in the discussion of incest, Gordon (1988) studied the records of child abuse, incest, and neglect cases from 1880 to 1960. Gordon's data came from the case records of "child saving" agencies in Boston. In earlier years the clients of the agencies were mainly poor Catholic immigrants from countries like Italy, Ireland, and Canada. In later years clients were likely to be African Americans and West Indians.

The cases revealed at least three sets of political relationships: (1) between husbands and wives; (2) between parents and children; and (3) between white Anglo-Saxon Protestant (WASP) social workers who represented the dominant culture and its agencies of control, and working-class racial ethnic people who were their clients.

Each of these three political relationships included oppression of the subordinate group. Men battered women and sexually abused children. Women abused and neglected children; throughout the period studied mothers and fathers were approximately equally likely to be the assailants in reported incidents of child abuse. Social workers degraded, disrespected, punished, and attempted to control their clients.

At the same time, the victims in each of these relationships were also likely to resist actively. As a way to draw attention to the importance of the resistance Gordon (1988) entitled her book *Heroes of Their Own Lives*. As we mentioned above, incest victims resisted their fathers' assaults. Battered women and abused children fought back against their husbands and parents. And communities, families, and individuals struggled against "the authorities" for the right to create their own definitions of family, morality, and childcare.

A key characteristic of Gordon's analysis is the political quality of the relationships that make up the issue of violence. Like the radical feminists discussed in the section on incest, Gordon centers her discussion around the concepts of oppression and control.

The picture Gordon presents of the history of family violence differs from the one presented by radical feminists, however, in its complexity. Gender inequality is at the heart of the violence she examines, but equally important are generation and class inequality. The complexity of the situation is further illustrated by the the active resistance to oppression by women and others that her work uncovers.

Another illustrative example of resistance work occurred in May, 1992. A national petition for sexually abused children's rights was presented to Congress during the Mother's Day Rally for Children's Rights (*Off Our Backs*, 1992, p. 1).

Child Victims of Sexual Abuse: A Bill of Rights

1. Each child victim should be entitled to have his or her case heard by a judge trained and educated about child development and the full effects of child sexual abuse.

2. Each child victim should be entitled to a lawyer GAL (trained Guardian Ad Litems) who would be that child's guardian angel advocate and oversee the work of the system including the lawyer GAL representing the child. The lawyer GAL should be trained and educated and limited in their work to being only a child advocate, not representing the parents or the offender.

3. Child protection issues separated from divorce court. Each child victim should be entitled to have legal matters pertaining to their protection heard by a different judge than the one hearing the divorce case. The child will not be a pawn in a property dispute.

4. Each child victim should have the legal right to hold any adult accountable for abusing them, participating in that abuse or for not protecting them from further abuse. This includes the legal right to file lawsuits against doctors, lawyers, GALs and any child protective service or government agency.

5. Burden of proof regarding visitation shifts to abuser. Once abuse has been demonstrated to the degree to warrant protection, the abuser should assume the burder of proof necessary to show that visitation would be good for the child, as opposed to the current system in which the protecting parent must show that visitation with the abuser would be bad for the child.

6. The right to confront their abuser but with protection. Each child victim should have the right to testify with permission from their therapist, so they can feel empowered to confront their abuser with the truth of the assault. However, the courts, prosecutors and GALs will be innovative via the use of TV and media technology to protect the child from further abuse and threat within the courtroom.

7. Child victims should have the same rights as adult victims to have lawbreakers punished and not to have their sentences lessened because the victim was a child or family member.

8. Child victims should be entitled to examination by doctors and experts who have extensive experience in child development and child sexual abuse issues.

9. Children should have the same legal and constitutional rights as adults have.

Although the effect of this petition has not yet been realized, the declaration shows the way in which the micro level continues to attempt to affect the macro level. This Children's Bill of Rights, along with activists in the Children's Defense Fund and radical feminists speaking out against child molesters, play an important role in shaping the social institutions that define and determine children's lives in the United States.

THE MICRO-MACRO CONNECTION

This discussion of children in families and children's families in society helps us to see the multiple layers of macro-level forces. Our society makes children primarily the responsibility of their families, but as we have noted neither families nor children exist in a vacuum. Families play an important role in many children's lives but they are not the only factor.

Other macro-level institutions that affect children's lives directly and through families include an economic system that does not provide sufficient access to food, shelter, and health care for many families. This makes children, who are economically dependent on those families, especially vulnerable and frequently unable to survive. Children's lives are also affected by a legal system that does not adequately protect them from sexual and physical abuse. The media, in addition, play a role in limiting children's opportunities, particularly in terms of their ability to see more gender-free choices.

Ideologies are a part of the macro forces in society. They help us to see or ignore what exists and they provide us a way of understanding what is happening, why it is happening, and what its significance is. Children are a large segment of our population and an important factor in many families but they are strangely invisible in much of the work done on families. When our ideas about children assume that children and childhood are something natural they need little explanation. But the historical evidence suggests that children and childhood are socially constructed and therefore need intense examination. Furthermore, that examination should pay attention to children not as just victims, learners, and threats but as social actors in all arenas of human life.

Macro-level forces in society are numerous and they are also contradictory. Ideologies that have paid attention to issues related to children, for example, are not in agreement. In this chapter we have examined competing sets of ideas about how gender is learned and maintained. We have also looked at two competing models of how to understand and address the problem of incest.

Political action by feminists and organizations like the Children's Defense Fund show us the ways in which the micro level can affect the macro level. These groups claim that in order to provide humane environments for children, our ideas about children and their relationships with families must be assessed and changed by activities ranging from scholarly work to running for office. CDF challenges fundamental ideas about who should be responsible for children—individual families or the broader community. Radical feminists, who have worked on issues such as incest, insist that we reexamine and change our assessment of relationships within families.

Gordon's uncovering of the resistance of children to child abuse and incest also provides an example of the way in which the micro level can put pressure on the macro level for social change. Bem's review of gender socialization in childrearing techniques and Peters' review of race socialization in childrearing also provide examples of resistance and the ongoing active struggle of people to reconstruct their societies in more just ways.

SUMMARY

Modern American Children

- The opening scenario provides a picture of the diverse and problematic issues that face children.
- Thorne proposes that three images dominate our society's view of children: children as learners of adult culture; children as threats to adult society; and children as victims of adults.
- Gender socialization is an important kind of social learning in children's lives.
- There are four models for understanding how gender socialization takes place: psychoanalytic, social learning, cognitive-developmental, and gender schema.
- Two other sources of gender are explored by social constructionists and social structural theorists.
- Racial socialization is another form of socialization that takes place in families.
- The second image of children is that they are threats. Children are threats to their families in three ways: parents are not as mentally healthy as nonparents, parents are not as happy with their marriages as nonparents, and parents have greater economic demands than nonparents.
- Children can also have positive effects on their parents by nurturing them, providing opportunities for parents to feel pride in taking responsibility and legitimating more open reactions to experiences.
- The third image of children is that they are victims. Children are victims of abuse, which can be physical or sexual, and of poverty.
- Two frameworks for understanding incest are a radical feminist analysis and a conservative psychoanalytic analysis.

Theoretical Problems in the Study of Children

- There are a number of theoretical issues related to understanding the experience of children in families. In addition to those that have already been reviewed there are two overriding theoretical issues: the socially constructed view of childhood and the use of theory to broaden our view of children and childhood.

Resistance Against Child Abuse and Incest

- Resistance movements have developed to fight the sexual and physical abuse of children, including both individual acts of resistance by victims and organized group efforts. One example of the latter is the Children's Bill of Rights presented to Congress in 1992.

The Micro-Macro Connection

- Macro-level institutions like the legal system, the economy, the media, and ideology affect children and their role in families.
- The activities of organizations like the Children's Defense Fund and race and gender socialization practices show how individuals have sought to alter at both the micro and the macro level of society.
- Gordon's discovery of the ways in which victims of abuse have resisted that abuse and actions like the development of the Children's Bill of Rights show further examples of the micro level effecting the macro level.

Marxists argue that the government works to serve business interests. This Gwich'in Athabascan family traveled 800 miles by boat to protest when the U.S. government allowed oil corporations to drill in the Artic Refuge.

14 FAMILIES, FAMILY POLICY, AND THE STATE

We want jobs, low income housing and our rights as true American citizens. We want to reinvent the social system of America from the bottom up, not from the top down. Remember "We the people"? We are following American tradition. It was the poor people that fled all those countries and came to America to make it wealthy and great and what it is today. The best story in my history book was the one about the Boston Tea Party where they weren't being represented and so they dumped all of the tea into the ocean. And I feel the same way. Like we're not being represented by the king and queen of America and we've gotten so far away from democracy. (Mary Uebelgunne, a homeless woman from North Carolina who organized a homeless organization called Home Street Home, speaking at a rally, quoted in Aulette & Fishman, 1991, p. 247)

Women are more aggressive when they see something harming their kids. . . . When the strikebreaker or the scab, he's taking her husband's job away, taking the food away from your kids, taking the shoes off their feet, I think women are the first to respond to that. . . . They're seeing the destruction of the family, and they are going to come out with tooth and nail. (Anna O'Leary, president of the Morenci Miners' Women's Auxiliary, an organization of wives of copper miners who struck against Phelps Dodge, quoted in Aulette & Mills, 1988)

Native women struggle against government and corporations, as well as against all individuals or interest groups who will stop at nothing in their dedication to obliterating the American Indian people. We must work to maintain tribal status; we must make certain that the tribes continue to be legally recognized entities, sovereign nations within the larger U.S., and we must wage this battle in a multitude of ways—political, educational, literary, artistic, individual, and communal. We are doing all we can . . . we daily demonstrate that we have no intention of disappearing, of being silent or of going along quietly with our extinction. (Paula Gunn Allen, a Native American woman writing in 1986, p. 410)

So what we did was we organized to fight back and to take what's ours because we are the sons and daughters, sisters and the brothers and the aunts and the uncles of the people who built this country and this is ours. And we challenge you, America, to take back and distribute the wealth among the people that need it. We're talking about the welfare movement. We're talking about the mothers with the babies. We're talking about childcare. We're talking about education. And that's why I'm here today and I'm pretty sure that's why you're here because we're tired. But we're not that tired and we're going to keep on struggling and the fight has just begun. (Leona Smith, a welfare mother, speaking out at Survival Summit in Philadelphia, quoted in Aulette & Fishman, 1991, p. 247)

These voices are those of people who want to see change in America and have called on the government as a place to begin making those changes. They have also faced the government as an adversary. Anna O'Leary's organization, for example, was attacked by the Arizona National Guard, called in by Governor Bruce Babbitt during demonstrations in support of the strike.

- *What do these people want for their families and what do they think the government can do for them?*
- *What do these people think is wrong with the government?*
- *What values do the voices express about democracy, freedom, rights, and support?*
- *What is a government and what is its relationship to the nation/state and social policy?*
- *The U.S. is one of a few wealthy industrialized countries that does not have an explicit, official family policy. Based on the statements made by these people, what kinds of issues do you think would need to be part of a family policy in the U.S.?*

In answering these questions you have had to think about what a government is and how it operates in a capitalist country like the United States. This chapter will discuss various points of view about the character and activities of our government, especially in regard to problems that families face.

Throughout this book, we have observed the way in which people have looked to the government as an institution to address the problems their families experience. For example, in Chapters 2 and 3 we examined the way in which people successfully changed public policy, laws, and even the government itself when the women's movement fought for the right to vote and own property, when the labor movement demanded that the government pass family wage laws, and when slaves and other antislavery forces toppled the slave system. In Chapters 4 through 7 we examined the more contemporary demands of housing for the homeless, family leave legislation, civil rights for racial ethnic people, and wages for housework. In Chapters 8 through 10, we added to this list the right to marriage for gay couples and divorce reform. Chapters 11 and 13 explored the need for changes in policy and law to protect children and adults from violence in families and Chapter 12 reviewed the growing movement around government support of childcare.

The discussions of these issues in the previous sections described the different groups and their points of view on what needs to be changed and how. The way in which the government is organized and the structural and ideological roots of that organization have not been emphasized. In this last chapter we will take a closer look at the organization of the government and its potential for supporting families making needed changes. The discussion of the government will center around one particular set of laws and policies that have been especially important for families, the welfare system.

The chapter is divided into five major sections. The first is a brief discussion of terms that are used to talk about public policy, like welfare state policy and family policy. The second section describes the current organization of the social policy that is the focus of this chapter, the welfare system. The third section reviews the historical roots of the welfare system and recent trends. The fourth section addresses the question of why the welfare system and family policy in general seem so inadequate in the U.S. by discussing three theoretical conceptualizations of the nation state. In the final section, the activities of people resisting current policies and seeking to reformulate welfare and the government are explored.

THE WELFARE STATE AND FAMILY POLICY

Since World War II, most of the industrial capitalist nations, including the United States, have come to identify themselves as welfare states. Atherton (1989) distinguishes between two types of welfare states: programmatic welfare states and redistributive welfare states. Programmatic welfare states are ones in which a "capitalist state devotes a portion of its gross national product, through taxation, to the solution of certain social problems without changing the basic nature of the economy" (Atherton, 1989, p. 169). Redistributive welfare states are those whose primary aim is to restructure the economy by redistributing wealth and resources. For example, suppose a company decides to close a plant because it wishes to move its operation to Mexico where labor is cheaper and profits higher. A programmatic welfare state would provide help to workers and their families in the region where the plant closed. Benefits like unemployment, AFDC, and food stamps would be provided to the families whose breadwinners had been laid off. A redistributive welfare state might take more radical action, for example, by making it illegal to close a plant. In this case the resource that would be redistributed is the power to decide whether a plant should move. The goal of the programmatic welfare state is to help people get by when the normal functioning of the economy causes them problems. The goal of the redistributive welfare state is to alter the "normal" functioning of the system. According to Atherton, the U.S. is a programmatic welfare state.

Within these two broad categories is a range of support for the welfare state. While most Western industrial nations would be considered programmatic welfare states like the U.S., they spend a larger proportion of their GNP on welfare and they have older and more extensive programs (Chelf, 1992).

Governments develop and implement these kinds of programs through policies. Policies are lines of action that allow countries to extract resources from and control people residing in them (Skocpol & Amenta, 1986). Policies include nearly everything the government does, from collecting taxes to imprisoning criminals.

Social policies are government "activities affecting the social status and life chances of groups, families and individuals" (Skocpol & Amenta, 1986, p. 132). Such policies would include, for example, the provision of public education, regulations affecting working conditions and the environment, and protection of vulnerable populations like children, the elderly, and the disabled. As a programmatic welfare state, the U.S. has an enormous array of social policies that are designed to help provide people with food, shelter, health care, and education. All of these programs have a powerful influence on families (Zimmerman, 1992). The United States, however, remains unusual among developed industrialized nations because it does not have an official family policy (Bane, 1980; Kamerman & Kahn, 1978).

The lack of an explicit family policy in the U.S. has been seen as problematic. First, without an overarching organized family policy, the policies that do touch families lack continuity, vision, and coherence. For example, one agency or program may provide work training for parents, while another provides day-care support and still another health care benefits. These programs may or may not be coordinated with one another.

Second, Alvin Schorr (1962) contends that problems can occur when services emerge for families by accident or as outgrowths of policies intended primarily for attaining some national objective other than serving those families. Because they were not intended to address the needs of families they may be distorted by the interests they were designed to serve and, therefore, less useful to families. For example, Schorr describes the Housing Act of 1949 as a policy that directly affected families. The intention of the act, however, was not to serve families but rather to guarantee housing starts in order to avoid postwar deflation (Moen & Schorr, 1987). Choices about where and for whom the housing would be built were based on the goal of providing economic support for communities and industries, not necessarily to provide shelter for individual families.

SOCIAL POLICY AND FAMILIES: THE CASE OF THE WELFARE SYSTEM

In this section we will look at one set of social policies that has been particularly important to families: welfare policy. One of the first observations we can make about social policy when we look at the welfare system is that it is sometimes as important to notice what is missing as what is included. In the United States most of the care of families is not provided by the government. An official social policy is missing in the provision of welfare even in cases where it is sorely needed.

One could say that the largest welfare system in the nation is the one provided by individual families. Moen and Schorr (1987) illustrate this view by quoting Reuben Hill, who wrote in 1949 but whose words are still valid:

> For too long the family has been called upon to take up the slack in a poorly integrated social order. If fiscal policies are bungled producing inflation, the family purse strings are tightened; if depressions bring sudden impoverishment, family savings and families' capacity to restrict consumption to subsistence levels are drawn upon; if real estate and building interest fail to provide housing, families must adapt themselves to filtered down, obsolete dwellings or double up into shoehorned quarters with other families. (Hill, 1949, p. ix)

The second largest welfare system in the U.S. is the one provided by the government (Moen & Schorr, 1987). Although, as we have observed, the United States does not have an explicit, official family policy, the welfare system is a primary example of what is considered to be a feature of family policy. The welfare system provides an example of a large and influential social policy that has had great importance for families. In 1986 a survey by the Census Bureau showed that within the previous thirty-two months, about one in every five Americans had used the welfare system for a month or more (Chelf, 1992). The Census Bureau further reported that about one-half of the households in the United States have used the welfare system at some time (Chisman & Pifer, 1987). This section describes the system before we move to a discussion of its historical roots and current direction.

Table 14-1 outlines the organization of social welfare services in the United States today. One distinction that can be made among services is that only some are means-tested. Means-tested programs are ones in which recipients must prove that they need the service and that they cannot provide it for themselves. For example, unemployment is an entitlement program that is not means-tested. Any person who qualifies by such criteria as length of employment and being laid off can collect unemployment. Unemployed people do not have to prove they are impoverished. They are not asked about how much money they have in savings accounts or whether they own property. They are entitled to the benefit because they had a job that was covered by the program and now they are unemployed. AFDC recipients, in contrast, are means-tested. They must prove that they are impoverished and that they have no other way to support themselves in order to receive a grant.

A second distinction among services is that they provide either cash assistance or in-kind support. Cash assistance refers to programs that provide grants to eligible persons who can then decide how best to use the money. In-kind

TABLE 14 • 1
The Classification of Social Welfare Services

SERVICES FOR PEOPLE WHO ARE ECONOMICALLY DEPENDENT
1. Cash Support Programs
Social Security
Public Assistance
AFDC for families with minor children
SSI for the aged, blind, or disabled
GA for indigent adults
VA benefits for veterans
Unemployment for workers who lose their job
Worker's Compensation for workers who are injured
2. In-Kind Programs
Medical services
Medicare for older people
Medicaid for poor people
VA for veterans
Community maternal and child services
Native American health services
Shelter
Public housing
Subsidized housing
Energy Assistance
Nutrition
Food stamps
Child nutrition, school lunches

Source: Popple and Leighninger, 1993, pp. 41–43.

programs provide specific services such as medical care and housing, and vouchers that can only be used to purchase specific items. For example, food stamps can only be used to purchase food.

The cash assistance programs for poor families and individuals are divided into two tiers. One tier is mandated and funded by the federal government. The second tier is designed and funded by state and local agencies. The result of this two-tiered system is dramatic variation from one state to another in terms of programs and amounts of grants. For example, the federal government demands that grants be paid to poor families with minor children. This is called Aid to Families with Dependent Children or AFDC; when the grant is to a family with two unemployed adults it is called AFDCU. Only seven states (Connecticut, Massachusetts, Minnesota, New Jersey, New York, Ohio, and Pennsylvania) *also* provide for adults who live in households without children, a benefit called General Assistance or GA.

At any given time, the amount of money granted to households also varies greatly by state. Although the federal government mandates that AFDC benefits be paid to eligible single-parent families with minor children, no minimum benefit is stipulated. Table 14-2 shows the variation among states in welfare benefits for a family of four. Maximum monthly grants for a three-person family in 1990 ranged from $121 in Alabama to $720 in Alaska. The median benefit for a single parent with two children was $359 per month.

Although grants vary from state to state they are low in all states. In forty-six states and the District of Columbia, the maximum monthly benefit for a

TABLE 14 • 2
Average Monthly AFDC Payment per Family, 1990, by State

DIVISION AND STATE OR OTHER AREA	Average monthly payment per family 1990
U.S.	$392
New England	396
ME	517
NH	425
VT	415
MA	518
RI	510
CT	508
Middle Atlantic	593
NY	463
NJ	530
PA	364
East North Central	391
OH	384
IN	323
IL	271
MI	329
WI	500
West North Central	459
MN	363
IA	496
MO	378
ND	275
SD	377
NE	290
KS	329
South Atlantic	328
DE	274
MD	293
DC	371
VA	398
WV	267
NC	243
SC	242
GA	209
FL	260
East South Central	269
KY	172
TN	219
AL	193
MS	121
West South Central	124
AR	183
LA	193
OK	169
TX	299

(continued)

TABLE 14 • 2
Average Monthly AFDC Payment per Family, 1990, by State (continued)

DIVISION AND STATE OR OTHER AREA	Average monthly payment per family 1990
Mountain	166
MT	294
ID	343
WY	282
CO	352
NM	316
AZ	264
UT	264
NV	353
Pacific	275
WA	610
OR	449
CA	368
AK	640
HI	720
	590

Source: U.S. Bureau of the Census. *Statistical Abstracts of the United States*, 1992, p. 371.

family of three with no income other than AFDC is 75 percent of the poverty line. The closest a grant gets to approaching the poverty line is in California, where a maximum grant for a family of three—$633—pays 84 percent of the poverty line.

The economic situation for welfare recipients has deteriorated over the past two decades. Between 1970 and 1987 the median decline in the maximum allowable benefit for a family of four was 31 percent. Recall that a median is the point at which half fall above and half fall below. For example, this median of 31 percent means that half the states had declines of less than 31 percent and half the states had declines of more than 31 percent. The main reason these declines have occurred is because there is no automatic adjustment for the grants to keep up with inflation.

In addition to changes in the welfare system that have resulted from changes in the economy like inflation, the welfare system also varies dramatically from one year to the next because of legislative changes. Street et al. (1979) write,

> Public assistance policies and practices are in almost perpetual flux, reflecting the complex permutations of the phasings of the federal, state, county, and local administrative and fiscal processes, the vagaries of political decisions, pressures, cycles of virtue and terror, local grassroots demands, and the organizational disarray induced by shifts in caseload size, runover of employees, and so on. (Street, et al., 1979, p. 40)

BOX 14 • 1 THREE PERSISTENT MYTHS ABOUT WELFARE MOTHERS

MYTH 1 Welfare mothers form single-parent households in order to obtain welfare.

If single-mother families were forming in order to become eligible we would expect to see a larger proportion participating in AFDC over time (Mishel & Bernstein, 1992). Figure 14-1 shows that the proportion of single-mother–headed families who receive AFDC peaked in the early 1970s and has declined since then.

MYTH 2 Welfare mothers purposely bear children in order to increase their grants.

Welfare families have an average of 2.2 children, which is smaller than the national average, and the longer a woman remains on welfare the less likely she is to have a baby. Mothers who are on AFDC have about one-fourth the births of mothers who are not on AFDC (Rank, 1989).

Furthermore, the economics of welfare grants make additional children highly "unprofitable." Welfare families increase their grant an average of $1,600 a year with the birth of a new child, not enough to provide for a child and certainly not likely to allow a mother to use childbirth as a money-making scheme. Although the empirical evidence refutes the myth that welfare mothers have babies as a way of enriching themselves, policy makers continue to create programs that are based on this false premise. "For example, in 1981 the Omnibus Budget and Reconciliation Act (OBRA) changed the rules so that pregnant women without other children are not eligible for AFDC and therefore for Medicaid coverage until the third trimester" (Sarvasy, 1988, p. 257). Conservative policy makers believed that this would discourage young women from having babies just to get on welfare. In a similar move in 1992, President Bush granted the state of Wisconsin a waiver to reduce welfare benefits to unmarried teenage women who had more children while receiving welfare. Before the changes Wisconsin paid a single mother $440 monthly if she had one child, $517 if she had two, and $617 if she had three. The proposed plan, which targeted teen mothers only, would pay $440 for one child and $479 for two or more children. California, Maine, New Jersey, and Illinois have considered similar changes (Green, 1992).

MYTH 3 Welfare accounts for a huge proportion of the national budget.

In fact, only about 1 percent of the federal budget is allocated to AFDC, a much smaller number than for defense, which is the largest single item and accounts for 28 percent of the federal budget (Children's Defense Fund, 1988b). In addition, at the same time that AFDC and other social welfare spending was cut, especially in recent decades military spending has increased, as shown in Figure 14-2.

Actually, even if we focus our attention only on social welfare programs, AFDC still accounts for a relatively small proportion of the money spent. Table 14-3 shows the division of spending. Social Security and Medicare, account for most of the spending. Only 3.5 percent of money for social welfare programs was spent on means-tested programs for poor people like AFDC and GA.

Another key characteristic of the welfare system is the differential manner in which needy people are treated by the system. From its inception, Social Security treated women and children less supportively than it did other needy groups. Social Security is a program for the people who are perceived to be the deserving poor, while AFDC is for women who are seen as having failed in their family responsibilities. For example, when the programs were initiated in the 1930s, widows (regarded as women who had become single mothers through no fault of their own) remained in the Social Security system and were not part of AFDC. Women who were in single female–headed households because of divorce or having never married were put into a separate program called ADC

FIGURE 14 • 1
AFDC Participation Rates, 1967–1987. Participation rate is the ratio of mother-only families on AFDC to mother-only families not participating in AFDC.

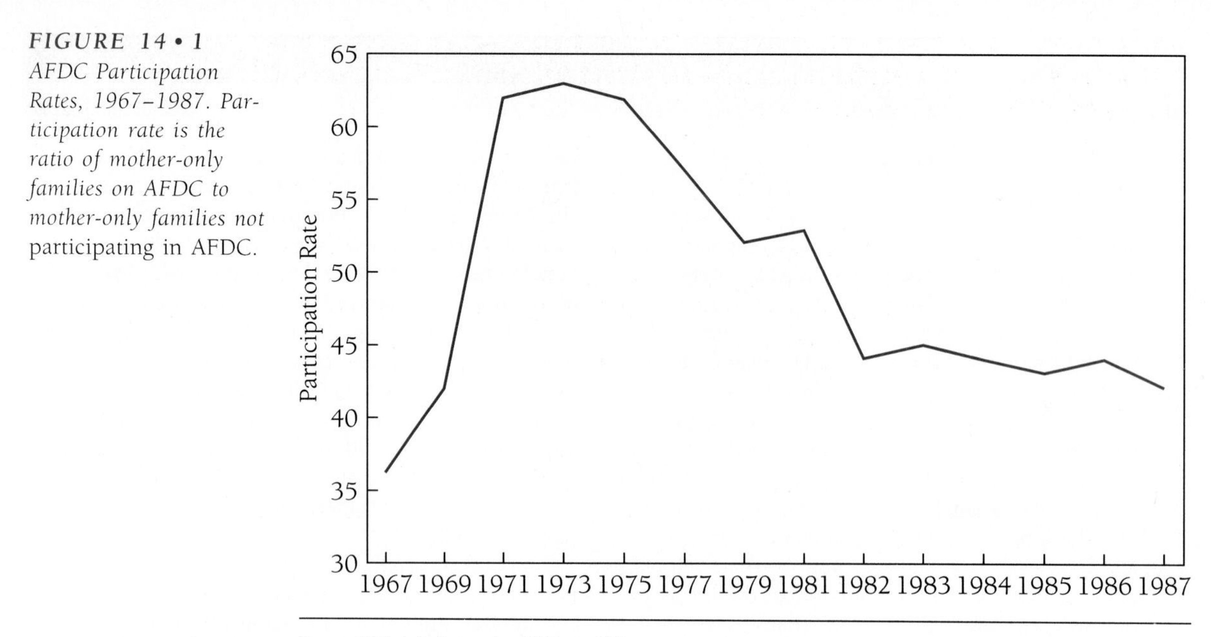

Source: Mishel & Bernstein, 1992, p. 304.

FIGURE 14 • 2
Military Spending, 1947–1991 (in billions of constant 1985 dollars)

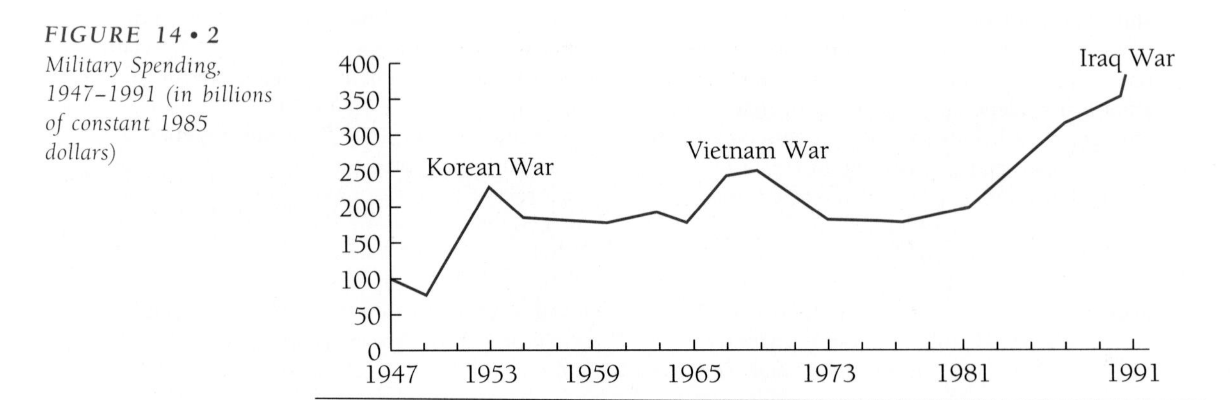

Source: Berberoglu, 1992, p. 89.

(which later became AFDC). This distinction still exists. Women who receive public money as a result of their relationship to a deceased husband remain better off and the gap between the two programs continues to grow. For example, between 1970 and 1980 the average payment to an AFDC recipient compared to a widow receiving Social Security declined from 58 percent to 40 percent (Folbre, 1987).

Welfare policy has always reflected a perception that AFDC mothers are bad women and need to have their personal lives monitored and controlled. Until the 1960s states cut aid to women who had relationships with men. For example, Arkansas did not allow aid to be given to mothers who were engaged in a "nonstable illegal union." Texas barred women from aid if they were in

TABLE 14 • 3
Social Welfare Expenditures, 1989

Social Services:	
AFDC, GA, Work incentive	3.5%
Social Insurance:	
Social Security, Medicare, Workers' Comp	68.8%
Medicaid	6.2%
Food stamps	2.4%
Supplemental Security Income	2.2%
Health care: civilian and defense hospital care and construction and medical research	4.3%
VA Programs	5.3%
Education	3.3%
Housing	2.7%
Other: for example, school lunches, vocational rehab	2.6%
Total in millions	$563,191

Source: U.S. Bureau of the Census. *Statistical Abstracts of the United States*, 1992, p. 355.

"pseudo–common law marriages." Michigan excluded women in households with "male boarders" (Folbre, 1987; Patterson, 1981). AFDC mothers are no longer monitored in this way but the perception of them as immoral and irresponsible remains.

The welfare system continues to have an influence on the organization of gender. Women are the major consumers of welfare and they are also overrepresented among workers in the welfare system: In 1980, 70 percent of the 17.3 million social service workers in the U.S. were women (Piven, 1990). Several authors have also examined the way the welfare system enforces a particular organization of gender, especially a particular role for women, one that is subordinate and oppressed (Pearce, 1989; Abramowitz, 1989; Miller, 1990; Cohen & Katzenstein, 1992). Dorothy Miller (1990), for example, argues that the treatment of women by the welfare system pushes them into marriage as the only alternative for survival. Marriage in turn operates as a way to keep women dependent on and controlled by men. Those women who insist on remaining single usually cannot maintain themselves financially because of inequality in the labor market and are therefore punished and controlled by the state through its welfare system. In Chapter 10 we explored the impoverishment of women associated with divorce and single parenting. One place divorced women might look for financial assistance is the welfare system. Our review of the system, however, reveals that it is inadequate to the needs of poor families.

In the 1970s, Johnny Tillmon, a welfare mother and president of the National Welfare Rights Organization, observed the relationship between AFDC mothers and the welfare system and the way in which that relationship represented and enforced gender inequality and the oppression of women. She summed it up this way:

> AFDC is a supersexist marriage. You trade in "a" man for "the" man. But you can't divorce him if he treats you bad. He can divorce you of course, cut you off anytime he want. But in that case "he" keeps the kids, not you.

Many people believe that welfare payments make up a large part of the federal budget. This belief is not well-founded since welfare accounts for only a small proportion while other expenses, such as defense, social security, and insurance for bankrupt savings and loans, create much larger public expenses.

> "The" man runs everything. In the ordinary marriage, sex is supposed to be for your husband. On AFDC, you're not supposed to have any sex at all. You give up control over your body. It's a condition of aid. . . . "The" man, the welfare system controls your money. He tells you what to buy, and what not to buy, where to buy it, and how much things cost. If things—rent, for instance—really costs more than he says they do, it's too bad for you. (Tillmon, 1976, p. 356)

Race ethnicity also plays a role in the welfare system (Mink, 1990). African American women are even more disadvantaged by the system than white women are. One reason African American women receive fewer benefits than white women is because African American women who receive AFDC are concentrated in the states with the lowest payments. Discrimination has also played a role in diminishing African American women's access to welfare benefits. For example, although the provision of economic assistance for single mothers and their children was developed as a national system in the 1930s, it was not until the 1960s and the civil rights movement that African American women in the south were allowed to participate. Caseworkers in the southern states assumed that "black mothers had always worked in the past, that they had more job opportunities than did white mothers, and that they had their children's grandmothers to provide childcare" and so they were systematically denied ADC (Sarvasy, 1988, p. 255).

In summary, the welfare system is a huge, complex system of programs that vary widely from one state to another, from one program to another, and from

one year to the next. The programs are inadequate for the needs of poor people and they are decreasing their ability to meet those needs. The welfare system plays a particularly important role in women's lives because women are the majority of the system's workers as well as the clients. In addition, the welfare system creates a mythical image of poor women as undeserving people who should be held responsible for their poverty and who are a serious drain on the nation's resources. Finally, the degrading treatment women receive in the welfare system may influence their lives by forcing them to stay in marriages or to enter marriages because they have no other means of survival.

HISTORY AND TRENDS FOR THE FUTURE IN WELFARE

This chapter has described the current organization of the welfare system. Next we move to a brief history of its emergence and evolution and discuss how it continues to develop in the 1990s. We will see that the welfare system is continuing to deteriorate and the insufficient funds that are provided seem to be headed toward further decline.

The Social Security system was set up during the Great Depression in the 1930s as a way to address the potentially volatile impoverishment of large sections of the American population (Abramowitz, 1989). Before the Depression, most relief for the poor came from private sources. For example, in 1928 only 12 percent of the money for relief in the fifteen largest cities in the U.S. came from public funds (Chelf, 1992). As the Depression deepened these private agencies were increasingly unable to obtain funds or provide services for the growing numbers of people who needed them. Local public agencies and eventually the federal government were called on to play a larger role. The 1935 Social Security Act established a federally funded and publicly administered welfare system. This act, regarded as the cornerstone of the federal welfare state, was composed of two components. One was a system of social insurance funded by participants'—employee and employer—contributions. The other was a program of social assistance funded by taxes from everyone (Chelf, 1992).

World War II pulled the U.S. economy out of the Depression and concern for poverty waned, but in the 1950s and 1960s, writers like Michael Harrington (1962) and John Kenneth Galbraith (1958) called political leaders' attention to the poverty that still existed in the prosperous United States (Gilbert & Kahl, 1993). These books, along with the civil rights and poor people's movements and the rebellions in the impoverished inner cities of large urban areas like Los Angeles, Detroit, and Newark, sparked a rediscovery of poverty and inequality. A new wave of policies appeared in the Kennedy and Johnson administrations, which came to be known as the War on Poverty. The War on Poverty introduced food stamps, Medicare and Medicaid, Head Start, Job Corps, Legal Services, and many other welfare programs. This period is regarded as one of tremendous welfare expansion. The numbers of people covered and served by government assistance grew during these years.

The 1980s and 1990s have been marked by further changes in the provision of welfare with the introduction of the Family Support Act and the trend toward privatization. Unlike the changes in the 1930s and 1960s that expanded the welfare programs, especially those that served low-income people, recent changes have been in the direction of cutbacks in programs that serve the poor and expansion of programs that benefit middle and upper-income households (Palmer & Sawhill, 1984).

One of the ways in which these cuts in federal support for low-income people took place was to create policies that transfered the responsibility for the provision of welfare services from the government to individuals and private institutions like churches and community organizations (Brown, 1988; Eisenstein, 1984; Starr, 1989). This process is called privatization (Smith & Stone, 1988; Kamerman & Kahn, 1989).

In 1987, President Reagan created a commission to investigate and propose changes in the division of responsibility between the federal government and the private sector in nine areas (President's Commission on Privatization, 1988). The report that was issued the following spring recommended that the federal government divest itself as much as possible from its welfare responsibilities by substituting the private sector as the provider for welfare needs. Housing topped the list of areas in which it was recommended that the government pull out and let private business and humanitarian community organizations take over (Stoesz, 1987).

One proposal that illustrates privatization in housing is a program called Stepping Stone Housing. Stepping Stone Housing also provides some good examples of the kinds of problems low-income families face when welfare programs are privatized (Aulette, 1991).

Stepping Stone Housing identifies people currently living in public housing whose income, according to the government, is adequate enough—over $12,500—to allow them to move into unsubsidized housing. These people are asked to participate in a seven-year transition program to make that move. These are households with incomes that average $16,000. Households that enter the Stepping Stone program are allowed to stay in public housing, with rents of $250 for a one-bedroom and $300 for a two-bedroom apartment, for two years. For the third through seventh year their rent is raised to 30 percent of their monthly income. The Housing Authority, however, continues to keep only $250 to $300 of those payments. The rest of the money is placed in a savings account for the renter. At the end of the seven years the participants must leave their subsidized apartments and, according to the Housing Authority, the households would have about $2,700 in their savings, which they could then use as a down payment on a house or as a security deposit on a privately owned apartment (Martin, 1988). The advantages to the participant are cheaper rent for two years, regardless of income, and the savings account created for them during the third through seventh years. The program, however, entails some risk because once a person decides to enter Stepping Stone, he or she cannot quit and is required to leave public housing at the end of seven years, regardless of any other changes that might have occurred in the seven-year period. Getting back into public housing is next to impossible. In most cities, the waiting lists are enormous and have been closed for years (Reyes & Waxman, 1986).

I interviewed people who had been identified as targets of this program to see what they thought about the idea of moving from government-subsidized apartments to private houses and apartments (Aulette, 1991). All of the respondents said they had thought about moving and all had criticisms of the housing provided in the project. They did not, however, agree with the government about their ability to handle the finances of a transition to private housing. Some people noted that even though their income appeared to be above the poverty line, their ability to provide for their families was barely adequate, even while living in subsidized housing. For example, one woman said, "My children were pressuring me to move, but I set them down and showed them what my check was, all the little charges and things I had to buy to make them look halfway decent when they go to school. I couldn't afford it" (Aulette, 1991, p. 155).

Others pointed out that even if they could afford it now, they might run into unforeseen difficulties such as losing their job or becoming ill. One person explained,

> I was looking for a house and I found one I could afford but then I looked at my paycheck and said: "Oh! What if my little girls get sick?" . . . like my mama had to go to the emergency room and I could do it but I can't afford a house too. Being in the projects, it helps you a lot. (Aulette, 1991, p. 155)

For the people in the study, public housing operated like insurance, allowing them to feel more secure knowing that if they ran into economic difficulties their rent would fall on the sliding scale and they would be less likely to become homeless. In addition, the incomes identified by Stepping Stone as adequate to finding housing in the local private market are unrealistic. The average price of a house in the city in which they live is $104,633. Stepping Stone participants would be seeking housing with an average annual income of about $16,000 and savings of about $3,000. Even finding a rental would be difficult. The average monthly rent on a two-bedroom is $415. This is a little less than their 30 percent of their average income but with no provision for a decline in income due to illness or job loss (Aulette, 1991).

Privatization plans like Stepping Stone Housing frequently fail to provide for people's needs. In Chapter 4, we observed how the Great U-Turn was characterized by a decline in the incomes of working-class and middle-class people. The policy makers who designed Stepping Stone Housing are assuming that people will see their incomes rise or at least remain the same. The people in the study are more cognizant of the Great U-Turn and fear that the risk of losing their place in public housing will cost them everything.

Family Support Act

A second example of recent changes in welfare policies is the introduction of the Family Support Act. Privatization indicates a change in thinking on the part of policy makers on the question of *who should provide welfare* for needy people. The Family Support Act indicates a change in thinking on the question of *what that support should look like.*

The Family Support Act was signed into law by President Reagan in 1988. It makes a number of changes, including increasing the ability of the government to more aggressively obtain child support from noncustodial parents who fail to make payments. The centerpiece of the legislation, however, is its emphasis on making recipients of welfare work for their grants. The word used to describe this policy is workfare (Katz, 1989).

Workfare is the requirement that welfare recipients must work in exchange for receiving their grants (Block & Noakes, 1988). Workfare exists under several names, like Community Work Experience Program (CWEP), Job Opportunity and Basic Skills (JOBS), Work Incentive Program (WIN), and Jobs Training Partnership Act (JTPA). The concept began to be introduced to welfare policy at the federal level in the late 1960s and was increasingly promoted throughout the 1980s and early 1990s. One particularly important recent change for families is in what had been an exemption for mothers of young children. Women with children under age six are no longer exempt from the programs and in some states women with children as young as six weeks are required to participate (Sarvasy, 1988).

Included in workfare programs are a variety of supportive programs that are supposed to assist people with education, job searches, transportation, and childcare. Although the idea of helping welfare recipients to find good jobs is one that many support, the reality of the workfare programs has not lived up to expectations. Most of the workfare jobs provide either no wages—recipients receive their grants for working at unpaid jobs—or the minimum wage, with no benefits, and are insufficient to raise families above the poverty line (Katz, 1989; Harlan, 1989). Seventy-five percent of the programs reporting in 1987 averaged wages of less than $4.47 per hour (Abramowitz, 1989). A study of workfare workers in Massachusetts found that 45 percent had no health coverage at their jobs (Amott, 1990). Support for workfare employees like education, transportation, and child support was nearly nonexistent and most of the assistance provided was in small programs that taught how to search for a job. The issue of childcare is especially important because families that are targeted by workfare are frequently composed of single parents and children.

Because many workfare programs are mandatory, people who refuse to participate can be sanctioned. For example, a mother may refuse to participate because she cannot find adequate childcare or she may not have money to purchase special clothing necessary for the job (Sarvasy, 1988). (In Michigan, people were placed in outside construction crews in the winter, which meant they needed expensive insulated jackets and boots.) When an adult is sanctioned his or her entire family loses their grant. This means that if an adult does not participate adequately, children who have no control over the situation must suffer.

Furthermore, it appears that a large proportion of those who are sanctioned have been treated unfairly. One study in New York, for example, found that 8,000 cases had received sanctions in a six-month period, but 98 percent subsequently won reversal of the sanction because of administrative incompetence or outright harassment (Abramowitz, 1989).

Workfare is also problematic from the point of view of workers who are not receiving welfare. When the welfare system mandates that welfare recipients take jobs, they have no control over wages and no right to belong to or organize unions. This means they are forced to become low-wage, nonunion competitors for jobs, which eliminates job openings and union strength among "regular" workers.

In spite of these problems, by 1987 forty-two states operated optional work programs and twenty-six states required that able-bodied AFDC parents "work off" their grants at unpaid jobs.

WHY IS THERE NO EXPLICIT FAMILY POLICY IN THE U.S.?

The United States is unusual among Western industrialized countries because it does not have a formal family policy. Scholars have attempted to determine what has prevented this country from developing such a policy. Phyllis Moen and Alvin Schorr (1987) argue that the lack of a family policy in the United States is a result of three factors that have shaped policy: an individualist ideology, an economy based on free enterprise, and a heterogeneous population.

A major tradition in the U.S. is the belief in individualism. The individual is perceived to be the fundamental element in society and policy has tended to be directed toward individuals rather than families (Bellah, et al., 1985). Bellah et al. (1985) write,

> Individualism is at the very core of American culture. . . . We believe in the dignity, indeed the sacredness, of the individual. Anything that would violate our right to think for ourselves, judge for ourselves, make our own decisions, live our lives as we see fit, is not only morally wrong, it is sacreligious. (Bellah, et al., 1985, p. 142)

This belief in the value of individualism has had an effect on public policy. First, it has made social policy itself difficult to develop because calling on the community through public tax systems and the government to provide for others goes against the ideal of individual responsibility. Second, even when this hurdle has been overcome and the public and the government agree to provide for some members of the society, the focus is on providing for them as individuals and not as members of families. Moen and Schorr (1987) note, for example, that the Social Security Act of 1935 was implemented to provide economic security for elderly individuals and to encourage retirement to create job openings for the unemployed. The notion that Social Security might serve families of individuals developed later.

An economic system based on free enterprise is a second factor that has inhibited the development of a family policy. The commitment to free enterprise has made any governmental action that interferes with the free market subject to scrutiny and criticism. All of us do not agree as to whether the free market should be interfered with or not, but this question always emerges in national discussions about social policy. Bellah et al. (1985) write that our ambivalence as a nation is expressed in the debate between those who are welfare liberalists and those who are neocapitalists. Both of these groups are concerned about how much the government should "interfere" with business. The neocapitalists argue that interference should be minimal. Welfare liberalists argue that capitalism cannot be maintained unless it is compassionate and therefore provides for the less fortunate in the system (Hewlett, 1991).

The third inhibition to developing a family policy in the U.S. is the heterogeneity of our large and diverse nation. Variability by region, state, family organization, and race ethnicity has made it difficult to agree on what a family is, as well as what the government should be doing for families. Throughout this book we have seen both the diversity of families and the problems scholars and policy makers have in trying to agree what a "real" family is. For example, we reviewed in this chapter recent proposals to cut welfare grants to young, single, poor women who have babies. These proposals imply a belief that these women and their children are not real families or good families, or at least not families the government wishes to help support.

In addition to the lack of an explicit family policy, the U.S. is characterized by a lack of understanding about the impact of the policies that do touch families. Schorr (1962) notes that much of the effect is accidental. *Family impact analysis* is the term used to describe assessments of how a policy might affect families. Family impact analysis would take into consideration issues like the intensity of consequences of policies on families and whether those consequences would be short or long term. It would also assess the range of families that would be affected and the perception of those effects by the families themselves. Moen and Schorr (1987) assert that although much discussion has promoted the idea that family impact analysis is essential, little has actually been accomplished.

Moen and Schorr (1987) conclude their discussion of families and social policy with four orientations they believe must be adopted by policy makers if we are to see any improvement in family policy. First, policy makers must acknowledge families as active agents, not just reactors, and therefore participants in finding and implementing solutions. Those who make policy related to families are usually not those whose lives are most affected by it. For example, Congress and the presidential cabinet do not include middle- or low-income people, single parents, or welfare recipients.

Second, policy makers must be informed about the diverse and changing nature of families. Much of the reaction of the public to Vice President Quayle's comment on the television character Murphy Brown, an employed single mother, was one of surprise and anger. The vice president implied that single female–headed households were not a valid family form. Research shows, however, that this is a growing family type and one not likely to be eliminated. Rather than ignoring single-parent families or trying to make them go away, policy makers should come to grips with the problems single-parent families face and the support the government might provide them.

Third, policy makers should emphasize families as their unit of analysis rather than focusing on individuals in families as their targets. For example, poverty among children cannot be solved by programs for children alone.

Fourth, family impact orientation is necessary to foresee potential inadvertent results of policies. For example, cutting programs that provide food and medical care for pregnant women and infants may result in the worse problems of very ill mothers and children.

What Is a State?

This chapter has provided a description of social policy through the case example of welfare policy. We have seen some of the problems and shortcomings of the policies, their historical evolution, and future trends. We have observed the way in which the macro level of social organization, in the form of social policy, has important effects on the micro-level experience of people who seek support from that system. The macro level of government and policy making, however, exists within an even broader social system, the state. In this section we will review the organization of the state and the theoretical debates that have developed around questions concerning its organization.

Throughout the chapter I have used the term *government* to describe the public institutions that make laws, provide for welfare, and use force. A government is part of the state, which is the broad political organization that organizes and wields power in our society.

In the U.S. the word state is commonly used to describe the fifty political and geographic units that make up the country. For example, we talk about the states of California, Iowa, and Georgia. When social scientists use the term *state* they mean something different. Sociologists refer to the state to indicate a social institution that can operate at several levels within a country. For example, in the U.S. the state would include the Supreme Court, the president, the armed forces, the IRS, the Census Bureau, and Congress, at the federal level, as well as governors and legislatures and more local bodies like city councils, school systems, welfare offices, and local police departments. Theda Skocpol and Edwin Amenta offer the following definition of states: "States are organizations that extract resources through taxation and attempt to extend coercive control and political authority over particular territories and the people residing in them" (Skocpol & Amenta, 1986, p. 131).

This definition of the state makes certain assumptions about the kind of society in which a state may exist. First, it assumes that people are living in some kind of settled community associated with a particular geographic site. Nomadic hunter-gatherers, for example, could not be subjects of a state. The emergence of states in human history is associated with the development of permanent settlements of people, first into cities and then into larger areas or nations.

A second factor that is associated with the emergence of the state in human history is the development of economic productivity through the invention of agriculture, manufacturing and trade, and population growth to a level that allowed for establishing a state. Skocpol and Amenta's definition assumes that economic productivity is at a level that can both maintain the survival of at least a substantial portion of the population and allow for some surplus that can be extracted through taxes. A society that is small and able to provide only subsistence for its members would not have enough that could be given to state officials without jeopardizing the existence of the group. States demand material support in order to run their affairs. States also need to be able to maintain people in authority positions to run the state, and these people would not have time to produce their own food and shelter.

For most of human history people lived in stateless societies. There may have been rulers, but there was no organized system of political and military support to maintain those rulers and to give them authority. About 10,000 years ago, when the agricultural revolution became firmly established in the Middle East, city states like Mesopotamia, which is in the region now known as Iraq, were created.

Modern states, which are called nation states, came about with the emergence of nations and developed with the growth of industrialization. Modern nation states are characterized by sovereignty (authority over a given area), citizenship (consciousness of people that they have certain rights and obligations by virtue of living in a state), and nationalism (identification of individuals with a state as a unified political community). In earlier states, most of the people had little conscious connection to their state. In a modern nation state like the

United States, people are aware of the authority of the state and understand themselves to be citizens of a particular nation state (Giddens, 1991). For example, in the quotes at the beginning of the chapter the last speaker says this is our country. She is identifying herself as a member of the nation state. The Native American speaker, on the other hand, reminds us that the absorption of Native American nations by the nation state known as the U.S. occurred by force, was not accepted by many Native Americans, and continues to be an unsettled issue.

States, then, are a relatively new phenomenon in human society. Their emergence is tied to changes that took place a few thousand years ago. Since then, states have extended their reach into our lives and through the use of policy play an important role in nearly all social interactions, including those related to families.

The State and Inequality

As you read the section on welfare policy, you may have begun to wonder why the welfare system is organized as it is. Why is it so inadequate? Why is it so complex? Why does it treat poor people like criminals? Why is there a recent push toward workfare and privatization? The answers to these questions depend on the point of view you use to examine them. In this section, we will explore some of the theoretical frameworks that currently propose explanations for the organization of the welfare system and in particular we will try to uncover whose interests are served by its current organization.

In the previous discussion, I described two features of the historical development of human society that allowed/necessitated the emergence of the state. I noted that in order for a state to exist a society must include permanent settlements of people in specific geographic sites. In addition, a society must be capable of producing, on the average at least, a surplus that can be taken in the form of taxes to maintain the state. In this section we will examine a third feature of the emergence of the state, inequality. The state became necessary because of a diversity of interests in society that necessitated some institution either negotiate a compromise among the adversarial parties or act on behalf of one party to control the others.

There are many perspectives on the question of whether, in a capitalist nation state like the U.S., the state acts as a negotiator on behalf of everyone or on behalf of certain strata. Two of the most popular points of view are the liberal democratic and the Marxist positions (Pupo, 1988; Abramowitz, 1989), shown in Table 14-4.

According to liberal democrats, the state is a neutral arbitrator among the numerous interest groups in a pluralist democracy like the United States. This model sees the United States as a nation of many different people, classes, and organizations. Each group works to influence the state through activities like lobbying, gaining media coverage, and running candidates. The state acts as vehicle by which these opposing interests can come together to create a fair and humane compromise. The state operates with the vague mission of providing for the social good (Pupo, 1988). Liberal democrats would argue that the welfare system is a product of much negotiation among many different parties. They would assert that the character of the system is a result of the ability of different interests to implement their will. Those who are better able to make their wishes known and build political support for their proposals will have a larger say in

TABLE 14 • 4
Theories of the State

LIBERAL DEMOCRATS

The state operates as a neutral arbitrer among all interest groups in the country. Different values and ideas create critical delineations among interest groups, but other factors like gender, race ethnicity, and region are also important.

MARXISTS

Instrumentalists: The state serves the interest of the capitalist class.
Structuralists: The state negotiates among different interests in the capitalist class.

FEMINISTS

Class Focus: The state controls/oppresses women to better serve the interests of the capitalist class.
Gender Focus: The state controls/oppresses women as a way to maintain male dominance.
Combination: The state controls/oppresses women to maintain male dominance and the capitalist class.

how the system operates. Liberal democrats would contend that the complexity of the welfare system is due to the complex sets of competing interests. For example, there are debates among federal, state, and local authorities about how best to spend money. There are different sets of interests among taxpayers and those who receive welfare grants. There are also differences in ideology among various factions. Shirley Zimmerman (1992) has been particularly interested in this factor.

Zimmerman (1992) argues that political culture plays a pivotal role in the formulation of policy. She defines political culture as the values and attitudes that people hold toward government and toward each other. Some examples of key values that influence policy making are family well-being, freedom, rights, justice, social intergration, efficiency, and equality of opportunity.

To test the relationships among family well-being, per-capita spending on public welfare, and political culture Zimmerman (1992) statistically analyzed these variables for all fifty states for four time periods: 1960, 1970, 1980, and 1985. She found that political culture influences public spending, which in turn affects family well-being. Zimmerman defined well-being as lower rates of divorce, births to teens, poverty, and suicide. Political culture was measured by percent voting Republican. For public spending she measured not only how much money was spent in each state per capita, but also how much taxes were paid per capita and what the fiscal capacity of the state—the financial ability of the state to provide services—was by obtaining the per capita income.

Zimmerman found that spending for public welfare was either insignificant or inversely related to family break-up, teen birthrates, poverty rates, and suicide rates. Spending money on public welfare tended to enhance family well-being as she defined it. She also found that political culture was a significant predictor in explaining spending for public welfare. During the four years she sampled, the percent voting Republican was related to a diminished spending on public welfare. Furthermore, this relationship between political culture and spending increased between 1960 and 1985. The proportion of people who voted Republican became an increasingly stronger predictor as the years went by.

Another factor that might play an important role in determining spending for public welfare is the fiscal capacity of the fifty states, which Zimmerman measured by determining the per capita income. States' fiscal capacity did not predict the level of spending for public welfare. Whether a state spent money on public welfare was not influenced by the poverty of its citizenry but it was strongly influenced by political culture, especially if that political culture had been around for a long time. Those states that had a history of a political culture that emphasized spending on public welfare were most likely to continue that spending even when the economy was declining.

Zimmerman concludes that the state is an arena of struggle among people who have differing ideologies about what a government should do. When those who have more moralistic and less individualistic ideas are able to dominate, more money is spent on public welfare, thereby enhancing the well-being of individuals and families. When those whose ideas are more individualistic, like those who predominated during the 1980s, have control, less money is spent on public welfare and well-being is diminished. Zimmerman's view fits in the pluralist framework because she believes that the state is a neutral arbiter among different ideological types.

The second major framework that seeks to understand the organization of the state is the Marxist position. Marxists maintain that the state is not a neutral arbiter but works to serve the interest of the ruling class, those who own the wealth. Wealth refers to cash, property, stock, buildings, factories and mines. Wealth can take the form of personal property, which means it is consumed by those who own it. This would include mansions, yachts, cars, jewels, and spending money. Wealth can also take the form of private property, which means that it is used to generate more wealth. This would include stocks, money invested in loans, factories, rental buildings, and mines. The people who own the majority of the wealth in a capitalist system like the U.S. are called capitalists. Remember from Chapter 5 that this includes a small number of people but they own a large proportion of all the wealth in the U.S. Furthermore, because they own the places of business where other classes of people work, they have a class relationship with working-class and middle-class people. In contrast to the liberal democrats who emphasize ideology, Marxists emphasize the class position as a source of motivation and power for the contending forces vying for power in the state.

The Marxist position on the state can be divided into two points of view, the instrumentalist perspective and the structuralist perspective. Instrumentalists contend that the state accomplishes its mission of serving the capitalist class by operating directly under its instruction. Instrumentalists, for example, would point out that the majority of members of Congress come from wealthy families and that some individuals float back and forth between leadership positions in business and leadership positions in the government. Structuralists, on the other hand, believe that there is so much conflict in the capitalist class that the state must act somewhat independently of any particular capitalist and instead behaves as an arbitrator among various factions. For example, the structuralist would point out that those capitalists who produce goods in the U.S. to sell here would advocate high tariffs on goods produced elsewhere. Capitalists with branches in other nations who wish to sell to the U.S. market would advocate lower tariffs for those same imported goods. The state must negotiate this difference of opinion by coming to some agreement about what is best for the maintenance of the political economy as a whole.

Both instrumentalist and structuralist Marxists assert that within a capitalist society, the state has two main tasks: to facilitate accumulation of capital by the capitalists and to maintain the system by either legitimating it in the eyes of the populace or by providing for some other form of social control like the police (O'Connor, 1973). To facilitate the accumulation of capital by the capitalists, the state, for example, provides for a properly educated work force (Bowles & Gintis, 1976). It also designs tax laws to keep money in the hands of the wealthy and it builds and maintains an infrastructure of roads, and airports, and mail, water, and sewage systems to allow the capitalists to produce and distribute commodities.

The maintenance of the system is accomplished by a number of state agencies. Information dissemination, education, and the welfare system help to inhibit those who are not in the ruling class from understanding that they are oppressed and taking action to change the system. If that does not work, however, the state can also use the courts, the police, and the army to keep "the peace."

Marxists would argue that the welfare system is organized to support the two major tasks of the state, accumulation and legitimation. Accumulation is facilitated by paying as little benefits as possible in order to keep more money available to the capitalists. Accumulation is also facilitated if the welfare system keeps more people seeking work in order to increase competition among workers and thereby lower wages. Legitimation is facilitated by stigmatizing welfare recipients and holding a victim-blaming image of poverty that deflects criticism away from the economic system and the political organization, which are actually the root of the problem.

A third school of thought on the state is a recently emerging one that incorporates the notion of gender inequality in its assessment of state activity (Connell, 1987; Zaretsky, 1982) and that could be called a feminist theory. This framework can be divided into three types. The first group of theorists contend that the state seeks to control and shape gender and sexuality to accomplish its primary mission of maintaining the capitalist system. For example, this approach would argue that the state does not demand equal or comparable pay for women primarily because keeping women as a cheap source of labor lowers production costs for business interests, thereby improving profitability and strengthening the capitalists.

The second group argues that the state operates to protect the patriarchy (MacKinnon, 1982). Theorists in this school of thought emphasize the way an insufficient welfare system—one that is means-tested, inadequate, constricting, and stigmatizing—coupled with unequal wages for women helps to preserve male dominance by providing men with more money and by keeping women more dependent on men. For example, consider a women with children who wishes to leave an unhappy marriage. In Chapter 10, we saw that she is likely to retain custody of the children but unlikely to receive child support from her husband. If she finds employment she is likely to be paid less than her husband and certainly less than they made as a two-income household. If she cannot find employment and childcare she can apply for AFDC but will find that her financial status falls below the poverty level. In short, she can choose to be poor or she can choose to stay with her husband. If the welfare system provided sufficient support it would create an avenue of independence for women. Advocates of this framework insist that our analysis of the state must fully appreciate the gender politics of the state and the coercive manner in which the state controls women and benefits men. This framework, furthermore, asserts that the state has actually grown to take on the role of the maintenance of male dominance as the classical patriarchal family has waned.

Boris and Bordaglio (1983) argue that familial patriarchy declined during the nineteenth and twentieth centuries while male dominance in the society at large was maintained. The maintenance of male dominance was accomplished by a transformation of patriarchy from a familial to a state form. Recall that in Chapter 2, we observed the disappearance of the classical patriarchal family of colonial America as the rise of industrial capitalism changed the organization of families. At the same time, patriarchy in a broader sense (referring to male dominance in society), continued to thrive. Theorists like Boris and Bordaglio (1983) argue that this occurred because patriarchal ideas and mechanisms were increasingly built into the laws and institutions, like the welfare system, that are a part of the state.

The third group of theorists argues that the state is used to promote both the dominance of capitalists over the working class and the dominance of men over women. Furthermore, the state serves as an arbiter between these two competing ruling strata: those who dominate by gender—men—and those who dominate by class—capitalists (Eisenstein, 1984b; Petchesky, 1990).

RESISTANCE TO THE WELFARE SYSTEM

This chapter has explored the way in which the macro level of social policy and the government create problems for people and the way in which policy makers and the government exist within an even broader system called the state. In this section we will turn our attention to the way people at the micro level respond to and resist the problems that are created for them by the organization of the macro level of society.

The welfare system appears to be fraught with problems. It is discriminating, inadequate, cumbersome, and deteriorating. How do recipients respond? Some people have created ways to try to transform the system. Others have attempted to cope with the system by devising ways to combine welfare with other resources in order to patch together a survival package. These two methods of resistance are the subject of this section, beginning with those who cope with the system.

Kathryn Edin (1991) examined this issue by interviewing fifty Chicago-area mothers who receive welfare. Edin (1991) found that all the women in her study supplemented their grants from AFDC and food stamps by using at least one source of unreported income: income from work or assistance from family, friends, boyfriends, or the fathers of their children.

Because their budgets were so tight, the women were able to give accurate accounts of how much money they spent and how much money they took in. The average woman spent a total of $864, with $501 going for housing and food. The remaining $363 was spent on items like phone, clothing, school sup-

The 1980s and 1990s were marked by cuts in social spending. Families have responded by protests aimed at government officials, like this protest at the state capitol in Maryland.

plies, transportation, toiletries, health care, diapers, burial insurance, and furniture. The average AFDC grant amounted to $324 and the average food stamps were $197. The mothers also received income from other sources totaling $376—an average of $166 from work and $210 from family and friends. This left a shortfall of about $343 per month.

The women explained to Edin (1991) how this constant shortfall affected them. One woman said, "I don't ever pay off all my bills, so there isn't ever anything left over. As soon as I get my check, it's gone, and I don't have anything left" (Edin, 1991, p. 465).

If they had not received money from sources other than welfare the women would have been even deeper in debt. Some of the money they obtained was from family and friends and some was from jobs. Receiving these funds and not reporting them to the welfare office, which would result in a cut in the grant received, is illegal. One woman explained that although she felt bad about having to lie, she saw no other way to survive. She said,

> Public Aid is an agency that I believe can teach a person how to lie. If you tell them the truth, you won't get any help. But if you go down there and tell them a lie, you get help. And I can't understand it, and every woman on Public Aid will tell you the same thing. It teaches you to lie. It won't accept the truth. So when you deal with Public Aid, you have to tell them a tale. (Edin, 1991, p. 469)

According to the respondents in Edin's work, many people who receive welfare resist the system by not complying with the regulations. A second form of resistance is through organizations like the National Welfare Rights Organization

(NWRO), which works to transform the system itself in order to make it more effective for the people who seek assistance (Piven, 1990).

Guida West (1981) wrote about the NWRO, which flourished in the 1960s and 1970s. During that time the organization mobilized thousands of welfare recipients and their advocates and millions of dollars and other resources. The movement gained national recognition after a ten-day march on Columbus, Ohio in 1966, which was coordinated with protests across the country. NWRO activities included lobbying, courtroom contests, public protests, education, and advocacy with individual welfare cases. NWRO's strategy was to flood the welfare system with eligible applicants in order to expose the inadequacy of the system and force change. Their goal was to obtain a guaranteed income system in the United States. Johnnie Tillmon, an African American woman from Los Angeles who eventually became chairwoman of NWRO and played a central role in its success, said,

> If I were President, I would solve this so-called welfare crisis in a minute, and go a long way toward liberating every woman. I'd just issue a proclamation that women's work is real work. In other words, I'd start paying women a living wage for doing the work we are already doing—child raising and housekeeping. And the welfare crisis would be over. Just like that. (West, 1981, p. 91)

Although NWRO did not obtain a guaranteed income it was successful in some of its efforts (Amott, 1990). NWRO educated potential applicants about available services and helped to guide people through the complex system. The number of AFDC families tripled between 1965 and 1976, the period in which NWRO was largest and most visible (Miller, 1990). In addition, some of the most punitive features of welfare, such as the midnight raids on AFDC mothers, were contested in court by NWRO and eventually ruled illegal. NWRO declined in the late 1970s as a national organization, although a few local Welfare Rights Organization groups remained strong throughout the 1980s and into the 1990s (Welfare Mothers Voice, 1992). During the late 1970s and into the 1980s welfare rights advocates took another tactic and sought positions within the welfare system, where they continued to work for change, an issue that Nancy Naples has investigated.

Naples (1991) interviewed forty-two women who had been unpaid community activists and had become paid workers in the state system, working in Community Action Programs (CAPs). CAPs are publicly funded programs that provide information, education, and other services to residents of low-income neighborhoods. Naples found that the women she interviewed had successfully challenged the organization of CAPs in four ways. First, they broadened the range of activities that were seen as part of their job. For example, traditional social service professionals worked from 9 to 5 and then went home. The community activists were in paid jobs from 9 to 5 but their work continued "after hours" because of their membership and participation in the community. Second, they contested the reliance on credentials as the most important or only criteria for being able to provide quality service. Third, they demanded that those who worked for the state should also be involved in challenging the state instead of playing the role of neutral observer. The supervisors attempted to stop paid workers from participating in political actions but workers ignored or resisted these regulations. Fourth, they insisted that their work with the people in their communities emerge from their relationship with their neighbors and that the services they provided and the way in which they were provided be appropriate to their particular community. For example, program administrators who attempted to manage centers from a central office were met with resistance by workers at various branches. Naples concludes, however, that despite the importance of their work in the community and the influence of their resistance on the bureaucracy of the CAPs, the cuts in welfare spending have eliminated many of the positions the women held.

In 1988 the National Welfare Rights Union (NWRU) emerged with its First Annual Convention in Detroit. Its leadership had come out of the National Welfare Rights Organization. The conference was entitled "Up and Out of Poverty." The NWRU has taken a more militant stand than the women in the Naples study. NWRU members maintain that poor people must openly confront the government for failing to provide a society in which all people can attain social justice and a decent standard of living. They believe that the victims of poverty, not their advocates, must take the leadership in the welfare rights movement. They also believe they must build coalitions with homeless unions, labor unions, students, and churches in order to wage their fight to "educate, organize and empower" (Kramer, 1988).

In their statement of purpose, NWRU writes,

> The current welfare system destroys families by keeping the husband/father out of the household, penalizes the parents who are trying to work themselves out of welfare, and punishes people, especially the children, for being poor. All people should have adequate income, regardless of their work, a guaranteed annual income so that no one in this nation should live in poverty. All people should be able to live a life of dignity with full freedom and respect for human rights. All low-income people and public assistance recipients should enjoy a fair and open system which guarantees the full protection of the U.S. Constitution. All low-income people and public assistance recipients should participate directly in the formation of decisions affecting their lives. (National Welfare Rights Union, 1988, p. 11)

Since 1988, NWRU has continued its effort through national and local Survival Summits:

> The NWRU will take our movement into the streets of this nation from the farms of rural America to the tenements of the inner city, from the homeless ghettos to the college campuses, from the Halls of Congress to the hearing rooms of our state houses. We are building a movement to insure a better world for our children and ourselves. We are building a movement that cannot be stopped. (National Welfare Rights Union, 1988, p. 12)

THE MICRO-MACRO CONNECTION

In this chapter our focus has been on the way in which the government uses social policy, especially welfare policy, to influence families. The welfare system directly affects poor families but it also affects other families as well, by defining what a family is and what a "proper" relationship should be between women and men and adults and children.

The book has discussed a number of other ways in which the state influences families. For example, one prominent aspect of the state is legislation. We have explored the way in which the laws shape and define the family and issues related to families such as sexuality, reproduction, marriage, divorce, relationships between parents and children, and violence in families. We have also noted how the financial activities of the state, such as collecting taxes and setting a minimum wage, have affected families. The state also includes the institutions of coercive control and here we have seen how the police have played a role in families on questions like sexuality and violence.

We have also observed the way in which various factors at the macro level may influence each other. For example, the welfare system is influenced by the government, which is influenced by the organization of the state, and the state is influenced by cultural values as well as the stratification systems of class, gender, and race. A variety of macro-level forces play upon each other.

In addition, the micro level of society affects the macro level. Families and individuals at the micro level have sought to resist and transform the macro level of social organization. For example, welfare mothers have responded to the welfare system by attempting to cope with the inadequacies of the services. In addition, the micro level has also worked to alter the organization of the welfare system. For example, community activists attempt to try to influence the provision of services to low-income communities by working within the system. The National Welfare Rights Organization was able to win concessions in the 1960s and 1970s and its members, now in the National Welfare Rights Union, continue their efforts to transform the system into a more effective and humane one.

SUMMARY

- The opening scenario presents the voices of activists who have confronted the government in their efforts to gain economic and social justice. Their comments call into question our understanding of the way in which the state works, how it has changed over time, and how it must be altered to address issues that affect families.

The Welfare State and Family Policy

- The United States, like most other industrial capitalist nations, is a welfare state, which means that it attempts to solve social problems like poverty by using government funds to pay for social service programs.
- Social policy refers to the activities in which the government is engaged to try to offer those programs.
- Unlike most other industrial capitalist nations, the U.S. does not have an explicit family policy, and this lack creates problems in providing adequate services for families.

Social Policy and Families: The Case of the Welfare System

- Individual families are the largest welfare system in the U.S., but the government has also created a welfare system that can be considered a prime example of family policy.
- The welfare system can be divided into a number of different levels and agencies that vary from one state to another.

- The efficacy of the welfare system has deteriorated in recent years.
- The welfare system is fraught with myths that center around the belief that AFDC mothers are undeserving.
- Racism and sexism are factors in the character of the system.

History and Trends for the Future in Welfare

- The modern welfare system in the U.S. emerged in 1935 and was marked by expansion during the 1960s War on Poverty.
- The welfare system has changed and contracted in recent decades.
- One aspect of the contraction is the promotion of the policy of privatization, which refers to the transference of responsibility for providing services for low-income people from the government to private agencies.
- The Family Support Act, passed in 1988, is another example of a recent change in welfare policy that has served to cut benefits.
- The centerpiece of the act is the problematic policy of workfare.

Why Is There No Explicit Family Policy in the U.S.?

- Moen and Schorr (1987) assert that the U.S. has no explicit family policy because of the strength of individualism, an economy based on free enterprise, and a heterogeneous population.
- In addition to having no family policy, little is known about the impact of those policies that touch families.
- Policy makers need to improve policies for families by changing their orientation.
- The state is the term used by social scientists to indicate the institution that extracts money through taxes and attempts to control specific territories and the people who live in them.
- States are a relatively new feature of human society.
- There are three major frameworks for understanding the organization of the state and the policies implemented by it: the liberal democratic framework, the Marxist framework, and the feminist framework.
- Liberal democrats emphasize the negotiating role played by the state among the various groups within our society who attempt to make the state meet their concerns.
- Liberal democrats are interested in the role that ideology plays in motivating different people to take varying positions on welfare policy.
- Marxists do not see the state as a neutral factor and emphasize the role the state plays in maintaining one class—the capitalists—in control.
- A third framework has recently developed, which pays particular attention to the way the state maintains the system of gender dominance as well as the system of class dominance.

Resistance to the Welfare System

- The inadequacy of the welfare system has caused people to use a number of creative ways in their attempts to survive. Welfare recipients often have to break the law in order to patch together a survival package.
- The National Welfare Rights Organization is an example of a political group that has attempted to change the organization of the welfare system.

- Others have also attempted to alter the system by working within welfare agencies like Community Action Programs.
- The National Welfare Rights Union emerged in the late 1980s as an organization committed to transforming not only the welfare system but the government in its efforts to win justice and a decent standard of living.

The Micro-Macro Connection

- The state, a macro-level institution, has an important effect on the micro level of families, especially on poor families with the creation and implementation of policies like AFDC.
- The macro level is complex with several layers of interacting forces including the government, ideology, and stratifications systems of race, class, and gender.
- The micro level has responded by trying to devise ways to survive in spite of the inadequacy of grants.
- The micro level has also responded by attempting to reshape the system through political organizing inside and outside of the welfare offices.

GLOSSARY

Abroad wives The name given to the wives of slaves who lived on other plantations.

Absorptiveness The extent to which a job draws in an employee and his or her family members.

Adjunct, support, and double duty role Three roles Kanter believed described the activities of corporate wives as they helped run businesses, provided emotional support, and took care of other jobs to leave their husbands free to pursue their careers.

Agency The notion that people create, maintain, and change the organization of their society.

Aid to Families with Dependent Children (AFDC) Federally mandated program to provide grants to eligible poor families with children.

Alimony Money the court orders one spouse to pay another in a divorce settlement.

Allopaths One sect among many health care providers in the nineteenth century. They eventually came to dominate and now are called regular M.D.s.

Alternative insemination Depositing sperm in a woman's vagina or uterus by means other than sexual intercourse.

Amniocentesis Sampling fetal cells by extracting them from amniotic fluid to determine abnormalities.

Baby M The daughter born to Mary Beth Whitehead as a result of alternative insemination by the sperm of William Stern. An important custody case developed when Whitehead refused to abide by the contract she had with Stern to turn the baby over to him.

Battered husband syndrome Controversial idea that men are as likely or almost as likely to be physically abused by their spouse as women are.

Battered women's movement Political organization of battered women and their advocates to develop ways to end the violence and to provide support for women who need to leave a batterer.

Birth rate The number of births per 1000 women between the ages of 14 and 44.

Black matriarchy False characterization of African American families as dominated by powerful women who inhibit the ability of African American men to behave in an adequately "masculine" manner.

Blueprints of love Cancian's (1987) typology for tracing the historical change in love in the U.S. through four stages: family duty, companionship, independence, and interdependence.

Calling Courtship practice in the nineteenth century in which the man came to the woman's house to visit.

Cash assistance Programs that provide grants to eligible persons who can then decide how best to spend the money.

Chicano Mexican American man or people. Chicana means Mexican American woman.

Child abuse Overt acts of aggression such as excessive verbal derogation, beating, or inflicting physical injury or sexual abuse.

Child custody Court decision as to where the children will live in a divorce settlement. Joint physical custody means the child lives intermittently with one parent and then the other. Joint legal custody means the child lives mostly with one parent but both parents are responsible for the child.

Child neglect Failing to provide adequate physical and emotional care.

Child support Money the court orders a noncustodial parent to pay to the parent with whom the child lives.

Children's Bill of Rights National petition for sexually abused children's rights, presented to Congress on Mother's Day, 1992.

Chinese Exclusion Act, 1882 Prohibited Chinese women from entering the United States even if their husbands were already living here.

Chorionic villi sampling Sampling fetal cells from a pregnant woman's cervix to determine abnormalities or the sex of the child.

Citizen wage Proposal to distribute goods and services produced by robots and computers on the basis of need.

Class A system of inequality based on differing economic position and relationships in society.

Cognitive development theory Posits that children learn how to be masculine or feminine by first identifying themselves and then positively responding to behaviors they associate with the proper gender.

Cohabitation Couples living together without being legally married.

Commodification of sexuality Selling sex, for example, prostitution or pornography, or using sex to sell other goods and services.

Compadrazgo System of shared childraising in Chicano community in which godparents celebrate rites of passage and provide support in times of need.

Companionate Family The second stage of the Modern Family, lasting from the early 1900s until about 1970.

Comstock Laws Legislation passed by Congress in 1873, named after Anthony Comstock, which made abortion and birth control illegal in the U.S.

Conservatives A political group that (1) stresses a narrow and specific morality; (2) believes it is important to maintain hierarchies; and (3) assumes that the interests of all family members can be collapsed into one. See also liberals.

Consciousness raising group Small groups of women who met during the early years of the women's liberation movement to discuss feminist literature, to talk about their lives, and to plan ways to fight for equality.

Content analysis Research methodology that involves reviewing cultural artifacts that were not originally intended to provide information for a researcher.

Corporate wives The wives of upper-middle-class men who help them to keep their jobs and obtain promotions.

Correlates of divorce Factors associated with divorce as causes, results, or both.

Courtship The period between singlehood and marriage.

Crime against nature Phrase used in laws that restrict sexuality; usually refers to anal intercourse.

Cult of True Womanhood Image of women prevalent in the nineteenth century that judged women by how well they exhibited four virtues: piety, purity, submissiveness, and domesticity.

Cultural artifacts Objects that are made by humans in different cultures, which are sometimes used as sources of information about that culture. For example, researchers might look at toys to try to understand gender socialization.

Culture of poverty The notion that the roots of poverty exist in the faulty values and practices of those people who are poor. See also social structural model.

De facto segregation The separation of blacks and whites that is not written into the laws but exists nevertheless. For example, the segregation of children into white or black schools because of the organization of school districts by neighborhoods that are segregated. See de jure segregation.

Deindustrialization Transition from an industrial-based economy to an electronically based one in which computers and robots increasingly take over manufacturing and other jobs. See electronics revolution.

De jure segregation The legal separation of blacks and whites. For example, the segregation of children into white or black schools that was mandated by law before the civil rights movement. See de facto segregation.

Democratic Family The first stage of the Modern Family, lasting from the late 1700s until the early 1900s.

Demography The study of the size, composition, and distribution of human populations.

Deprivation model of childless couples One in which people who choose to remain childless are preceived to be maladjusted because they have not had the experience of having children. See also diagnostic model, labeling model.

Diagnostic model of childless couples One in which people who choose to remain childless are by virtue of that fact alone perceived to be maladjusted. See also deprivation model, labeling model.

Dialectical materialism The philosophical basis of Marxism. Materialism refers to the assumption that the material realities, especially economic ones, of human life shape the ways in which people think and structure their society. Dialectics refers to a complex set of ideas about how social phenomena are related to each other and how change takes place.

Disaffiliate One of four strategies men use to avoid doing housework. They passively avoid taking responsibility by forgetting, doing the tasks in an unskilled way, and waiting to be asked. See also selective encouragement, substitute offering, reduce the need.

Divorce rate The number of divorces over some period of time. Sociologists use a variety of ways to calculate the incidence of divorce, including comparing the number of divorces to other factors such as population size or the number of marriages.

Doctrine of couverture Defined marriage in eighteenth and early nineteenth centuries as a unity in which husband and wife are one and that one is the husband. A married woman could not own property, keep her own wages, or be sued.

Domestice partners Unmarried couples who live together in committed, sexually intimate relationships.

Economic institution The organization and distribution of the goods and services people use to survive.

Economy of gratitude Term used to describe the exchanges of "gifts," which might be material objects, services, or emotional support.

Electronics Revolution Perhaps a third major turning point in human history, which allows computers and robots to potentially take over nearly all productive and service work, leaving humans to emphasize the work only they can do: creating, nurturing, exploring, and learning. See also deindustrialization.

Emotional climate The social-psychological influences at work that affect employees and their relationships within families.

Emotional labor The work people do to support and nurture others. Romero (1992) observed that maids were often called on to do this kind of work for their employers.

Essentialism Theoretical model that posits that social behavior is an immutable part of a person's character.

Ethnography A research methodology that involves the collection of artifacts, interviews, and observations of communities over a relatively long period of time.

Ethnomethodology Research methodology that focuses attention on the way people work together to understand and react to their everyday social experience.

Exchange theory Posits that social interaction involves exchange of rewards between people, including husbands and wives.

Expressive Term used by functionalists to describe the nurturant, domestic, child-care-providing activities they believe are best performed by women in families.

Extended family Relatives by blood or marriage other than spouse and children. In postmodern families extended families have come to include others like ex-spouses, spouses, and children of ex-spouses, and ex-spouses' new mates. These new extended families are composed of people who are not related by blood or marriage and frequently do not get along well.

Families we choose Social arrangements that include intimate relationships between couples and close familial relationships among other couples and other adults and children, often found in the gay community (Weston 1991).

Family and nonfamily households Terms the U.S. Bureau of the Census uses to distinguish among households. A family household includes those people who live together with others who are related by blood, marriage, or adoption.

Family impact analysis Term used to describe assessments of how a policy might affect families.

Family Leave Bill Signed by President Clinton in 1993. Makes it illegal to fire employees who must take unpaid time off to care for babies or ill family members.

Family policy Policies that affect families. The United States is unusual among Western nations because it does not have an explicit family policy. A variety of policies, however, could be considered family policy because of the important impact they have on families.

Family Support Act Federal law signed by President Reagan in 1988 whose centerpiece is its emphasis on making recipients of welfare work for their grant.

Family wage A wage that could be earned by a man to support his entire family without his wife and children having to work.

Feminist enlightenment Term used by Jessie Bernard to describe the period beginning in the late 1960s in regard to the important effect a feminist perspective has had on the discipline of sociology.

Feminist theory Highly variable among a range of different schools of thought, including radical feminists, liberal feminists, socialist feminists, Third Wave feminists, and others. According to Chafetz (1988) they all share the following characteristics: focus on gender; see gender inequality and gender relations as problematic; and believe that gender is socially constructed.

Feminization of poverty Term coined by Pearce to describe the trend toward greater impoverishment among women and their children.

Fetal monitoring Recording and evaluating the heartbeat of a fetus, especially during delivery.

First Wave feminists Organized the movement for equality for women that emerged in the nineteenth century in the U.S. and waned after the passage of the Nineteenth Amendment giving women the right to vote in 1920.

Flextime Flexible work schedules in which workers vary the time they arrive at work and leave, the number of hours they work, and the permanency of their employment.

French letters Condoms. In the nineteenth century, people in the U.S. used the word French to indicate an association of some product with birth control.

Gamete donor A person who donates sperm or egg cells to be used for alternative insemination.

Gender Socially constructed categories that imply expectations about what men and women and boys and girls should be like. Distinguishable from sex, which is a relatively discernible distinction based on biological differences between males and females.

Gender factory Term coined by Berk to describe the way the division of housework in families creates and reinforces gender.

Gender ideologies Ideas about differences and similarities between women and men or boys and girls.

Gender schema theory Similar to cognitive development theory except for one important factor—acknowledging the importance of social forces telling children they must identify with one sex or the other.

Gender socialization The process by which people learn what it is to be masculine and feminine.

General Assistance (GA) State-supported program in some states that provides assistance to poor individuals or families without minor children.

Geographic mobility Expectation that people, especially middle-class professionals, are willing to move their families around in order to further their career.

Godly Family Patriarchal family dominant in preindustrial agricultural economy of Puritan society.

Government subsidization The provision of publicly funded support by the government. Includes not only welfare for poor people, but benefits for the middle class and the wealthy as well.

Gramm-Rudman-Hollings Act Signed in 1985 with the intention of balancing the budget by stopping "government handouts."

Great Migration Movement of Southerners, especially African Americans, from the rural South to Southern cities and the North between 1910 and 1930.

Great U-Turn Term coined by Bennett Harrison and Barry Bluestone to describe the reversal in the United States economy in the mid-1970s from growth and increasing affluence to growing unemployment and impoverishment.

Heterosexuality Sexual attraction or interaction with those of the opposite sex.

Homelessness According to Congress: individuals who lack a fixed regular nighttime residence or who have a primary nighttime residence in a shelter or a place not ordinarily used for sleeping by humans such as a cave or abandoned building.

Homestead Act of 1862 Provided 160 acres of government-owned land for $10 for any citizen or anyone declaring their intention to become a citizen.

Homophobia Hatred and fear of homosexuals and lesbians.

Homosexuality Sexual attraction or interaction with those of the same sex; usually refers to men. Lesbian is the word used for women.

Husband care Term coined by Hartmann to describe the difference between the amount of housework husbands do and the larger amount they generate.

Hyde Amendment Federal amendment passed in 1977 prohibiting the use of federal funds—Medicaid—for abortion.

Hypergamy Marrying a person in a higher social class.

Incest Sexual contact between close relatives.

Indian removal Euphemistic term for the United States government's policy of forcibly evicting Native Americans from their land.

Individualism The belief that the individual is the fundamental element of society and that all actions should take place for the benefit of individuals.

Industrial Revolution Second major turning point in human history, which began in the late 1700s when people began to use machines driven by water, steam, and later electricity.

Infant mortality The number of children who die before their first birthday among every 1,000 babies born.

In-kind support Programs that provide specific services such as medical care and housing and vouchers that can only be used to purchase specific items. Food stamps are an example.

Instrumental Term used by functionalists to describe the rational decision-making, money-earning activities they believe are best performed by men in families.

Instrumentalist Marxists Assert that the state serves the capitalist class by operating directly under its direction.

Integrated marriages Marriages in which husbands and wives share relatively equally in tasks and decision making. See also segregated, modified segregated, wife-shaped.

Inuit Native Americans living in Alaska and northern Canada, commonly called Eskimos.

Issei, Nisei, Sensei First-, second-, and third-generation Japanese Americans.

Jim Crow Laws and social norms that segregated blacks and whites in the South during the first half of the twentieth century.

Jumping the broom Although slaves were not allowed to legally marry each other, women and men would publicly declare their relationship to the community through symbolic rituals like jumping over a broom.

Labeling model of childless couples One in which people who choose to remain childless are perceived to be not maladjusted but at a serious disadvantage because of the way in which others see childless people. See also diagnostic model, deprivation model.

League of the Iroquois Organization of six Native American nations that lived in the area that is now the state of New York prior to the immigration of Europeans to North America.

Learned helplessness The notion that women who are abused by their parents and/or their husbands eventually become passive, unresourceful victims.

Learning theory Posits that children learn how to be masculine or feminine through rewards and punishments.

Liberal feminism Focuses on the issue of inequality between women and men and posits that the organization of society is largely responsible for differences between women and men.

Liberals A political grouping that (1) is concerned about equality; (2) focuses on the individual; and (3) bases its ideology about families on reason. See also conservatives.

Life cycle approach Research that pays attention to changes over time in a family's or an individual's history.

Live-in maids Maids were required to live with their employers and were only allowed to visit their own families a few days a year.

Machismo Spanish word for masculine, used to describe the supposed domination of overbearing men in Latino families. This concept has been questioned in terms of its existence, its expression relative to other racial ethnic groups, and its roots if it does in fact exist.

Macro level Focuses on social structure, which includes ideologies, technologies, and social institutions like government, the economy, and social classes.

Male revolt Movement in the 1950s, 1960s, and 1970s, which sought to change ideas about masculinity and free men from the requirement of marrying in order to prove themselves adults and heterosexuals.

Mandatory arrest Recently passed laws requiring that police arrest batterers regardless of whether the battered person requests it or not.

Marital dissolution Marriages ending by legal action such as divorce or by death of a spouse.

Marriage contract Legal document to which husband and wife agree when they marry. Most of the duties and rewards in a marriage contract are specified by the government.

Marriage squeeze Shortage of eligible men for women seeking husbands.

Married Women's Property Laws Laws passed in mid-1800s allowing married women to own property.

Maternal thinking The point of view people (often but not exclusively biological mothers) develop as a result of their participation in mothering activities.

Matriarchy Rule by mothers. Some scholars identify the Iroquois as a matriarchy because of the power Iroquois women wielded in their society.

Mean The average of a set of numbers. See also Median.

Means-tested programs Participants must prove they are impoverished and have no other way to support themselves.

Medicalized childbirth Organization of birthing that makes pregnant women patients of physicians, who deliver babies in hospitals, surrounded by health care professionals and subject to a number of technical medical procedures.

Median The point at which half fall above and half below. See also Mean.

Micro level Focuses on the family life of individual families or individual family members in face-to-face relationships.

Modern Family Term historians use to describe nuclear families with a breadwinning husband and a housewife, dominant among the white middle class after the Industrial Revolution.

Modified segregated marriages Marriages in which wives act as junior partners to their husbands.

Mommy Track Term used for the policy of judging employed women who have children less rigorously on the job in exchange for not promoting them or giving them raises.

Monogamy The practice of marrying only one person at a time and remaining sexually exclusive with that one person.

Monolithic Something that can be described as totally uniform. A monolithic view of families implies that only one kind of family exists or is valid.

Mon Valley Unemployment Council Organization of steelworkers laid off in the late 1970s and early 1980s in Pennsylvania.

Mortality rate Proportion of people dying over some period of time.

Motherhood mystique An idealized version of motherhood that ignores the problems that real mothers have.

Mothers of East Los Angeles (MELA) Political organization of Mexican American women working to improve the quality of life in their community and to defend it from hazardous industries.

Moynihan Report Report submitted by Senator Moynihan in the 1960s, which claimed that the root of the poverty in the black community was the faulty organization of the black family.

Naming ceremony Ritual among Native Americans in which children are given a name and adults, other than biological parents, are chosen to help to raise the child.

National Organization for Women (NOW) Political organization of feminists formed in 1966.

National Union of the Homeless Political organization established in the early 1980s to publicize and develop solutions to the problem of homelessness.

National Welfare Rights Organization (NWRO) An organization of welfare recipients and their advocates that emerged during the 1960s.

National Welfare Rights Union (NWRU) An organization of welfare recipients that emerged from the Welfare Rights Movement in the 1980s.

Neocapitalists Believe that the government should not interfere with free enterprise. See also welfare liberalists.

Neolithic Revolution First major turning point in human history, about 8000 years ago, when people began to develop agriculture and to establish permanent settlements.

Neo-Traditionalist One of three types Sidel found in her research. These young women prioritized being good wives and mothers, although most of them also wished to work outside the home. See also Outsider, New American Dreamer.

New American Dreamer One of three types Sidel found in her research. These young women prioritized success on the job and believed that if they worked hard enough they would succeed. See also Neo-Traditionalist, Outsider.

New reproductive technologies (NRT) Application of advanced technology on reproduction that may inhibit the development of new life, monitor it, or create it.

No-fault divorce Divorce reforms passed in the mid-1970s allowing a couple to divorce without having to prove one spouse did something wrong.

Norplant Contraceptive lasting for several years after being implanted in a woman's arm.

Oedipal process According to Freudian theory the process by which children come to be properly gendered adults as a result of resolving their fear (boys) or disappointment (girls) when they discover the genital differences between males and females.

Oral history Research methodology that involves interviewing people about their experience over a long period of time.

Other mothers Term used to describe people other than biological mothers who frequently take responsibility for raising children.

Outsider One of three types Sidel found in her research. These young women did not see themselves as being either good wives and mothers or successful employees. See also New American Dreamer, Neo-Traditionalist.

The Passage An initiation rite for African American youth designed to socialize them into the black community.

Paterollers Patrols of whites whose task was to see that no slaves traveled about at night without passes.

Patriarchy This word has two meanings: (1) male-dominated extended households like those in Puritan society; (2) general dominance of men in families and in society.

Pedophobia Fear, dread, aversion, or hatred of children.

Policy Lines of action that allow countries to extract resources from and control people residing in them. Policies include nearly everything the government does, from collecting taxes to imprisoning criminals.

Political culture The values and attitudes that people hold toward government and toward each other.

Political economy The manner in which a society organizes its political and economic institutions.

Political institution The organization and distribution of power and decision making.

Portuguese female pills Abortifacients. In the nineteenth century, people in the U.S. used the word Portuguese to indicate a product's association with terminating pregnancy.

Potter's Field Pauper's cemetery on an island used as a prison in New York City.

Power Weberian: Process by which individuals gain or maintain capacity to impose their will on others through punishment or rewards. Berger and Luckmann: Ability to impose one's definition of reality on others.

Privatization Relatively recent move to create policies that transfer the responsibility for the provision of welfare services from the government to individuals and private institutions.

Pro-choice Advocates keeping abortion legal.

Programmatic welfare state "Capitalist states that devote a portion of their gross national product through taxation to the solution of certain social problems without changing the basic nature of the economy" (Atherton, 1989, p. 169). See also redistributive welfare state.

Pro-life Advocates making abortion illegal.

Psychoanalytic theory Theory developed by Freud, which posits that children are biologically predestined to move through certain stages of psychosexual development.

Qualitative methodology Methods for doing social research that emphasize the meaning of the responses to researchers' questions or observations, especially their meaning from the point of view of those being studied.

Quantitative methodology Methods for doing social research that stress the compilation of information that can be quantified—turned into numbers.

Quickening Moment at which fetus is believed to gain life. In nineteenth century, marked by the mother feeling movement.

Race A nonscientific and arbitrary division of people on the basis of physical characteristics. Although race is entirely "made up" it becomes an important social distinction in many societies.

Racial ethnic Social category of people who identify themselves on the basis of their ancestry and culture.

Racial socialization Learning what racism is and how best to confront it.

Radical feminism Theoretical model that is based on an intense positive valuation of women and a belief that men often control and oppress women through the use of violence.

Random sample Research technique in which subjects are chosen without regard to any of their individual characteristics in order to argue that they are representative of a larger population.

Rational choice Theoretical model that posits that people make decisions about social relationships on the basis of how well those decisions help them to maximize their interests.

Recipient donor A person who donates sperm or egg cells to impregnate a surrogate mother with the intention of receiving the child.

Redistributive welfare state Those states whose primary aim is to restructure the economy by redistributing wealth and resources. See also programmatic welfare state.

Reduce the need One of four strategies men use to avoid doing housework; they redefine the task by arguing it does not need to be done or can be done with less effort, for example, going out to a restaurant instead of making a meal at home. See also disaffiliate, selective encouragement, substitute offering.

Research methodology Techniques used by social scientists to gather and analyze data. They include interviews, observations, case studies, historical documents, experiments, surveys and content analysis.

Restraining order Also called protective order and temporary restraining order. Directive given to abusive person to stay away from another person or to discontinue some abusive behavior.

Roe v. Wade Legal case decided by the Supreme Court in 1973 making abortion legal in the U.S.

Rosie the Riveter Name given to large numbers of women who were hired at manufacturing plants during World War II.

RU486 Steroid that can be used with progesterone within 49 days of conception to cause a woman's body to expel a fetus.

Second Wave feminists Organized the movement for women's liberation in the 1960s. The focus of their work was on identifying the similarities among women in the problems they faced in a male-dominated society.

Segmented labor market theory Asserts that jobs create workers' characteristics. Some activities we associate with women may in fact be more closely linked to women's jobs than to the women who occupy those jobs.

Segregated marriage Marriages in which husbands and wives participate in different sets of tasks. See also integrated, modified segregated, and wife-shaped.

Selective encouragement One of four strategies men use to avoid doing housework; they profusely praise their wife for the way in which she so adequately takes care of the work. See also disaffiliation, substitute offering, reduce the need.

Self-disclosure The expression of one's feelings and especially one's vulnerabilities.

Separate spheres The notion that work and family are two distinct and separate sets of activities done by two distinct and separate groups of people—employed men and housewives.

Sexual hierarchy Ranking of sex acts from good, normal, natural, and blessed to bad, unnatural, abnormal, and damned.

Sexual standards Standards telling us how to feel and act sexually. In the contemporary U.S. they include: heterosexual, coital, orgasmic, two-person, romantic, and marital.

Sharecropping The economic system established in the South after the Civil War, which served to maintain much of the oppressive exploitation that characterized the slave system. Former slaves and poor whites were forced to work on the plantations of the old slavemasters.

Shared partnership doctrine Currently dominant basis of marriage in the U.S. designating husbands and wives as equals in duties and responsibilities. See also doctrine of couverture.

Sibling violence Abuse such as kicking, biting, or punching among brothers and sisters.

Social construction of childhood Thinking about children and treating them in a way that differentiates them from adults. For example, although children are smaller than adults and must wear smaller clothing we socially construct childhood by dressing children in different styles than those adults wear.

Social constructionism Theoretical model that posits that social behaviors are a product of social interaction and ideologies rather than being inborn, unchanging phenomena.

Social interaction theory Posits that gender is not so much what we learn as what we do and that gender pervades nearly every social activity.

Social policy Government "activities affecting the social status and life chances of groups, families and individuals" (Skocpol & Amenta, 1986, p. 132).

Social Security Act of 1935 Established for the first time in the United States a federally funded and publicly administered welfare system.

Social Security Amendment of 1983 Included the provision that whenever possible Medicare and Medicaid patients should use outpatient care and that reimbursement for in-home care should be sharply curtailed. This resulted in a shift in health care from paid professionals in health care facilities to unpaid women in families.

Social stigmatization Perceiving people as different or less worthy than others because of their appearance or behavior.

Social structural model In contrast to the culture of poverty model, the social structural model asserts that economic decline causes poverty, which may then alter family structure.

Socialist feminism Theoretical model that asserts that the oppression of women is rooted both in the capitalist system and the patriarchal organization of society.

Socialization Process by which people learn to behave in an acceptable manner in their society—what behaviors are considered appropriate or inappropriate and what people might expect in response to those activities.

Socializing housework The proposal that all work done to feed, shelter, and care for people be equally divided throughout society rather than only within each individual household.

Sociobiology Theoretical model that posits that much of human behavior, especially gender and parenting, are largely determined by biological differences between males and females.

Sociological Imagination C. Wright Mills coined this term to describe the activity he argued was the task of sociology: bridging the gap between society and the individual social experience.

Sociologists for Women in Society (SWS) Professional organization that grew out of the feminist enlightenment in the 1970s and continues to work to further equality for women.

Spending down capital Upper-class children are taught that in order to maintain the family's fortune they should spend only the interest on an estate rather than spending down the principal.

Split household Family form dominant in the late 1800s among Chicanos and Chinese Americans. Men were forced to live a long distance from their wives and children in order to earn a living.

State "Organizations that extract resources through taxation and attempt to extend coercive control and political authority over particular territories and the people residing in them" (Skocpol & Amenta, 1986, p. 131).

Statistical Significance The differences found in the research were probably not a result of the fact that a relatively small sample of people was chosen to represent a large population.

Stratification Systems of social inequality among different groups of people.

Structural functionalism A theoretical perspective in sociology that examines society as a whole system and assumes that the organization of American society is essentially stable and beneficial for nearly all its members.

Structural Marxists Assert that the state serves the capitalist class by acting as an arbitrator among various factions in the ruling class.

Structuration theory Theoretical model that posits that social structure plays a critical role in determining people's day-to-day activities while individual social interaction simultaneously influences, shapes, and determines the character of social structure.

Student Non-Violent Coordinating Committee (SNCC) Student organization involved in sit-ins and other political actions in the civil rights movement.

Substitute offering One of four strategies men use to avoid doing housework; they try to alleviate the tension by patiently listening to their wife and offering her advice while she struggles to do it all. See also disaffiliate, selected encouragement, reduce the need.

Surrogate motherhood Becoming pregnant and delivering a child with the purpose of giving it to another.

Swapping Term used by Carol Stack to describe the circulation of goods and services among poor families.

Symbolic interactionism Theoretical model that emphasizes the way people work together to socially construct an understanding and a way to react to social experience.

Theory "Gives us a description of the problems we face, provides an analysis of the forces which maintain social life, defines the problems we should concentrate on, and acts as a set of criteria for evaluating the strategies we develop" (Hartsock, 1983, p. 8).

Third Wave feminists Emphasize the diversity of experience among women, especially by class and race ethnicity.

Tithingmen People who were appointed by the court in Puritan society to watch over households to make sure that family members were behaving properly.

Underground Railroad Political organization of slaves and other abolitionists that helped thousands of slaves to run away.

Value free Sociologists who support value-free research argue they must be detached and objective about their work.

Value committed Sociologists who support value-committed research contend that maintaining a value-free stance is impossible, disrespectful of participants, and impractical.

Wealth Assets owned, for example, money, houses, stocks, bonds, and land.

Webster Decision The 1989 Supreme Court decision that seemed to challenge the legality of abortion in the U.S.

Welfare hotels Hotels paid by the government to house poor people as an alternative to shelters.

Welfare liberalists Believe that the government should provide for its less fortunate citizens by creating policies that control or alter free enterprise. See also neocapitalists.

White House Conference on Families Conference called by President Carter to fulfill his campaign promise to address problems of American families. The conference name became a center of controversy because conservative participants wanted to call it the White House Conference on the Family.

Wife-shaped marriage Marriages in which wives are more active than their husbands in decision-making and financial activities. See also segregated, modified segregated, and integrated.

Woman battering "The use of physical force by a man against his intimate cohabiting partner. This force can range from pushes and slaps to coerced sex to assaults with deadly weapons" (Bograd, 1988, p. 12).

Women's liberation Movement in the 1960s that sought to end discrimination against and oppression of women.

Workfare The requirement that welfare recipients must work in exchange for receiving their grant.

World historic defeat of women Engels used this phrase to describe the "moment" in history at which monogamy was established, tying women to male-dominated and oppressive relationships of marriage.

World view The values and ideas promoted at work that influence an employee's behavior at home.

BIBLIOGRAPHY

Abel, Emily. (1986). Adult daughters and care for the elderly. *Feminist Studies 12* (3): 479–493.

Abramowitz, Mimi. (1989). *Regulating the lives of women: Social welfare policy from colonial times to the present.* Boston: South End Press.

Achilles, Rona. (1990). Desperately seeking babies: New technologies of hope and despair. In K. Arnup, A. Levesque and R. Pierson (eds.), *Delivering motherhood: Maternal ideologies and practices in the 19th and 20th centuries* (pp. 284–312). New York: Routledge.

Ahrons, Constance and Roy Rodgers. (1987). *Divorced families: A multidisciplinary view.* New York: Norton and Norton.

Allan, Graham. (1985). *Family life: Domestic roles and social organization.* New York: Basil Blackwell.

Allen, Paula Gunn. (1986). *The sacred hoop.* Boston: Beacon Press.

Alliance Housing Council. (1988). *Housing and homelessness.* Washington, D.C.: Alliance to End Homelessness.

Altman, Dennis. (1971). *Homosexual: Oppression and liberation.* New York: Outerbridge and Dientsfrey.

———. (1982). *The homosexualization of America.* Boston: Beacon Press.

American Association for Protecting Children. (1979). *National study on neglect and abuse reporting.* Denver, CO: American Humane Association.

———. (1985). *National study on neglect and abuse reporting.* Denver, CO: American Humane Association.

Amott, Teresa. (1990). Black women and AFDC: Making entitlement out of necessity. In L. Gordon (ed.), *Women, the state and welfare* (pp. 280–300). Madison: University of Wisconsin Press.

Amott, Teresa and Julie Matthaei. (1991). *Race, gender and work: A multicultural economic history of women in the United States.* Boston: South End Press.

An act for the better ordering and governing of negroes and slaves, South Carolina, 1712. (1992). In P. Rothenberg (ed.), *Race, Class and Gender in the U.S: an Integrated Study* (pp. 258–264), 2d ed. New York: St. Martin's Press.

Anderson, Margaret. (1988). *Thinking about women: Sociological perspectives on sex and gender.* 2d ed. New York: Macmillan.

Andler, Judy and Gail Sullivan. (1980). The price of government funding. *Journal of Alternative Human Services* 6(2): 15–18.

Appelbaum, Richard. (1989). The affordability gap. *Society 26* (May/June): 6–8.

Appelbaum, Richard and Peter Dreier. (1992). Census count no help to the homeless. In P. Baker, L. Anderson and D. Dorn (eds.), *Social*

problems: A critical thinking approach (pp. 336–337). Belmont, CA: Wadsworth.

Aptheker, Herbert. (1943). *American Negro slave revolts.* New York: International Publishers.

Arditti, Joyce. (1992). Differences between fathers with joint custody and noncustodial fathers. *American Journal of Orthopsychiatry 62*(2): 186–195.

Arendell, Terry. (1986). *Mothers and divorce: Legal, economic and social dilemmas.* Berkeley: University of California Press.

Aries, Philippe. (1962). *Centuries of childhood: A social history of family life.* New York: Knopf.

Aronson, Jane. (1992). Women's sense of responsibility for the care of old people: "But who else is going to do it?" *Gender and Society 6.* (1): 8–29.

Atherton, Charles. (1989). The welfare state: Still on solid ground. *Social Service Review 63.* (June): 169–178.

Aulette, Judy. (1991). The privatization of housing in a declining economy: The case of Stepping Stone Housing. *Sociology and Social Welfare 18*(1): 149–164.

Aulette, Judy and Walda Katz Fishman. (1991). Working class women and the women's movement. In B. Berberoglu (ed.), *Critical perspectives in sociology* (pp. 241–252). Dubuque, IA: Kendall Hunt.

Aulette, Judy and Trudy Mills. (1988). Something old, something new: Auxiliary work in the 1983–1986 copper strike. *Feminist Studies 14* (2): 251–268.

Baca Zinn, Maxine. (1980). Employment and education of Mexican American women: The interplay of modernity and ethnicity in eight families. *Harvard Educational Review 50* (1): 47–62.

———. (1982). Qualitative methods in family research: A look inside Chicano families. *California Sociologist* (Summer): 58–79.

———. (1987). *Minority families in crisis: The public discussion.* Memphis, TN: Memphis State University, Center for Research on Women.

———. (1989a). Chicano men and masculinity. In M. Kimmel and M. Messner (eds.), *Men's lives* (pp. 67–76). 2d ed. New York: Macmillan.

———. (1989b). Family race and poverty in the eighties. *Signs 14.* (4): 856–874.

Baca Zinn, Maxine and Stanley Eitzen. (1990). *Diversity in American families.* 2d ed. New York: Harper and Row.

Bahr, Stephen. (1974). Effects of power and division of labor in the family. In L. Hoffman and F. Nye (eds.), *Working mothers* (pp. 167–185). San Francisco: Jossey-Bass.

Baldwin, Wendy and Christine Nord. (1984). Delaying childbirth in the U.S.: Facts and fictions. *Population Bulletin 39*: 1–37.

Balswick, Jack. (1979). How to get your husband to say "I love you." *Family Circle.*

Balswick, Jack and Charles Peek. (1971). The inexpressive male: A tragedy of American society. *The Family Coordinator 20*: 363–368.

Bane, Mary Jo. (1980). Toward a description and evaluation of U.S. family policy. In J. Aldous and W. Dumon (eds.), *The politics and programs of family policy* (pp. 155–191). Notre Dame: UND and Leuven University Press.

———. (1986). Household composition and poverty. In S. Danziger and D. Weinberg (eds.), *Fighting poverty: What works and what doesn't* (pp. 209–231), Cambridge, MA: Harvard University Press.

Barak, Gregg. (1992). *Gimme shelter: A social history of homelessness in contemporary America.* New York: Praeger.

Barash, David. (1977). *Sociobiology and behavior.* New York: Elsevier.

Barerra, Mario. (1979). *Race and class in the Southwest.* Notre Dame: University of Notre Dame Press.

Barret, Robert and Bryan Robinson. (1992). *Gay fathers.* Lexington, MA: Lexington.

Barry, Kathleen. (1981). *Female sexual slavery.* Englewood Cliffs, NJ: Prentice-Hall.

Bart, Pauline. (1983). Review of Chodorow's "The reproduction of mothering." In J. Trebilcot (ed.), *Mothering: Essays in feminist theory* (pp. 147–152). Totowa, NJ: Rowman and Allanheld.

Basow, Susan. (1992). *Gender stereotypes and roles.* 3d ed. Pacific Grove, CA: Brooks/Cole.

Bassuk, E., L. Rubin and A. Lauriat. (1986). Characteristics of sheltered homeless families. *American Journal of Public Health 76* (September): 1097–1101.

Beals, Ralph, Harry Hoijer and Alan Beals. (1977). *An introduction to anthropology.* New York: Macmillan.

Becker, Gary. (1965). A theory of the allocation of time. *Economic Journal 75* (299): 493–517.

———. (1981) *A treatise on the family*. Cambridge, MA: Harvard University Press.

Becker, Howard. (1979). What's happening to sociology? *Society 15* (5): 19–24.

Bedard, Marcia. (1992). *Breaking with tradition: Diversity, conflict and change in contemporary American families*. Dix Hills, NY: General Hall.

Begus, Sarah and Pamela Armstrong. (1982). Daddy's right: Incestuous assault. In I. Diamond (ed.) *Family Politics and Public policy: A feminist dialogue on women and the state* (pp. 236–249). New York: Longman.

Bellah, Robert, Richard Madsen, William Sullivan, Ann Swidler and Steven Tipton. (1985). *Habits of the heart: Individualism and commitment in American life*. New York: Harper and Row.

Belle, Deborah. (1982). *Lives in stress: Women and depression*. Beverly Hills, CA: Sage.

Bem, Sandra Lipsitz. (1981). Gender schema theory: A cognitive account of sex typing. *Psychological Review 88*. (4): 354–364.

———. (1982). Gender schema theory and self-schema theory compared: A comment on Markus, Crane, Bernstein, and Siladi's "self schemas and gender." *Journal of Personality and Social Psychology 43* (6): 1192–1194.

———. (1983). Gender schema theory and its implications for child development: Raising gender-aschematic children in a gender-schematic society. *Signs 8* (4): 598–616.

Benenson, Harold. (1985). The community and family bases of U.S. working class protest, 1880–1920. In L. Kriesberg (ed.), *Research in social movements, conflicts and change* (pp. 112–126). Greenwich, CT: JAI Press.

Bennett, Neil, David Bloom and Patricia Craig. (1989). The divergence of black and white marriage patterns. *American Journal of Sociology 95*: 692–722.

Benson, Susan. (1978). The clerking sisterhood: Rationalization and the work culture of saleswomen in American department stores, 1890–1960. *Radical America 12* (2): 41–55.

Benston, Margaret. (1969). The political economy of women's liberation. *Monthly Review 21*: 13–27.

Berberoglu, Berch. (1988). Labor, capital and the state: Economic crisis and class struggle in the U.S. in the 1970s and 1980s. *Humanity and Society 12* (1): 1–20.

———. (1992). *The legacy of empire: Economic decline and class polarization in the United States*. New York: Praeger.

Berger, Peter and Hansfried Kellner. (1964). Marriage and the construction of reality. *Diogenes 46*: 1–32.

Berger, Peter and Thomas Luckmann. (1966). *The social construction of reality*. New York: Random House.

Berheide, Catherine. (1984). Women's work in the home: Seems like old times. *Marriage and Family Review 7* (Fall/Winter): 37–50.

Berheide, Catherine, Sarah Fenstermaker Berk and Richard Berk. (1976). Household work in the suburbs: The job and its participants. *Pacific Sociological Review 19* (4): 491–518.

Berk, Richard and Sarah Fenstermaker Berk. (1979). *Labor and leisure at home: Content and organization of the household day*. Beverly Hills, CA: Sage.

———. (1983). Supply side sociology of the family: The challenge of the new home economics. *Annual Review of Sociology 9*: 375-395.

———. (1988). Women's unpaid labor: Home and community. In A. Stromberg and S. Harkess (eds.), *Women working* (pp. 287–302), 2d ed. Mountain View, CA: Mayfield.

Berk, Richard, Sarah Berk, Donileen Loseke, and David Rauma. (1983). Mutual combat and other family violence myths. In D. Finkelhor et al. (eds.), *The dark side of families* (pp. 197–212). Beverly Hills, CA: Sage.

Berk, Sarah Fenstermaker. (1985). *The gender factory: The apportionment of work in American households*. New York: Plenum.

Bernard, Jessie. (1972). *The future of marriage*. New Haven: Yale University Press.

———. (1989). The dissemination of feminist thought: 1960–1988. In R. Wallace (ed.), *Feminism and sociological theory* (pp. 23–33). Newbury Park, CA: Sage.

Bernard, M. and V. Bernard. (1983). Violent intimacy: The family as a model for love relationships. *Family Relations 32* (April): 283–286.

Billingsley, Andrew. (1968). *Black families in white America*. Englewood Cliffs, NJ: Prentice-Hall.

Billington, Ray. (1949). *Westward expansion: A history of the American frontier*. New York: Macmillan.

Binder, Arnold and James Meeker. (1992). Arrest as a method to control spouse abuse. In E. Buzawa and C. Buzawa (eds.), *Domestic violence: The changing criminal justice response* (pp. 129–140). Westport, CT: Auburn House.

Blackstone, William. (1803). *Commentaries on the laws of England.* 14th ed. Book 1. London: Strahan.

Blackwood, Evelyn. (1984). Sexuality and gender in certain Native American tribes: The case of cross-gender females. *Signs 10* (1): 27–42.

Blassingame, John. (1977). *Slave testimony: Two centuries of letters, speeches, interviews and autobiographies.* Baton Rouge: Louisiana State University Press.

Blau, Peter and Otis Dudley Duncan with Andrea Tyree. (1967). *The American occupational structure.* New York: Wiley.

Blauner, Robert. (1964). *Alienation and freedom.* Chicago: University of Chicago Press.

Bleier, Ruth. (1984). *Science and gender: A critique of biology and its theories on women.* Elmsworth, NY: Pergamon Press.

Block, Fred. (1984). Technological change and employment: New perspectives on an old controversy. *Economia and Lavora 18*: 3–21.

———. (1988). Rethinking responses to economic distress: A critique of full employment. In P. Voydanoff and L. Majka (eds.), *Families and economic distress: Coping strategies and social policy* (pp. 190–206). Beverly Hills, CA: Sage.

Block, Fred, Richard Cloward, Barbara Ehrenreich and Frances Fox Piven. (1987). *The mean season: The attack on the welfare state.* New York: Pantheon.

Block, Fred and J. Noakes. (1988). The politics of new style workfare. *Socialist Review 18* (3): 31–58.

Blood, Robert and Donald Wolf. (1960). *Husbands and wives: The dynamics of married living.* New York: Free Press.

Blumstein, Phillip and Pepper Schwartz. (1983). *American couples: Money, work and sex.* New York: William Morrow.

———. (1990a). Getting and spending money among American couples. In J. Heeren and M. Mason (eds.), *Windows on society* (pp. 124–129). Los Angeles: Roxbury Publishing.

———. (1990b). Intimate relationships and the creation of sexuality. In D. McWhirter, S. Sanders and J. Reinisch (eds.), *Homosexuality/heterosexuality: Concepts of sexual orientation* (pp. 96–109). New York: Oxford University Press.

Bograd, Michele. (1988). Feminist perspectives on wife abuse: An introduction. In K. Yllo and M. Bograd (eds.), *Feminist perspectives on wife abuse* (pp. 11–27). Beverly Hills, CA: Sage.

Bonomo, Thomas. (1987). Working class movements in the Reagan era: The potential for progressive change. *Humanity and Society 11* (1): 12–39.

———. (1993). Personal communication.

Bookman, Ann. (1991). Parenting without poverty: The case for funded parental leave. In J. Hyde and M. Essex (eds.), *Parental leave and children: Setting a research and policy agenda* (pp. 66–89). Philadelphia: Temple University Press.

Boris, Eileen and Peter Bordaglio. (1983). The transformation of patriarchy: The historic role of the state. In I. Diamond (ed.), *Families, politics and public policy: A feminist dialogue on women and the state* (pp. 70–93). New York: Longman.

Bose, Christine. (1987). Dual spheres. In B. Hess and M. Ferree (eds.), *Analyzing gender* (pp. 267–285). Newbury Park, CA: Sage.

Boulding, Elise. (1980). The nurturance of adults by children in family settings. In H. Lopata (ed.), *Research in the interweave of social roles.* Vol. 1 (pp. 167–189). Greenwich, CT: JAI Press.

Bourne, Patricia and Norma Wikler. (1982). Commitment and the cultural mandate: Women in medicine. In R. Kahn-Hut et al. (eds.), *Women and work: Problems and perspectives* (pp. 111–112). New York: Oxford University Press.

Bovee, Tim. (1993). Interracial marriage on rise: "It's a very normal thing." *Charlotte Observer* (Feb. 12): 1A, 5A.

Bowker, Lee. (1986). *Ending the violence: A guidebook based on the experience of 1000 wives.* Holmes Beach, FL: Learning Publications.

Bowles, Samuel and Herbert Gintis. (1976). *Schooling in capitalist America.* New York: Basic Books.

Bowman, Madonna and Constance Ahrons. (1985). Impact of legal custody status on fathers' parenting post divorce. *Journal of Marriage and the Family 47*: 481–488.

Boyle, James. (1913). *The minimum wage and syndicalism.* Cincinnati: Stewart and Kidd.

Bozett, Frederick. (1984). Parenting concerns of gay fathers. *Topics in Clinical Nursing 6*: 60–71.

———. (1988). Gay fatherhood. In P. Bronstein and C. Cowan (eds.), *Fatherhood today: Men's changing role in the family* (pp. 60–71). New York: Wiley.

———. (1993). Children of gay fathers. In C. Brettell and C. Sargent (eds.), *Gender in cross-cultural perspective* (pp. 191–200). Englewood Cliffs, NJ: Prentice-Hall.

Brake, Mike. (1982a). Sexuality as praxis—A consideration of the contribution of sexual theory to the process of sexual being. In M. Brake (ed.), *Human sexual relations: Towards a redefinition of sexual politics* (pp. 13–34). New York: Pantheon.

Brake, Mike (ed.). (1982b). *Human sexual relations: Towards a redefinition of sexual politics*. New York: Pantheon.

Braun, Denny. (1991). *The rich get richer: The rise of income in inequality in the U.S. and the world.* Chicago: Nelson Hall.

Braverman, Harry. (1974). *Labor and monopoly capital: The degradation of work in the twentieth century.* New York: Monthly Review Press.

Breines, Wini, Margaret Cerullo and Judith Stacey. (1978). Socio biology, family studies and antifeminist backlash. *Feminist Studies 4* (1): 43–67.

Brewer, Rose. (1988). Black women in poverty: Some comments on female-headed families. *Signs 13* (2): 331–339.

Bridenthal, Renate and C. Koonz (eds.). (1977). *Becoming visible: Women in European history.* Boston: Houghton Mifflin.

Broder, David. (1983). Phil Gramms' free enterprise. *Washington Post.* (2/16): A1.

Brody, Jane. (1989). Who's having sex? Data are obsolete, experts say. *New York Times* 2/18: A1.

Brown, Judith. (1977). Economic organization and the position of women among the Iroquois. *Ethnohistory 17*: 151–167.

Brown, Michael (ed.). (1988). *Remaking the welfare state: Retrenchment and social policy in America and Europe.* Philadelphia: Temple University Press.

Brown, Rita Mae. (1976). The shape of things to come. *Plain Brown Rapper.* Baltimore: Diana Press.

Browne, Angela. (1987). *When battered women kill.* New York: Free Press.

Bullough, Vern. (1976). *Sexual variance in society and history.* New York: John Wiley.

Bumpass, Larry and James Sweet. (1989). National estimates of cohabitation. *Demography 26*: 615–625.

———. (1991). The role of cohabitation in declining rates of marriage. *Journal of Marriage and the Family 53* (November): 913–927.

Burch, E.S. (1970). Marriage and divorce among North American Eskimos. In P. Bohannon (ed.), *Divorce and after* (pp. 152–181). New York: Doubleday.

Burgess, Ernest and L. Cottrell. (1939). *Predicting success or failure in marriage.* New York: Prentice-Hall.

Burgess, Ernest and Harvey Locke. (1945). *The family: From institution to companionship.* New York: American Book.

Burris, Beverly. (1991). Employed mothers. *Social Science Quarterly 72* (March): 50–66.

Bushnell, Don and Robert Burgess. (1969). Some basic principles of behavior. In R. Burgess and D. Bushnell (eds.), *Behavioral sociology* (pp. 27–48). New York: Columbia University Press.

Buss, D. M. (1985). Human mate selection. *American scientist 73*: 48.

Butler, Sandra. (1979). *Conspiracy of silence: The trauma of incest.* San Francisco: New Glide.

Cahill, Spencer. (1983). Reexamining the acquisition of sex roles: A symbolic interactionist approach. *Sex Roles 9* (1): 1–15.

Camarillo, Albert. (1979). *Chicanos in a changing society: From Mexican pueblos to American barrios in Santa Barbara and Southern California, 1848–1930.* Cambridge, MA: Harvard University Press.

Cancian, Francesca. (1987). *Love in America: Gender and self development.* New York: Cambridge University Press.

———. (1993). Feminist science: Methodologies that challenge inequality. *Gender and Society 6* (4): 623-642.

Cargan, Leonard and Matthew Melko. (1982). *Singles: Myths and realities.* Beverly Hills, CA: Sage.

Carlson, D., R. Labarba, J. Sclafani and C. Bowers. (1986). Cognitive and motor development in infants of adolescent mothers: A longitudinal analysis. *International Journal of Behavior Development 9* (1): 1–14.

Carnegie-Mellon University, School of Urban and Public Affairs. (1983). *Milltowns in the Pittsburgh region: Conditions and prospects.* Pittsburgh: Carnegie-Mellon University.

Carter, Hugh and Paul Glick. (1976). *Marriage and divorce: A social and economic study.* Cambridge, MA: Harvard University Press.

Cate, Rodney. (1992). *Courtship.* Beverly Hills, CA: Sage.

Caulfield, Mina. (1974). Imperialism, the family and cultures of resistance. *Socialist Revolution 20*: 67-85.

Cavan, Ruth and Katherine Ranck. (1938). *The family and the Depression: A study of one hundred Chicago families.* Chicago: University of Chicago Press.

Chaduri, Molly and Kathleen Daly. (1992). Do restraining orders help? Battered women's experience with male violence and legal process. In E. Buzawa and C. Buzawa (eds.) *Domestic violence: The changing criminal justice response* (pp. 227–252). Westport, CT: Auburn House.

Chafetz, Janet. (1988). *Feminist sociology: An overview of contemporary theories.* Itasca, IL: Peacock.

Chambers, David. (1979). *Making fathers pay.* Chicago: University of Chicago Press.

Charlotte Observer. (1990). Teenage mothers studied: Babies healthier, moms better off. February 17, p. 16A.

Chelf, Carl. (1992). *Controversial issues in social welfare policy: The government and the pursuit of happiness.* Newbury Park, CA: Sage.

Cherlin, Andrew. (1982). The trends: Marriage, divorce, remarriage. In A. Skolnick and J. Skolnick (eds.), *Family in transition* (pp. 128–137), 4th ed. Boston: Little Brown.

———. (1992a). *Marriage, divorce and remarriage.* Cambridge, MA: Harvard University Press.

———. (1992b). The strange career of the "Harvard-Yale Study." In A. Skolnick and J. Skolnick (eds.), *Family in transition* (pp. 553–559). New York: Harper Collins.

Chesler, Phyllis. (1986). *Mothers on trial: The battle for children and custody.* New York: McGraw-Hill.

Childe, J. Gordon. (1948). *The dawn of European civilization.* London: Kegan Paul.

Children's Defense Fund. (1988a). *A children's defense budget: An analysis of our nation's investment in children.* Washington, D.C.: Children's Defense Fund.

———. (1988b). *Vanishing dream: The growing economic plight of America's young families.* Washington, D.C.: Children's Defense Fund.

———. (1990). *Children 1990: A report card briefing book and action primer.* Washington, D.C.: Children's Defense Fund.

Chisman, F. and A. Pifer. (1987). *Government for the people: The federal social role, what it is, what it should be.* New York: Norton.

Chodorow, Nancy. (1976). Oedipal asymetries and heterosexual knots. *Social Problems 23* (4): 454–468.

———. (1978b). Considerations on a biosocial perspective on parenting. *Berkeley Journal of Sociology 22*: 179–197.

———. (1978b). *The reproduction of mothering: Psychoanalysis and the sociology of gender.* Berkeley: University of California Press.

Chow, Esther. (1987). The development of feminist consciousness among Asian American women. *Gender and Society 1* (September): 284–299.

Chow, Esther and Katherine Berheide. (1988). The interdependence of family and work. *Family Relations 37* (January): 23–28.

Clay, Phil. (1987). At risk of loss: The endangered future of low-income rental housing resources. *Safety Network* (May): 1-7.

Cobb, S. and S. Kasl. (1977). *Termination: The consequences of job loss.* Cincinnati: NIOSH.

Cohen, Susan and Mary Katzenstein. (1992). The war over the family is not over the family. In M. Hutter (ed.), *The family experience: A reader in cultural diversity* (pp. 101–120). New York: Macmillan.

Cole, Jonathan. (1979). *Fair science: Women in the scientific community.* New York: Free Press.

Collier, Jane, Michelle Rosaldo, and Sylvia Yanagisako. (1992). Is there a family? New anthropological views. In B. Thorne with M. Yalom (eds.), *Rethinking the family: Some feminist questions* (pp. 31–48). New York: Longman.

Collins, Patricia Hill. (1991). The meaning of motherhood in black culture and black mother-daughter relationships. In P. Scott et al. (eds.), *Double stitch: Black women write about mothers and daughters* (pp. 42–60). Boston: Beacon Press.

———. (1989). A comparison of two works on black family life. *Signs 14* (4): 875–884.

———. (1990). *Black feminist thought: Knowledge, consciousness and the politics of empowerment.* New York: Harper Collins.

Collins, Randall and Scott Coltrane. (1991). *Sociology of marriage and the family: Gender, love and property*. 3d ed. New York: Nelson Hall.

Committee on Health Care for Homeless People (CHCHP). (1988). *Homelessness, health and human needs*. Washington, D.C.: National Academy Press.

Condry, J. (1989). *The psychology of television*. Hillsdale, NJ: Lawrence Erlbaum Associates.

Connell, Robert. (1987). *Gender and power: Society, the person and sexual politics*. Stanford, CA: Stanford University Press.

Connor, J. and L. Serbin. (1978). Children's responses to stories with male and female characters. *Sex Roles 4*: 637–645.

Conway, Elizabeth. (1990). Women and contingent work. In S. Rix (ed.), *The American woman, 1990–1991: A status report* (pp. 203–211). New York: W. W. Norton.

Coontz, Stephanie. (1988). *The social origins of private life: A history of American families, 1600–1900*. London: Verso.

———. (1992). *The way we never were: American families and the nostalgia trap*. New York: Basic Books.

Costello, C. (1985). "We're worth it." Work culture and conflict at the Wisconsin Education Association Insurance Trust. *Feminist Studies 11* (Fall): 497–518.

Coverman, Shelley. (1989). Women's work is never done. The division of domestic labor. In Jo Freeman (ed.), *Women: A feminist perspective* (pp. 356–370). Mountain View, CA: Mayfield.

Cowan, Ruth Schwartz. (1983). *More work for mothers: The ironies of household technology from the open hearth to the microwave*. New York: Basic Books.

———. (1989). More work for mothers. In A. Skolnick and J. Skolnick (eds.), *Family in transition* (pp. 57–67). 6th ed. New York: Scott, Foresman.

Coyle, Laurie, Gail Hershatter and Emily Honig. (1980). Women at Farah: An unfinished story. In M. Mora and A. del Castillo (eds.), *Mexican women and Chicano families* (pp. 117–144). Los Angeles: University of California, Chicano Research Center Publications.

Cronan, Sheila. (1971). Marriage. In *Notes from the third year: Women's liberation* (pp. 62–66). New York: Notes from the Second Year, Inc.

Crosby, John. (1980). A critique of divorce statistics and their interpretation. *Family Relations 29*: 51–68.

Cullen, Countee. (1947). *On these I stand*. New York: Harper and Row.

Curie-Cohen, Martin, Lesleigh Luttrell and Sander Shapiro. (1979). Current practice of artificial insemination by donor in the U.S. *New England Journal of Medicine 300* (11): 589.

Damon, William. (1977). *The social world of the child*. San Francisco: Jossey-Bass.

D'Andrea, Ann. (1983). Joint custody as related to paternal involvement and paternal self-esteem. *Conciliation Courts Review 21*: 81–87.

Daniels, Arlene Kaplan. (1988). *Invisible careers: Women civic leaders from the volunteer world*. Chicago: University of Chicago Press.

Daniels, Roger. (ed.). (1978). *Anti-Chinese violence in America*. New York: Arno Press.

Danziger, Sheldon and Peter Gottschalk. (1985). The poverty of *Losing Ground*. *Challenge 28* (May/June): 32–38.

Darity, William and Samuel Meyers. (1984). Does welfare dependency cause female headship? The case of the black family. *Journal of Marriage and the Family 46* (4): 765–779.

Darling, Carol, David Kallen, and Joyce VanDusen. (1992). Sex in transition, 1900–1980. In A. Skolnick and J. Skolnick (eds.), *Family in transition* (pp. 151–160). 7th ed. New York: Harper Collins.

Dassbach, Carl. (1986). Industrial robots in the American automobile industry. *Insurgent Sociologist 13* (Summer): 53–61.

Davis, Angela. (1981). *Women, race and class*. New York: Random House.

———. (1991). Racism, birth control and reproductive rights. In M. Fried (ed.), *From abortion to reproductive freedom: Transforming a movement* (pp. 15–26). Boston: Southend Press.

Davis, Margaret R. (1982). *Families in a working world: The impact of organizations on domestic life*. New York: Praeger.

Davis, S. (1990). Men as success objects and women as sex objects: A study of personal advertisements. *Sex Roles 23*: 43–50.

Degler, Carl. (1980). *At odds: Women and the family in America: From the Revolution to the present*. Oxford: Oxford University Press.

Deitch, Cynthia. (1984). Collective action and unemployment: Responses to job loss by workers

and community groups. *International Journal of Mental Health 13* (1–2): 139–153.

D'Emilio, John and Estelle Freedman. (1988). *Intimate matters: A history of sexuality in America.* New York: Harper.

Demos, John. (1970). *A little commonwealth: Family life in the Plymouth Colony.* New York: Oxford University Press.

———. (1986). *Past, present and personal: The family and life course in American history.* New York: Oxford University Press.

Devault, Marjorie. (1987). Doing housework: Feeding and family life. In N. Gerstel and H. Gross (eds), *Families and work* (pp. 178–191). Philadelphia: Temple University Press.

———. (1991). *Feeding the family: The social organization of caring as gendered work.* Chicago: University of Chicago Press.

Diamond, Irene. (1983). *Families, politics and public policy.* New York: Longman.

Dibble, Ursula and Murray Straus. (1980). Some social structure determinants of inconsistency between attitudes and behavior: The case of family violence. *Journal of Marriage and the Family 42* (February): 71–80.

Dickerson, R. and L. Bean. (1915). *The single woman.* Baltimore: Williams and Wilkins.

Dill, Bonnie Thornton. (1983). Race, class and gender: Prospects for an all-inclusive sisterhood. *Feminist Studies 9* (1): 131–150.

———. (1986). *Our mother's grief: Racial ethnic women and the maintenance of families.* Memphis, TN: MSU Center for Research on Women.

———. (1988). Making your job good yourself: Domestic service and the construction of personal dignity. In A. Bookman and S. Morgen (ed.), *Women and the politics of empowerment* (pp. 33–52). Philadelphia: Temple University Press.

Dinnerstein, Dorothy. (1977). *The mermaid and the minotaur.* New York: Harper and Row.

D'Iorio, Judith. (1982). Feminist fieldwork in a masculinist setting: Personal problems and methodological issues. Paper presented at annual meetings of the North Central Sociological Association, Detroit.

Dobash, R. Emerson and Russell Dobash. (1979). *Violence against wives: A case against the patriarchy.* New York: Free Press.

———. (1988). Research as social action: The struggle for battered women. In K. Yllo and M. Bogard (eds.), *Feminist Perspectives on Wife Abuse* (pp. 28–50). Beverly Hills, CA: Sage.

———. (1992). *Women, violence and social change.* New York: Routledge.

Dolbeare, Cushing. (1983). The low income housing crisis. in C. Hartman (ed.). *America's housing crisis: What is to be done?* (pp. 32-47). Boston: Routledge and Kegan.

Domhoff, William. (1970). *The higher circles: The governing class in America.* New York: Random House.

Douvan, Elizabeth and Joseph Adelson. (1966). *The adolescent experience.* New York: Wiley.

Dubois, W. E. B. (1969/1908). *The Negro American family.* New York: New American Library.

Duncan, Otis Dudley, B. Featherman and Beverly Duncan. (1972). *Social change in a metropolitan community.* New York: Russell Sage.

Dutton, Donald. (1986). Wife assaulters' explanations for assault: The neutralization of self-punishment. *Canadian Journal of Behavioral Science 18* (4): 381–390.

Dutton, Donald, Stephen Hart, Les Kennedy and Kirk Williams. (1992). Arrest and the reduction of repeat wife assault. In E. Buzawa and C. Buzawa (eds.), *Domestic violence: The changing criminal justice response* (pp. 111–127). Westport, CN: Auburn House.

Dworkin, Andrea. (1981). *Pornography: Men possessing women.* New York: Putnam.

———. 1987). *Intercourse.* New York: Free Press.

Eckenrode, John and Susan Gore (eds.). (1990). *Stress between work and family.* New York: Plenum.

Edin, Kathryn. (1991). Surviving the welfare system: How AFDC recipients make ends meet in Chicago. *Social Problems 38* (4): 462–474.

Ehrenreich, Barbara. (1983). *Hearts of men: American dreams and the flight from commitment.* Garden City, NY: Anchor.

Ehrenreich, Barbara and Dierdre English. (1989). Blowing the whistle on the mommy track. *MS. 18*: 56–58.

Ehrenreich, Barbara, Beth Hess and G. Jacobs. (1986). *Remaking love: The feminization of sex.* Garden City, NY: Anchor.

Ehrensaft, Diane. (1980). When women and men mother. *Socialist Review 49*: 37–73.

———. (1987). *Parenting together: Men and women sharing the care of children.* New York: Free Press.

———. (1990). Feminists fight (for) fathers. *Socialist Review 59*. 57–80.

Eisenstein, Zillah. (1984a). *Feminism and sexual equality: Crisis in liberal America*. New York: Monthly Review Press.

———. (1984b). The patriarchal relations of the Reagan state. *Signs 10* (2): 329–337.

Eisenstein, Zillah (ed.). (1979). *Capitalist patriarchy and the case of socialist feminism*. New York: Monthly Review Press.

Eitzen, D. Stanley. (1985). *In conflict and order: Understanding society*. 3d. ed. Boston: Allyn and Bacon.

Eitzen, Stanley and Maxine Baca Zinn (eds.). (1989). *The reshaping of America*. Englewood Cliffs, NJ: Prentice-Hall.

———. (1992). *Social problems*. 5th ed. Boston: Allyn and Bacon.

Elder, Glen. (1969). Appearance and education in marriage mobility. *American Sociological Review 34*: 519–533.

———. (1974). *Children of the Great Depression*. Chicago: University of Chicago Press.

———. (1986). Military times and turning points in men's lives. *Developmental Psychology 22*: 233–245.

Ellis, Ellen. (1983). Abortion: Is a woman a person? In. A. Snitow, C. Stansell and S. Thompson (eds.), *Powers of desire: The politics of sexuality* (pp. 471–476). New York: Monthly Review Press.

Ellwood, David and Lawrence Summers. (1986). Poverty in America: Is welfare the answer or the problem? In. S. Danziger and D. Weinberg (eds.), *Fighting poverty: What works and what doesn't* (pp. 78–105). Cambridge, MA: Harvard University Press.

Emery, R. (1988). *Marriage, divorce and children's adjustment*. Newbury Park, CA: Sage.

Engels, Frederick. (1884/1970). *Origins of the family, private property and the state*. New York: International Publishers.

England, Paula. (1989). A feminist critique of rational choice theories: Implications for sociology. *The American Sociologist 19* (Spring): 14–28.

Epstein, Cynthia. (1983). *Women in law*. New York: Anchor Press.

Erlichman, Karen. (1989). Lesbian mothers: Ethical issues in social work practice. In E. Rothblum and E. Cole (eds), *Lesbianism: Affirming nontraditional roles* (pp. 207–244). New York: Haworth Press.

Etaugh, C. (1980). The effects of non-maternal care on children: Research evidence and popular views. *American Psychologist 35*: 309–319.

Ettlebrick, Paula. (1992). Since when is marriage the path to liberation? In G. Bird and M. Sporakowski (eds.), *Taking sides: Clashing views on controversial issues in family and personal relationships* (pp. 80–84). Guilford, CT: Dushkin Publishing.

Evans, Sara. (1979). *Personal politics: The roots of women's liberation in the civil rights movement and the new left*. New York: Knopf.

———. (1989). *Born for liberty: A history of women in America*. New York: Free Press.

———. (1991). The first American women. In L. Kerber and J. De Hart (eds.), *Woman's America: Refocusing the past* (pp. 31–40). 3d ed. New York: Oxford University Press.

Exner, M. (1915). *Problems and principles of sex education: A study of 948 college men*. New York: Association Press.

Eyer, Diane. (1992). *Mother-infant bonding: A scientific fiction*. New Haven, CT: Yale University Press.

Fagan, Jeffrey, Douglas Stewart and Karen Valentine. (1983). Violent men or violent husband? Background factors and situational correlates. In D. Finkelhor, et al. (eds), *The dark side of families: Current family violence reserach* (pp. 49–68). Beverly Hills, CA: Sage.

Falk, P. (1989). Lesbian mothers: Psychological assumptions in family law. *American Psychologist 44*: 941–947.

Faludi, Susan. (1991). *Backlash: The undeclared war against American women*. New York: Crown.

Farel, A. (1980). Effects of preferred maternal roles, maternal employment and sociographic status on school adjustment and competence. *Child Development 50*: 1179–1186.

Farrar, Eliza. (1837). *The young lady's friend*. Boston: Ticknor and Fields.

Fass, Paula. (1977). *The damned and the beautiful: American youth in the 1920s*. New York: Oxford University Press.

Fassinger, Polly. (1989). The impact of gender and past marital experience on heading a household alone. In B. Risman and P. Schwartz (eds.), *Gender in intimate relationships* (pp. 165–180). Belmont, CA: Wadsworth.

Feagin, Joe and Clarice Feagin. (1990). *Building American cities*. 2d ed. Englewood Cliffs, NJ: Prentice-Hall.

Feldberg, Roslyn and Evelyn Nakano Glenn. (1979). Male and female: Job versus gender models in the sociology of work. *Social Problems 26*: 524–538.

Ferber, Marianne and Brigid O'Farrell (eds.). (1988). *Work and family*. Washington, D.C.: National Academy.

Ferman, Lewis and Mary Blehar. (1983). Family adjustment to unemployment. In A. Skolnick and J. Skolnick (eds.), *Family in transition* (pp. 587–600). 4th ed. Boston: Little, Brown.

Ferraro, Kathleen. (1981). *Battered women and the shelter movement*. Ph.D. dissertation, Arizona State University.

———. (1989). Policing woman battering. *Social Problems 36* (1) 61–74.

Ferraro, Kathleen and John Johnson. (1990). How women experience battering: The process of victimization. In J. Heeren and M. Mason (eds.), *Sociology: Windows on society* (pp. 109–115). Los Angeles: Roxbury.

Ferree, Myra Marx. (1990). Beyond separate spheres: Feminism and family research. *Journal of Marriage and the Family 52* (November): 866–884.

Ferree, Myra Marx and Beth Hess. (1985). *Controversy and Coalition: The new feminist movement*. Boston: Twayne.

Finkelhor, David. (1979). *Sexually victimized children*. New York: Basic Books.

Finkelhor, David and Kersti Yllo. (1983). Rape in marriage: A sociological view. In D. Finkelhor et al. (eds.), *The dark side of families* (pp. 119–130). Beverly Hills, CA: Sage.

———. (1985). *License to rape: Sexual abuse of wives*. New York: Free Press.

———. (1989). Marital rape: The myth versus the reality. In J. Henslin (ed.), *Marriage and the family in a changing society* (pp. 382–291). New York: Free Press.

Firestone, Shulamith. (1970). *The dialectic of sex: The case for feminist revolution*. New York: William Morrow.

Fishman, Pamela. (1978). Interaction: The work women do. *Social Problems 25*: 308–406.

Flax, Jane. (1982). The family in contemporary feminist thought: A critical review. In J. Elshtain (ed.), *The family in political thought* (pp. 223–253). Amherst, MA: University of Massachusetts.

———. (1993). Women do theory. In M. Pearsall (ed.), *Women and values: Readings in recent feminist philosophy* (pp. 3–7). Belmont, CA: Wadsworth.

Fleming, Jeanne. (1988). Public opinion on changes in women's rights and roles. In S. Dornbusch and M. Strober (eds.), *Feminisim, children and the new families* (pp. 47–66). New York: Guilford.

Folbre, Nancy. (1987). The pauperization of motherhood: Patriarchy and public policy in the U.S. In N. Gerstel and H. Gross (eds.), *Families and work* (pp. 491–511). Philadelphia: Temple University Press.

Forbes. (1992). Billionaires. (July): 16–18.

Foucault, Michael. (1978). *The history of sexuality*. New York: Pantheon.

Fowlkes, Martha. (1987). The myth of merit and male professional careers: The roles of wives. In N. Gerstel and H. Gross (eds.), *Families and work* (pp. 347–361). Philadelphia: Temple University Press.

Franklin, John Hope. (1988). A historical note on black families. In H. McAdoo (ed.), *Black families* (pp. 23–26). 2d ed. Beverly Hills, CA: Sage.

Freedman, Estelle and Barrie Thorne. (1984). Introduction to feminist sexuality debates. *Signs 10* (Autumn): 102–105.

Freeman, Jo. (1975). *The politics of women's liberation*. New York: Longman.

———. (1989). Feminist organization and activities from suffrage to women's liberation. In J. Freeman (ed.), *Women: A feminist perspective* (pp. 541–555). 4th ed. Mountain View, CA: Mayfield.

Freud, Sigmund. (1954). *The origins of psychoanalysis: Letters to Wilhelm Fliess, drafts and notes, 1887–1902*. New York: Basic Books.

———. (1963). *Introductory lectures on psychoanalysis*. New York: W. W. Norton.

Fried, Marlene (ed.). (1991). *From abortion to reproductive freedom: Transforming a movement*. Boston: Southend Press.

Friedan, Betty. (1963). *The feminine mystique*. New York: Dell.

Friedman, Dana. (1988). Estimates from the Conference Board and other national monitors of employer-supported child care. Unpublished memo. New York: The Conference Board.

Frisch, Michael and Dorothy Watts. (1980). Oral history and the presentation of class consciousness: The *New York Times* vs. the Buffalo unemployed. *International Journal of Oral History*. 1: 89–110.

Frye, Marilyn. (1983). *The politics of reality*. Trumansburg, NY: Crossing Press.

Furstenburg, Frank. (1990). Divorce and the American family. *Annual Review of Sociology 16*: 379–403.

Furstenburg, F., J. Brooks-Gunn and S. Morgan. (1987). *Adolescent mothers in later life*. Cambridge: Cambridge University Press.

Furstenburg, Frank, S. Phillip Morgan, and Paul Allison. (1987). Paternal participation and children's well-being after marital dissolution. *American Sociological Review 52*: 695–701.

Galbraith, John. (1958). *The affluent society*. Boston: Houghton, Mifflin.

Gallagher, E. (1986). *No place like home: The tragedy of homeless children and their families in Massachusetts*. Boston: Massachusetts Committee for Children and Youth.

Gallup, George and F. Newport. (1991). Babyboomers seek more family time. *Gallup Poll Monthly* (April): 31–38.

Garcia, Mario. (1980). La familia: The Mexican immigrant family 1900–1930. In M. Barrera, A. Camarillo and F. Hernandez (eds.), *Work, family, sex roles, language* (pp. 117–140). Berkeley: Tonatius-Quinto Sol International.

Gaylord, Maxine. (1984). Relocation and the corporate family. In P. Voydanoff (ed.), *Work and family: Changing roles of women and men* (pp. 144–152). Palo Alto, CA: Mayfield.

Geerken, Michael and Walter Gove. (1983). *At home and at work: The family's allocation of labor*. Beverly Hills, CA: Sage.

Gelles, Richard. (1977). Violence in the American family. In J. Martin (ed.), *Violence and the family* (pp. 169–182). New York: Wiley.

Gelles, Richard and Claire Cornell. (1985). *Intimate violence in families*. Beverly Hills, CA: Sage.

Gelles, Richard and Murray Straus. (1987). Is violence toward children increasing? A comparison of 1975–1985 national survey rates. *Journal of Interpersonal Violence* 2: 212–222.

———. (1976). Abused wives: Why do they stay? *Journal of Marriage and the Family 38*: 659–668.

Genevie, L. and E. Margolies. (1987). *The motherhood report: How women feel about being mothers*. New York: Macmillan.

Gerson, Kathleen. (1987). How women choose between employment and family: A developmental perspective. In N. Gerstel and H. Gross (eds.), *Families and work* (pp. 270–288). Philadelphia: Temple University Press.

Gerstel, Naomi. (1987). Divorce and stigma. *Social Problems 34* (2): 172–186.

Gerstel, Naomi and Hannah Gross (eds.). (1987). *Families and work*. Philadelphia: Temple University Press.

Giddens, Anthony. (1977). *Studies in social and political theory*. New York: Basic Books.

———. (1984). *The constitution of society: Outline of the theory of structuration*. Berkeley: University of California Press.

———. (1991). *Introduction to sociology*. New York: W. W. Norton.

Giddings, Paula. (1984). *When and where I enter: The impact of black women on race and sex in America*. New York: Bantam Books.

Gilbert, Dennis and Joseph Kahl. (1993). *The American class structure: A new synthesis*. Belmont, CA: Wadsworth.

Gilderbloom, John and Richard Appelbaum. (1988). *Rethinking Rental Housing*. Philadelphia: Temple University Press.

Gillespie, Dair. (1971). Who has the power? The marital struggle. *Journal of Marriage and the Family 31*: 445–558.

Gilligan, Carol. (1982). *In a different voice: Psychological theory and women's development*. Cambridge, MA: Harvard University Press.

Gilman, Carolina. (1834). *Recollections of a housekeeper*. New York:

Gilman, Charlotte Perkins. (1898–1966). *Women and Economics*. New York: Harper and Row.

Ginsberg, Faye. (1990). *Contested lives: The abortion debate in an American community*. Berkeley: University of California Press.

Ginsberg, Faye and Anna Lowenhaupt Tsing. (1990). *Uncertain terms: Negotiating gender in American culture*. Boston: Beacon Press.

Glass, Jennifer. (1992). Gender, family and job-family compatibility. *American Journal of Sociology 98* (1): 131–151.

Glazer, Nona. (1987). Servants to capital: Unpaid domestic labor and paid work. In N. Gerstel and H. Gross (eds.), *Families and work* (pp. 236–255). Philadelphia: Temple University Press.

———. (1990). The home as workshop: Women as amateur nurses and medical care providers. *Gender and Society 4* (4): 479–499.

Glenn, Evelyn Nakano. (1987). Gender and the family. In P. Hess and M. M. Ferree (eds.), *Analyzing gender: A handbook of social science research* (pp. 348–380). Newbury Park, CA: Sage.

———. (1990). The dialectics of wage work: Japanese-American women and domestic service, 1905–1940. In E. Dubois and V. Ruiz (eds.), *Unequal sisters: A multi-cultural reader in U.S. women's history* (pp. 345–372). New York: Routledge.

———. (1991). Racial ethnic women's labor: The intersection of race, gender and class oppression. In R. Blumberg (ed.), *Gender, family and economy: The triple overlap* (pp.173–201). Newbury Park, CA: Sage.

Glick, Paul. (1977). Updating the life cycle of the family. *Journal of Marriage and the Family 39*: 5–13.

———. (1988). A demographic picture of black families. In H. McAdoo (ed.), *Black families* (pp. 111–133). 2d ed. Beverly Hills, CA: Sage.

Glick, Paul and Sung-Ling Lin. (1986). Recent changes in divorce and remarriage. *Journal of Marriage and the Family 48* (4): 737–747.

Glick, Paul and Arthur Norton. (1979). Marrying, divorcing and living together in the U.S. today. *Population Bulletin 32* (February): 1–41.

Gluck, Sherna. (1987). *Rosie the Riveter revisited: Women, the war and social change*. New York: Twayne.

Glueck, W. (1979). Changing hours of work: A review and analysis of the research. *Personnel Administrator 3*: 44–47.

Goffman, Erving. (1963). *Stigma: Notes on the management of spoiled identity*. Englewood Cliffs, NJ: Prentice-Hall.

———. (1977). The arrangement between the sexes. *Theory and Society 40*: 301–331.

———. (1979). *Gender advertisements*. New York: Harper and Row.

Goldin, Cynthia. (1981). Family strategies and the family economy in the late nineteenth century: The role of secondary works. In T. Hershberg (ed.), *Philadelphia* (pp. 183–201). New York: Oxford University Press.

Goldman, Emma. (1910). *Anarchism and other essays*. Port Washington, NY: Kennikat Press.

Goldscheiter, Frances and Linda Waite. (1991). *New families, no families: The transformation of the American home*. Berkeley: University of California Press.

Goode, William. (1964). *The family*. Englewood Cliffs, NJ: Prentice-Hall.

Goodman, Ellen. (1992). The White House and your house. *Charlotte Observer* (October 10): 2c.

Googins, Bradley. (1991). *Work/family conflicts: Private lives, public responses*. New York: Auburn House.

Gordon, Linda. (1977). *Woman's body, woman's right: A social history of birth control in America*. New York: Penguin.

———. (1982). Why nineteenth century feminists did not support birth control and twentieth century feminists do. In B. Thorne (ed.), *Rethinking the family: Some feminist questions* (pp. 40–53). New York: Longman.

———. (1988). *Heroes of their own lives: The politics and history of family violence, Boston 1880–1960*. New York: Viking.

Gordon, Linda, and Paul O'Keefe. (1984). Incest as a form of family violence: Evidence from historical case records. *Journal of Marriage and the Family*. 49: 27–34.

Gordon, Michael (ed.). (1978). *The American family in social-historical perspective*. 2d ed. New York: St. Martin's Press.

Gorelick, Sherry. (1991). Contradictions of feminist methodology. *Gender and Society 5* (4): 459–478.

Gould, Meredith. (1984). Lesbians and the law: Where sexism and heterosexism meet. In T. Darty and S. Potter (eds.), *Women identified women* (pp. 149–162). Palo Alto, CA: Mayfield.

Gouldner, Alvin. (1970). *The coming crisis of Western sociology*. New York: Basic Books.

Gove, Walter. (1972). The relation between sex roles, marital status and mental illness. *Social Forces 51*: 34–44.

Gramsci, Antonio. (1971). *Selections from the prison notebooks*. New York: International Publishers.

Grant, Linda, Layne Sampson and Xue Lai Rong. (1990). Development of work and family commitments: A study of medical students. *Journal of Family Issues 8*: 176–198.

Grant, Linda, Kathryn Ward, Donald Broom, and William Moore. (1987). Gender, parenthood and work hours of physicians. *Journal of Marriage and the Family* 52 (February): 39–49.

Greeley, Andrew, Robert Michael and Tom Smith. (1990). *A most monogamous people: Americans and their sexual partners.* Chicago: NORC.

Green, Charles. (1992). Bush says test welfare limits plan. *Charlotte Observer* (April 10): 1A.

Green. S. and P. Sandos. (1983). Perceptions of male and female initiators of relationships. *Sex Roles* 9: 849–852.

Greenberger, Ellen. (1987). Children's employment and families. In N. Gerstel and H. Gross (eds.), *Families and Work* (pp. 396–406). Philadelphia: Temple University Press.

Greer, Germaine. (1970). *The female eunuch.* London: MacGibbon and Kee.

Greif, Geoffrey. (1979). Fathers, children and joint custody. *American Journal of Orthopsychiatry* 49: 311–319.

———. (1985). *Single Fathers.* Lexington, MA: Lexington Books.

Gresham, Jewell. (1989). White patriarchal supremacy: The politics of family in America. *Nation* 249(4): 116–121.

Greven, Philip. (1978). Family structure in the 17th century. Andover, MA: *William and Mary Quarterly* 23: 234–256.

Griscom, Joan. (1992). The case of Sharon Kowalski and Karen Thompson: Ableism, heterosexism and sexism. In P. Rothenberg (ed.), *Race, class and gender in the U.S.: An integrated study* (pp. 215–224). 2d ed. New York: St. Martin's Press.

Gutman, Herbert. (1976). *The black family in slavery and freedom, 1750–1925.* New York: Pantheon.

———. (1992). Americans and their sexual partners. In G. Bird and M. Sporakowski (eds.), *Taking sides: Clashing views on controversial issues and family and personal relationships* (pp. 254–262). Guilford, CT: Dushkin.

Guttentag, Marcia and Paul Secord. (1983). *Too many women: The sex ratio problem.* Beverly Hills, CA: Sage.

Gwartney-Gibbs, Patricia. (1986). The institutionalization of premarital cohabitation: Estimates from marriage license applications, 1970–1980. *Journal of Marriage and the Family* 48: 423–434.

Hacker, Andrew. (1992). *Two nations: Black and white, separate, hostile and unequal.* New York: Charles Scribner's Sons.

Halle, Robert. (1984). *America's working man: Work, home and politics among blue collar property owners.* Chicago: University of Chicago Press.

Halsted, James. (1992). Domestic violence: Its legal definitions. In S. Buzawa and C. Buzawa (eds.), *Domestic violence: The changing criminal justice response* (pp. 143–160). Westport, CT: Auburn House.

Hansen, D. and V. Johnson. (1979). Rethinking family stress theory. In W. Burr et al. (eds.), *Contemporary theories about the family.* Vol. 1 (pp. 582–603). New York: Free Press.

Hansen, Karen and Ilene Philipson (eds.). (1990). *Women, class and the feminist imagination: A socialist-feminist reader.* Philadelphia: Temple University Press.

Harding, Susan. (1981). Family reform movements: Recent feminism and its opposition. *Feminist Studies* 7 (1): 58–75.

———. (1987). *Feminism and methodology.* Bloomington: University of Indiana Press.

Hare, Nathan and Julia Hare. (1985). *Bringing the black boy to manhood: The passage.* San Francisco: Black Think Tank.

Harlan, S. (1989). Introduction to welfare, workfare and training. In S. Harlan and R. Steinberg (eds.), *Job training for women* (pp. 359–364). Philadelphia: Temple University Press.

Harrington, Michael. (1962). *The other America: Poverty in the U.S.* New York: Penguin.

Harris, Diana. (1992). You're pregnant? You're out. *Working Woman* November: 48–51.

Harrison, Althea. (1987). Images of black women. *The American woman, 1987–1988, a report in depth.* New York: Norton.

Harrison, Bennett and Barry Bluestone. (1988). *The great U-turn: The corporate restructuring and the polarizing of America.* New York: Basic Books.

Harry, Joseph. (1983). Gay male and lesbian relationships. In E. Macklin and R. Rubin (ed.), *Contemporary families and alternative lifestyles: Handbook of research and theory* (pp. 216–234). Beverly Hills, CA: Sage.

Hartmann, Betsy. (1987). *Reproductive rights and wrongs: The global politics of population control and contraceptive choice.* New York: Harper & Row.

Hartmann, Heidi. (1981). The family as the locus of gender, class and political struggle: The example of housework. *Signs* 6 (3): 366–394.

———. (1991). Women's work and diversity and employment stability: Public policy responses to new realities. Testimony before U.S. Senate Committee of Labor and Human Resources. Washington, D.C.: Institute for Women's Policy Research.

Hartmann, Heidi and Diana Pearce. (1989). *High skill and low pay: The economics of child care work. Executive summary.* Washington, D.C.: IWPR.

Hartsock, Nancy. (1983). *Money, sex and power: Toward a feminist historical materialism.* New York: Longman.

———. (1993). Feminist theory and the development of revolutionary strategy. In M. Pearsall (ed.), *Women and values: Readings in recent feminist philosophy* (pp. 8–17). Belmont, CA: Wadsworth.

Haskins, R. (1988). Child support: A father's view. In S. Kamerman and A. Kahn (eds.), *Child support: From debt collection to public policy* (pp. 306–327). Beverly Hills, CA: Sage.

Hayden, Dolores. (1981). *The grand domestic revolution: History of feminist designs for American homes, neighborhoods, and cities.* Cambridge: MIT Press.

Hedges, J. and J. Barnett. (1972). Working women and the division of household tasks. *Monthly Labor Review 95* (1): 9–14.

Heller, Celia. (1966). *Mexican American youth: Forgotten youth at the crossroads.* New York: Random House.

Hendrick, Susan and Clyde Hendrick. (1992). *Liking, loving and relating.* 2d ed. Belmont, CA: Wadsworth.

Henley, Nancy. (1977). *Body politics: Power, sex and nonverbal communication.* Englewood Cliffs, NJ: Prentice-Hall.

Henshaw, S. and J. Van Vort (eds.). (1992). *Abortion Services in the U.S. Each State and metropolitan area, 1967–1988.* New York: Alan Guttmacher Institute.

Herman, Judith. (1981). *Father-daughter incest.* Cambridge, MA: Harvard University Press.

———. (1992). *Trauma and recovery.* New York: Basic Books.

Herman, Judith and Lisa Hirschman. (1977). Father-daughter incest. *Signs 2* (4): 735–756.

Hess, Beth and Myra Marx Ferree. (1987). *Analyzing gender: A handbook of social science research.* Beverly Hills, CA: Sage.

Hetherington, E. (1979). Divorce: A child's perspective. *American Psychologist 34*: 851–858.

Hetherington, E., Martha Cox and Roger Cox. (1976). Divorced fathers. *The Family Coordinator 25*: 417–428.

Hewitt, John. (1984). *Self and society: A symbolic interactionist social psychology.* 3d ed. Boston: Allyn and Bacon.

Hewlett, Sylvia. (1991). *When the bough breaks: The cost of neglecting our children.* New York: Basic Books.

Higgenbotham, Elizabeth. (1981). Is marriage a priority: Class differences in marital options of educated black women. In P. Stein (ed.), *Single life: Unmarried adults in social context* (pp. 259–267). New York: St. Martin's Press.

Hill, Martha S. (1985). Patterns of time use. In F. Thomas Juster and Frank P. Stafford (eds.), *Time, goods, and well-being* (pp. 133–176). Ann Arbor, MI: Institute for Social Research, UM.

Hill, Martha and Michael Ponza. (1983). Poverty and welfare dependence across generations. *Economic Outlook, USA* (Summer): 61–64.

Hill, Richard and Cynthia Negrey. (1989). Deindustrialization and racial minorities in the Great Lakes region, USA. In D. Stanley Eitzen and M. Baca Zinn (eds.), *The reshaping of America: Social consequences of a changing economy.* Englewood Cliffs, NJ: Prentice Hall.

Hill, Robert. (1972). *The strengths of black families.* New York: Emerson Hall.

———. (1977). *Informal adoption among black families.* Washington, D.C.: National Urban League Research Department.

Hill, Robert et al. (eds.). (1989). *Research on the African American family: A holistic perspective and assessment of the status of African Americans.* Vol. II. Boston: Robert Trotter Institute, University of Massachusetts.

Hill, Rueben. (1949). *Families under stress.* New York: Harper and Row.

Hirschel, J. David and Ira Hutchison. (1989). The theory and practice of spouse abuse arrest policies. Paper presented at the annual meetings of the ASC, Reno, NV.

Hochschild, Arlie. (1971). Inside the clockwork of male careers. In F. Howe (ed.), *Women and the power to change* (pp. 47–80). New York: McGraw-Hill.

———. (1983). *The managed heart: Commercialization of human feeling.* Berkeley: University of California Press.

———. (1989). *The second shift: Working parents and the revolution at home*. New York: Viking.

———. (1991). The economy of gratitude. In M. Hutter (ed.), *The family experience: A reader in cultural diversity* (pp. 499–512). New York: Macmillan.

Hofferth, Sandra and Deborah Phillips. (1987). Childcare in the United States, 1970–1995. *Journal of Marriage and the Family 49*: 559–571.

Hoffman, Lois. (1961). Effects of maternal employment on the child. *Child Development 32*: 187–97.

———. (1984). Maternal employment and the young child. In M. Perlmutter (ed.), *Minnesota symposium in child psychology*. Hillsdale, NJ: Erlbaum.

Hoffman, Lois and J. Manis. (1979). The value of children in the U.S.: A new approach to the study of fertility. *Journal of Marriage and the Family 41*: 583–596.

Hoffman, Saul and Greg Duncan. (1988). What are the economic consequences of divorce? *Demography 25* (4): 641–645.

Hoffnung, Michele. (1989). Motherhood: Contemporary conflict for women. In J. Freeman (ed.), *Women: A feminist perspective* (pp. 147–175). Mountain View, CA: Mayfield.

Hole, Judith and Ellen Levine. (1971). *Rebirth of feminism*. New York: Random House.

Homans, George. (1958). Social behavior as exchange. *American Journal of Sociology 63*: 597–606.

Hooks, Bell. (1984). *Feminist theory: From margin to center*. Boston: Southend Press.

Hope, Marjorie and James Young. (1986). *The faces of homelessness*. Lexington, MA: Lexington Books.

Horowitz, Gad and Michael Kaufman. (1987). Male sexuality: Toward a theory of liberation. In M. Kaufman (ed.), *Beyond patriarchy: Essays by men on pleasure, power and change* (pp. 81–119). New York: Oxford University Press.

Houseknecht, Sharon. (1987). Voluntary childlessness. In M. Sussman and S. Steinmetz (eds.), *Handbook of marriage and the family* (pp. 369–396). New York: Plenum.

Houseknecht, Sharon, Suzanne Vaughn, and Anne Macke. (1984). Marital disruption among professional women: The timing of career and family events. *Social Problems 31* (1): 273–284.

Huaco, George. (1986). Ideology and general theory: The case of sociological functionalism. *Comparative Studies in Society and History 28*: 34–54.

Hughes, Everett. (1945). Dilemmas and contradictions of status. *American Journal of Sociology 50*: 353–359.

———. (1958). *Men and their work*. Chicago: Univeristy of Chicago Press.

Hunt, D. (1985). Parents and children in history. In P. Worsley (ed.), *Modern sociology* (pp. 195–199). New York: Penguin.

Hunt, Morton. (1974). *Sexual behavior in the seventies*. Chicago: Playboy Press.

———. (1983). Marital sex. In A. Skolnick and J. Skolnick (eds.), *Family in transition* (pp. 219–234). 4th ed. Boston: Little, Brown.

Hyde, Janet and M. Essex. (1991). *Parental leave and child care*. Philadelphia: Temple University Press.

Hymowitz, Carol and Michaele Weissman. (1978). *A history of women in America*. New York: Bantam Books.

Jacob, Herbert. (1989). Women and divorce reform. In L. Tilly and P. Gurin (eds.), *Women, politics and change* (pp. 482–502). New York: Russell Sage.

Jacobs, Janet. (1990). Reassessing mother blame in incest. *Signs 15* (3): 500–514.

Jagger, Alison. (1983). *Feminist politics and human nature*. Totowa, NJ: Rowman and Allanheld.

Janiewski, Doris. (1983). Sisters under the skin: Southern working women 1880–1950. In J. Hauks and S. Skemp (eds.), *Sex, race and the role of women in the South*. Jackson: University of Mississippi Press.

Jensen, Joan. (1991). Native American women and agriculture: A Seneca case study. In E. Dubois and V. Ruiz (eds.), *Unequal Sisters* (pp. 51–65). New York: Routledge.

Jessor, S. and R. Jessor. (1975). Transition from virginity to non-virginity among youth: A social-psychological study over time. *Developmental Psychology 11* (4): 473–484.

Joffe, Carol. (1986). *The regulation of sexuality: Experiences of family planning workers*. Philadelphia: Temple University Press.

Johnson, John. (1981). Program enterprise and official cooptation in the battered women's shelter movement. *American Behavioral Scientist 24* (6): 827–842.

———. (1991). The changing concept of child abuse and its impact on the integrity of family life. In M. Hutter (ed.), *The family experience: A reader in cultural diversity* (pp. 671–685). New York: Macmillan.

Jones, Barry. (1985). *Sleepers, wake! Technology and the future of work.* New York: Oxford University Press.

Jones, Elise et al. (eds.). (1986). *Teenage pregnancy in industrialized countries.* New Haven, CT: Yale University Press.

Jones, Jacqueline. (1985). *Labor of love, labor of sorrow: Black women, work, and family from slavery to the present.* New York: Basic Books.

———. (1989). The public dimensions of the "private" life: Southern women and their families, 1865–1965. In P. Cortelyou Little and R. Vaughn (eds.), *A new perspective: Southern women's cultural history from the Civil War to civil rights* (pp. 29–40). Charlottesville, VA: Virginia Foundation for the Humanities.

Journal of the American Medical Association (*JAMA*). (1989). Abortion frequency before and after Roe v. Wade. *JAMA 262* (October 10): 2076.

Judd, Ted. (1978). Naturizing what we do: A review of the film "Sociobiology: Doing what comes naturally." *Science for the People 10* (1): 16–19.

Kain, Edward. (1990). *The myth of family decline: Understanding families in a world of rapid social change.* Lexington, MA: Lexington.

Kamerman, Sheila. (1991). Parental leave and infant care: U.S. and international trends and issues, 1978–1988 . In J. Hyde and M. Essex (eds.), *Parental leave and children: Setting a research and policy agenda* (pp. 11–23). Philadelphia: Temple University Press.

Kamerman, Sheila and Arthur Kahn. (1978). *Family policy, government and families in 14 countries.* New York: Columbia University Press.

———. (1988). *Mothers alone: Strategies for a time of change.* Dover, MA: Auburn House.

———. (eds.). (1989). *Privatization and the welfare state.* Princeton: Princeton University Press.

———. (eds.). (1991). *Child care, parental leave and the under three's: Policy innovation in Europe.* New York: Auburn.

Kandel, Denise and Gerald Lesser. (1972). *Youth in two worlds.* San Francisco: Jossey-Bass.

Kanowitz, Leo. (1969). *Women and the law.* Albuquerque: University of New Mexico Press.

Kanter, Rosabeth Moss. (1977a). *Men and women of the corporation.* New York: Basic Books.

———. (1977b). *Work and family in the United States: A critical review and agenda for research and policy.* New York: Russell Sage Foundation.

———. (1986). Wives. In J. Cole (ed.), *All American women: Lines that divide, ties that bind* (pp. 155–171). New York: Free Press.

Kasarda, John. (1985). Urban change and minority opportunities. In P. Peterson (ed.), *The New Urban Reality* (pp. 33–67). Washington, D.C.: Brookings Institution.

Kasinitz, Phillip. (1984). Gentrification and homelessness: The single room occupant and inner city revival. *The Urban and Social Change Review 17*: 9–14.

Kassoff, Elizabeth. (1989). Nonmonogamy in the lesbian community. In E. Rosenblum and E. Cole (eds.), *Lesbianism: Affirming nontraditional roles* (pp. 167–182). New York: Haworth Press.

Katchadourian, Herant. (1985). *Fundamentals of human sexuality.* New York: Holt Rinehart and Winston.

Katz, Michael. (1989). *The undeserving poor: From the war on poverty to the war on welfare.* New York: Pantheon.

———. (1990). The invention of heterosexuality. *Socialist Review 59*: 7–34.

Keller, John. (1983). *Power in America.* Chicago: Vanguard.

Kellman, Jeff. (1989). Decoding MTV: Values, views and videos. *Media and Values 46*: 15–16.

Kelly, Gary. (1992). *Sexuality today: The human perspective.* Guilford, CT: Dushkin.

Kelly, Joan. (1979). The doubled vision of feminist theory: A postscript to the Women of Power Conference. *Feminist Studies 5* (1): 221–240.

Kelly, Liz. (1988). How women define their experience of violence. In K. Yllo and M. Bograd (eds.), *Feminist Perspectives on wife abuse* (pp. 114–132). Newbury Park, CA: Sage.

Kephart, William. (1967). Some correlates of romantic love. *Journal of Marriage and the Family 29*: 470–474.

Kessler-Harris, Alice. (1982). *Out to Work: A history of wage-earning women in the United States.* New York: Oxford University Press.

Kimmel, Michael and Michael Messner. (1989). *Men's lives.* New York: Macmillan.

———. (1990). Men as "gendered beings." In S. Ruth (ed.), *Issues in feminism: An introduction to women's studies* (pp. 56–58). Mountain View, CA: Mayfield.

Kincaid, P. (1982). *The omitted reality: Husband-wife violence in Ontario and policy implications for education.* Maple, Ontario: Learner's Press.

Kinsey, Alfred. (1948). *Sexual behavior in the human male.* Philadelphia: Saunders.

———. (1953). *Sexual behavior in the human female.* Philadelphia: Saunders.

Kitano, Harry. (1976). *Japanese Americans: The evolution of a sub-culture.* 2d. ed. Englewood Cliffs, NJ: Prentice-Hall.

Kitano, Harry and Roger Daniels. (1988). *Asian Americans: Emerging minorities.* Englewood Cliffs, NJ: Prentice Hall.

Kloby, Jerry. (1991). Increasing class polarization in the United States: The growth of wealth and income inequality. In B. Berberoglu (ed.), *Critical perspectives in sociology: A reader* (pp. 39–54). Dubuque, IO: Kendall/Hunt.

Kohlberg, Lawrence. (1966). A cognitive developmental analysis of children's sex role concepts and attitudes. In E. Maccoby (ed.), *The development of sex differences* (pp. 82–173). Stanford, CA: Stanford University Press.

Komarovsky, Mirra. (1940). *The unemployed man and his family: The effect of unemployment upon the status of the man in fifty-nine families.* New York: Dryden Press.

———. (1953). *Women in the modern world.* Boston: Little Brown.

Kolder, V., J. Gallagher and M. Parsons. (1987). Court-ordered obstetrical interventions. *New England Journal of Medicine 316* (19): 1192–1196.

Kozol, Jonathan. (1988). *Rachel and her children: Homeless families in America.* New York: Crown.

Kramer, Marian. (1988). Report on National Welfare Rights Union to Members and Friends. *Convention Proceedings.* Detroit, MI: September 3–5.

Kurz, Demie. (1989). Social science perspectives on wife abuse: Current debates and future directions. *Gender and Society 3* (4): 489–505.

Kuznets, Simon. (1971). *The economic growth of nations.* Cambridge, MA: Harvard University Press.

Lader, Lawrence. (1991). *RU486: The pill that could end the abortion wars and why American women don't have it.* New York: Addison-Wesley.

Ladner, Joyce. (1971). *Tomorrow's tomorrow: The black woman.* Garden City, NY: Doubleday.

Lamanna, Marianne and Agnes Reidman. (1991). *Marriages and families: Making choices and facing change.* Belmont, CA: Wadsworth.

Lamb, E. and A. Elster. (1986). Parental behavior of adolescent mothers and fathers. In A. Elster and M. Lamb (eds.), *Adolescent fatherhood* (pp. 89–106). Hillsdale, NJ: Lawrence Erlbaum.

Lamb, Michael. (1987). *The father's role: Cross-cultural perspectives.* Hillsdale, NJ: Erlbaum.

Lamphere, Louise. (1985). Bringing the family to work: Women's culture on the shop floor. *Feminist Studies 11* (3): 519–540.

LaRossa, Ralph. (1992). Fatherhood and social change. In M. Kimmel and M. Messner (eds.), *Men's lives* (pp. 521–534). 2d ed. New York: Macmillan.

LaRossa, Ralph and Maureen LaRossa. (1981). *Transition to parenthood.* Beverly Hills, CA: Sage.

Lasch, Christopher. (1977). *Haven in a heartless world: The family besieged.* New York: Basic.

Lazarre, Jane. (1986). *The mother knot.* New York: McGraw-Hill.

Leacock, Eleanor. (1973). Introduction to F. Engels *Origins of the family.* New York: International Publishers.

Lee, Dwight. (1986). Government policy and the distortions in family housing. In J. Peden and F. Glahe (eds.), *American family and the state.* San Francisco: Pacific Research Institute, pp. 310–320.

Lee, John. (1974). *Colours of love.* Toronto: New Press.

Lein, Laura, M. Durham, M. Pratt, M. Schudson, R. Thomas and H. Weiss. (1974). *Work and family life: Final report to the National Institute of Education.* Cambridge, MA: Center for the Study of Public Policy.

Lembcke, Jerry. (1991). New approaches to the problems of labor. In B. Berberoglu (ed.), *Critical perspectives in sociology* (pp. 93–103). Dubuque, IO: Kendall Hunt.

Lengermann, Patricia Madoo and Jill Niebrugge-Brantley. (1992). Contemporary feminist theory. In G. Ritzer (ed.), *Sociological theory* (pp. 447–496). 3d ed. New York: McGraw-Hill.

Lengermann, Patricia and Ruth Wallace. (1985). *Gender in America: Social control and social change.* Englewood Cliffs, NJ: Prentice-Hall.

Leontief, Wassily and Faye Duchen. (1986). *The future impact of automation on workers.* New York: Oxford University Press.

Lerner, Gerda. (1973). *Black women in white America: A documentary history.* New York: Vintage.

———. (1982). *An economic history of women in America: Women's work, the sexual division of labor, and the development of capitalism.* New York: Schocken.

———. (1986). *The creation of patriarchy.* New York: Oxford University Press.

Levitan, Sarah, Richard Belous and Frank Gallo. (1988). *What's happening to the American family? Tensions, hopes, realities.* Baltimore: Johns Hopkins University Press.

Lewin, Ellen. (1984). Lesbianism and motherhood: Implications for child custody. In T. Darty and S. Potter (eds.), *Women-identified women* (pp. 163–183). Palo Alto, CA: Mayfield.

Lewin, Miriam. (1992). Coping with catastrophe. *Women's Review of Books 10* (2): 16–17.

Lewin, Tamar. (1991). Shadows cloud early optimism about Norplant. *New York Times* 11/29: A1.

Lewontin, Richard, Steven Rose and Leon Kamin. (1984). *Not in our genes: Ideology and human nature.* New York: Pantheon.

Liem, Ramsay. (1988). Unemployed workers and their families: Social victims or social critics. In P. Voydanoff and L. Majka (eds.), *Families and economic distress: Coping strategies and social policy* (pp. 135–152). Beverly Hills, CA: Sage.

Lipman-Blumen, Jean. (1984). *Gender roles and power.* Englewood Cliffs, NJ: Prentice-Hall.

Littlefield, Alice. (1989). The B.I.A. boarding school: Theories of resistance and social reproduction. *Humanity and Society 13* (4): 428–441.

Look. (1956). American women at home. October 16, p. 35.

Lopata, Helena Znaniecki. (1971). *Occupation housewife.* New York: Oxford University Press.

Lorber, Judith. (1984). *Women physicians.* New York: Tavistock.

Lorber, Judith, Rose Coser, Alice Rossi and Nancy Chodorow. (1981). On the reproduction of mothering: A methodological debate. *Signs 6* (3): 482–514.

Loseke, Donileen. (1992). *The battered woman and shelters: The social construction of wife abuse.* Albany, NY: SUNY Press.

Loseke, Donileen and Spencer Cahill. (1984). The social construction of deviance: Experts on battered women. *Social Problems 31*: 296–310.

Lott, Bernice. (1987). *Women's lives: Themes and variation in gender learning.* Belmont, CA: Wadsworth.

Lowenberg, Bert and Ruth Bogin (eds.). (1976). *Black women in 19th century American life.* University Park, PA: Pennsylvania State University.

Luepnitz, Deborah. (1982). *Child custody: A study of families after divorce.* Lexington, MA: Heath.

Luker, Kristin. (1984). *Abortion and the politics of motherhood.* Berkeley: University of California Press.

———. (1992). Dubious conceptions: The controversy over teen pregnancy. In A. Skolnick and J. Skolnick (eds.), *Family in transition* (pp. 160–172). 7th ed. New York: Harper Collins.

Lund, Kristina. (1990). A feminist perspective on divorce therapy for women. *Journal of Divorce 13* (3): 57–67.

Lyle, Jack and Heidi Hoffman. (1972). Children's use of television and other media, pp. 145–181. In U.S. Surgeon General's Scientific Advisory Committee on Television and Social Behavior. *Reports and paper. vol. 4. Television in day-to-day life: Patterns of use.* Washington, DC: USGPO.

Maccoby, Eleanor and Carol Jacklin. (1974). *The psychology of sex differences.* Stanford, CA: Stanford University Press.

MacDermid, Shelley, Ted Huston and Susan McHale. (1990). Changes in marriage associated with the transition to parenthood: Individual differences as a function of sex-role attitudes and changes in the division of household labor. *Journal of Marriage and the Family 52*: 475–486.

Machung, Anne. (1984). Word processing: Forward for business, backward for women. In K. Sacks and D. Remy (eds.), *My troubles are going to have troubles with me* (pp. 124–139). New Brunswick, NJ: Rutgers University Press.

Mack, Delores. (1978). The power relations in black families and white families. In R. Staples (ed.), *The black family* (pp. 144–149). Belmont, CA: Wadsworth.

MacKinnon, Catherine. (1982). Feminism, Marxism, method and the state: An agenda for theory. *Signs 7* (Spring): 515–544.

Mainardi, Pat. (1970). The politics of housework. In. D. Scott and B. Wishy (eds.), *America's Families: A documentary history* (pp. 516–518). New York: Harper.

Mandell, Nancy. (1988). The child question: Links between women and children in families (pp. 49–84). In N. Mandell and A. Duffy (eds). *Reconstructing the Canadian family: Feminist perspectives.* Toronto: Butterworths.

Mandle, Jay. (1978). *The roots of black poverty: Southern plantation economy after the Civil War.* Durham: Duke University Press.

Mann, Susan Archer. (1986). *Social change and sexual inequality: The impact of the transition from slavery to sharecropping on black women.* Memphis, TN: Memphis State University Center for Research on Women.

Marcuse, Herbert. (1966). *Eros and civilization.* New York: Beacon Press.

Marks, Carole. (1985). Black workers and the great migration north. *Phylon 46* (2): 148–161.

Marmor, Judd (ed.). (1980). *Homosexual behavior.* New York: Basic Books.

Martin, E. (1988). 150 apply for plan permitting tenants to save for homes. *Charlotte Observer*, August 8, p. 1.

Martin, Judith. (1983). Maternal and paternal abuse of children: Theoretical and research perspectives. In D. Finkelhor et al. (eds.), *The dark side of families: Current family violence research* (pp. 293–304). Beverly Hills, CA: Sage.

Martin, Teresa and Larry Bumpass. (1989). Recent trends in marital disruption. *Demography 26*: 37–52.

Martineau, Harriet. (1834). *Illustration of political economy, the moral of many fables.* 6 vols. London: Charles Fox.

Marx, Fern. (1985). Child care. In H. McAdoo and T. Parkham (eds.), *Services to young families: Program review and policy recommendations.* Washington, D.C.: American Public Welfare Association.

Marx, Fern and Michelle Seligson. (1990). Child care in the U.S. In S. Rix (ed.), *The American woman 1990–1991: A status report* (pp. 132–169). New York: W. W. Norton.

Marx, Karl. (1869/1963). *The 18th Brumaire of Louis Bonaparte.* New York: International Publishers.

Matsumoto, Valerie. (1990). Japanese American women during World War II. In E. Dubois and V. Ruiz (eds.), *Unequal sisters: A multicultural reader in U.S. women's history* (pp. 373–386). New York: Routledge.

Matsuoko, Jon. (1990). Differential acculturation among Vietnamese refugees. *Social Work 35*: 341–345.

Mattessich, Paul and Rueben Hill. (1987). Life cycle and family development. In M. Sussman and S. Steinmetz (eds.), *Handbook of Marriage and the Family* (pp. 437–469). New York: Plenum.

Matthaei, Julie. (1982). *An economic history of women in America.* New York: Schocken.

May, Martha. (1982). The historical problem of the family wage: The Ford Motor Company and the five dollar day. *Feminist Studies 8* (2): 399–424.

McChesney, K. (1986). New findings on homeless families. *Family Professional 1* (2):

McCubbin, Hamilton, Constance Joy, A. Cauble, Joan Comeau, Joan Patterson and Richard Needle. (1980). Family stress and coping: A decade review. *Journal of Marriage and the Family 42*: 855–871.

McKinlay, D. (1964). *Social class and family life.* New York: Free Press.

McLanahan, Sara. (1983). Family structure and stress: A longitudinal study of two-parent and female-headed families. *Journal of Marriage and the Family 45*: 347–357.

McLanahan, Sara and Julia Adams. (1987). Parenthood and psychological well-being. *Annual Review of Sociology 13*: 237–257.

McLaurin, Melton. (1992). *Celia: A slave.* Athens, GA: University of Georgia Press.

McNall, Scott and Sally McNall. (1983). *Plains families: Exploring sociology through social history.* New York: St. Martin's.

Mead, George Herbert. (1934/1962). *Mind, self and society: From the standpoint of a social behaviorist.* Chicago: University of Chicago Press.

Mead, Margaret. (1935). *Sex and temperament in three primitive societies.* New York: Dell.

Mednick, Martha. (1992). Single mothers: A review and critique of current literature. In A. Skolnick and J. Skolnick (eds.), *Family in transition* (pp. 363–378). 7th ed. New York: Harper Collins.

Medrich, Elliott, Judith Roizan, Victor Rubin and Stuart Buckley. (1982). *The serious business of* *growing up*. Berkeley: University of California Press.

Mehl, Lewis, Gail Peterson, Michael Whitt and Warren Hawes. (1980). Evaluation of outcomes of non-nurse midwives: Matched comparisons with physicians. *Women and Health 5*: 17–29.

Meissner, Martin, E. W. Humphries, S. M. Meis, and W. J. Scheu. (1975a). No exit for wives: Sexual division of labour and the cumulation of household demands. *Canadian Review of Sociology and Anthropolgy 12*: 424–439.

———. (1975b). *The sociology of housework*. New York: Pantheon Books.

———. (1980). Prologue: Reflections on the study of household labor. In S. Berk (ed.), *Women and household labor* (pp. 7–14). Beverly Hills, CA: Sage.

Mercer, R. (1986). *First time motherhood: Experiences from the teens to the forties*. New York: Springer.

Meyer, Agnes. (1950). Women aren't men. *Atlantic Monthly 186*: 32.

Meyers, Daniel and Seven Garasky. (1991). *Custodial fathers: Myths, realities and child support policy*. U.S. Department of Health and Human Services. Washington, D.C.: USGPO.

Meyersohn, Rolf. (1963). Changing work and leisure routines. In E. Smigel (ed.), *Work and leisure: A contemporary social problems* (pp. 97–106). New Haven, CT: College and University Press.

Middleton, R. and S. Putney. (1960). Dominance in decisions in the family: Race and class differences. *American Journal of Sociology 65* (6): 605–609.

Mies, Maria. (1983). Toward a methodology for feminist research. In G. Bowles and R. Klein (eds.), *Theories of women's studies* (pp. 117–139). London: Routledge and Kegan Paul.

Milkman, Ruth. (1976). Women's work and economic crisis: Some lessons of the Great Depression. *Radical Review of Political Economy 8*: 73–97.

———. (1991). Gender at work: The sexual division of labor during World War II. In L. Kerber and J. De Hart (eds.), *Women's America* (pp. 437–450). 3d ed. New York: Oxford University Press.

Miller, C. (1987). Qualitative differences among gender-stereotyped toys: Implications for cognitive and social development in girls and boys. *Sex Roles 16*: 473–488.

Miller, Dorothy. (1990). *Women and social welfare: A feminist analysis*. New York: Praeger.

Millett, Kate. (1970). *Sexual politics*. Garden City, NY: Doubleday.

Millman, Marcia. (1991). *Warm hearts and cold cash: The intimate dynamics of families and money*. New York: Free Press.

Mills, C. Wright. (1956). *The power elite*. London: Oxford University Press.

———. (1959). *The sociological imagination*. London: Oxford University Press.

Mincer, Jacob. (1962). Labor force participation of married women. In *National Bureau of Economic Research, Aspects of Labor Economics* (pp. 63–97). Princeton, NJ: Princeton University Press.

Mink, Gwendolyn. (1990). The lady and the tramp: Gender, race and the origins of the welfare state. In L. Gordon (ed.), *Women, the state and welfare* (pp. 92–122). Madison: University of Wisconsin Press.

Mintz, Stephen and Susan Kellogg. (1988). *Domestic revolutions: A social history of American family life*. New York: Free Press.

Mirowsky, John and Catherine Ross. (1989). *Social causes of psychological distress*. New York: Aldine DeGruyter.

Mischel, Harriet and Robert Fuhr. (1988). Maternal employment: Its psychological effects on children and their families. In S. Dornbusch and M. Strober (eds.), *Feminism, children and the new families* (pp. 191–211). New York: Guilford Press.

Mishel, Lawrence and Jared Bernstein. (1992). *The state of working America, 1992–1993*. Washington, D.C.: Economic Policy Institute.

Mitchell, Juliet. (1984). *Women: The longest revolution*. New York: Pantheon.

———. (1986). Reflections on twenty years of feminism. In J. Mitchell and A. Oakley (eds.), *What is feminism: A re-examination* (pp. 34–48). New York: Pantheon.

Modell, John and Tamara Haraven. (1978). Urbanization and the malleable household: An examination of boarding and lodging in American families. In M. Gordon (ed.), *The American Family in social historical perspective* (pp. 51–68). New York: St. Martin's Press.

Moen, Phyllis. (1989). *Working parents: Gender roles and public policies in Sweden*. Madison: University of Wisconsin Press.

———. (1992). *Women's two roles*. Westport, CT: Auburn House.

Moen, Phyllis and Alvin Schorr. (1987). Families and social policies. In M. Sussman and S. Steinmetz (eds.), *Handbook of Marriage and the Family* (pp. 795–813). New York: Plenum.

Moore, Charles, David Sink, and Patricia Hoban-Moore. (1988). The politics of homelessness. *PS Political Science and Politics 21*: 57–63.

Moore, Kristin and Isabel Sawhill. (1984). Implications of women's employment for home and family life. In A. Stromberg and S. Harkess (eds.), *Women working: Theories and facts in perspective* (pp. 201–223). Palo Alto, CA: Mayfield.

Morgan, David. (1975). *Social theory and the family*. Boston: Routledge and Kegan Paul.

———. (1985). *The family, politics and social theory*. Boston: Routledge and Kegan Paul.

Morgan, Edmund. (1978). The puritans and sex. In M. Gordon (ed.), *The American family in social-historical perspective* (pp. 363–373). 2nd ed. New York: St. Martin's Press.

Morgan, L. (1965). *Houses and houselife of the American aborigines*. Chicago: University of Chicago Press.

Mortimer, J. (1976). Social class, work and the family: Some implications of the father's occupation for familial relationships and son's career decisions. *Journal of Marriage and the Family 38*: 241–256.

Mortimer, J. And D. Kumka. (1982). A further examination of the occupational linkage hypothesis. *Sociological Quarterly 23* (Winter): 3–16.

Moynihan, Daniel. (1965). *The Negro family: The case for national action*. Office of Policy Planning and Research, U.S. Department of Labor. Washington, D.C.: GPO.

Murdock, George. (1949). *Social Structure*. New York: Macmillian.

Murray, Charles. (1984). *Losing ground: American social policy: 1950–1980*. New York: Basic.

Murstein, Bernard and C. Holden. (1979). Sexual behavior and correlates among college students. *Adolescence 4* (56): 625–639.

Myers, S. (1979). Child abuse and the military community. *Military Medicine 144*: 23–25.

Nabakov, Peter. (1991). *Native American testimony: A chronicle of Indian-white relations from prophecy to the present, 1492–1992*. New York: Viking.

Nakano, Mei. (1990). *Japanese American women: Three generations, 1890–1990*. Berkeley: Mina Press.

Nagata, Donna. (1991). Transgenerational impact of the Japanese-American internment: Clinical issues in working with children of former internees. *Psychotherapy 28*: 121–128.

Naples, Nancy. (1991). Contradictions in the gender subtext of the war on poverty: The community work and resistance of women from low income communities. *Social Problems 38* (3): 316–332.

———. (1992). Activist mothering: Cross-generational continuity in the community work of women from low-income urban neighborhoods. *Gender and Society 6* (3): 441–463.

National Council for the Prevention of Child Abuse (NCPCA). (1991). *Current trends in child abuse reporting and fatalities: The results of the 1990 annual fifty state survey*. Chicago: NCPCA.

National Welfare Rights Union. (1988). Convention Report, First Annual Convention. *Up and out of poverty*. Detroit, September 3–5.

NC NOW. (1992). Sharon Kowalski is home. *Nownews* (Spring): 6.

Nee, Victor and Herbert Wong. (1985). Asian Americans socioeconomic achievement: The strength of the family bond. *Sociological Perspectives 28*: 281–306.

Negrey, Cynthia. (1990). Contingent work and the rhetoric of autonomy. *Humanity and Society 14* (1): 16–33.

New York Times. (1992). Excerpts from vice president's speech on cities and poverty. *New York Times* (5/20): A20.

Newman, Katherine. (1988). *Falling from grace: The experience of downward mobility in the American middle class*. New York: Random House.

Newman, Maria. (1992). Lawmaker from riot zone insists on a new role for black politicians. *New York Times* (5/19): A18.

Nock, Stephen. (1987). *Sociology of the family*. Englewood Cliffs, NJ: Prentice-Hall.

Norlicht, S. (1979). Effects of stress on the police officer and family. *New York State Journal of Medicine 79*: 400–401.

Norsigian, Judy. (1990). RU-486. In M. Fried (ed.), *From abortion to reproductive freedom* (pp. 197–203). Boston: South End Press.

Nulty, Peter. (1987). Pushed out at 45: Now what? *Fortune 115* (5): 26–30.

Nye, F. Ivan. (1974). Emerging and declining family roles. *Journal of Marriage and the Family 36*: 238–245.

Oakley, Ann. (1974). *The sociology of housework.* New York: Pantheon.

———. (1981a). Interviewing women: A contradiction in terms. In H. Roberts (ed.), *Doing feminist research* (pp. 30–61). London: Routledge and Kegan Paul.

———. (1981b). *Subject women.* New York: Pantheon.

O'Connor, James. (1973). *Fiscal crisis of the state.* New York: St. Martin's Press.

Off Our Backs. (1992). Child victims of sexual abuse: A bill of rights. *Off Our Backs 22* (5): 1.

Ogbu, John. (1978). *Minority education and caste: The American system in cross cultural perspective.* New York: Academic Press.

Ollman, Bertell. (1976). *Alienation.* 2d ed. Cambridge: Cambridge University Press.

Oppenheimer, Valerie. (1969). *The female labor force in the U.S.: Demographic and economic factors governing its growth and changing composition.* Berkeley: University of California Population Monograph Series 5.

Osmond, Marie. (1987). Radical-critical sociology. In M. Sussman and S. Stienmetz (ed.) *Handbook of Marriage and the family* (pp. 103–124. New York: Plenum Press.

Ostrander, Susan. (1984). *Women of the upper class.* Philadelphia: Temple University Press.

Oxford Analytica. (1986). *America in perspective.* Boston: Houghton Mifflin .

Padgug, Robert. (ed.). (1979). Special Issue on Sexuality. *Radical History Review* Spring/Summer.

Pagelow, Mildred. (1981). *Woman battering: Victims and their experience.* Beverly Hills, CA: Sage.

Palmer, John and Isabel Sawhill. (1984). *The Reagan record.* Cambridge, MA: Ballinger.

Pardo, Mary. (1990). Mexican American women grassroots community activists: The Mothers of East Los Angeles. *Frontiers 11* (1): 1–7.

Parker, Robert. (1991). Urban social problems in the United States: Issues in urban political economy. In B. Berberoglu (ed.), *Critical perspectives in sociology: A reader* (pp. 169–184). Dubuque, IA: Kendall/Hunt.

Parsons, Talcott. (1955). The American family: Its relation to personality and to the social structure. In T. Parsons and R. Bales (eds.), *Family socialization and interaction process* (pp. 3–33). Glencoe, IL: Free Press.

Pascal, Gillian. (1986). *Social policy: A feminist analysis.* New York: Tavistock.

Patterson, James. (1981). *America's struggle against poverty, 1900–1980.* Cambridge: Harvard University Press.

Patzer Gordon. (1985). *The phsyical attractiveness phenomenon.* New York: Plenum Press.

Pearce, Diana. (1978). The feminization of poverty: Women, work and welfare. *Urban and Social Change 2*: 24–36.

———. (1989). Farewell to alms. In J. Freeman (ed.), *Women: A feminist perspective* (pp. 493–506). 4th ed. Mountain View, CA: Mayfield.

Peplau, Letitia and Susan Campbell. (1989). The balance of power in dating and marriage. In J. Freeman (ed.), *Woman: A feminist perspective* (pp. 121–137). Mountain View, CA: Mayfield.

Perucci, Carolyn and Dena Targ. (1988). Effects of plant closings on marriage and family life. In P. Voydanoff and L. Majka (eds.), *Families and economic distress: Coping strategies and public policy* (pp. 55–72). Beverly Hills, CA: Sage.

Petchesky, Rosalind. (1990). *Abortion and woman's choice: The state, sexuality and reproductive freedom.* Boston: Northeastern University Press.

Peters, Marie. (1985). Racial socialization of young black children. In H. McAdoo and J. McAdoo (eds.), *Black children: Social, educational and parental environments* (pp. 159–173). Newbury Park, CA: Sage.

———. (1988). Parenting in black families with young children: A historical perspective. In H. McAdoo (ed.), *Black families* (pp. 228–241). 2nd ed. Newbury Park, CA: Sage.

Peterson, Paul. (1985). Introduction, technology, race and urban policy. In P. Peterson (ed.), *The New Urban Reality* (pp. 1–35). Washington, D.C.: Brookings Institute.

Philipson, Ilene. (1982). Heterosexual antagonisms and the politics of mothering. *Socialist Review 66*: 55–77.

Phoenix, Ann. (1991). *Young mothers.* Cambridge, MA: Basil Blackwell.

Piven, Francis Fox. (1990). Ideology and the state: Women, power and the welfare state. In L. Gordon (ed.), *Women, the state and welfare* (pp. 250–264). Madison: University of Wisconsin Press.

Piven, Francis Fox and Richard Coward. (1982). *The new class war: Reagan's attack on the welfare state and its consequences.* New York: Pantheon.

Pizzey, Erin. (1974). *Scream quietly or the neighbors will hear you.* New York: Penguin.

Pleck, Elizabeth. (1991). *An historical overview of American gender roles and relations from precolonial times to the present.* No. 242. Wellesley, MA: Wellesley College Center for Research on Women.

Pleck, Elizabeth, Joseph Pleck, M. Grossman and Pauline Bart. (1977–1978). The battered data syndrome: A comment on Steinmetz' article. *Victimology: An International Journal* 2: 680–684.

Pleck, Joseph. (1983). Husband's paid work and family roles: Current research issues. In H. Lopata and J. Pleck (eds.), *Research in the interweave of social roles.* Vol. 3. Greenwich, CT: JAI Press.

———. (1987). The theory of male sex role identity: Its rise and fall 1936, to the present. In H. Brod (ed.), *The making of masculinities* (pp. 21–38). Boston: Allen and Unwin.

Pleck, Joseph, Graham Staines and Linda Lang. (1980). Conflicts between work and family. *Monthly Labor Review* (March): 29–31.

Pogrebin, Letty. (1983). *Family politics: Love and power on an intimate frontier.* New York: McGraw-Hill.

Pohlman, E. (1969). *The psychology of birth planning.* Cambridge, MA: Shenkman.

Polatnick, Margaret. (1983). Why men don't rear children. In J. Trebilcot (ed.), *Mothering* (pp. 21–40). Totawa, NJ: Rowman and Allanheld.

Polikoff, Nancy. (1983). Gender and child custody determinations: Exploding the myths. In I. Diamond (ed.), *Families, politics and public policy: A feminist dialogue on women and the state* (pp. 183–202). New York: Longman.

Pollin, Robert. (1990). Borrowing more, buying less. *Dollars and Sense 156* (May): 7.

Popple, Philip and Leslie Leighninger. (1993). *Social work, social welfare, and American society.* Boston: Allyn and Bacon.

President's Commission on Privatization. (1988). *Privatization: Toward more effective government,* Washington, D.C.: President's Commission on Privatization.

Ptacek, James. (1988). Why do men batter wives? In K. Yllo and M. Bograd (eds.), *Feminist perspectives of wife abuse* (pp. 133–157). Beverly Hills, CA: Sage.

Pupo, Norene. (1988). Preserving patriarchy: Women, the family and the state. In N. Mandell and A. Duffy (eds.), *Reconstructing the Canadian family* (pp. 207–238). Toronto: Buttersworth.

Purrington, Beverly. (1980). Effects of children on their parents: Parents' perceptions. Ph.D. dissertation. Michigan State University.

Radhill, Samual. (1968). A history of child abuse and infanticide. In C. Kempe (ed.), *The Battered Child.* Chicago: University of Chicago Press.

Rainwater, Lee and William Yancey. (1967). *The Moynihan report and the politics of the controversy.* Cambridge: MIT Press.

Ramey, C., D. Bryant, and T. Suarez. (1985). Preschool compensatory education and the modifiability of intelligence: A critical review. In D. Detterman (ed.), *Current Topics in Human Intelligence* (pp. 247–296). Norwood, NJ: Ablex.

Ramos, Reyes. (1979). The Mexican American: Am I who they say I am? In A. Trejo (ed.), *The Chicanos as we see ourselves* (pp. 49–66). Tucson, AZ: University of Arizona Press.

Randle, M. (1951). Iroquois women, then and now. In W. Fenton (ed.) *Symposium on local diversity in Iroquois culture.* Bureau of American Ethnology, Bulletin 149. Washington, D.C.

Rank, Mark. (1989). Fertility among women on welfare. *American Sociological Review 54* (April).

Rapp, Rayna. (1982). Family and class in contemporary America: Notes toward an understanding of ideology. In B. Thorne with M. Yalom (eds.), *Rethinking the family: Some feminist questions* (pp. 168–187). New York: Longman.

Rapp, Rayna, Ellen Ross, and Renate Bridenthal. (1979). Examining family history. *Feminist Studies 5* (1): 172–200.

Red Horse, John. (1980). Family structure and value orientation in American Indians. *Social Casework: The Journal of Contemporary Social Work 59*: 462–467.

Reich, Wilhelm. (1945). *The sexual revolution.* New York: Farrar, Straus and Giroux.

Reinharz, Shulamit. (1983). Experiential analysis: A contribution to feminist research. In G. Bowles and R. Klein (eds.), *Theories of women's studies* (pp. 162–191). London: Routledge and Kegan Paul.

———. (1992). *Feminist methods in social research.* New York: Oxford University Press.

Reinisch, June. (1990). *The Kinsey Institute new report on sex: What you must know to be sexually literate.* New York: St. Martin's Press.

Reiss, Ira. (1986). *Journey into sexuality: An exploratory voyage.* Englewood Cliffs, NJ: Prentice-Hall.

Reissman, Catherine. (1991). *Divorce talk: Women and men make sense of personal relationships.* New Brunswick, NJ: Rutgers University Press.

Reiter, Rayna (ed.). (1975). *Toward an anthropology of women.* New York: Monthly Review Press.

Renvoize, Jean. (1985). *Going solo: Single mothers by choice.* Boston: Routledge and Kegan Paul.

Renzetti, Claire and Daniel Curran. (1992). *Social problems: Society in crisis.* 2d ed. Boston: Allyn and Bacon.

Reyes, L. and L. Waxman. (1986). *The continued growth of hunger, homelessness and poverty in America's cities, 1986.* U.S. Conference on Mayors. Washington, D.C.: USGPO.

Rheingold, H. and K. Cook. (1975). The content of boys' and girls' rooms as an index of parents' behavior. *Child Development 46*: 459–463.

Rhode, Deborah. (1989). *Justice and gender, sex discrimination and the law.* Cambridge: Harvard University Press.

Rich, Adrienne. (1976). *Of woman born: Motherhood as experience and institution.* New York: Norton.

———. (1980). Compulsory heterosexuality and lesbian existence. *Signs 5* (Summer): 631–660.

Rich, Ruby. (1986). Feminism and sexuality in the 1980s. *Feminist Studies 12* (3): 525–561.

Richardson, B. (1981). *Racism and child-rearing: A study of black mothers.* Ph.D. dissertation. Claremont Graduate School.

Ries, Paula and Anne Stone. (1992). *The American woman: 1992–1993, a status report.* New York: Norton.

Riley, Glenda. (1991). *Divorce: An American tradition.* New York: Oxford.

Risman, Barbara. (1987). Intimate relationships from a microstructural perspective: Men and women who mother. *Gender and Society 1* (1): 6–32.

Ritzer, George. (1992). *Sociological theory.* 3d. ed. New York: McGraw-Hill.

Robinson, Jon. (1988). Who's doing the housework? *American Demographics 10*: 24–28, 63.

Robson, Ruthann. (1992). *Lesbian (out)law: Survival under the rule of law.* Ithaca, NY: Firebrand Books.

Rodgers, William and Arland Thornton. (1985). Changing patterns of first marriage in the U.S. *Demography 22*: 265–279.

Rodriguez, Julia. (1988). Labor migration and familial responsibilities: Experience of Mexican women. In M. Melville (ed.), *Mexicanas at work in the U.S.* (pp. 47–63). Houston, University of Houston Press.

Rodrique, Jessie. (1990). The black community and the birth control movement. In E. Dubois and V. Ruiz (eds.), *Unequal sisters: A multicultural reader in U.S. women's history* (pp. 333–344. New York: Routledge.

Rogers, Susan. (1978). A woman's place: A critical review of anthropological theory. *Comparative Studies in Society and History 20*: 123–162.

Rohrbaugh, J. (1979). Femininity on the line. *Psychology Today* (August): 30.

Rollins, Judith. (1985). *Between women: Domestics and their employers.* Philadelphia: Temple University Press.

Rollins, Boyd and Kenneth Cannon. (1974). Marital satisfaction over the family life cycle. *Journal of Marriage and the Family 36*: 271–284.

Romero, Mary. (1988). Sisterhood and domestic service: Race, class and gender in the mistress-maid relationship. *Humanity and Society 12* (4): 318–346.

———. (1992). *Maid in the USA.* New York: Routledge.

Rosaldo, Michele and Louise Lamphere (eds.). (1974). *Women, culture and society.* Stanford: Stanford University Press.

Rose, S. and I. Frieze. (1989). Young singles' scripts for a first date. *Gender and Society 3*: 258–268.

Rosen, Ellen. (1987). *Bitter choices.* Chicago, University of Chicago Press.

——— (1991). Beyond the factory gates: Blue collar women at home. In L. Kraem (ed.), *The sociology of gender* (pp. 233–254). New York: St. Martin's Press.

Rosenblatt, Roger. (1983). The baby in the factory. *Time 14* (February): 72.

Rosenthal, Andrew. (1992). Quayle says riots sprang from lack of family values. *New York Times* (5/20): A1, 20.

Rossi, Alice. (1964). Equality between the sexes: An immodest proposal. *Daedalus* (Spring): 1–19.

———. (1977). A bio-social perspective on parenting. *Daedalus 106* (2): 1–31.

———. (1992). Transition to parenthood. In A. Skolnick and J. Skolnick (eds.), *Family in transition* (pp. 453–463). 7th ed. New York: Harper Collins.

Rossi, Alice (ed.). (1973). *The feminist papers.* New York: Bantam.

Rothberg, B. (1983). Joint custody: Parental problems and satisfactions. *Family Process 22*: 43–52.

Rothman, Barbara Katz. (1982). *In labor: Women and power in the birthplace.* New York: W. W. Norton.

———. (1989a). *Recreating motherhood: Ideology and technology in a patriarchal society.* New York: W. W. Norton.

——— (1989b). Women, health and medicine. In J. Freeman (ed.), *Women: A feminist perspective* (pp. 76–86). 4th ed. Mountain View, CA: Mayfield.

Rothman, Sheila and Emily Marks. (1989). Flexible work schedules and family policy. In F. Gerstel and H. Gross (eds), *Families and Work* (pp. 469–477). Philadelphia: Temple University Press.

Rotundo, E. (1985). American fatherhood: A historical perspective. *American Behavioral Scientist 29*: 7–23.

Rouse, L., R. Breen and M. Howell. (1988). Abuse in intimate relationships. A comparison of dating and married college students. *Journal of Interpersonal Violence 3*: 414–429.

Rubel, Arthur. (1966). *Across the tracks: Mexican Americans in Texas City.* Austin: University of Texas Press.

Rubin, Gayle. (1975). The traffic in women. In R. Reiter (ed.), *Toward an anthropology of women* (pp. 157–211). New York: Monthly Review Press.

———. (1984). Thinking sex: Notes for a radical theory of the politics of sexuality. In C. Vance (ed.), *Pleasure and danger: Exploring female sexuality* (pp. 267–319). Boston: Routledge and Kegan Paul.

Rubin, Lillian. (1976). *Worlds of pain: Life in working class families.* New York: Basic Books.

———. (1981). *Intimate Strangers.* New York: Harper Row.

———. (1991). *Erotic wars: What happened to the sexual revolution.* New York: Harper.

Rubin, Zick. (1973). *Liking and loving: An invitation to social psychology.* New York: Holt, Rinehart and Winston.

Ruddick, Sara. (1982). Maternal thinking. In B. Thorne with M. Yalom (eds.), *Rethinking family: Some feminist questions* (pp. 76–94). New York: Longman.

Ruiz, Vicki. (1990). A promise fulfilled: Mexican cannery workers in Southern California. In D. Dubois and V. Ruiz (eds.), *Unequal sisters: A multicultural reader in U.S. women's history* (pp. 264–274). New York: Routledge.

Rupp, Leila and Verta Taylor. (1987). *Survival in the doldrums: The American women's rights movement, 1945 to the 1960s.* New York: Oxford University Press.

Rush, F. (1980). *Best kept secrets: Sexual abuse of children.* Englewood Cliffs, NJ: Prentice-Hall.

Russell, Diana. (1986). *The secret trauma: Incest in the lives of girls and women.* New York: Basic Books.

Ryan, William. (1971). *Blaming the victim.* New York: Random House.

Ryscavage, P. (1979). More wives in the labor force have husbands with "above average" incomes. *Monthly Labor Review 102* (6): 40–42.

Sacks, Karen. (1984). Generations of working-class families. In K. Sacks and D. Remy (eds.), *My troubles are going to have trouble with me: Everyday triumphs of women workers* (pp. 15–38). New Brunswick, NJ: Rutgers.

Safilios-Rothschild, Constantina. (1970) The study of family power structure: A review 1960–1969. *Journal of Marriage and the Family 31*: 539–543.

——— (1977). *Love, sex and sex roles.* Englewood Cliffs, NJ: Prentice-Hall.

Sanger, Margaret. (1931). *An autobiography.* New York: Dover.

Sanik, Margaret Mietus. (1981). Divisions of household work: A decade comparison, 1967–1977. *Home Economics Research Journal 10*: 175–180.

Sarvasy, Wendy. (1988). Reagan and low-income mothers: A feminist recasting of the debate. In M. Brown (ed.), *Remaking the welfare state* (pp. 253–276). Philadelphia: Temple University Press.

Sattel, Jack. (1976). The inexpressive male: Tragedy or sexual politics. *Social Problems 23* (April): 469–477.

Scanzoni, John. (1971). *The black family in modern society.* Boston: Allyn and Bacon.

Scarr, Sandra. (1984). *Mother care/other care.* New York: Basic Books.

Scarr, Sandra, Deborah Phillips, and Kathleen McCartney. (1992). Working mothers and their families. In A. Skolnick and J. Skolnick (eds.), *Family in transition* (pp. 414–430). 7th ed. New York: Harper Collins.

Schechter, Susan. (1982). *Women and male violence: The visions and struggles of the battered women's movement*. Boston: South End Press.

———. (1988). Building bridges between activists, professionals and researchers. In K. Yllo and M. Bogard (eds.), *Feminist perspectives on wife abuse* (pp. 299–312). Beverly Hills, CA: Sage.

Schneider, Beth and Meredith Gould. (1987). Female sexuality: Looking back into the future. In B. Hess and M. Ferree (eds.), *Analyzing gender* (pp. 120–153). Newbury Park, CA: Sage.

Schoen, Robert. (1983). Measuring the tightness of the marriage squeeze. *Demography 20*: 61–78.

Schorr, Alvin. (1962). Family policy in the United States. *International Social Science Journal 14*: 452–467.

Schwartz, Felice. (1989). Management, women and the new facts of life. *Harvard Business Review 67* (January–February): 65–76.

Schwartz, L. and W. Markham. (1985). Sex stereotyping in children's toy advertisements. *Sex Roles 12*: 157–170.

Schwartz, Martin. (1987). Gender and injury in spousal assault. *Social Forces 20* (1): 61–75.

Scott, Anne Firor. (1989). Conclusions, trends and future directions. In C. Little and R. Vaughn (eds.), *A new perspective: Southern women's cultural history from the Civil War to civil rights* (pp. 77–83). Charlottesville, VA: Virginia Foundation for the Humanities.

Scott, Jerome. (1988). Four stages of African American history. Paper presented at Conference for Black history month, Howard University Washington, D.C.

Scott, Rebecca. (1985) The battle over the child: Child apprenticeship and the Freedman's Bureau in North Carolina. In N. Hiner and J. Hawes (eds.), *Growing up in America: Children in historical perspective* (pp. 193–207). Chicago: University of Chicago Press.

Seidenberg, R. (1973). *Corporate wives—Corporate casualties?* New York: Amacon.

Senour, Maria and Lynda Warren. (1976). Sex and ethnic differences in masculinity, femininity, and anthropology. Paper presented at the annual meetings of the Western Psychological Association, Los Angeles, CA.

Shaiken, Harley. (1984). *Work transformed: Automation and labor in the computer age*. New York: Holt Rinehart and Winston.

Shapiro, Isaac and Robert Greenstein. (1988). *Holes in the safety nets: Poverty programs and policies in the states*. Washington, D.C.: Center on Budget and Policy Priorities.

Shapiro, L. (1990). Guns and dolls. *Newsweek 28* (May): 56–65.

Shaw, Anna Howard. (1915). *Story of a pioneer*.

Shaw, Nancy. (1974). *Forced labor: Maternity care in the United States*. New York: Pergamon.

Sheak, Robert. (1990). Corporate and state attacks on the material conditions of the working class. *Humanity and society 14*: 105–127.

Sheppard, Annamary. (1982). Unspoken premises in custody litigation. *Women Rights Law Reporter 7* (Spring): 229–234.

Sherman, Lawrence and Richard Berk. (1984). The specified deterrent effects of arrest for domestic assault. *American Sociological Review 49*: 261–272.

Sidel, Ruth. (1990). *On her own: Growing up in the shadow of the American dream*. New York: Viking.

Sigelman, Lee and Susan Welch. (1991). *Black Americans' views of racial inequality: The dream deferred*. Cambridge: Cambridge University Press.

Simmel, Georg. (1907/1978). *The philosophy of money*. London: Routledge and Kegan Paul.

Simon, Rita and Gloria Danziger. (1991). *Women's movements in America: Their successes, disappointments, and aspirations*. New York: Praeger.

Skocpol, Theda and Edwin Amenta. (1986). States and social policies. In R. Turner and J. Short, (eds.), *Annual review of sociology* (pp. 131–157). Vol. 12. Palo Alto, CA: Annual Reviews.

Smith, Dorothy. (1974). Women's perspective as a radical critique of sociology. *Sociological Inquiry 44*: 7–13.

———. (1979). A sociology for women. In J. Sherman and E. Beck (eds.), *The prism of sex: Essays in the sociology of knowledge* (pp. 16–35). Madison: University of Wisconsin Press.

———. (1987). Women's inequality and the family. In N. Gerstel and H. Gross (eds.), *Families and work* (pp. 23–54). Philadelphia: Temple University Press.

Smith, Steven and Deborah Stone. (1988). The unexpected consequences of privatization. In M. Brown (ed.), *Remaking the welfare state: Retrenchment and social policy in America and Europe* (pp. 232–252). Philadelphia: Temple University Press.

Smith, Tom. (1990). Adult sexual behavior in 1989. Unpublished paper presented at the annual meetings of the American Association for the Advancement of Science.

Smith-Rosenberg, Carroll. (1985). *Disorderly conduct.* New York: Knopf.

Snitow, Ann. (1978). Thinking about the mermaid and the minotaur. *Feminist Studies 4* (2): 190–198.

Sobel, Herman. (1962). Can I help my husband avoid a heart attack? *Reader's Digest* (September): 69.

Sokoloff, Natalie. (1980). *Between money and love: The dialectics of women's home and market work.* New York: Praeger.

Song, Young. (1991). Single Asian American women as a result of divorce: Depressive affect and changes in social support. *Journal of Divorce and Remarriage 14*: 219–230

Spanier, Graham. (1983). Married and unmarried cohabitation in the U.S. 1980. *Journal of Marriage and the Family 45* (May): 277–288.

Spanier, Graham and Paul Glick. (1981). Marital instability in the U.S.: Some correlates and recent changes. *Family Relations 30* (July): 329–338.

Spenner, K. (1981). Occupational role characteristics and intergenerational transmission. *Work and Occupations 8*: 89–112.

Spinetta, J. and Rigler, D. (1972). The child-abusing parent: A psychological review. *Psychological Bulletin 77*: 296–304.

Stacey, Judith. (1988). Can there be a feminist ethnography? *Women Studies International Forum 11*: 21–27.

———. (1990). *Brave new families: Stories of domestic upheaval in late twentieth century America.* New York: Basic Books.

Stacey, Judith and Barrie Thorne. (1985). The missing feminist revolution in sociology. *Social Problems 32*: 301–316.

Stack, Carol. (1974). *All our kin: Strategies for survival in the black community.* New York: Harper and Row.

Stafford, Frank and Greg Duncan. (1978). Market hours, real hours and labor productivity. *Economic Outlook USA 5*: 103–119.

Stanford Center for the Study of Families, Children, and Youth. (1991). *The Stanford studies of homeless families, children, and youth.* Stanford, CA: Stanford University Press.

Stanton, Elizabeth Cady. (1889/1990). Speech before the legislature 1860. In S. Ruth (ed.), *Issues in Feminism* (pp. 465–470). Toronto: Mayfield.

Stanworth, Michelle. (1990). Birth pangs: Conceptive technologies and the threat to motherhood. In M. Hirsch and E. Keller (eds.), *Conflicts in Feminism* (pp. 288–304). New York: Routledge.

Staples, Robert. (1973). *The Black Woman in America: Sex, marriage and the family.* Chicago: Nelson Hall.

———. (1981). The myth of the black matriarchy. *Black scholar.* (December): 32.

Staples, Robert and T. Jones. (1985). Culture, ideology and black television images. *The Black Scholar 16*: 10–20.

Stark, Evan, and Anne Flitcraft. (1979). Medicine and patriarchal violence: The social construction of a "private" event. *International Journal of Health Services 98*: 461–491.

———. (1983). Social knowledge, social policy and the abuse of women: The case against patriarchal benevolence. In D. Finkelhor et al. (eds.), *The dark side of families: Current family violence research* (pp. 330–348). Beverly Hills, CA: Sage.

Starr, Paul. (1989). The meaning of privatization. In S. Kamerman and A. Kahn (eds.), *Privatization and the welfare state* (pp. 15–48). Princeton: Princeton University Press.

Stein, Peter. (ed.). (1981). *Single life: Unmarried adults in social context.* New York: St. Martin's Press.

Steinberg, Lawrence et al. (1982). Effects of early work experience on adolescent development. *Developmental Psychology 18*: 385–395.

Steiner, Jerome. (1972). What price success? *Harvard Business Review.* (March): 69–74.

Steinmetz, Suzanne. (1977 –1978). The battered husband syndrome. *Victimology: An International Journal 2*: 499–509.

Steinmetz, Suzanne, Sylvia Clavin and Karen Stein. (1990). *Marriage and the family.* New York: Harper and Row.

Stetson, Dorothy. (1991). *Women's rights in the U.S.A.: Policy debates and gender roles.* Pacific Grove, CA: Wadsworth.

Stier, Haya. (1991). Immigrant women go to work: Analysis of immigrant wives' labor supply for six Asian groups. *Social Science Quarterly 72*: 67–82.

Stockard, Jean and Miriam Johnson. (1992). *Sex and gender in society.* 2nd ed. Englewood Cliffs, NJ: Prentice-Hall.

Stoesz, D. (1987). Privatization: Reforming the welfare state. *Journal of Sociology and Social Welfare 14*: 3–20.

Straus, Murray. (1980). Sexual inequality and wife beating. In M. Straus and G. Hotaling (eds.), *The social causes of husband-wife violence* (pp. 86–93). Minneapolis: UM Press.

———. (1986). Societal change and change in family violence from 1975 to 1985 as revealed by two national surveys. *Journal of Marriage and the Family 48*: 465–479.

Straus, Murray, Richard Gelles, and Suzanne Steinmetz. (1980). *Behind closed doors: Violence in the American family.* Garden City, NY: Doubleday.

———. (1988). The marriage license as a hitting license. In A. Skolnick and J. Skolnick (eds.), *Family in transition* (pp. 301–314). 6th ed. Boston: Scott, Foresman.

Street, David, George Martin, and Laura Gordon. (1979). *The welfare industry.* Beverly Hills, CA: Sage.

Strober, Myra. (1988). Two-earner families. In S. Dornbusch and M. Strober (eds.), *Feminism, children and new families* (pp. 161–190). New York: Guilford Press.

Strober, Myra and Sanford Dornbusch. (1988). Public policy alternatives. In S. Dornbusch and M. Strober (eds.), *Feminism, children and the new families* (pp. 327–356). New York: Guilford Press.

Sudarkasa, Niara. (1988). Interpreting the African heritage in Afro-American family organization. In H. McAdoo (ed.), *Black families* (pp. 27–43). 2d ed. Beverly Hills, CA: Sage.

Sullivan, Andrew. (1992). Here comes the groom. In G. Bird and M. Sporakowski (eds.), *Taking sides: Clashing views on controversial issues in family and personal relationships* (pp. 76–79). Guilford, CT: Dushkin.

Sullivan, Deborah and Rose Weitz. (1988). *Labor pains: Modern midwives and home birth.* New Haven: Yale University Press.

Sullivan, Joyce. (1984). Family support systems paychecks can't buy. In P. Voydanoff (ed.), *Work and family: Changing roles of men and women* (pp. 310–319). Palo Alto, CA: Mayfield.

Suro, Roberto. (1992). For women, varied reasons for single motherhood. *New York Times* (May 26): A12.

Surra, Catherine. (1991). Research and theory on mate selections and premarital relationships in the 1980s. In A. Booth (ed.), *Contemporary families: Looking forward, looking back* (pp. 54–75). Minneapolis: National Council on Family Relations.

Swanstrom, Todd. (1992). Homeless a product of policy. In P. Baker, L. Anderson and D. Dorn (eds.), *Social problems: A critical thinking approach* (pp. 335–336). Belmont, CA: Wadsworth.

Swart, J. (1979). Flextime's debit and credit option. *Harvard Business Review 1*: 10–12.

Sweet, Larry, James Bumpass and Vaughn Call. (1988). *The design and contact of the survey of families and households.* NSFH Working Paper No. 1. Madison: University of Wisconsin Center for Demography and Ecology.

Sydie, Rosalnd. (1987). *Natural woman, cultured men: A feminist perspective on sociological theory.* New York: New York University Press.

Szasz, Margaret. (1985). Federal boarding schools and the Indian child, 1920–1960. In N. Hiner and J. Hawes (eds.), *Growing up in America: Children in historical perspective* (pp. 209–218). Chicago: University of Chicago Press.

Tafel, Selma, Paul Placek and Mary Moien. (1985). One fifth of U.S. births by cesarean section. *American Journal of Public Health 75*: 190.

Takaki, Ronald. (1989). *Strangers from a different shore: A history of Asian Americans.* Boston: Little Brown.

Tanzer, Michael. (1974). *The energy crisis: World struggle for power and wealth.* New York: Monthly Review Press.

Taylor, Paul. (1991). Day care: Soaring popularity, stable cost. *Washington Post* (November 7): A20.

Taylor, Robert, Linda Chatters, M. Belinda Tucker, and Edith Lewis. (1992). Developments in research on black families: A decade review. In A. Skolnick and J. Skolnick (eds.), *Family in*

Transition (pp. 439–471). 7th ed. New York: Harper Collins.

Taylor, Verta. (1983). The future of feminism in the 1980s: A social movement analysis. In V. Taylor and L. Richardson (eds.), *Feminist frontiers: Rethinking sex, gender and society* (pp. 434–451). Palo Alto, CA: Mayfield.

Teachman, Jay. (1991). Contributions to children by divorced fathers. *Social Problems 38* (3): 358–371.

TenHouten, W. (1970). The black family: Myth and reality. *Psychiatry 25*: 145–173.

Testa, Mark, N.M. Astone, Marilyn Krogh, and Kathryn Neckerman. (1989). Employment and marriage among inner-city fathers. *Annals of the American Academy of Political and Social Science 501*: 79–91.

Thoits, Peggy. (1985). Multiple identities: examining gender and marital status differences in distress. *American Sociological Review* 51: 259–272.

Thompson, Sharon. (1984). Search for tomorrow: On feminism and the construction of teen romance. In C. Vance (ed.), *Pleasure and danger: Exploring female sexuality* (pp. 350–357). Boston: Routledge and Kegan Paul.

Thorne, Barrie. (1987). Revisioning women and social change: Where are the children? *Gender and Society 1*: 85–109.

Thorne, Barrie with Marilyn Yalom. (1992). *Rethinking the family: Some feminist questions.* 2d ed. New York: Longman.

Thornton, Arland. (1988). Cohabitation and marriage in the 1980s. *Demography 25*: 497–508.

Thornton, Arland and Deborah Freedman. (1983). The changing American family. *Population Bulletin 38.* Washington, D.C.: Reference Bureau.

Thrall, C. (1978). Who does what? Role stereotyping, children's work, and continuity between generations in the household division of labor. *Human Relations 31*: 249–265.

Tierney, Kathleen. (1982). The battered women's movement and the creation of the wife beating movement. *Social Problems 29* (February): 207–220.

Tillmon, Johnnie. (1976). Welfare is a woman's issue. In R. Baxandall, L. Gordon and S. Reverby (eds.), *America's working women: A documented history, 1600 to the present* (pp. 354–356). New York: Vintage Books.

Tilly, Louise and Joan Scott. (1978). *Women, work and family.* New York: Holt, Rinehart and Winston.

Timmer, Doug and D. Stanley Eitzen. (1992). The root causes of urban homelessness in the United States. *Humanity and Society 16* (2): 159–175.

Traveler's Aid Program and Child Welfare League. (1987). *Study of homeless children and families: Preliminary findings.* Conducted by P. Maza and J. Hall.

Trennert, Robert. (1979). Peaceably if they will, forcibly if they must: The Phoenix Indian school, 1890-1901. *Journal of Arizona History 20* (Autumn): 297–322.

———. (1990). Educating Indian girls at non-reservation boarding schools, 1878–1920. In E. DuBois and V. Ruiz (eds.), *Unequal sisters: A multicultural reader in U.S. women's history* (pp. 224–237). New York: Routledge.

Tucker, M. Belinda and Robert Taylor. (1989). Demographic correlates of relationship status among black Americans. *Journal of Marriage and the Family 51*: 655–665.

Turnbull, C. (1972). *The mountain people.* New York: Simon and Schuster.

Udry, J. (1977). The importance of being beautiful: A re-examination and racial comparison. *American Journal of Sociology 83*: 154–160.

Ullmann, Owen. (1993). U.S. future relies on adapting to global economy. *Charlotte Observer* (4/25): 1e, 4e.

Uhlenberg, John. (1989). Death and the family. In A. Skolnick and J. Skolnick (eds.), *Family in transition* (pp. 87–96). 6th ed. Boston: Scott Foresman.

U.S. Bureau of the Census. (1975). *Historical statistics of the United States, colonial times to 1970.* Washington, D.C.: USGPO.

———. (1983). *Contribution of wives' earnings to family income.* Current Population Reports, Series P-60. Washington, D.C.: GPO.

———. (1985). *Money income and poverty status of families and persons in the U.S.: 1984.* Current Population Reports, Series P-60, No. 149. Washington, D.C.: GPO.

———. (1991). *Money income of households, families and persons in the United States, 1991.* Current Population Reports, Series P-60, Washington, D.C.: USGPO.

———. (1988). *Households, families, marital status and living arrangements, March 1988.* Current

Population Reports, Series P-20, No. 432, Washington, D.C.: USGPO.
———. (1989a). *Child support and alimony, 1985.* Current Population Reports, Series P-23, No. 154. Washington, D.C.: USPGO.
———. (1989b). *Population profile of the United States.* Current Population Reports, Series P-23, No. 129. Washington, D.C.: USGPO.
———. (1989c). *Marital status and living arrangements: March 1988.* Current Population Reports, Series P-20, No. 418. Washington, D.C.: Government Printing Office.
———. (1989d). *Changes in America family life.* (Current Population Reports P-23, No. 163).
———. (1990a). *Household and family characteristics: 1990 and 1989.* P-20, No. 447. Washington, D.C.: USGPO.
———. (1990c). *Money Income and Poverty Status of Families and Persons in the U.S.: 1989.* (Current Population Reports, Series P-60). Washington, D.C.: GPO.
———. (1991a). *Fertility of American women: June 1990.* Washington, D.C.: USGPO
———. (1991b). *Marital status and living arrangements: March 1991.* Current Population Reports, Series P-20, No. 450. Washington, D.C.: USGPO.
———. (1991c). *Statistical abstract of the U.S., 1989.* Washington, D.C.: USGPO.
———. (1991d). Studies in household and family formation.
———. (1992a). *Census and You* 27:7
———. (1992b). *Census and You* May: 10
———. (1992c) *Statistical Abstracts of the U.S.* Washington, DC: USGPO.
U.S. Commission on Civil Rights. (1981). *Child care and equal opportunity for women.* Clearinghouse Publication 67, June. Washington, D.C.: USGPO.
U.S. Congress House Committee on Ways and Means. (1987). *Background material and data on programs within the jurisdiction of the Committee on Ways and Means.* 100th Congress, 1st Session, March 6. Washington, D.C.: USGPO.
———. (1987b). *Stewart P. McKinney Homeless Assistance Act.* Conference Report to accompany H.R. 558. 100th Congress, 1st Session. Washington, D.C.: USGPO.
U.S. Department of Agriculture. (1991). *The cost of raising a child.* Washington, D.C.: USGPO.
U.S. Department of Health and Human Services. (1991). *Cohabitation, marriage, marital dissolution and remarriage in the United States, 1988.* Washington, D.C.: USGPO.
U.S. Department of Justice. (1985). *FBI uniform crime reports.* Washington, D.C.: GPO.
Upson, Norma. (1974). *How to survive as a corporate wife.* Garden City, NY. : Doubleday & Co.
Vance, Carol. (1984). *Pleasure and danger: Exploring female sexuality.* Boston: Routledge and Kegan Paul.
VanEvery, Dale (1976). *The disinherited: The lost birthright of the American Indian.* New York: Morrow.
Veevers, Jean. (1980). *Childless by choice.* Toronto: Butterworth.
Veroff, Joseph, Elizabeth Douvan, and Richard Kukla. (1981). *The inner American: A self-portrait from 1957–1976.* New York: Basic books.
Vickery, C. (1979). Women's economic contribution to the family. In R. Smith (ed.), *The subtle revolution: Women at work* (pp. 159–200). Washington, D.C.: The Urban Institute.
Voydanoff, Patricia. (1963). *The Influence of Economic Security on Morale.* Unpublished thesis, Wayne State University.
———. (1984). Unemployment: Family strategies for adaptation. In P. Voydanoff (ed.), *Work and family: Changing roles of men and women* (pp. 61–72). Palo Alto, CA: Mayfield.
Voydanoff, Patricia and Brenda Donnelley. (1988). Economic distress, family coping and the quality of life. In P. Voydanoff and L. Majka (eds.), *Families and economic distress: Coping strategies and public policy* (pp. 97–116). Beverly Hills, CA: Sage.
Walby, Sylvia. (1986). *Patriarchy at work: Patriarchal and capitalist relations in employment.* Minneapolis: University of Minnesota Press.
Walker, Alice. (1990). The right to life: What can the white man say to the black woman? In M. Fried (ed.), *From abortion to reproductive freedom: Transforming a movement* (pp. 65–70). Boston: South End Press.
Walker, K. and M. Woods. (1976). *Time use: A measure of household production of family goods and services.* Washington, D.C.: American Home Economics Association.
Walker, Lenore. (1979). *The battered woman.* New York: Harper.

———. (1983. The battered woman syndrome study. In D. Finkelhor et al. (eds.), *The dark side of families: Current family violence research* (pp. 31–48). Beverly Hills, CA: Sage.

Wallerstein, Judith and Sandra Blakeslee. (1989). *Second chances: Men, women and children a decade after divorce.* New York: Ticknor and Field.

Washburne, Carolyn. (1983). A feminist analysis of child abuse and neglect. In D. Finkelhor et al. (eds.), *The dark side of families: Current family violence research* (pp. 289–292). Beverly Hills, CA: Sage.

Washington Post. (1992). Violence against women. September 1, 115: WH5.

Weber, Max. (1969). *The theory of social and economic organization.* New York: Free Press.

Weeks, Jeffrey. (1985). *Sexuality and its discontents: Meanings, myths and modern sexualities.* London: Routledge and Kegan Paul.

Weeks, John. (1993). *Population: Introduction to concepts and issues.* 5th ed. Belmont, CA: Wadsworth.

Weissman, M. and S. Paykel. (1972). *The depressed woman: A study of social relationships.* Chicago: University of Chicago Press.

Weitz, Rose and Deborah Sullivan. (1986). The politics of childbirth: The re-emergence of midwifery in Arizona. *Social Problems 33* (3): 163–175.

Weitzman, Lenore. (1981). *The marriage contract: Spouses, lovers and the law.* New York: Free Press.

———. (1985). *The divorce revolution: The unexpected social and economic consequences for women and children in America.* New York: Free Press.

Weitzman, Lenore and Ruth Dixon. (1992). The transformation of legal marriage through no-fault divorce. In A. Skolnick and J. Skolnick (eds.), *Family in Transition* (pp. 217–230). 7th ed. New York: Harper Collins.

Welfare Mother's Voice. (1992). California moms fight back. Spring: 1.

Welter, Barbara. (1978). The cult of true womanhood, 1820–1860. In M. Gordon (ed.), *The American family in social-historical perspective* (pp. 313–333). 2d ed. New York: St. Martin's Press.

West, Candace and Don Zimmerman. (1987). Doing gender. *Gender and Society 1*: 125–151.

West, Guida. (1981). *The national welfare rights movement: The social protest of poor women.* New York: Praeger.

Weston, Kath. (1991). *Families we choose.* New York: Columbia University Press.

Westwood, Sallie. (1984). *All day, every day: Factory and family in the making of women's lives.* London: Pluto Press.

Wharton, Carol. (1987). Establishing shelters for battered women: Local manifestations of a social movement. *Qualitative Sociology 10* (2): 146–163.

White, Deborah. (1985). *Aren't I a woman: Female slaves in the plantation South.* New York: W. W. Norton.

White, Lynn. (1991). Determinants of divorce: Review of research in the 1980s. In A. Booth (ed.), *Contemporary families: Looking forward, looking back* (pp. 141–149). Minneapolis: National Council on Family Relations.

White, Lynn, Alan Booth and John Edwards. (1986). Children and marital happiness: Why the negative relationship. *Journal of Family Issues 7*: 131–148.

White, Lynn and David Brinkerhoff. (1987). Children's work in families: Its significance and meaning. In N. Gerstel and H. Gross (eds.), *Families and Work* (pp. 204–219). Philadelphia: Temple University Press.

Wiggins, David. (1985). The play of slave children in plantation communities in the old south, 1820–1860, pp. 173–192. In N. Hiner and J. Hawes, *Growing up in America: Children in historical perspectives.* Chicago: University of Illinois Press.

Wiley, Norbert. (1979). The rise and fall of dominating theories in sociology. In W. Snizek, E. Fuhrman and M. Miller (eds.), *Contemporary Issues in theory and research* (pp. 47-49). Westport, CT: Greenwood Press.

Wilkinson, Doris. (1984). Afro-American women and their families. *Marriage and Family Review 7* (Fall): 459–467.

William T. Grant Foundation Commission on Work, Family and Citizenship. (1988). *The forgotten half: Pathways to success for America's youth and young families.* Washington, D.C.: William T. Grant Foundation Commission on Work, Family and Citizenship.

Williams, Constance. (1990). *Black teenage mothers.* New York: Lexington.

Williams, Norma. (1990). Role making among married Mexican American women: Issues of class and ethnicity. In C. Carlson (ed.), *Perspec-*

tives on the family: History, class and feminism (pp. 186–204). Belmont, CA: Wadsworth.

Williams, Terry and William Kornblum. (1991). Sneaker Mothers. In M. Hutter (ed.), *The family experience* (pp. 589–600). New York: MacMillan.

Willie, Charles. (1983). *Race, ethnicity and socio-economic status: A theoretical analysis of their interrelationship.* Dix Hills, NY: General Hall.

———. (1985). *Black and white families: A study in complementarity.* Bayside, NY: General Hall.

———. (1988). *A new look at black families.* 3d ed. Bayside, NY: General Hall.

Willis, Ellen. (1983). Abortion: Is a woman a person? In A. Snitow, C. Stansell and S. Thompson (eds.), *Powers of desire: The politics of sexuality* (pp. 471–476). New York: Monthly Review Press.

Wilson, E.O. (1975). *Sociobiology: The new synthesis.* Cambridge, MA: Harvard University Press.

Wilson, William Julius and Kathryn Neckerman. (1986). Poverty and family structure: The widening gap between evidence and public policy issues. In S. Danziger and D. Weinberg (eds.), *Fighting poverty: What works and what doesn't* (pp. 240–264). Chicago: University of Chicago Press.

Winnick, Andrew. (1988). The changing distribution of income and wealth in the U.S., 1960–1985: An examination of the movement toward two societies separate but equal. In P. Voydanoff and L. Majka (eds.), *Economic distress and families: Coping strategies and public policy* (pp. 232–260). Beverly Hills, CA: Sage.

Wolf, Wendy and Neil Fligstein. (1979). Sex and authority in the workplace. *American Sociological Review 44*: 235–252.

Women's Institute for Freedom of the Press (WIFP). (1986). 1955 to 1985: Women in prime time TV still traditional, but new treatment of women's rights themes. *Media Report to Women* (November–December): 7.

———. (1990). TV portrayal of the childless black female: Superficial, unskilled, dependent. *Media Report to Women* (March–April): 4.

World Almanac and Book of Facts, 1992. (1991). Child care arrangements in the U.S., p. 945.

Wright, Eric Olin. (1985). *Classes.* London: Verso.

———. (1990). *A debate on classes.* London: Verso.

Wright, Eric Olin, Karen Shire, Shu-Ling Hwang, Maureen Dolan, and Janeen Baxter. (1992). The non-effects of class on the gender division of labor in the home: A comparative study of Sweden and the U.S. *Gender and Society 6* (2): 252–282.

Yangisako, Sylvia. (1985). *Transforming the Past: Tradition and kinship among Japanese Americans.* Stanford: Stanford University Press.

Ybarra, Leonarda. (1977). *Conjugal role relationships in the Chicano family.* Ph.D. dissertation. University of California, Berkeley.

———. (1982a). Marital decision making and the role of machismo in the Chicano family. *De Colores, Journal of Chicano Expression and Thought 6*: 32–47.

———. (1982b). When wives work: The impact on the Chicano family. *Journal of Marriage and the Family 44.* (1): 169–178.

Yllo, Kersti. (1988). Political and methodological debates in wife abuse research. In K. Yllo and M. Bograd (eds.), *Feminist perspectives on wife abuse* (pp. 28–51). Newbury Park, CA: Sage.

Zaretsky, Eli. (1982). *Capitalism, the family and personal life.* New York: Harper.

———. (1987). The place of family in the origins of the welfare state. In B. Thorne with M. Yalom (eds.), *Rethinking the family: Some feminist questions* (pp. 188–224). New York: Longman.

Zavella, Patricia. (1987). *Women's work and Chicano families: Cannery workers of the Santa Clara Valley.* Ithaca: Cornell University Press.

Zelnick, Melvin, John Kanter and Kathleen Ford. (1981). *Sex and pregnancy in adolescence.* Beverly Hills, CA: Sage.

Zelizer, Viviana. (1985). *Pricing the priceless child: The changing social value of children.* New York: Basic Books.

Zigler, Edward and Meryl Frank. (1988). *The parental leave crisis.* New Haven: Yale University Press.

Zimmerman, Shirley. (1992). *Family policies and family well-being: The role of political culture.* Newbury Park, CA: Sage.

Zinn, Howard. (1980). *A people's history of the U.S.* New York: Harper & Row.

NAME INDEX

SUBJECT INDEX